# Protein Engineering and Design

# Protein Engineering and Design

Edited by
## Sheldon J. Park
## Jennifer R. Cochran

CRC Press
Taylor & Francis Group
Boca Raton   London   New York

CRC Press is an imprint of the
Taylor & Francis Group, an **informa** business

First edition published 2024

First published in 2010
by CRC Press
2385 NW Executive Center Drive, Suite 320, Boca Raton FL 33431

and by CRC Press
4 Park Square, Milton Park, Abingdon, Oxon, OX14 4RN

© 2010, 2024 by Taylor and Francis Group, LLC

*CRC Press is an imprint of Taylor & Francis Group, LLC*

No claim to original U.S. Government works

**Library of Congress Cataloging-in-Publication Data**

Protein engineering and design / editors, Sheldon J. Park, Jennifer R. Cochran.
    p. ; cm.
  Includes bibliographical references and index.
  ISBN 978-1-4200-7658-5 (hardcover : alk. paper)
  1. Protein engineering. I. Park, Sheldon J. II. Cochran, Jennifer R. III. Title.
  [DNLM: 1. Protein Engineering. 2. Models, Molecular. 3. Proteins--chemistry. QU 450 P9668 2010]

TP248.65.P76P738 2010
660.6'3--dc22
    2009024409

ISBN: 978-1-4200-7658-5 (hbk)
ISBN: 978-1-03-283668-3 (pbk)
ISBN: 978-0-429-19163-3 (ebk)

DOI: 10.1201/9781420076592

**Visit the Taylor & Francis Web site at**
**http://www.taylorandfrancis.com**

**and the CRC Press Web site at**
**http://www.crcpress.com**

# Contents

# Preface

Proteins possess a broad range of structural and functional properties that are unmatched by any other class of biological molecules. Amazingly, nature has arranged simple atoms and chemical bonds in such a way to facilitate complex biological processes like molecular recognition and catalysis. Nature has also inspired many scientists and engineers to design and create their own customized proteins. These engineered proteins can serve as novel molecular tools for scientific, medical, and industrial applications, thus addressing many needs unmet by naturally occurring proteins.

Protein engineering requires identification of particular amino acid sequences that will result in desired structural and functional properties. Despite recent advances in the field, however, protein engineering remains as much an art as it is a science. Engineering an arbitrary protein structure or function remains a formidable challenge, because the rules defining sequence-structure-function relationships are still not well understood. Even with refined quantitative models, the large degrees of freedom present in a typical protein do not easily allow identification of optimal sequences using currently available computational techniques. Furthermore, the complexity of proteins present engineering challenges whose solutions will most likely require a combination of experimental and computational approaches.

This book discusses two general strategies commonly used to engineer new proteins: diversity-oriented protein engineering and computational protein design. Diversity-oriented protein engineering, or directed evolution, identifies protein variants with desired properties from a large pool of mutants. As such, its success depends on generating sufficient sequence diversity and employing sensitive high-throughput assays. Computational protein design, on the other hand, generates and screens protein sequences *in silico* before synthesizing them in the laboratory. This is still an unfamiliar concept to many, so an important goal of this book is to demystify the subject by describing its development and current implementations. Structure-based protein engineering similarly uses computation to facilitate the discovery of interesting protein sequences. However, computational protein design places emphasis on both engineering new, useful proteins and on testing sequence-structure relationships. In this regard, it shares a deep philosophical root with protein folding, which similarly seeks to understand the relationship between protein sequence and tertiary structure.

The book is organized into two sections. The first half of the book discusses experimental approaches to protein engineering and starts by describing several high-throughput protein engineering platforms (Chapters 1–3). This is followed by a chapter on key techniques used for diversity generation (Chapter 4). The next few chapters present examples of therapeutics, enzymes, biomaterials, and other molecules that were engineered by rational or combinatorial-based approaches (Chapter 5–8). The section finishes with a chapter on the use of unnatural amino acids in protein engineering (Chapter 9).

The second half of the book introduces computational protein design, which designs new sequences by quantitatively modeling sequence-structure relationships. Despite their unique approaches, protein engineering and design are increasingly developing a synergistic relationship. To that end, more and more experimentalists are recognizing computation as an important molecular tool for protein engineering, and vice versa. These days, it is routine for those planning a protein engineering project to first perform sequence analysis and to visualize protein structures in a molecular viewer. It is thus appropriate to start this section with a chapter on the common use of computers and informatics in protein engineering (Chapter 10). Examples of heuristic protein design are described in Chapter 11, before the core components of computational protein design are discussed in detail in Chapters 12–14. Subsequent chapters present examples of computationally designed proteins that played critical roles in advancing the use of computers in protein engineering (Chapters 15–17). The field has not yet fully matured and there are difficulties that remain to be resolved; these challenges are discussed in the last chapter of the book (Chapter 18).

Modern biology has provided a deep understanding of the molecular nature of biological processes. In particular, we now have a variety of tools that can be used to analyze and control key biological processes with molecular precision. Protein engineering and design are attempts to accomplish exactly these goals. As examples throughout the book show, certain categories of problems have attracted attention from scientists and engineers with a diverse range of technical expertise. We hope these studies will help the reader identify potential opportunities to bridge experimental protein engineering and computational protein design and will lead to exciting breakthroughs in biotechnology and medicine.

**Sheldon J. Park**
**Jennifer R. Cochran**

# Editors

**Sheldon Park** holds a B.A. in math and physics from the University of California (Berkeley), an M.S. in physics from Massachusetts Institute of Technology, and a Ph.D. in biophysics from Harvard University. He studied protein engineering and design while working as a postdoc for Dr. Jeffery Saven and Dr. Eric Boder at the University of Pennsylvania. Since 2006, he has been a professor of chemical and biological engineering at University at Buffalo. In his research, Dr. Park uses modeling and simulation to analyze protein molecules and uses high-throughput screening to engineer protein molecules of various structure and function. He is particularly interested in developing efficient methods of engineering complex protein molecules with potential biotechnological and biomedical applications.

**Jennifer Cochran** holds a B.S. in biochemistry from the University of Delaware and a Ph.D. in biological chemistry from Massachusetts Institute of Technology (MIT). She studied and developed combinatorial protein engineering methods while a postdoctoral fellow in the lab of K. Dane Wittrup in the Department of Biological Engineering at MIT. Since 2005, she has been a professor of bioengineering at Stanford University. Dr. Cochran's laboratory uses interdisciplinary approaches in chemistry, engineering, and biophysics to study complex biological systems and to create designer protein therapeutics and diagnostic agents for biomedical applications. She is interested in elucidating molecular details of receptor-mediated cell signaling events and at the same time developing protein and polymer-based tools that will allow manipulation of cell processes on a molecular level.

# Contributor List

**Pamela A. Barendt**
Department of Bioengineering
University of Pennsylvania
Philadelphia, Pennsylvania

**Subhayu Basu**
Codon Devices, Inc.
Cambridge, Massachusetts

**Brian M. Baynes**
Codon Devices, Inc.
Cambridge, Massachusetts

**Izhack Cherny**
Department of Chemistry
Princeton University
Princeton, New Jersey

**Eun Jung Choi**
Department of Biochemistry and
    Biophysics
University of North Carolina
Chapel Hill, North Carolina

**A. B. Chowdry**
Biophysics Graduate Group
University of California
Berkeley, California
and
Skaags School of Pharmacy and
    Pharmaceutical Science
University of California
San Diego, California

**Patrick C. Cirino**
Department of Chemical Engineering
Pennsylvania State University
University Park, Pennsylvania

**Frank V. Cochran**
Biophysics Program
Stanford University
Stanford, California

**Jennifer R. Cochran**
Department of Bioengineering
Stanford University
Stanford, California

**Andreas Ernst**
Banting and Best Department for
    Medical Research
and
Terence Donnelly Center for Cellular
    and Biomolecular Research
University of Toronto
Toronto, Ontario, Canada

**Michael A. Fisher**
Department of Molecular Biology
Princeton University
Princeton, New Jersey

**Cheryl Wong Po Foo**
Department of Materials Science and
    Engineering
Stanford University
Stanford, California

**Christopher S. Frei**
Department of Chemical Engineering
Pennsylvania State University
University Park, Pennsylvania

**Menachem Fromer**
School of Computer Science and
    Engineering
Hebrew University of Jerusalem
Jerusalem, Israel

**David F. Green**
Department of Applied Mathematics
    and Statistics
Stony Brook University (SUNY)
Stony Brook, New York

**Gurkan Guntas**
Department of Biochemistry and
    Biophysics
University of North Carolina
Chapel Hill, North Carolina

**T. M. Handel**
Skaags School of Pharmacy and
    Pharmaceutical Science
University of California
San Diego, California

**M. S. Hanes**
Biophysics Graduate Group
University of California
Berkeley, California
and
Skaags School of Pharmacy and
    Pharmaceutical Science
University of California
San Diego, California

**Michael H. Hecht**
Departments of Chemistry and
    Molecular Biology
Princeton University
Princeton, New Jersey

**Sarah C. Heilshorn**
Department of Materials Science and
    Engineering
Stanford University
Stanford, California

**Daniel Hsieh**
Department of Biochemistry
Robert Wood Johnson Medical School
University of Medicine and Dentistry of
    New Jersey
Piscataway, New Jersey
and
Center for Advanced Biotechnology and
    Medicine
Piscataway, New Jersey

**Yulia Ivanova**
Department of Chemistry
Tufts University
Medford, Massachusetts

**Patrice Koehl**
Department of Computer Science and
    Genome Center
University of California
Davis, California

**Shohei Koide**
Department of Biochemistry and
    Molecular Biology
The University of Chicago
Chicago, Illinois

**Hidetoshi Kono**
Computational Biology Group
Japan Atomic Energy Agency
Kyoto, Japan
and
PRESTO, Japan Science and
    Technology

**Brian Kuhlman**
Department of Biochemistry and
    Biophysics
University of North Carolina
Chapel Hill, North Carolina

**Krishna Kumar**
Department of Chemistry
Tufts University
Medford, Massachusetts
and
Cancer Center
Tufts Medical Center
Boston, Massachusetts

**Daša Lipovšek**
Codon Devices, Inc.
Cambridge, Massachusetts

**Shaun M. Lippow**
Codon Devices, Inc.
Cambridge, Massachusetts

**Loren L. Looger**
Howard Hughes Medical Institute
Janelia Farm Research Campus
Ashburn, Virginia

**Jonathan S. Marvin**
Howard Hughes Medical Institute
Janelia Farm Research Campus
Ashburn, Virginia

**Marco Mena**
Codon Devices, Inc.
Cambridge, Massachusetts

**Sarah J. Moore**
Department of Bioengineering
Stanford University
Stanford, California

**Vikas Nanda**
Department of Biochemistry
Robert Wood Johnson Medical School
University of Medicine and Dentistry of
    New Jersey
Piscataway, New Jersey
and
Center for Advanced Biotechnology and
    Medicine
Piscataway, New Jersey

**Mark J. Olsen**
Pharmaceutical Sciences
Midwestern University
Glendale, Arizona

**Diren Pamuk**
Department of Chemistry
Tufts University
Medford, Massachusetts

**Shona C. Patel**
Department of Chemical Engineering
Princeton University
Princeton, New Jersey

**Sheryl B. Rubin-Pitel**
Department of Chemical and
    Biomolecular Engineering
University of Illinois at Urbana-
    Champaign
Urbana, Illinois

**Casim A. Sarkar**
Department of Bioengineering
University of Pennsylvania
Philadelphia, Pennsylvania

**Julia M. Shifman**
Department of Biological Chemistry
The Alexander Silberman Institute of
    Life Sciences
Hebrew University of Jerusalem
Jerusalem, Israel

**Sachdev S. Sidhu**
Banting and Best Department for
    Medical Research
and
Terence Donnelly Center for Cellular
    and Biomolecular Research
University of Toronto
Toronto, Ontario, Canada

**Fei Wen**
Department of Chemical and
    Biomolecular Engineering
University of Illinois at Urbana-
    Champaign
Urbana, Illinois

**Fei Xu**
Department of Biochemistry
Robert Wood Johnson Medical School
University of Medicine and Dentistry of
    New Jersey
Piscataway, New Jersey
and
Center for Advanced Biotechnology and
    Medicine
Piscataway, New Jersey

**Deniz Yüksel**
Department of Chemistry
Tufts University
Medford, Massachusetts

**Huimin Zhao**
Departments of Chemical and
    Biomolecular Engineering and
    Chemistry
University of Illinois at Urbana-
    Champaign
Urbana, Illinois

# 1 Phage Display Systems for Protein Engineering

## *Andreas Ernst and Sachdev S. Sidhu*

## CONTENTS

Information in biological systems is contained and passed on through genes, while the traits of a biological system are given by the functions of encoded proteins and other macromolecules. In molecular biology, the focus of research is the function of biological macromolecules, especially proteins, which are studied in an isolated setting, allowing us to develop our understanding of biological systems. However, these studies are usually limited to a small set of states that can be established in the experiment. With phage display, we can study proteins and protein–protein interactions on a combinatorial scale. This means that it is possible to probe billions of different protein variants simultaneously. Phage display can be applied to the generation of affinity reagents, which are invaluable tools for diagnostics and therapeutic development. The technology can also be used for improving protein stability and for identifying and mapping natural protein–protein interactions in detail.

## THE PHAGE DISPLAY CONCEPT

Phage display technology provides an *in vitro* version of classical Darwinian evolution by establishing a physical linkage between a polypeptide and the encoding genetic information. By mutating DNA encoding for a displayed polypeptide, it is possible to generate a multitude of related variants. This library of variants can be displayed on the surfaces of phage particles, while the DNA encoding for each

1

variant is encased in the phage capsid (Figure 1.1). Consequently, the technology allows for the handling of diverse protein variants and enables the selection of those variants with properties defined by the selection process. The selection process sifts through the pool of variants and selects those proteins with a particular property of interest, which is usually a specific binding activity. Because the function of a protein is linked to its folding and stability, phage display also enables selections for these additional properties.

Phage display was first developed with the *Escherichia coli*–specific filamentous bacteriophage M13 (Smith 1985), and the success of M13 phage display prompted the subsequent development of numerous alternative display systems. These include systems that utilize other *E. coli*–specific phage, such as λ-phage (Santini et al. 1998) and T4 phage (Ren and Black 1998), and also systems that use eukaryotic viruses (Possee 1997). In addition, as discussed in Chapter 2, polypeptides have been displayed on the surfaces of bacteria and yeast (Georgiou et al. 1997). Although these newer systems provide useful alternative approaches, filamentous phage display remains the dominant technology and is the focus of this chapter.

## PHAGE STRUCTURE AND ASSEMBLY

The filamentous bacteriophage particle is 65 Å in diameter and 9300 Å in length, and it consists of a single-stranded DNA (ssDNA) genome of approximately 6400 bases coated with approximately 2700 copies of the major coat protein, pVIII (Figure 1.2) (Vanwezenbeek et al. 1980; Marvin 1998). One end of the particle is capped with

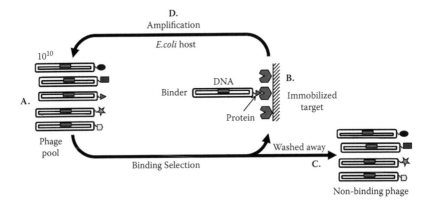

**FIGURE 1.1**   Phage display cycle for the selection of binding proteins. A library of proteins is displayed on the phage particle as fusions to coat proteins. Each phage particle displays a unique protein and encapsulates the encoding DNA (A). Highly diverse libraries with more than $10^{10}$ unique members can be represented as phage pools. Binding clones are captured with an immobilized target (B), nonbinding phage are washed away (C), and bound phage are amplified by passage through bacteria (D). The amplified pool can be cycled through additional rounds of selection to further enrich for binding clones. Positive clones are indentified by enzyme-linked immunosorbent assay (ELISA) and are subjected to DNA sequencing to decode the sequences of the displayed proteins.

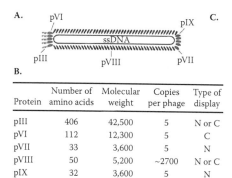

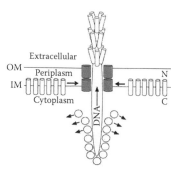

| Protein | Number of amino acids | Molecular weight | Copies per phage | Type of display |
|---------|----------------------|------------------|------------------|-----------------|
| pIII | 406 | 42,500 | 5 | N or C |
| pVI | 112 | 12,300 | 5 | C |
| pVII | 33 | 3,600 | 5 | N |
| pVIII | 50 | 5,200 | ~2700 | N or C |
| pIX | 32 | 3,600 | 5 | N |

**FIGURE 1.2** Filamentous phage structure and assembly. (A) An illustration of the filamentous phage particle depicts the ssDNA genome surrounded by a protein coat. The coat is mainly composed of pVIII molecules that cover the length of the phage particle. One end of the particle is capped by minor coat proteins pIII and pVI, and the other end is capped by pVII and pIX. (B) Characteristics of the phage coat proteins. Each of the five coat proteins has been used for the functional display of polypeptides either as N- or C-terminal fusions. (C) Phage assembly at the pore complex (gray cylinders) formed by nonstructural proteins (pIV, pXI, and pI) spanning the inner (IM) and outer membranes (OM) of the *E. coli* host. Coat proteins (white cylinders) are imbedded in the inner membrane with their N- and C-termini pointing into the periplasm or cytoplasm, respectively. Genomic ssDNA is extruded through the pore complex and is concomitantly stripped of pV (white circles) and encapsulated by coat proteins. In this way, assembling phage particles are extruded in a nonlytic manner from the host cell into the extracellular environment.

five copies each of the minor coat proteins pVII and pIX, and the other end is capped with five copies each of the minor coat proteins pIII and pVI (Barbas et al. 2001). With the exception of pIII, the coat proteins are small and simple. In contrast, pIII is a 42 kDa polypeptide consisting of three distinct structural domains (N1, N2, and CT). The first two domains, N1 and N2, are necessary for infection (Russel et al. 1988; Krebber et al. 1997) and have been crystallized (Lubkowski et al. 1998). The C-terminal (CT) domain is involved in contacts with the phage coat and is necessary for the incorporation of pIII into phage particles (Rakonjac and Model 1998).

During infection of host bacteria, the N2 domain binds to the F-pilus of *E. coli* (Lubkowski et al. 1999), the pilus retracts to the membrane surface, and the N1 domain binds to the outer membrane protein TolA (Riechmann and Holliger 1997). Upon contact with the membrane, the phage particle dissembles and the genome is injected into the host cell (Nakamura et al. 2003), where the ssDNA viral genome is converted into double-stranded DNA (dsDNA) by host proteins. The dsDNA serves as a template for the synthesis of new ssDNA genomes and also viral proteins, which constitute all the components necessary for the assembly of new phage particles.

Phage structural proteins insert spontaneously into the inner membrane, and once a critical concentration of these proteins is reached, phage assembly commences (Barbas et al. 2001). In parallel to this process, the ssDNA-binding protein pV is expressed, and, once it reaches a critical concentration, it coats newly synthesized ssDNA genomes and thereby prevents their conversion to dsDNA (Figure 1.2C)

(Konings et al. 1995). A specific hairpin structure in the pV-associated ssDNA is recognized by a complex of three assembly proteins (pIV, pXI, and pI), which form a pore spanning the inner and outer membranes (Marciano et al. 1999; Opalka et al. 2003). At the assembly site, the ssDNA is extruded through the pore, and pV is concomitantly stripped and replaced by phage structural proteins (Marciano et al. 2001). pVII and pIX cap the initiating end of the assembling phage, pVIII covers the length, and assembly is terminated by a cap composed of pIII and pVI (Feng et al. 1997). At the termination point, pIII dissociates from the membrane, and the assembled phage is released from the host cell (Rakonjac and Model 1998).

## VECTORS AND PLATFORMS

Filamentous phage display is predominantly achieved by fusion of heterologous polypeptides to either pVIII or pIII, although display has also been demonstrated with the other coat proteins (Gao et al. 1999; Hufton et al. 1999; Fuh and Sidhu 2000; Weiss and Sidhu 2000; Gao et al. 2002). The earliest phage display systems relied on fusions to the N-terminus of either pVIII or pIII in the viral genome (Il'Ichev et al. 1989; Cwirla et al. 1990; Devlin et al. 1990; Scott and Smith 1990; Felici et al. 1991; Greenwood et al. 1991), but utility was limited because polypeptides that compromised coat protein function could not be efficiently displayed. The development of hybrid phage display systems largely alleviated these limits.

In hybrid phage, the displayed fusion protein is an additional component of a natural phage coat that contains all five wild-type coat proteins. This can be achieved by introducing the fusion gene as an addition to a complete phage genome (Smith 1993), or alternatively, with a phagemid-based system (Bass et al. 1990; Lowman et al. 1991). A phagemid is a dsDNA plasmid that is also capable of being converted into ssDNA that can be packaged into phage particles (Figure 1.3) (Barbas et al. 2001). A phagemid carries an antibiotic resistance marker, bacterial (dsDNA) and phage (ssDNA) origins of replication, a phage-packaging signal, and a gene encoding for the heterologous protein of interest fused to a phage coat protein. To package phagemid DNA into phage particles, it is necessary to co-infect *E. coli* with helper phage virions that provide the genes encoding for all the phage proteins needed to assemble functional phage particles. During the assembly process, copies of the phagemid-encoded coat protein fusion are incorporated into phage particles that encapsulate the phagemid DNA, and thus, functional and selectable phage display particles are obtained.

In hybrid phage, the deleterious effects of fusion proteins are attenuated by the presence of wild-type coat proteins, and this enables the display of polypeptides that could not be displayed with earlier phage-based systems. In addition, fusions to pIII or pVIII typically provide different levels of display that can be exploited for different applications. Fusion of polypeptides to the high copy pVIII typically results in polyvalent display, and, therefore, binding selections generally yield low affinity ligands because the avidity effect of multiple pVIII copies present on the surface of the phage amplifies intrinsic affinities (Chappel et al. 1998). In contrast, display of polypeptides on the low copy pIII typically results in monovalent display, which enables the selection of ligands on the basis of intrinsic affinity (Barbas et al.

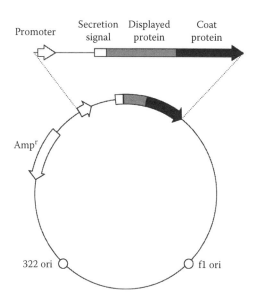

**FIGURE 1.3** A phagemid vector designed for phage display. A phagemid vector contains the origins of replication for single-strand DNA (ssDNA ori) and double-strand DNA (dsDNA ori), as well as a selective marker, such as the β-lactamase gene (Amp^r), which confers resistance to ampicillin. For phage display, the vector also contains a cassette consisting of a promoter-regulated gene that encodes for a secreted fusion of the displayed protein and a phage coat protein.

1991). In general, polypeptide fusions to pIII are used to generate high affinity binders against a particular target, while fusions to pVIII are used to select binders with weak affinities.

For protein libraries, researchers typically use monovalent display to avoid the effects of binding avidity. To increase the stringency of selection for high affinity binders, it is possible to further reduce display levels by introducing an amber stop codon between the displayed protein and pIII. When expressed in an *E. coli* suppressor strain, only a fraction of the translations result in the production of full-length fusion proteins that can be inserted in the phage coat, and thus the displayed copy number is reduced (Lowman et al. 1991). The level of display can also be attenuated by using tightly regulated promoters, which allow only low-level expression of the fusion protein (Huang et al. 2000).

Whether fused to the N-terminus of pIII or pVIII, proteins to be displayed on phage must be secreted into the host periplasm. Although underappreciated, the pathway used for secretion can play a critical role in the efficiency with which proteins are displayed on phage particles. For conventional phage display, secretion of proteins occurs through the post-translational secretion pathway (Barbas et al. 2001). This pathway requires that the protein be threaded through the inner membrane of *E. coli* in an unfolded state. Refolding occurs in the periplasm, and this leads to the display of a folded and functional protein. Somewhat paradoxically, it has been shown that display is compromised for proteins that are highly stable and fold rapidly in the cytoplasm (Kohl et al. 2003). This is because highly stable proteins are unable to undergo the unfolding process that is necessary prior to secretion, and consequently levels of display are severely reduced (Steiner et al. 2006). This secretion bottleneck has been overcome by use of the cotranslational secretion pathway mediated by the signal recognition particle (Steiner et al. 2006). Compared to the post-translational pathway, use of the cotranslational pathway increased display levels of highly stable proteins by almost three orders of magnitude (Koide et al. 2007).

## THE SELECTION PROCESS

In comparison with other display platforms, phage particles are very robust (Kristensen and Winter 1998). For example, although ribosome display allows access to very large libraries (described in Chapter 3), the selection conditions are restricted to low temperatures due to the low stability of the ternary complex of the ribosome, mRNA, and displayed polypeptide (Hanes and Plückthun 1997). Alternatively, yeast cell-surface display allows for fine control over selections using fluorescence-activated cell sorting, but the libraries are rather small and the yeast cells are also sensitive to high temperature and buffer conditions (Gai and Wittrup 2007). In contrast, filamentous phage remains infectious even after extended incubation at temperatures up to 65°C. Also, phage particles are tolerant of chemical denaturants, and it is possible to work under denaturing conditions to enable, for example, the selection of antibodies with improved thermodynamic stabilities (Jung et al. 1999).

In phage display, the displayed proteins that will ultimately be selected from a library are present in a very low copy number, as typically only a few thousand copies of each library variant are present in the unselected phage pool. Furthermore, in phage display, the encoded sequence space in a library does not change during the course of the experiment, and thus the selection process operates on a "static library." This is one of the main differences between phage display and other display techniques, such as ribosome display or mRNA display. In those methods, it is possible to further alter the outcome of each selection cycle by introducing additional mutations during the amplification between rounds, thereby providing access to an expanded sequence space beyond the diversity encoded in the naive library (Hanes et al. 1998). Thus, in ribosome display, the library itself is responsive to selective pressure and subject to dynamic changes, while in phage display, the library is considered fixed.

Library selection experiments are conceptually simple, but they can contain many technical challenges. Most experiments require immobilization of the target on a surface, and subsequently phage pools are added and extensively washed to remove non-binding particles. The selection round is concluded by an elution step, followed by infection of *E. coli* to amplify a new generation of phage for further selection or analysis (Figure 1.1). This process of binding, washing, elution, and amplification is repeated several times to eventually obtain a population dominated by binding clones.

To isolate phage-displayed polypeptides with desired phenotypes, two major obstacles have to be overcome within a library selection experiment. First, it is crucial to efficiently couple the target to a solid support in a homogenous fashion; the target must be in its native form, stable throughout the selection experiment, and reproducibly coated on the surface. Furthermore, the target must be solvent-exposed to allow for efficient enrichment of binding phage. Coupling the target directly to the solid support is not always advisable, because proteins can unfold due to adsorption to the surface (Engel et al. 2002; Engel et al. 2004; Gray 2004). To achieve efficient coupling, it is possible to use fusion tags that can be selectively captured by standard affinity reagents; for example, fusions to glutathione S-transferase can be captured with glutathione. Also, it is possible to crosslink proteins *in vitro* with biotin, which can subsequently be captured and immobilized by binding to streptavidin. Alternatively, proteins can be biotinylated *in vivo* by the enzyme BirA, which adds biotin to a short peptide epitope tag (Cull and Schatz 2000).

A second major factor affecting the success of a selection is that, apart from design issues, the library must be large and diverse enough to contain binding clones. In a phage display selection experiment, library members that have the desired properties are gradually enriched in consecutive rounds of screening. By the final screening round, functional proteins that share the same properties dominate the phage pool and their sequences can be revealed by sequencing of the encoding DNA.

## EVOLUTION OF BINDING AGENTS

In the past decade, several different protein frameworks have been harnessed to derive novel binding agents for use in biotechnology and drug discovery (Figure 1.4). Antibodies represent by far the most abundant framework under development. Within antibodies, six hypervariable loops or complementarity determining regions (CDRs) define the combined site responsible for antigen recognition (Knappik et al. 2000;

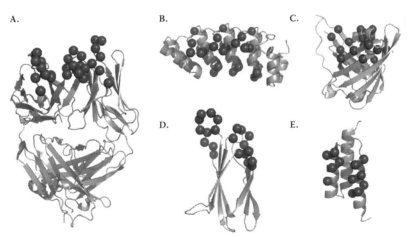

**FIGURE 1.4** Structures of scaffolds used for developing binding agents. Positions in the binding sites are shown as spheres, and the frameworks are shown as ribbons. (A) An antigen-binding fragment (Fab) of an antibody (PDB entry 1tzi) (Fellouse et al. 2004). (B) A designed ankyrin repeat protein (PDB 1svx) (Binz et al. 2004; Steiner et al. 2006). (C) A lipocalin scaffold based on a retinol binding protein (PDB 1lnm) (Schlehuber et al. 2000). (D) A fibronectin type III domain (PDB 2obg) (Koide et al. 2007). (E) An affibody based on the Z domain of protein A (PDB 2b87) (Lendel et al. 2006).

Krebs et al. 2001; Hoet et al. 2005). These CDR loops, which are supported by a more conserved structural framework, provide the diversity responsible for the vast repertoire of binding molecules in the human immune system. Although bacterial expression is possible, antibodies are difficult to express in *E. coli* in their full-length form (Simmons et al. 2002). Instead, phage display relies on the expression of antibody fragments, such as single-chain variable fragments (scFv) or antigen-binding fragments (Fab) (Figure 1.4A) (Sheets et al. 1998; de Haard et al. 1999). Tight affinities in the low- to sub-nanomolar range have been selected from phage-displayed antibody fragment libraries without further affinity maturation (Röthlisberger et al. 2004; Sidhu and Fellouse 2006; Fellouse et al. 2007).

Despite the considerable success of phage display, the most common method for deriving antibodies is still animal immunization, which leads to the production of polyclonal antibodies directly from sera or monoclonal antibodies from cell lines derived by hybridoma methods (Winter and Milstein 1991). However, these popular technologies involve lengthy and expensive processes and also impose several limitations. For example, animal immunization does not allow for the selection of antibodies against molecules that are toxic to the host, and the vertebrate immune system might not produce high affinity antibodies against some antigens due to immunological tolerance. In addition, the DNA encoding the antibodies is not obtained in a recombinant form, and thus it is not possible to identify the sequence of the antibody or to improve affinity, specificity, or stability. In contrast, phage display and other *in vitro* display methods (described in Chapters 2 and 3) do not suffer from these limitations, and thus they possess considerable advantages over animal immunization and hybridoma technologies.

Furthermore, *in vitro* display methods are not restricted to antibody repertoires, and many studies have investigated alternative scaffolds as substitutes for antibodies (described in detail in Chapter 5). In general, alternative scaffolds should have a stable protein core, which allows for the efficient formation of an antigen-binding site presented on the surface of the protein. As is the case for antibodies, the binding site must be tolerant to mutations and should not affect the overall folding of the protein. Additionally, these proteins should be well suited for high-level production in *E coli*. To date, several protein folds have been used as alternative scaffolds, and the most extensively studied include designed ankyrin repeat proteins (Figure 1.4B) (Binz et al. 2004), lipocalins (4C) (Beste et al. 1999), fibronectins (4D) (Koide and Koide 2007), and "affibodies" based on the Z domain of protein A (4E) (Lendel et al. 2006). In addition, other domains have also been used for more specialized applications, and these include kunitz domains (Ley et al. 1996), bovine pancreatic trypsin inhibitor (BPTI) (Roberts et al. 1992), and insect defensin (Dennis and Lazarus 1994). This list is far from complete, and we refer the interested reader to several recent reviews (Skerra 2000; Binz et al. 2005; Sidhu and Koide 2007; Skerra 2007).

Some alternative scaffolds (e.g., kunitz domains, BPTI, and insect defensin) are too small to form a substantial hydrophobic core and instead are stabilized mainly by disulfide bridges. These observations suggested that it might be possible to select small, structured binding peptides stabilized by disulfide bonds. In confirmation of this supposition, many studies have used random peptide libraries to select disulfide-constrained binding peptides (Sidhu et al. 2000; Szardenings 2003; Mori 2004).

These peptides have been used to target a remarkably diverse array of targets, including cell-surface receptors, hormones, and enzymes, and structural analyses have revealed that the peptides usually form discrete, compact structures (Figure 1.5) (Dennis and Lazarus 1994; Livnah et al. 1996; Wiesmann et al. 1998; Eigenbrot et al. 2001; Deshayes et al. 2002; Schaffer et al. 2003). Although the affinities of these binding peptides are typically lower than those of antibodies, peptide ligands can be readily synthesized and may be useful for specialized applications (Lowman 1997; Schaffer et al. 2003; Uchiyama et al. 2005).

## IMPROVEMENT OF PROTEIN STABILITY AND FOLDING

As mentioned previously, the secretion, folding, and stability of a protein all contribute to the efficiency with which it is displayed on phage particles, and thus these parameters play an important role in the success of phage display experiments. In addition, the dependence of phage display on these factors can be used to select for improved stability of a protein, and although phage display is not a true selection method for folding, it is nonetheless sensitive to folding.

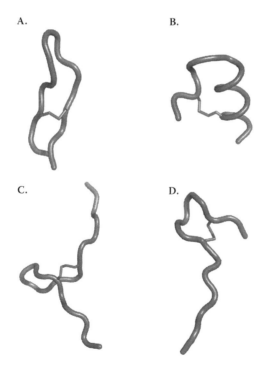

**FIGURE 1.5**  Structures of disulfide-constrained binding peptides selected from naive libraries. The peptides bind to (A) the erythropoietin receptor (PDB entry 1ebp) (Livnah et al. 1996), (B) insulin-like growth factor 1 (PDB 1lb7) (Deshayes et al. 2002; Schaffer et al. 2003), (C) vascular endothelial growth factor (PDB 1vpp) (Wiesmann et al. 1998), or (D) factor VIIa (PDB 1jbu) (Dennis and Lazarus 1994; Eigenbrot et al. 2001).

Since the phage particle is highly stable, it can sustain infectivity under relatively harsh conditions, such as high temperature, presence of denaturant, and challenge with proteolytic enzymes. Indeed, Kristensen and Winter have shown that treatment with proteases can be used to select stably folded proteins from phage-displayed libraries (Kristensen and Winter 1998). The rationale behind this type of selection is that an unfolded or molten globule protein exposes protease-sensitive sites. For this particular study, the protein of interest was cloned between the second (N2) and third (CT) domains of pIII. While the CT domain anchors pIII to the phage coat, the preceding two domains are necessary for host recognition and infection. Thus, proteolytic cleavage of the displayed protein results in the loss of the N-terminal domains and renders the phage noninfective. The validity of this approach was demonstrated in a control experiment using a cleavable linker, and subsequently the strategy was used to distinguish between proteins with differing stabilities. Furthermore, it was convincingly demonstrated that the stability of filamentous phage particles is not compromised under a variety of harsh conditions, including a wide range of pH, temperature, and concentrations of denaturants such as urea and guanidine (Kristensen and Winter 1998).

In the previously described study, the selection pressure is for phage infectivity and not directly for the functionality of the protein of interest. However, the methodology can be expanded by focusing directly on binding activity of the displayed protein. Jung et al. applied stress in the form of elevated temperature or presence of denaturant to a phage display library based on an antifluorescein scFv, which was selected for antigen binding (Jung et al. 1999). Five distinct variants exhibiting at least twofold improvements in affinity were identified, and additionally two of the variants also exhibited significant improvements in stability when selected under high temperature in the presence of denaturants. Interestingly, while the wild-type scFv exhibited significant attenuation of affinity in the presence of denaturant, the two variants with improved stability showed only a moderate reduction of affinity under the same conditions.

Alternatively, conformation-specific affinity reagents can be used to probe a protein domain for proper folding at a location distal to the antigen-binding site. Certain antibody heavy-chain variable ($V_H$) domains bind to protein A through a conserved binding site that is distinct from the antigen-binding site (Graille et al. 2000). This interaction depends on a correctly folded $V_H$ domain, and thus binding to protein A can be used to select stable antibody fragments. In one application, this selection strategy was used to evolve autonomous $V_H$ domains that fold and function in the absence of a light chain (Bond et al. 2003; Bond et al. 2005; Barthelemy et al. 2008).

In another approach, Georgiou and coworkers have shown that the twin-arginine translocation (Tat) pathway of *E. coli*, which selectively secretes only folded proteins, can also be used for the detection of stably folded proteins (DeLisa et al. 2003). They examined the relationship between protein folding and export competence in *E. coli* by analyzing the subcellular localization of proteins containing disulfide bonds. The *E. coli* strains used in their study either allowed or disfavored oxidative protein folding in the cytoplasm. As a result, they observed the Tat-dependent accumulation of active alkaline phosphatase and other multidisulfide proteins, including scFv and Fab antibody fragments. Successful export into the periplasm was observed only in *E. coli* strains that allowed the formation of disulfide bonds in the cytosol. Interestingly, the ability to distinguish between folded and unfolded proteins seems to be localized in one of the four proteins forming the Tat translocation machinery.

Subsequently, this feature of the *E. coli* machinery was applied to phage display (Paschke and Hohne 2005). A system was developed in which the protein of interest was secreted through the Tat pathway, while the pIII coat protein used for display was exported through the Sec protein-translocation pathway. This was necessary because pIII does not fold in the cytosol and thus would not be secreted by the Tat pathway. The protein of interest and pIII were fused to heterodimer leucine zippers such that they would pair in the periplasm to enable phage display of the heterologous protein. The system was used to display circularly permuted green fluorescent protein (GFP), and as a proof of concept, affinity selection was demonstrated for a GFP variant recognized by a monoclonal antibody. In another use of the Tat pathway, the efficient display of an immunoglobulin family protein (CD147) was demonstrated by direct fusion to pVIII (Thammawong et al. 2006).

## IDENTIFYING NATURAL PROTEIN–PROTEIN INTERACTIONS

Genome sequencing projects have provided vast databases at the DNA level and have identified thousands of open reading frames (Venter et al. 2001). Consequently, the demand for technologies that enable the functional analysis of gene products is at a premium. Furthermore, the identification and subsequent characterization of protein–protein interactions have become central to drug discovery and biotechnology. Because phage display provides a physical linkage between proteins and encoding DNA, the technology has proven useful for the screening of protein interaction pairs. Indeed, the very first report of phage display suggested that the method could be used to search DNA libraries for interactions at the protein level (Smith 1985).

Highly diverse display libraries have been constructed by fusing either cDNA (Crameri and Suter 1993; Crameri et al. 1994; Jespers et al. 1995; Dunn 1996; Rhyner et al. 2002) or genomic DNA fragments (Palzkill et al. 1998; Jacobsson and Frykberg 2001) to the genes encoding for pIII or pVIII. Conventional fusions to the N-termini of phage coat proteins are suboptimal for these applications because display is impaired by the early termination of protein translation due to the presence of stop codons at the ends of open reading frames (Crameri et al. 1996; Crameri and Kodzius 2001; Walter et al. 2001). However, it is possible to functionally display polypeptides fused to the C-termini of phage coat proteins (Kang et al. 1991; Jespers et al. 1995; Fuh and Sidhu 2000; Gao et al. 2002). Although not widely used, C-terminal fusions can provide functional display of cDNA libraries despite the presence of stop codons. A more widely used indirect strategy that also overcomes the problem of stop codons relies on heterodimer leucine zippers fused to the N- or C-termini of phage proteins or displayed proteins, respectively (Crameri and Suter 1993). This method has been successfully applied to the analysis of IgE antigens from various allergenic sources (Crameri and Kodzius 2001; Rhyner et al. 2002).

## SPECIFICITY PROFILING OF PEPTIDE-BINDING MODULES

Many intracellular proteins are composed of small, structurally discrete modules. In particular, scaffolding proteins that assemble signaling complexes often contain multiple peptide-binding modules. Thousands of peptide recognition domains are embedded in the human proteome, and these have been classified into over 70 distinct families (Pawson and Nash 2003). These families include domains that recognize phosphorylated peptides (SH2 and PTB), polyproline stretches (SH3 and WW) or C-terminal sequences (PDZ) (Figure 1.6). Phage-displayed peptide libraries have been used by many groups for specificity profiling of ligand binding to these types of domains (Sidhu et al. 2003a; Han et al. 2005).

Some of the earliest identified and characterized peptide-binding domains are SH3 domains, which due to their pivotal role in protein complexes have attracted considerable attention from molecular and cell biologists. SH3 domains consist of approximately 60 amino acids and are found in a wide variety of membrane-associated and cytoskeletal proteins, as well as in proteins with enzymatic activity and in adaptor proteins without catalytic activity (Macias et al. 2002). In general, SH3

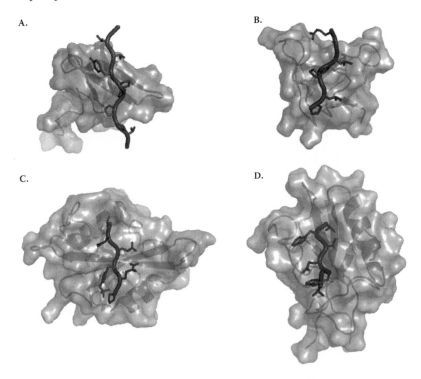

**FIGURE 1.6**  Structures of peptide-binding domains and peptide ligands. The domains are shown as gray surfaces, and the peptides are shown in dark gray. (A) The YAP65 WW domain bound to the peptide GTPPPPYTVG (PDB 1jmq) (Pires et al. 2001). (B) The Src SH3 domain bound to the peptide RALPPLP (PDB 1rlp) (Feng et al. 1994). (C) The Src SH2 domain bound to the phosphorylated peptide PQpYEEI (PDB 1sps) (Waksman et al. 1993). (D) The Erbin PDZ domain bound to the peptide TGWETWV (PDB 1n7t) (Skelton et al. 2003).

domains bind proline-rich peptides, and studies on their ligand binding specificities have provided a fascinating glimpse into the principles governing subcellular signaling architecture.

Phage-displayed peptide libraries have proven to be a powerful tool for the profiling of a wide array of SH3 domains. Several studies have provided significant insights into the mechanisms responsible for recognition of proline-rich sequences by SH3 domains (Rickles et al. 1994; Rickles et al. 1995; Sparks et al. 1996; Cestra et al. 1999; Mongiovi et al. 1999; Tong et al. 2002; Landgraf et al. 2004). Phage display has been used to identify proline-rich peptides that resemble natural ligands but in many cases have higher SH3 binding affinities. In addition, some studies have identified binding motifs that differ from the canonical SH3 binding motifs, suggesting that the range of SH3 ligand specificities may be more diverse than previously appreciated (Cestra et al. 1999; Mongiovi et al. 1999).

In theory, the specificity profiles defined by phage display could be used to search for natural binding partners in genomic databases, but such searches usually yield an excessively large number of false positives. However, these false positives can be reduced by combining phage display analysis with other experimental methods. For example, Tong et al. used an approach that combined data from yeast two-hybrid with data from phage display (Tong et al. 2002). In the first step, phage display was used to identify binding sequences for 20 yeast SH3 domains, and this data set was used to computationally predict putative natural interaction partners in the yeast genome. Independently, the yeast SH3 domains were used as "baits" in yeast two-hybrid screens to identify interacting partners from cDNA libraries. Comparison of the predicted partners from the phage display and yeast two-hybrid analyses revealed a common set of predicted partners that was enriched for bona fide interactions and depleted of false positives.

PDZ domains are modules of approximately 90 amino acids, which predominantly recognize the C-termini of various intracellular and membrane-associated proteins. They act as scaffolding modules to assemble multiprotein complexes at subcelluar sites (Craven and Bredt 1998; Fanning and Anderson 1999). PDZ domains are highly abundant in metazoans and are associated with numerous functions that are typical of multicellular organisms. Many PDZ domains share a common binding mode for ligand recognition, but nonetheless different domains are specific for different sets of ligands. The central question of how this domain family mediates specificity has been investigated by phage display.

Peptide libraries fused to the C-terminus of pVIII (Fuh and Sidhu 2000) were used to investigate the binding specificity of the PDZ domain of Erbin (Jung et al. 1999; Laura et al. 2002; Skelton et al. 2003), a protein originally identified as a putative natural ligand for the cell-surface receptor ErbB2 (Borg et al. 2000). Surprisingly, phage display revealed a binding consensus significantly different from the proposed binding sequence in ErbB2, and follow-up studies showed that the Erbin PDZ domain binds with much higher affinity to a family of catenins localized at cell-cell junctions (Laura et al. 2002). Subsequent studies have used a similar approach to profile the specificities of other PDZ domains involved in cell-cell junctions (Appleton et al. 2006; Zhang et al. 2006), and also domains involved in the regulation of intracellular and extracellular proteolysis (Runyon et al. 2007; Zhang et al. 2007). Recently, the technology has been applied to a large-scale analysis of close to 100 domains, and the resulting database has provided a specificity map for the PDZ domain family (Tonikian et al. 2008).

## MAPPING BINDING ENERGETICS

Although most phage display studies have focused on either improving natural protein functions or developing novel protein functions, the method can also be used to understand the molecular basis of protein structure and function. Libraries with restricted diversity have proven to be highly effective for studying protein function, because reducing the chemical complexity of the library greatly simplifies the identification of key interactions within a binding interface (Morrison and Weiss 2001; Avrantinis and Weiss 2005; Runyon et al. 2007; Zhang et al. 2007).

Shotgun alanine scanning (described in more detail in Chapter 4) was developed as a high-throughput, phage display version of traditional alanine scanning for mapping binding energetics at protein binding interfaces (Weiss et al. 2000). Libraries are constructed using binary degenerate codons that ideally encode for only the wild-type and alanine at each position to be scanned, and dozens of positions can be scanned in a single library. Following selection for either structure or function using appropriate immobilized ligands, individual clones are sequenced, and the importance of each residue for the selected trait is determined from the ratio of wild-type relative to alanine at each position. The wild-type sequence is conserved at positions that are important for the selected trait, and the analysis is quantitative, so the energetic contributions of individual side chains can be assessed with accuracies approaching those for traditional biophysical analyses with purified proteins.

Shotgun alanine scanning was first applied to study interactions between human growth hormone (hGH) and its receptor (Weiss et al. 2000). This combinatorial method successfully reiterated previous results obtained by conventional alanine scanning, and enabled subsequent homolog and serine scans that provided deeper insights into the molecular details of the interaction (Pal et al. 2005). The method was also used to establish the structural basis for the improved affinity of an hGH variant derived from phage display selections (Opalka et al. 2003; Pal et al. 2005). In addition, the hGH system showed that shotgun alanine scanning data can be used to detect cooperative interactions between interface side chains by analysis of the frequencies of double-alanine mutations (Pal et al. 2005). Following the success of shotgun alanine scanning, a "quantitative saturation" scanning approach was developed where diversity is restricted spatially rather than chemically (Pal et al. 2006). This method allowed for the quantitative assessment of the effects of all possible mutations at all positions in the hGH interface for receptor binding. The interface was found to be highly tolerant to mutations, but the nature of the tolerated mutations challenged generally accepted views about the biophysical and evolutionary pressures governing protein–protein interactions.

In recent years, shotgun scanning approaches have been applied to many other proteins. The method was applied to the phage display platform itself to determine how coat proteins assemble into phage particles (Roth et al. 2002; Weiss et al. 2003). Several antibodies have been scanned to shed light on the details of antigen recognition (Vajdos et al. 2002; Fellouse et al. 2006; Lee et al. 2006), and peptide-protein interactions have also been analyzed (Murase et al. 2003; Schaffer et al. 2003; Skelton et al. 2003). The method has been extended beyond protein–protein interactions in studies of proteins that recognize DNA (Sato et al. 2004) or small molecules (Avrantinis et al. 2002). Shotgun scanning principles also apply in systems other than phage display, and combinatorial alanine scanning has been adapted for *in vivo* selections with protein complementation assays (Phillips et al. 2006). Overall, these numerous studies validate the use of combinatorial libraries as effective tools for the rapid yet quantitative analysis of the principles underlying protein structure and function.

## FUTURE PERSPECTIVES

Phage display is the longest-standing platform for the selection of protein structure and function from combinatorial libraries. The robustness of the method has made phage display a powerful tool for protein engineering, and we anticipate that automation and robotics will exponentially expand the power of the technology and enable high-throughput protein engineering. The developing fields of functional genomics and systems biology are concerned with the analysis of complete processes rather than isolated proteins (Cesareni et al. 2005; Neduva and Russell 2006; Bader et al. 2008), and such studies demand tools that can allow for the efficient and rapid analysis of protein–protein interactions on a much larger scale than is possible with traditional biophysical methods. In such an environment, phage display and other combinatorial methods will play a critical role in enabling large-scale studies of protein function (Sidhu et al. 2003b). In the near future, we envision an integrated system for protein engineering that utilizes phage display, automation, and high-throughput screening for the rapid analysis of natural proteins and the facile engineering of synthetic proteins with novel functions.

## REFERENCES

Appleton, B. A., Y. N. Zhang, P. Wu, J. P. Yin, W. Hunziker, et al. 2006. "Comparative structural analysis of the Erbin PDZ domain and the first PDZ domain of ZO-1—Insights into determinants of PDZ domain specificity." *J Biol Chem* 281: 22312–20.

Avrantinis, S. K., R. L. Stafford, X. Tian and G. A. Weiss. 2002. "Dissecting the streptavidin-biotin interaction by phage-displayed shotgun scanning." *Chem Bio Chem* 3: 1229–34.

Avrantinis, S. K. and G. A. Weiss (2005). Mapping protein functional epitopes. *Phage display in biotechnology and drug discovery.* S. S. Sidhu. Boca Raton, CRC Press. 3: 441–60.

Bader, S., S. Kühner and A. C. Gavin. 2008. "Interaction networks for systems biology." *FEBS Lett* 582: 1220–24.

Barbas, C. F., III, D. R. Burton, J. K. Scott and G. J. Silverman. 2001. "Phage display: A laboratory manual." *Phage display: A laboratory manual,* i–xvi, 1.1–23.32, A1.1–A6.3, I.1–I.22.

Barbas, C. F., A. S. Kang, R. A. Lerner and S. J. Benkovic. 1991. "Assembly of the combinatorial antibody libraries on phage surfaces—The gene-III site." *Proc Natl Acad Sci U S A* 88: 7978–82.

Barthelemy, P. A., H. Raab, B. A. Appleton, C. J. Bond, P. Wu, et al. 2008. "Comprehensive analysis of the factors contributing to the stability and solubility of autonomous human V-H domains." *J Biol Chem* 283: 3639–54.

Bass, S., R. Greene and J. A. Wells. 1990. "Hormone phage—an enrichment method for a variant proteins with altered binding-properties." *Proteins* 8: 309–14.

Beste, G., F. S. Schmidt, T. Stibora and A. Skerra. 1999. "Small antibody-like proteins with prescribed ligand specificities derived from the lipocalin fold." *Proc Natl Acad Sci U S A* 96: 1898–1903.

Binz, H. K., P. Amstutz, A. Kohl, M. T. Stumpp, C. Briand, et al. 2004. "High-affinity binders selected from designed ankyrin repeat protein libraries." *Nat Biotechnol* 22: 575–82.

Binz, H. K., P. Amstutz and A. Plückthun. 2005. "Engineering novel binding proteins from nonimmunoglobulin domains." *Nat Biotechnol* 23: 1257–68.

Bond, C. J., J. C. Marsters and S. S. Sidhu. 2003. "Contributions of CDR3 to VHH domain stability and the design of monobody scaffolds for naive antibody libraries." *J Mol Biol* 332: 643–55.

Bond, C. J., C. Wiesmann, J. C. Marsters and S. S. Sidhu. 2005. "A structure-based database of antibody variable domain diversity." *J Mol Biol* 348: 699–709.

Borg, J. P., S. Marchetto, A. Le Bivic, V. Ollendorff, F. Jaulin-Bastard, et al. 2000. "ERBIN: a basolateral PDZ protein that interacts with the mammalian ERBB2/HER2 receptor." *Nat Cell Biol* 2: 407–14.

Cesareni, G., A. Ceol, C. Gavrila, L. M. Palazzi, M. Persico, et al. 2005. "Comparative interactomics." *FEBS Lett* 579: 1828–33.

Cestra, G., L. Castagnoli, L. Dente, O. Minenkova, A. Petrelli, et al. 1999. "The SH3 domains of endophilin and amphiphysin bind to the proline-rich region of synaptojanin 1 at distinct sites that display an unconventional binding specificity." *J Biol Chem* 274: 32001–7.

Chappel, J. A., M. He and A. S. Kang. 1998. "Modulation of antibody display on M13 filamentous phage." *J Immunol Methods* 221: 25–34.

Crameri, R., S. Hemmann and K. Blaser. 1996. "pJuFo: A phagemid for display of cDNA libraries on phage surface suitable for selective isolation of clones expressing allergens." *Adv Exp Med Biol* 409: 103–10.

Crameri, R., R. Jaussi, G. Menz and K. Blaser. 1994. "Display of expression products of cDNA libraries on phage surfaces—A versatile screening system for selective isolation of genes by specific gene-product ligand interaction." *Eur J Biochem* 226: 53–58.

Crameri, R. and R. Kodzius. 2001. "The powerful combination of phage surface display of cDNA libraries and high throughput screening." *Comb Chem High Throughput Screen* 4: 145–55.

Crameri, R. and M. Suter. 1993. "Display of biologically active proteins in the surface of Filamentous Phage—A cDNA cloning system for the selection of functional gene-products linked to the genetic information responsible for their production." *Gene* 137: 69–75.

Craven, S. E. and D. S. Bredt. 1998. "PDZ proteins organize synaptic signaling pathways." *Cell* 93: 495–8.

Cull, M. G. and P. J. Schatz (2000). Biotinylation of proteins in vivo and in vitro using small peptide tags. *Applications of Chimeric Genes and Hybrid Proteins, Pt A.* 326: 430–40.

Cwirla, S. E., E. A. Peters, R. W. Barrett and W. J. Dower. 1990. "Peptides on phage—a vast library of peptides for identifying ligands." *Proc Natl Acad Sci U S A* 87: 6378–82.

de Haard, H. J., N. van Neer, A. Reurs, S. E. Hufton, R. C. Roovers, et al. 1999. "A large nonimmunized human Fab fragment phage library that permits rapid isolation and kinetic analysis of high affinity antibodies." *J Biol Chem* 274: 18218–30.

DeLisa, M. P., D. Tullman and G. Georgiou. 2003. "Folding quality control in the export of proteins by the bacterial twin-arginine translocation pathway." *Proc Natl Acad Sci U S A* 100: 6115–20.

Dennis, M. S. and R. A. Lazarus. 1994. "Kunitz domain inhibitors of tissue factor-factor VIIA. 1. Potent inhibitors selected from libraries by phage display." *J Biol Chem* 269: 22129–36.

Deshayes, K., M. L. Schaffer, N. J. Skelton, G. R. Nakamura, S. Kadkhodayan, et al. 2002. "Rapid identification of small binding motifs with high-throughput phage display: Discovery of peptidic antagonists of IGF-1 function." *Chem Biol* 9: 495–505.

Devlin, J. J., L. C. Panganiban and P. E. Devlin. 1990. "Random peptide libraries—a source of specific protein-binding molecules." *Science* 249: 404–6.

Dunn, I. S. 1996. "Phage display of proteins." *Curr Opin Biotechnol* 7: 547–53.

Eigenbrot, C., D. Kirchhofer, M. S. Dennis, L. Santell, R. A. Lazarus, et al. 2001. "The factor VII zymogen structure reveals re-registration of beta strands during activation." *Structure* 9: 627–36.

Engel, M. F. M., C. P. M. van Mierlo and A. Visser. 2002. "Kinetic and structural characterization of adsorption-induced unfolding of bovine alpha-lactalbumin." *J Biol Chem* 277: 10922–30.

Engel, M. F. M., A. Visser and C. P. M. van Mierlo. 2004. "Conformation and orientation of a protein folding intermediate trapped by adsorption." *Proc Natl Acad Sci U S A* 101: 11316–21.

Fanning, A. S. and J. M. Anderson. 1999. "Protein modules as organizers of membrane structure." *Curr Opin Cell Biol* 11: 432–9.

Felici, F., L. Castagnoli, A. Musacchio, R. Jappelli and G. Cesareni. 1991. "Selection of antibody ligands form a large library of oligopeptides expressed on multivalent exposition vector." *J Mol Biol* 222: 301–10.

Fellouse, F. A., P. A. Barthelemy, R. F. Kelley and S. S. Sidhu. 2006. "Tyrosine plays a dominant functional role in the paratope of a synthetic antibody derived from a four amino acid code." *J Mol Biol* 357: 100–14.

Fellouse, F. A., K. Esaki, S. Birtalan, D. Raptis, V. J. Cancasci, et al. 2007. "High-throughput generation of synthetic antibodies from highly functional minimalist phage-displayed libraries." *J Mol Biol* 373: 924–40.

Fellouse, F. A., C. Wiesmann and S. S. Sidhu. 2004. "Synthetic antibodies from a four-amino-acid code: A dominant role for tyrosine in antigen recognition." *Proc Natl Acad Sci U S A* 101: 12467–72.

Feng, J. N., M. Russel and P. Model. 1997. "A permeabilized cell system that assembles filamentous bacteriophage." *Proc Natl Acad Sci U S A* 94: 4068–73.

Feng, S. B., J. K. Chen, H. T. Yu, J. A. Simon and S. L. Schreiber. 1994. "2 Binding orientations for peptides to the Src SH3 domain—development of a general-model for SH3-ligand interactions." *Science* 266: 1241–7.

Fuh, G. and S. S. Sidhu. 2000. "Efficient phage display of polypeptides fused to the carboxy-terminus of the M13 gene-3 minor coat protein." *FEBS Lett* 480: 231–4.

Gai, S. A. and K. D. Wittrup. 2007. "Yeast surface display for protein engineering and characterization." *Curr Opin Struct Biol* 17: 467–73.

Gao, C. S., S. L. Mao, G. Kaufmann, P. Wirsching, R. A. Lerner, et al. 2002. "A method for the generation of combinatorial antibody libraries using pIX phage display." *Proc Natl Acad Sci U S A* 99: 12612–16.

Gao, C. S., S. L. Mao, C. H. L. Lo, P. Wirsching, R. A. Lerner, et al. 1999. "Making artificial antibodies: A format for phage display of combinatorial heterodimeric arrays." *Proc Natl Acad Sci U S A* 96: 6025–30.

Georgiou, G., C. Stathopoulos, P. S. Daugherty, A. R. Nayak, B. L. Iverson, et al. 1997. "Display of heterologous proteins on the surface of microorganisms: From the screening of combinatorial libraries to live recombinant vaccines." *Nat Biotechnol* 15: 29–34.

Graille, M., E. A. Stura, A. L. Corper, B. J. Sutton, M. J. Taussig, et al. 2000. "Crystal structure of a Staphylococcus aureus protein A domain complexed with the Fab fragment of a human IgM antibody: Structural basis for recognition of B-cell receptors and superantigen activity." *Proc Natl Acad Sci U S A* 97: 5399–5404.

Gray, J. J. 2004. "The interaction of proteins with solid surfaces." *Curr Opin Struct Biol* 14: 110–15.

Greenwood, J., A. E. Willis and R. N. Perham. 1991. "Multiple display of foreign peptides on a filamentous bacteriophage—peptides from Plasmodium-Falciparum circumsporozoite proteins as antigens." *J Mol Biol* 220: 821–7.

Han, Z., E. Karatan and B. Kay (2005). Mapping intracellular protein networks. *Phage Display in Biotechnology and Drug Discovery*. S. Sidhu, CRC: 322–46.

Hanes, J., L. Jermutus, S. Weber-Bornhauser, H. R. Bosshard and A. Plückthun. 1998. "Ribosome display efficiently selects and evolves high-affinity antibodies in vitro from immune libraries." *Proc Natl Acad Sci U S A* 95: 14130–5.

Hanes, J. and A. Plückthun. 1997. "In vitro selection and evolution of functional proteins by using ribosome display." *Proc Natl Acad Sci U S A* 94: 4937–42.

Hoet, R. M., E. H. Cohen, R. B. Kent, K. Rookey, S. Schoonbroodt, et al. 2005. "Generation of high-affinity human antibodies by combining donor-derived and synthetic complementarity-determining-region diversity." *Nat Biotechnol* 23: 344–8.

Huang, W. Z., M. McKevitt and T. Palzkill. 2000. "Use of the arabinose p(bad) promoter for tightly regulated display of proteins on bacteriophage." *Gene* 251: 187–97.

Hufton, S. E., P. T. Moerkerk, E. V. Meulemans, A. de Bruine, J. W. Arends, et al. 1999. "Phage display of cDNA repertoires: the pVI display system and its applications for the selection of immunogenic ligands." *J Immunol Methods* 231: 39–51.

Il'Ichev, A. A., O. O. Minenkova, S. I. Tat'Kov, N. N. Karpyshev, A. M. Eroshkin, et al. 1989. "Production of a viable variant of phage M13 with a foreign peptide inserted into the coat basic protein." *Doklady Akademii Nauk SSSR* 307: 481–3.

Jacobsson, K. and L. Frykberg. 2001. "Shotgun phage display cloning." *Comb Chem High Throughput Screen* 4: 135–43.

Jespers, L. S., J. H. Messens, A. Dekeyser, D. Eeckhout, I. Vandenbrande, et al. 1995. "Surface expression and ligand-based selection of cDNAs fused to filamentous Phage Gene VI." *Bio-Technology* 13: 378–82.

Jung, S., A. Honegger and A. Plückthun. 1999. "Selection for improved protein stability by phage display." *J Mol Biol* 294: 163–80.

Kang, A. S., C. F. Barbas, K. D. Janda, S. J. Benkovic and R. A. Lerner. 1991. "Linkage of recognition and replication functions by assembling combinatorial antibody Fab libraries along the phage surfaces." *Proc Natl Acad Sci U S A* 88: 4363–6.

Knappik, A., L. M. Ge, A. Honegger, P. Pack, M. Fischer, et al. 2000. "Fully synthetic human combinatorial antibody libraries (HuCAL) based on modular consensus frameworks and CDRs randomized with trinucleotides." *J Mol Biol* 296: 57–86.

Kohl, A., H. K. Binz, P. Forrer, M. T. Stumpp, A. Plückthun, et al. 2003. "Designed to be stable: Crystal structure of a consensus ankyrin repeat protein." *Proc Natl Acad Sci U S A* 100: 1700–05.

Koide, A., R. N. Gilbreth, K. Esaki, V. Tereshko and S. Koide. 2007. "High-affinity single-domain binding proteins with a binary-code interface." *Proc Natl Acad Sci U S A* 104: 6632–7.

Koide, A. and S. Koide. 2007. "Monobodies—antibody mimics based on the scaffold of the fibronectin type III domain." *Methods Mol Biol* 352: 95–109.

Konings, R. N. H., R. H. A. Folmer, P. J. M. Folkers, M. Nilges and C. W. Hilbers. 1995. "3-Dimensional structure of the single-stranded DNA-binding protein encoded be gene-V of the filamentous bacteriophage-M13 and a model of its complex with single-stranded DNA." *Fems Microbiology Reviews* 17: 57–72.

Krebber, C., S. Spada, D. Desplancq, A. Krebber, L. M. Ge, et al. 1997. "Selectively-infective phage (SIP): A mechanistic dissection of a novel in vivo selection for protein–ligand interactions." *J Mol Biol* 268: 607–18.

Krebs, B., R. Rauchenberger, R. Silke, C. Rothe, M. Tesar, et al. 2001. "High-throughput generation and engineering of recombinant human antibodies." *J Immunol Methods* 254: 67–84.

Kristensen, P. and G. Winter. 1998. "Proteolytic selection for protein folding using filamentous bacteriophages." *Fold Des* 3: 321–8.

Landgraf, C., S. Panni, L. Montecchi-Palazzi, L. Castagnoli, J. Schneider-Mergener, et al. 2004. "Protein interaction networks by proteome peptide scanning." *PLoS Biol* 2: 94–103.

Laura, R. P., A. S. Witt, H. A. Held, R. Gerstner, K. Deshayes, et al. 2002. "The Erbin PDZ domain binds with high affinity and specificity to the carboxyl termini of delta-catenin and ARVCF." *J Biol Chem* 277: 12906–14.

Lee, C. V., S. G. Hymowitz, H. J. A. Wallweber, N. C. Gordon, K. L. Billeci, et al. 2006. "Synthetic anti-BR3 antibodies that mimic BAFF binding and target both human and murine B cells." *Blood* 108: 3103–11.

Lendel, C., J. Dogan and T. Hard. 2006. "Structural basis for molecular recognition in an Affibody: Affibody complex." *J Mol Biol* 359: 1293–1304.

Ley, A. C., W. Markland and R. C. Ladner. 1996. "Obtaining a family of high-affinity, high-specificity protein inhibitors of plasmin and plasma kallikrein." *Mol Divers* 2: 119–24.

Livnah, O., E. A. Stura, D. L. Johnson, S. A. Middleton, L. S. Mulcahy, et al. 1996. "Functional mimicry of a protein hormone by a peptide agonist: The EPO receptor complex at 2.8 angstrom." *Science* 273: 464–71.

Lowman, H. B. 1997. "Bacteriophage display and discovery of peptide leads for drug development." *Annual Rev Biophys Biomol Struct* 26: 401–24.

Lowman, H. B., S. H. Bass, N. Simpson and J. A. Wells. 1991. "Selecting high-affinity binding-proteins by monovalent phage display." *Biochemistry* 30: 10832–8.

Lubkowski, J., F. Hennecke, A. Plückthun and A. Wlodawer. 1998. "The structural basis of phage display elucidated by the crystal structure of the N-terminal domains of g3p." *Nat Struct Biol* 5: 140–47.

Lubkowski, J., F. Hennecke, A. Plückthun and A. Wlodawer. 1999. "Filamentous phage infection: crystal structure of g3p in complex with its coreceptor, the C-terminal domain of TolA." *Structure* 7: 711–22.

Macias, M. J., S. Wiesner and M. Sudol. 2002. "WW and SH3 domains, two different scaffolds to recognize proline-rich ligands." *FEBS Lett* 513: 30–37.

Marciano, D. K., M. Russel and S. M. Simon. 1999. "An aqueous channel for filamentous phage export." *Science* 284: 1516–9.

Marciano, D. K., M. Russel and S. M. Simon. 2001. "Assembling filamentous phage occlude pIV channels." *Proc Natl Acad Sci U S A* 98: 9359–64.

Marvin, D. A. 1998. "Filamentous phage structure, infection and assembly." *Curr Opin Struct Biol* 8: 150–8.

Mongiovi, A. M., P. R. Romano, S. Panni, M. Mendoza, W. T. Wong, et al. 1999. "A novel peptide-SH3 interaction." *EMBO J* 18: 5300–9.

Mori, T. 2004. "Cancer-specific ligands identified from screening of peptide-display libraries." *Curr Pharm Des* 10: 2335–43.

Morrison, K. L. and G. A. Weiss. 2001. "Combinatorial alanine-scanning." *Curr Opin Chem Biol* 5: 302–7.

Murase, K., K. L. Morrison, P. Y. Tam, R. L. Stafford, F. Jurnak, et al. 2003. "EF-Tu binding peptides indentified, dissected, and affinity optimized by phage display." *Chem Biol* 10: 161–8.

Nakamura, M., K. Tsumoto, I. Kumagai and K. Ishimura. 2003. "A morphologic study of filamentous phage infection of Escherichia coli using biotinylated phages." *FEBS Lett* 536: 167–72.

Neduva, V. and R. B. Russell. 2006. "Peptides mediating interaction networks: New leads at last." *Curr Opin Biotechnol* 17: 465–71.

Opalka, N., R. Beckmann, N. Boisset, M. N. Simon, M. Russel, et al. 2003. "Structure of the filamentous phage pIV multimer by cryo-electron microscopy." *J Mol Biol* 325: 461–70.

Pal, G., S. Y. Fong, A. A. Kossiakoff and S. S. Sidhu. 2005. "Alternative views of functional protein binding epitopes obtained by combinatorial shotgun scanning mutagenesis." *Protein Sci* 14: 2405–13.

Pal, G., J.-L. K. Kouadio, D. R. Artis, A. A. Kossiakoff and S. S. Sidhu. 2006. "Comprehensive and quantitative mapping of energy landscapes for protein–protein interactions by rapid combinatorial scanning." *J Biol Chem* 281: 22378–85.

Palzkill, T., W. Z. Huang and G. M. Weinstock. 1998. "Mapping protein–ligand interactions using whole genome phage display libraries." *Gene* 221: 79–83.

Paschke, M. and W. Hohne. 2005. "A twin-arginine translocation (Tat)-mediated phage display system." *Gene* 350: 79–88.

Pawson, T. and P. Nash. 2003. "Assembly of cell regulatory systems through protein interaction domains." *Science* 300: 445–2.

Phillips, K. J., D. M. Rosenbaum and D. R. Liu. 2006. "Binding and stability determinants of the PPARγ nuclear receptor-coactivator interface as revealed by shotgun alanine scanning and in vivo selection." *J Am Chem Soc* 128: 11298–306.

Pires, J. R., F. Taha-Nejad, F. Toepert, T. Ast, U. Hoffmuller, et al. 2001. "Solution structures of the YAP65 WW domain and the variant L30 K in complex with the peptides GTPPPPYTVG, N-(n-octyl)-GPPPY and PLPPY and the application of peptide libraries reveal a minimal binding epitope." *J Mol Biol* 314: 1147–56.

Possee, R. D. 1997. "Baculoviruses as expression vectors." *Curr Opin Biotechnol* 8: 569–72.

Rakonjac, J. and P. Model. 1998. "Roles of pIII in filamentous phage assembly." *J Mol Biol* 282: 25–41.

Ren, Z. J. and L. W. Black. 1998. "Phage T4 SOC and HOC display of biologically active, full-length proteins on the viral capsid." *Gene* 215: 439–44.

Rhyner, C., R. Kodzius and R. Crameri. 2002. "Direct selection of cDNAs from filamentous phage surface display libraries: Potential and limitations." *Curr Pharm Biotechnol* 3: 13–21.

Rickles, R. J., M. C. Botfield, Z. G. Weng, J. A. Taylor, O. M. Green, et al. 1994. "Identification of Src, Fyn, Lyn, PI3K and Abl SH3-domain ligands using phage display libraries." *EMBO J* 13: 5598–5604.

Rickles, R. J., M. C. Botfield, X. M. Zhou, P. A. Henry, J. S. Brugge, et al. 1995. "Phage display selection of ligand residues important for Src-homology-3 domain binding-specificity." *Proc Natl Acad Sci U S A* 92: 10909–13.

Riechmann, L. and P. Holliger. 1997. "The C-terminal domain of ToIA is the coreceptor for filamentous phage infection of E-coli." *Cell* 90: 351–60.

Roberts, B. L., W. Markland, K. Siranosian, M. J. Saxena, S. K. Guterman, et al. 1992. "Protease inhibitor display M13-phage—Selection of high-affinity neutrophil elastase inhibitors." *Gene* 121: 9–15.

Roth, T. A., G. A. Weiss, C. Eigenbrot and S. S. Sidhu. 2002. "A minimized M13 coat protein defines the requirements for assembly into the bacteriophage particle." *J Mol Biol* 322: 357–67.

Röthlisberger, D., K. M. Pos and A. Plückthun. 2004. "An antibody library for stabilizing and crystallizing membrane proteins—selecting binders to the citrate carrier CitS." *FEBS Lett* 564: 340–8.

Runyon, S. T., Y. G. Zhang, B. A. Appleton, S. L. Sazinsky, P. Wu, et al. 2007. "Structural and functional analysis of the PDZ domains of human HtrA1 and HtrA3." *Protein Sci* 16: 2454–71.

Russel, M., H. Whirlow, T. P. Sun and R. E. Webster. 1988. "Low-Frequency of infection of the F-bacteria by transducing particles of filamentous bacteriophages." *J Bacteriol* 170: 5312–6.

Santini, C., D. Brennan, C. Mennuni, R. H. Hoess, A. Nicosia, et al. 1998. "Efficient display of an HCV cDNA expression library as C-terminal fusion to the capsid protein D of bacteriophage lambda." *J Mol Biol* 282: 125–35.

Sato, K., M. D. Simon, A. M. Levin, K. M. Shokat and G. A. Weiss. 2004. "Dissecting the engrailed homedomain-DNA interaction by phage-displayed shotgun scanning." *Chem Biol* 11: 1017–23.

Schaffer, M. L., K. Deshayes, G. Nakamura, S. Sidhu and N. J. Skelton. 2003. "Complex with a phage display-derived peptide provides insight into the function of insulin-like growth factor I." *Biochemistry* 42: 9324–34.

Schlehuber, S., G. Beste and A. Skerra. 2000. "A novel type of receptor protein, based on the lipocalin scaffold, with specificity for digoxigenin." *J Mol Biol* 297: 1105–20.

Scott, J. K. and G. P. Smith. 1990. "Searching for peptide ligands with an epitope library." *Science* 249: 386–90.

Sheets, M. D., P. Amersdorfer, R. Finnern, P. Sargent, E. Lindqvist, et al. 1998. "Efficient construction of a large nonimmune phage antibody library: The production of high-affinity human single-chain antibodies to protein antigens." *Proc Natl Acad Sci U S A* 95: 6157–62.

Sidhu, S. S., G. D. Bader and C. Boone. 2003a. "Functional genomics of intracellular peptide recognition domains with combinatorial biology methods." *Curr Opin Chem Biol* 7: 97–102.

Sidhu, S. S., W. J. Fairbrother and K. Deshayes. 2003b. "Exploring protein–protein interactions with phage display." *Chembiochem* 4: 14–25.

Sidhu, S. S. and F. A. Fellouse. 2006. "Synthetic therapeutic antibodies." *Nat Chem Biol* 2: 682–8.

Sidhu, S. S. and S. Koide. 2007. "Phage display for engineering and analyzing protein interaction interfaces." *Curr Opin Struct Biol* 17: 481–7.

Sidhu, S. S., H. B. Lowman, B. C. Cunningham and J. A. Wells. 2000. "Phage display for selection of novel binding peptides." *Applications of Chimeric Genes and Hybrid Proteins, Pt C.* 328: 333–63.

Simmons, L. C., D. Reilly, L. Klimowski, T. S. Raju, G. Meng, et al. 2002. "Expression of full-length immunoglobulins in Escherichia coli: rapid and efficient production of agly-cosylated antibodies." *J Immunol Methods* 263: 133–47.

Skelton, N. J., M. F. T. Koehler, K. Zobel, W. L. Wong, S. Yeh, et al. 2003. "Origins of PDZ domain ligand specificity: Structure determination and mutagenesis of the Erbin PDZ domain." *J Biol Chem* 278: 7645–54.

Skerra, A. 2000. "Engineered protein scaffolds for molecular recognition." *J Mol Recognit* 13: 167–87.

Skerra, A. 2007. "Alternative non-antibody scaffolds for molecular recognition." *Curr Opin Biotechnol* 18: 295–304.

Smith, G. P. 1985. "Filamentous fusion phage—novel expression vectors that display cloned antigens on the virion surface." *Science* 228: 1315–7.

Smith, G. P. 1993. "Surface display and peptide libraries." *Gene* 128: 1–2.

Sparks, A. B., J. E. Rider, N. G. Hoffman, D. M. Fowlkes, L. A. Quilliam, et al. 1996. "Distinct ligand preferences of Src homology 3 domains from Src, Yes, Abl, Cortactin, p53bp2, PLC gamma, Crk, and Grb2." *Proc Natl Acad Sci U S A* 93: 1540–4.

Steiner, D., P. Forrer, M. T. Stumpp and A. Plückthun. 2006. "Signal sequences directing cotranslational translocation expand the range of proteins amenable to phage display." *Nat Biotechnol* 24: 823–31.

Szardenings, M. 2003. "Phage display of random peptide libraries: Applications, limits, and potential." *J Recept Signal Transduct Res* 23: 307–49.

Thammawong, P., W. Kasinrerk, R. J. Turner and C. Tayapiwatana. 2006. "Twin-arginine signal peptide attributes effective display of CD147 to filamentous phage." *Appl Microbiol Biotechnol* 69: 697–703.

Tong, A. H. Y., B. Drees, G. Nardelli, G. D. Bader, B. Brannetti, et al. 2002. "A combined experimental and computational strategy to define protein interaction networks for peptide recognition modules." *Science* 295: 321–4.

Tonikian, R., Y. N. Zhang, S. L. Sazinsky, B. Currell, J. H. Yeh, et al. 2008. "A specificity map for the PDZ domain family." *Plos Biology* 6(9): 2043–2059.

Uchiyama, F., Y. Tanaka, Y. Minari and N. Toku. 2005. "Designing scaffolds of peptides for phage display libraries." *J Biosci Bioengin* 99: 448–56.

Vajdos, F. F., C. W. Adams, T. N. Breece, L. G. Presta, A. M. de Vos, et al. 2002. "Comprehensive functional maps of the antigen-binding site of an anti-ErbB2 antibody obtained with shotgun scanning mutagenesis." *J Mol Biol* 320: 415–28.

Vanwezenbeek, P., T. J. M. Hulsebos and J. G. G. Schoenmakers. 1980. "Nucleotide-sequence of the filamentous bacteriophage M13-DNA genome—comparison with phage-FD." *Gene* 11: 129–48.

Venter, J. C., M. D. Adams, E. W. Myers, P. W. Li, R. J. Mural, et al. 2001. "The sequence of the human genome." *Science* 291: 1304–51.

Waksman, G., S. E. Shoelson, N. Pant, D. Cowburn and J. Kuriyan. 1993. "Binding of a high-affinity phosphotyrosyl peptide to the SRC SH2 domain—crystal structures of the complexed and peptide-free forms." *Cell* 72: 779–90.

Walter, G., Z. Konthur and H. Lehrach. 2001. "High-throughput screening of surface displayed gene products." *Comb Chem High Throughput Screen* 4: 193–205.

Weiss, G. A., T. A. Roth, P. F. Baldi and S. S. Sidhu. 2003. "Comprehensive mutagenesis of the C-terminal domain of the M13 gene-3 minor coat protein: the requirements for assembly into the bacteriophage particle." *J Mol Biol* 332: 777–82.

Weiss, G. A. and S. S. Sidhu. 2000. "Design and evolution of artificial M13 coat proteins." *J Mol Biol* 300: 213–9.

Weiss, G. A., C. K. Watanabe, A. Zhong, A. Goddard and S. S. Sidhu. 2000. "Rapid mapping of protein functional epitopes by combinatorial alanine scanning." *Proc Natl Acad Sci U S A* 97: 8950–4.

Wiesmann, C., H. W. Christinger, A. G. Cochran, B. C. Cunningham, W. J. Fairbrother, et al. 1998. "Crystal structure of the complex between VEGF and a receptor-blocking peptide." *Biochemistry* 37: 17765–72.

Winter, G. and C. Milstein. 1991. "Man-made antibodies." *Nature* 349: 293–9.

Zhang, Y. N., B. A. Appleton, P. Wu, C. Wiesmann and S. S. Sidhu. 2007. "Structural and functional analysis of the ligand specificity of the HtrA2/Omi PDZ domain." *Protein Sci* 16: 1738–50.

Zhang, Y. N., S. Yeh, B. A. Appleton, H. A. Held, P. J. Kausalya, et al. 2006. "Convergent and divergent ligand specificity among PDZ domains of the LAP and zonula occludens (ZO) families." *J Biol Chem* 281: 22299–311.

# 2 Cell Surface Display Systems for Protein Engineering

*Sarah J. Moore, Mark J. Olsen, Jennifer R. Cochran, and Frank V. Cochran*

## CONTENTS

Cell surface display systems are effective tools for protein engineering by directed evolution, and show great promise in both medical and industrial applications. With this technology, combinatorial protein libraries are expressed on the surface of host cells and screened in a high-throughput manner to identify mutants with a desired phenotype. Numerous cell surface display systems have been developed using a range of host organisms, including Gram negative bacteria (Daugherty 2007), Gram positive bacteria (Wernerus and Stahl 2004), yeast (Gai and Wittrup 2007; Pepper et al. 2008), insect cells/baculovirus (Makela and Oker-Blom 2008), and mammalian cells (Beerli et al. 2008). These systems complement phage display technology and cell-free protein engineering systems, discussed in Chapters 1 and 3 of this volume, respectively. This chapter surveys the most common cell surface display formats, describes library screening methods, and discusses issues that should be considered when initiating a protein engineering project using cell surface display technology.

## OVERVIEW OF PROTEIN ENGINEERING USING CELL SURFACE DISPLAY

Currently, bacteria and yeast are the most commonly used cell surface display platforms for protein engineering. In addition to protein engineering, cell surface-displayed proteins have been used in applications such as vaccine development (Leclerc et al. 1991), bioabsorbants (Pazirandeh et al. 1998), biocatalysts (Shiraga et al. 2005), and biosensors (Shibasaki et al. 2001). Figure 2.1 outlines how cell surface display is commonly used to engineer proteins. A typical project starts with deciding on a suitable surface display platform, based on characteristics of the protein to be displayed and access to an appropriate screening method. A mutant DNA library of the protein of interest is then generated, incorporated into a display plasmid, and transformed into the host organism. The size and diversity of the resultant library is estimated, and protein expression on the cell surface is confirmed. Using standard techniques, library sizes of up to $10^{11}$ transformants have been reported with each cell displaying thousands of identical copies of a particular

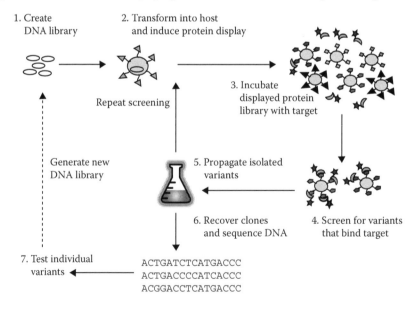

**FIGURE 2.1**  Overview of a protein engineering project using cell surface display.

mutant protein. High-throughput screening of the library is then performed by a number of methods. Cells that display proteins possessing the desired phenotype, such as increased binding to a target of interest, are isolated and propagated in culture. Compared to intracellular expression, cell surface display eliminates the need for the target to diffuse into the cell for binding interactions. Furthermore, library screening conditions, such as buffer composition, pH, and ionic strength, can easily be tuned. Several rounds of screening are typically required to obtain an enriched pool of mutants with the desired phenotype. Individual clones are then isolated, and their DNA is recovered and sequenced to determine the identity of the corresponding amino-acid sequences. This genotype-phenotype link is one of the most important components of a cell surface display platform, as without it there would be no easy way to identify protein variants isolated from these library screens. If necessary, this genetic material can serve as a starting point for the generation of subsequent DNA libraries for further rounds of evolution. Properties such as binding affinity, thermal stability, and catalytic efficiency can be measured while the proteins are still tethered to the cell surface, allowing rapid characterization of individual mutants without the need for soluble protein expression and purification, which is laborious and time consuming (Orr et al. 2003). The most promising mutants are then produced in soluble form for further characterization or use in the intended application.

## BACTERIAL SURFACE DISPLAY

### Overview

Bacterial surface display was first reported over two decades ago (Charbit et al. 1986). Bacterial surface display constructs typically include DNA encoding for a signal targeting the fusion protein to the bacterial membrane or cell wall, an anchor for embedding the fusion in the membrane or cell wall, and the protein to be engineered. Over the years, researchers have developed several bacterial surface display platforms, which have been used to engineer peptides (Lu et al. 1995), antibody fragments (Daugherty et al. 1998), and enzymes (Olsen et al. 2000; Varadarajan et al. 2005).

### Surface Display on *E. coli* Bacteria

*Escherichia coli* is currently the most commonly used host for protein engineering by bacterial surface display. As a Gram negative bacteria, *E. coli* has an inner membrane and an outer membrane separated by a periplasmic space. Membrane proteins native to *E. coli* have served as convenient targeting and anchoring portions in this system, but heterologous anchor proteins have been used as well. Numerous *E. coli* surface display systems exist (Lee et al. 2003; Daugherty 2007). Figure 2.2 provides an overview of the representative formats that are discussed in this chapter. Separate from the bacterial display formats discussed in this section, the *E. coli* endopeptidase OmpT was engineered for increased activity and substrate specificity by a novel screening strategy based on electrostatic trapping of a fluorescent product on the cell surface, which enabled flow cytometric sorting (Olsen et al. 2000; Varadarajan et al. 2005).

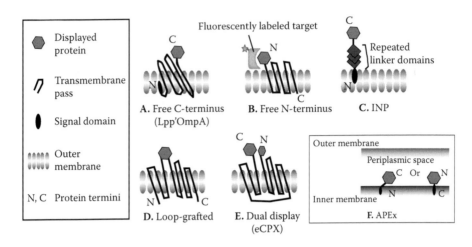

**FIGURE 2.2** Representative *E. coli* surface display platforms that have been used in protein engineering applications. An example of a binding interaction with a fluorescently labeled target is shown in B. Abbreviations: INP (ice nucleation protein), eCPX (enhanced circularly permuted OmpX), APEx (anchored periplasmic expression).

As these results are described in detail in Chapter 6 of this volume, they will not be discussed here.

Two early *E. coli* surface display formats tether the protein of interest as N-terminal fusions, such that the C-terminus is free. One of these, the Lpp'OmpA system developed by Georgiou and coworkers, is a three-part fusion protein anchored into the *E. coli* outer membrane (Figure 2.2A) (Francisco et al. 1992). The first component, from *E. coli* lipoprotein (Lpp), is a nine-amino-acid signal sequence responsible for targeting the protein to the outer membrane. The second component, from the *E. coli* outer membrane protein OmpA, is responsible for anchoring the fusion into the outer membrane (Stathopoulos et al. 1996). The third component is the protein of interest. The Lpp'OmpA format was the first bacterial display system combined with library screening by flow cytometry for use in protein engineering applications (Daugherty et al. 1998). In this study, a single-chain variable fragment (scFv) antibody library was screened to identify variants that bound the target molecule digoxin with a $K_D$ < 1 nM. However, Lpp'OmpA fusion proteins were found to be toxic to the host cells primarily due to membrane disruption, which resulted in lysed cultures following protein induction (Christmann et al. 1999; Daugherty et al. 1999). Because of this toxicity, optimization of promoter strength, media composition, and induction temperature are necessary with this platform. Furthermore, the effects of toxicity limit the size of proteins that can be successfully displayed with the Lpp'OmpA format to around 35 kDa (Lee et al. 2003). A second display construct that tethers the protein of interest as an N-terminal fusion is based on the ice nucleation protein (INP) (Figure 2.2C) (Wolber 1993; Jung et al. 1998). In its natural context, this protein is expressed by the bacterial plant pathogen *Pseudomonas syringe* to nucleate ice crystals for cutting open plant tissue, which enables infection. This heterologous system was adapted to *E. coli* surface display by replacing the repeat domains that nucleate water with the protein of interest. This format displays the fusion proteins at a further distance from the cell surface and minimizes disruption of the outer membrane. Consequently, lower levels of toxicity have been reported with INP fusions in comparison to the Lpp'OmpA format, and proteins of up to 60 kDa have been successfully displayed, including the enzyme levansucrase and an scFv that binds the human oncoprotein c-myc (Jung et al. 1998; Bassi et al. 2000). In one example, organophosphorous hydrolase was engineered to have improved activity for hydrolysis of various organophosphate nerve agents, including a 25-fold increase in activity for one particular agent (Cho et al. 2002).

In some cases, it is critical that a particular protein terminus remain free in order to permit binding to a target or other biological activity. It is therefore useful to have display platforms that present proteins of interest in orientations other than N-terminal fusions, as in the two systems described previously. Display with a free N-terminus is possible through autotransporter formats with the IgA1 protease and EstA systems (Figure 2.2B) (Maurer et al. 1997; Becker et al. 2005; Jose and Meyer 2007). EstA was used as an anchoring motif for the display of lipases and esterases, and a screening strategy based on covalent capture of biotinylated product has shown promise for engineering enzymes with improved activity (Becker et al. 2005; Becker et al. 2007).

Displayed polypeptides can also be inserted as a loop with both the N- and C-termini fused to the anchor component (Figure 2.2D), yet these loops are generally limited to less than 15 amino acids in length. One example of this format uses the outer membrane protein A (OmpA) as an anchor protein (Bessette et al. 2004). Using this format, a random 15-mer library of $5 \times 10^5$ clones was separately screened against five unrelated proteins to identify high affinity peptides that each retained binding to the respective target when transferred from the OmpA format and expressed as insertional fusions within a monomeric fluorescent protein. The commercially available FliTrx system (Invitrogen) displays random 12-mer peptides, which in this case are inserted into the active site loop of $E.\ coli$ thioredoxin. This protein in turn exists as an inserted component of the flagellar protein FliC (Lu et al. 1995; Lu et al. 2003). The FliTrx system has been used for epitope mapping (Lu et al. 2003), discovery of peptide epitopes with metal-dependent binding to an antibody (Tripp et al. 2001), and discovery of peptides that bind and inhibit aggregation of the Alzheimer's disease amyloid β-protein (Aβ) (Schwarzman et al. 2005). FliTrx has also been used as a scaffold for the engineering of self-assembling nanotube bundles (Kumara et al. 2006). One study that compared the FliTrx system to the Ph.D.-C7C phage-displayed cyclic peptide library (New England Biolabs) reported obtaining an expected consensus peptide motif from only the phage library when panning against surface-immobilized streptavidin (Lunder et al. 2005).

The OmpX outer membrane protein has been engineered into a circularly permuted scaffold to allow both C-terminal and N-terminal display of proteins of interest simultaneously (Figure 2.2E). The most current version of this system, referred to as enhanced circularly permuted OmpX, or eCPX, achieves display levels similar to the native OmpX protein (Rice et al. 2006; Rice and Daugherty 2008). Two epitope tags, one on each terminus of eCPX, were simultaneously detectable using dual-color flow cytometry (Rice and Daugherty 2008). Thus, in addition to displaying dimeric polypeptides, normalization of surface display levels is also possible by including an epitope tag on the N- or C-terminus. Library screening has yet to be reported with this new construct.

Displaying proteins of interest on the periplasmic side of the *E. coli* inner membrane is also a valuable format for protein engineering. Advantages of this format include the lack of complex carbohydrates on the inner membrane that could sterically interfere with protein binding to large targets, and that the displayed protein is required to traverse only one membrane, which avoids factors that may restrict export. The anchored periplasmic expression (APEx) system anchors the protein of interest to the inner membrane either as an N-terminal fusion to a six-residue sequence derived from the native *E. coli* lipoprotein NlpA, or as a C-terminal fusion to the phage gene three minor coat protein of M13 (Figure 2.2F) (Harvey et al. 2004). Selective permeabilization of the outer membrane allows targets of up to at least 240 kDa to interact with the displayed protein. This format was initially used to improve the affinity of a neutralizing scFv to the *Bacillus anthracis* protective antigen by more than 200-fold. In a follow-up report, scFvs were engineered by APEx for increased binding affinity to methamphetamine (Harvey et al. 2006). These improved scFvs showed low cross-reactivity to over-the-counter drugs and could be useful as methamphetamine class immunodiagnostics. Screening scFv libraries against surface-immobilized antigen is also a viable strategy for the APEx format (Jung et al. 2007).

Recently, whole antibody display on the surface of the *E. coli* inner membrane was demonstrated (Mazor et al. 2007; Mazor et al. 2008). Full-length heavy and light chains were secreted for assembly in the periplasm, followed by capture with an Fc-binding protein that was anchored to the inner membrane. An analogous display platform, referred to as APEx 2-hybrid, involves coexpression of an anchored protein along with a soluble binding partner, which colocalize in the periplasmic space (Jeong et al. 2007). The soluble periplasmic contents are released by disruption of the outer membrane, yet binding partners that interact with the anchored protein are retained. An epitope tag on the captured binding partner allows for the interaction to be detected by flow cytometry. An advantage of this format is that one does not need to separately prepare and purify the binding partner protein. The APEx 2-hybrid format was used to obtain scFvs with increased affinity to the protective antigen of *Bacillus anthracis* from cells that coexpressed libraries of scFv mutants along with endogenously expressed antigen, and for engineering fragment antigen binding (Fab) antibodies for improved expression.

Protein libraries displayed on the *E. coli* inner membrane have also been used to engineer a G protein-coupled receptor (GPCR) for increased expression and stability (Sarkar et al. 2008). The rat neurotensin receptor-1 (NTR1) was used as a model system to demonstrate the feasibility of this approach. An NTR1 mutant library was screened by flow cytometry to yield mutants with expression levels that were an order of magnitude greater than wild-type receptor. These mutants retained agonist binding, antagonist binding, and native signaling. A second library screen identified a single amino-acid substitution that abolished antagonist binding while retaining agonist binding. These achievements are exciting given the difficulties of studying and engineering integral membrane proteins, and the pharmaceutical relevance of GPCRs, which compose about 60% of current drug targets.

## Surface Display on Staphylococcus Bacteria

Gram positive bacteria have a single cell membrane and an outer peptidoglycan layer that forms a rigid cell wall. Because of this robust cell wall, Gram positive bacteria may be suitable for industrial applications of cell surface-displayed enzymes in which the bacteria would be subjected to the strong shear forces that exist within a bioreactor. These hosts could also be very useful in screens requiring relatively harsh conditions that may be incompatible with a more fragile organism.

Gram positive bacterial surface display has been investigated primarily by Stahl and coworkers using the *Staphylococcus aureus* Protein A system displayed on the *Staphylococcus xylosus* or *Staphylococcus carnosus* host (Hansson et al. 1992; Robert et al. 1996; Wernerus and Stahl 2004). The protein to be displayed is fused to the N-terminus of a polypeptide sequence that is recognized and processed by sortase (Mazmanian et al. 1999), which results in covalent attachment of the C-terminus to a free amino group within the peptidoglycan layer of the cell wall. This system has been successful in the display of scFvs (Gunneriusson et al. 1996), affibodies (Gunneriusson et al. 1999), and metal-binding peptides (Samuelson et al. 2000). Achieving sufficient transformation efficiencies and protein display levels has proven to be a significant hurdle in this system. Improvements in plasmid vectors and optimized host protocols have addressed these limitations (Wernerus and Stahl 2002; Lofblom et al. 2005; Lofblom et al. 2007a; Lofblom et al. 2007b). The first report of screening a mutant library to obtain a protein with improved binding affinity using Gram positive bacterial display has recently appeared (Kronqvist et al. 2008). A library of $3 \times 10^9$ affibody variants was initially enriched for binding human tumor necrosis factor-alpha (TNF-alpha) using one round of phage display. Afterwards, the isolated variants were transferred to the staphylococcal host. The resulting library of $\sim 10^6$ variants was then sorted three times by flow cytometry to yield TNF-alpha binders with affinities ranging from 95 pM to 2.2 nM.

## Advantages and Disadvantages of Bacterial Surface Display

A significant advantage of bacterial surface display platforms is that relatively high transformation efficiencies of *E. coli* allow libraries of $10^9$–$10^{11}$ members to be easily accessible. This is in contrast to yeast surface display where libraries of this size require conducting many parallel transformations due to poorer transformation efficiency. While *E. coli* display libraries are similar in size to that of phage display libraries, they are still smaller than library sizes routinely obtained with mRNA and ribosomal display ($10^{13}$–$10^{14}$) (Lipovsek and Pluckthun 2004).

A significant drawback to prokaryotic display systems is the lack of protein folding and quality control machinery that exists in eukaryotes. This limits the application of bacterial display systems to proteins with simpler folds relative to eukaryotic display systems. However, this system allows for identification of variants that express well in bacteria, which is useful when large-scale production in bacteria is planned for the engineered proteins.

When fused to membrane protein anchors, displayed proteins that are large and complex may be toxic, since they can disrupt the integrity of the host membrane.

This becomes especially problematic when display levels are high. Because cell viability may be compromised in bacterial surface display systems, slower growth rates of clones with toxic protein variants compared to clones with nontoxic protein variants can lead to biased libraries due to more rapidly growing clones dominating a culture. One way to account for the toxic effects of displayed proteins is to simultaneously screen the library by flow cytometry for the desired phenotype as well as cell viability using a dye such as propidium iodide. This permits only cells that are both viable and displaying a mutant protein with a desired phenotype to be collected for further propagation. However, new evidence suggests that toxicity may be eliminated by coexpression of periplasmic chaperones to assist in transport, folding, and targeting of the protein of interest, while in turn significantly improving display levels (Narayanan and Chou 2008). It is not currently possible to predict which fusion proteins in any given display construct will be toxic to a host. Instead, individual culture optimization is often necessary.

## YEAST SURFACE DISPLAY

### OVERVIEW

Yeast surface display physically couples the protein to be engineered to an anchor protein embedded within the yeast cell wall. The eukaryotic secretory pathway of yeast allows a wide variety of proteins to be displayed, including many with complicated folds. Proteins that have been engineered using yeast surface display include scFvs (Feldhaus and Siegel 2004; Chao et al. 2006), Fab antibody fragments (Weaver-Feldhaus et al. 2004), single-chain T-cell receptors (scTCRs) (Shusta et al. 1999), single-chain major histocompatibility complexes (scMHCs) (Esteban and Zhao 2004), human epidermal growth factor (EGF) (Cochran et al. 2006), cytokines (Rao et al. 2003), epidermal growth factor receptor (EGFR) extracellular domains (Kim et al. 2006), fibronectin type III domains (Lipovsek et al. 2007b; Hackel et al. 2008), cystine-knot peptides (Kimura et al. 2009; Silverman et al. 2009), and horseradish peroxidase (Lipovsek et al. 2007a). The ability of this eukaryotic system to display mammalian proteins has also allowed screening of human cDNA libraries (Wadle et al. 2005; Bidlingmaier and Liu 2006). Figure 2.3 illustrates representative yeast surface display platforms.

### YEAST SURFACE DISPLAY PLATFORMS

Early work in yeast surface display involved immobilization of antigens for vaccine development and surface display of enzymes for biocatalysts (Schreuder et al. 1993; Schreuder et al. 1996). While most efforts have focused on using *Saccharomyces cerevisiae,* other yeast such as *Pichia pastoris* have also been explored (Tanino et al. 2006). Progress in using yeast surface display formats for applications other than protein engineering has recently been reviewed (Kondo and Ueda 2004).

The vast majority of protein engineering efforts with yeast surface display have been performed using the Aga1-Aga2 display format in *S. cerevisiae,* a technology developed by Wittrup and coworkers over a decade ago (Figure 2.3A) (Boder and

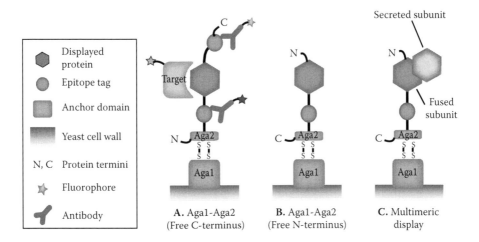

**FIGURE 2.3** Representative yeast surface display platforms that have been used in protein engineering applications. The presence, position, and identity of epitope tags are variable. An example of binding interactions with fluorescently labeled target and fluorescently labeled antibodies is shown in A.

Wittrup 1997; Gai and Wittrup 2007). The protein to be engineered is fused to the yeast agglutinin protein Aga2, whose expression is controlled by a galactose-inducible promoter. Aga2 is coexpressed and attached to the yeast Aga1 protein through disulfide bonds. Aga1 is covalently attached to the cell wall through phosphotidylinositol glycan linkages, serving as an anchor for protein display. While fusing the N-terminus of the protein to be engineered to Aga2 is the most common display format, tethering of the C-terminus to Aga2 is also possible (Figure 2.3B) (Wang et al. 2005b). Epitope tags allow immunofluorescent detection of displayed proteins, removal of truncated proteins, and normalization of activity or binding with protein expression levels. Additionally, the epitope tags may be excluded if their presence interferes with activity or a binding interaction (Bhatia et al. 2003).

Aga1-Aga2 display has been widely used for affinity maturation of antibody-antigen interactions (Feldhaus and Siegel 2004; Colby et al. 2004c; Chao et al. 2006; Miller et al. 2008). Using this format, one of the strongest reported antibody-antigen interactions was engineered, generating an scFv that recognized fluorescein with a $K_D = 48$ fM (Boder et al. 2000a). In other examples, an scFv was engineered to bind carcinoembryonic antigen (CEA) with a dissociation half-time of >4 days at 37 °C (Graff et al. 2004), and scFvs with bispecific reactivity against two botulinum neurotoxin subtypes have been engineered (Garcia-Rodriguez et al. 2007). In addition, scFvs can be identified from naive yeast-displayed antibody libraries (Feldhaus et al. 2003). An scFv specific for huntingtin protein was discovered from such a library, and a variable light chain domain derived from this scFv was engineered for high affinity binding to intracellular huntingtin protein for inhibition of disease-causing protein aggregation (Colby et al. 2004a; Colby et al. 2004b). Additionally, scFvs that bind fluorogenic dyes and activate them for fluorescence have also been discovered from these libraries for use in imaging applications (Ozhalici-Unal et al. 2008; Szent-Gyorgyi et al. 2008).

Other protein–protein interactions have also been affinity-matured using yeast surface display. Human EGF was engineered for up to a 30-fold improvement in binding affinity to EGFR for potential applications in wound healing (Cochran et al. 2006). The cytokine interleukin-2 has also been engineered for 30-fold tighter binding to one of its receptor subunits (Rao et al. 2003). This led to altered endocytic trafficking and enhanced T-cell proliferation for potential applications in cancer therapy. Interestingly, in both cases these affinity improved mutants showed a strong statistical bias toward orthologous substitutions at sites that varied among protein family members, confirming that nature has in most cases already identified optimal protein variants for a given fold and function (Cochran et al. 2006).

Yeast surface display has also been used to engineer proteins for improved stability and soluble expression yields. Some notable examples include scTCRs (Kieke et al. 1999; Shusta et al. 1999), scMHCs (Esteban and Zhao 2004; Jones et al. 2006), EGFR extracellular domain (Kim et al. 2006), the tumor necrosis factor receptor (TNFR) extracellular domain (Schweickhardt et al. 2003), and the tumor antigen NY-ESO-1 (Piatesi et al. 2006). This approach was used when initial attempts at displaying wild-type scTCRs on the surface of yeast were unsuccessful. Screening a mutant library for displayed variants yielded scTCRs with the correct fold (Kieke et al. 1999). Subsequent work demonstrated that, in general, thermostability and expression levels of proteins displayed on the surface of yeast correlated with soluble protein thermostability and expression (Shusta et al. 1999); moreover, using increased thermostability as a screening criteria can lead to proteins with enhanced expression levels (Shusta et al. 2000). However, recent work has shown there may be limits to this correlation. For example, in cases where the protein to be engineered already possesses high structural stability, the yeast quality control machinery may be unable to discriminate between proteins that are either well-folded or in a molten globule state (Park et al. 2006). Recently, a high-throughput flow cytometric assay for screening mutant yeast libraries for clones with improved secretion of heterologous proteins was reported (Rakestraw et al. 2006). This method was based on covalently coupling a cognate ligand to the yeast surface, which in turn captured the secreted protein, producing the required genotype-phenotype link. The resulting assembly was then labeled for sorting by flow cytometry.

Engineering scTCRs with increased stability and expression levels has laid a foundation for developing these proteins as a novel class of therapeutic agents (Richman and Kranz 2007). These efforts have also shed light on molecular details underlying the immune response. For example, scTCRs were engineered to have a low nanomolar binding affinity ($K_D \approx 9$ nM) for peptide/MHC complexes (Holler et al. 2000), which is much stronger than the weak peptide/MHC-TCR binding affinities commonly observed in nature, thus demonstrating that neither a genetic nor structural limitation exists for this immune system interaction.

Enzyme engineering using cell surface display is challenging because catalytic turnover often results in a product that diffuses away from the cell surface, breaking the critical genotype-phenotype link that is required for screening and subsequent identification of isolated mutants. However, horseradish peroxidase (HRP), a commercially useful enzyme, was engineered by yeast surface display for enhanced enantioselectivity using a novel fluorescent substrate that forms a product that is trapped on the yeast surface to maintain the genotype-phenotype link (Lipovsek et al. 2007a). Yeast-displayed HRP libraries were screened for enzymatic activity by flow cytometric sorting, and variants with up to an eightfold altered enantioselectivity were isolated. Mutants with proper enantioselectivity were obtained by alternating rounds of positive screening for activity toward the desired enantiomer and negative screening to remove library members with activity toward the opposite enantiomer. To increase screening efficiency, a subsequent study used D- and L-enantiomeric substrates that were each labeled with a different fluorescent dye to screen yeast-displayed HRP libraries by two-color flow cytometric sorting (Antipov et al. 2008). This strategy led to engineered enzyme variants with up to two orders of magnitude enantioselectively for either substrate.

In addition to protein engineering, the Aga1-Aga2 yeast display format has also proven successful for epitope mapping of protein–protein interactions (Chao et al. 2004; Cochran et al. 2004; Siegel et al. 2004; Oliphant et al. 2005; Chung et al. 2006; Oliphant et al. 2006; Levy et al. 2007). Information gained from epitope mapping studies is valuable for antibody engineering efforts aimed at developing improved modes of antibody-based therapy and diagnostics. In an initial study, domain-level epitopes of several clinically relevant anti-EGFR antibodies were mapped by expressing fragments of the EGFR extracellular domain in a native or denatured state on the surface of yeast and measuring antibody reactivity (Cochran et al. 2004). A follow-up study used yeast-displayed libraries for fine-level epitope mapping of several EGFR antibodies by identifying receptor mutations that resulted in loss of antibody binding (Chao et al. 2004). Another report demonstrated domain- and fine-level epitope mapping of antibodies that recognize botulinum neurotoxin, an extremely poisonous substance that is responsible for botulism (Levy et al. 2007).

Most cell surface display platforms are limited to displaying monomeric polypeptides. However, recent efforts enable yeast display of multimeric proteins in novel formats. In all of these systems, one protein subunit is anchored as a fusion to the yeast cell wall, while other protein subunits are targeted for secretion and associate with the anchored subunit, resulting in the display of a multimeric protein on the cell surface (Figure 2.3C). Using α-agglutinin as an anchor, a catalytic Fab antibody fragment (composed of a heavy and light chain) was the first reported example of a functional heterodimeric protein to be displayed on the surface of yeast (Lin et al. 2003). In addition, a platform based on Aga1-Aga2 has also been developed to display heterodimeric Fab fragments on the surface of yeast. In this format, a single vector carried distinct expression cassettes for the Fab heavy and light chain subunits (van den Beucken et al. 2003). The heavy chain subunit was fused to the Aga2 anchor protein, while the light chain subunit contained a signal sequence that targeted it for secretion. Expression of the heavy and light chains resulted in an assembled heterodimeric Fab fragment on the yeast surface. Another format has been established using the Aga1-Aga2 platform in which haploid yeast are mated to produce yeast-displayed Fab libraries (Weaver-Feldhaus et al. 2004). In this system, haploid yeast contained a vector encoding either heavy chain subunits fused to the Aga2 protein or light chain subunits that were targeted for soluble expression. Following yeast mating, diploid yeast contained one of each vector, allowing them to display libraries of heterodimeric Fab fragments. Moreover, this strategy produced library sizes greater than what is typically accessible with yeast through transformation-based methods. Yeast surface display of a noncovalent MHC class II α/β heterodimer complexed with antigenic peptide has also been reported (Boder et al. 2005). In this system, a single plasmid contained two distinct cassettes for simultaneous expression of a soluble MHC α-subunit and an MHC β-subunit plus antigenic peptide fused to Aga2, which assembled as a α/β heterodimer on the yeast surface. In a final example, a platform for yeast surface display of homo-oligomers has also been reported (Furukawa et al. 2006). One subunit of homo-tetrameric streptavidin was expressed as a fusion to the cell wall anchor protein Flo1. From a separate cotransformed vector, the other subunits were expressed in soluble form, and associated with the surface displayed component. The fully assembled streptavidin showed binding to biotinylated compounds, as demonstrated by fluorescence microscopy.

## ADVANTAGES AND DISADVANTAGES OF YEAST SURFACE DISPLAY

To date, there have been many reports using yeast surface display for protein engineering. Many bacterial display reports have been limited to antibody fragments or peptides of less than 20 amino acids, while yeast display studies have encompassed a much more diverse range of proteins. The protein folding machinery and quality control mechanisms of the eukaryotic secretory pathway allows complex proteins to be engineered with yeast. Furthermore, unlike some *E. coli* surface display formats, significant cellular toxicity resulting from displayed proteins has yet to be reported with yeast. Proteins of over 500 amino acids have been successfully displayed on the yeast surface with little consequential effect on cell density or growth (Cochran et al. 2004; Kim et al. 2006). In addition, evidence suggests that yeast display libraries are of higher quality than those produced in other noneukaryotic display formats. In a recent study, an HIV-1 immune scFv library, created in both phage- and yeast-displayed formats, was screened against the HIV-1 gp120 envelope glycoprotein (Bowley et al. 2007). The yeast and phage libraries both yielded the same six scFv variants. Yet, yeast display identified an additional twelve novel clones not found by phage display. Only five of these twelve clones could be displayed by phage, likely due to the greater protein folding capabilities of yeast.

Many eukaryotic proteins require glycosylation for folding and/or function (Mitra et al. 2006). Glycosylated proteins have been successfully displayed on yeast, despite the differences between mammalian and yeast glycosylation patterns. However, hyperglycosylation of proteins by yeast could be problematic by interfering with protein function. Therefore, point mutations to remove potential glycosylation sites may be required for surface display of properly folded and functional proteins. Recent work aimed at incorporating human-like glycosylation patterns into yeast opens the possibility of using yeast surface display for the engineering of proteins with native human glycosylation (Wildt and Gerngross 2005; Hamilton et al. 2006).

A potential disadvantage of yeast surface display is the smaller library size typically accessible relative to other display formats. While libraries of up to $10^9$ unique clones have been generated (Feldhaus et al. 2003), library sizes of $10^6$–$10^7$ are more typical due to low yeast transformation efficiencies. Yet, despite these smaller library sizes, yeast surface display has been very successful for a variety of protein engineering efforts. Moreover, there is also a slightly longer experimental time associated with yeast display relative to bacterial display due to slower culture growth and a separate protein induction step. These factors result in a 2–3 day lag time between rounds of screening, which increases the total time of the library screening process.

## EMERGING CELL SURFACE DISPLAY SYSTEMS: INSECT CELL AND MAMMALIAN CELL DISPLAY

Insect cell/baculovirus display is a well-established technology, yet unlike bacterial or yeast surface display, this system has not been extensively used for protein engineering applications (Makela and Oker-Blom 2006; Makela and Oker-Blom 2008). Despite the relatively few examples of protein engineering with insect cell/baculovirus systems, which require greater expertise in handling compared to microbial-based systems, recent studies have shown much potential. This format involves either the infection of insect cells with baculovirus and subsequent expression of displayed proteins on the cell wall, or display of proteins on the virus capsid. Display of heterologous proteins mainly uses the gp64 major envelope glycoprotein as an anchor (Boublik et al. 1995), but other proteins have also been used (Ernst et al. 1998). This eukaryotic system has the ability to process proteins with complex folds and the capability to perform post-translational modifications. The first demonstration of this system for the display and screening of a library involved the display of an HIV-1 gp41 epitope specific for a neutralizing monoclonal antibody (Ernst et al. 1998). Amino acids adjacent to the epitope were randomized to produce a library of 8000 variants on baculovirus-infected insect cells. Flow cytometric sorting led to a clone with significantly increased binding affinity to the antibody. Baculovirus-infected insect cells have also been used to display peptide libraries bound to class I and class II MHC proteins (Crawford et al. 2006). In this system, each insect cell was able to display a unique MHC-peptide complex on its surface. These libraries were screened with soluble, fluorescent T cell receptors to identify peptide mimotopes for MHC class I–specific (Wang et al. 2005a) and MHC class II–specific (Crawford et al. 2004) T cells. Furthermore, these mimotopes could be used to stimulate T cells expressing the same receptor. Follow-up studies demonstrated the ability to use baculovirus-infected cell-displayed libraries to examine the role of MHC-bound peptides in the interaction of the staphylococcal enterotoxin A superantigen with an MHC class II protein, and to identify peptide mimotopes that bound with high affinity to an MHC class I protein and an antigenic TCR for potential uses in tumor therapy (Crawford et al. 2006). Although there are limited examples of using insect cell/baculovirus technologies for protein engineering, this methodology should be generally applicable for studying other ligand/receptor interactions.

Cell surface display systems using mammalian hosts are also useful for engineering eukaryotic proteins. As in the case of insect cell display, mammalian surface display platforms have not yet found widespread use for protein engineering, but this may also be due to the high level of skill needed on the part of the researcher. Recent reports suggest that their use will become more prominent in the near future. As an initial proof of concept, a peptide library of ~$10^7$ members was displayed as fusions to the N-terminus of the CCR5 chemokine receptor in Jurkat-e cells, and screened to isolate an epitope mimetope (Wolkowicz et al. 2005). Beyond this study, most mammalian cell display efforts are motivated by the need to address problems that arise when translating antibodies engineered in nonmammalian hosts to diagnostic and therapeutic applications. With this in mind, human embryonic kidney 293T cells were used to display scFvs tethered to the transmembrane domain of human platelet-derived growth factor receptor (Ho et al. 2006). An scFv with higher affinity was enriched 240-fold by a single pass of cell sorting after being mixed with a large excess of cells expressing wild-type antibody with a slightly lower affinity. This study also reported significant enrichment of a mutant with increased binding affinity for CD22 after a single round of screening from an scFv library with a randomized antibody hotspot. The authors reported that transformation protocols with this system permit library sizes of ~$10^9$ variants to be relatively accessible. Antibody libraries of ~$10^8$ members of whole IgG molecules were displayed on the surface of a human embryonic kidney-derived cell line, and screened for binding to target by a combination of magnetic bead sorting and flow cytometry (Akamatsu et al. 2007). IgG antibodies that neutralize human and mouse interleukin-12 were successfully isolated from this library. In another report, human scFvs with specificity to virus antigen or to nicotine were isolated from a mammalian display library in BHK cells (Beerli et al. 2008). The therapeutic potential of monoclonal antibodies derived from the nicotine-specific scFvs was validated in a preclinical mouse model by inhibiting nicotine entry into the brain.

## LIBRARY SCREENING METHODS

### OVERVIEW

Cell surface display systems are valuable tools for protein engineering when coupled with a method for high-throughput screening of large libraries (Boder and Wittrup 1998; Daugherty et al. 1998). While flow cytometry remains a common method for screening cell surface display libraries, other methods such as magnetic bead sorting and panning have also been developed. The library screening strategies discussed in this section are generally applicable to multiple hosts and display formats.

### FLUORESCENCE-ACTIVATED CELL SORTING (FACS)

Flow cytometry, a mature technology analogous to fluorescence microscopy, is used in both research and clinical settings (Shapiro 2003). When used to isolate distinct populations from mixtures of cells, it is often referred to as FACS. Briefly, the central concept involves hydrodynamically focusing a cell suspension into a stream of single cells. Laser light is then directed onto the cells as they flow one at a time through a detection chamber, which allows simultaneous measurement of light scatter and fluorescence. Cells that match user-defined optical criteria are then isolated by charge deflection into a collection tube for further propagation.

Polypeptide libraries displayed on phage, mRNA, or ribosomes cannot be screened by FACS because these particles are too small to produce the light scattering necessary for detection. In contrast, bacteria, yeast, insect, and mammalian cells are all large enough for detection. For protein engineering projects aimed at improving binding affinity, such as with antibody-antigen interactions, cell surface-displayed libraries are first incubated with a fluorescently labeled target. FACS is then used to isolate fluorescently stained cells resulting from the desired binding interaction. Screening stringency can be controlled through altering ligand concentration, the number of wash steps, and incubation time in the presence of competitive binders. Both equilibrium and kinetic binding screens have been successfully employed using FACS (Boder and Wittrup 1998). For equilibrium binding screens, the yeast-displayed library is incubated with fluorescent ligand at a concentration below the $K_D$ that is desired for affinity enhancement. For kinetic binding screens, mutants with a decreased off-rate of binding can be isolated using an extended wash step, followed by incubation with excess unlabeled ligand prior to sorting. Optimization of these variables, in combination with multiple rounds of FACS, can be used to isolate rare library members with particularly interesting properties (Boder and Wittrup 1998). Library screening criteria can be adjusted based upon cell data collected in real-time. For example, in affinity maturation experiments, cell sorting criteria can be clearly defined to retain only the top 1% of clones with high binding affinity. In FACS, cells are analyzed at speeds of 20,000–50,000 cells/second, depending on the instrumentation used. This creates an upper practical limit of about $10^8$ cells that can be sorted in a typical session. A 10-fold excess of the library diversity should be sampled to ensure recovery of rare clones; thus, only library sizes of up to $10^7$ can reasonably be covered. While enrichment for high affinity clones has been observed after only one round of cell sorting, it is more common to perform several rounds of FACS to enable enrichment of extremely rare clones. Between sort rounds, the pool of collected yeast cells can be analyzed by flow cytometry to measure enrichment of clones with the desired phenotype. After sufficient enrichment, individual mutants are isolated and sequenced, and binding parameters are often analyzed by flow cytometry while the proteins are still tethered to the cell surface, removing the need for soluble protein expression and purification.

When sorting yeast-displayed libraries, it is common to perform dual-color FACS, in which two separate fluorescent signals are measured simultaneously (Boder and Wittrup 2000b; Colby et al. 2004c). Bacterial display systems that employ epitope tags are also compatible with dual-color FACS (Kenrick et al. 2007; Rice and Daugherty 2008). As shown in Figure 2.4, one signal results from a terminal epitope tag that is bound to a fluorescently labeled antibody (used to measure display levels of the protein of interest), while the other signal results from binding of a fluorescently labeled ligand (used to measure protein–ligand binding interactions). Simultaneous detection and, hence, normalization of ligand binding with protein expression enables high-affinity clones to be isolated regardless of their expression level. Moreover, this normalization prevents inadvertent collection of clones with apparent high binding resulting from high expression levels rather than strong binding to the target ligand.

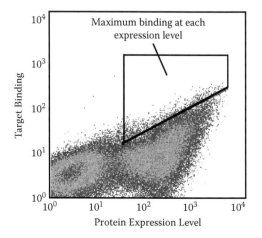

**FIGURE 2.4** Flow cytometry dot plot of a yeast surface-displayed protein library incubated with a fluorescently labeled antibody that binds an epitope tag (x-axis) and a fluorescently labeled target (y-axis). Protein expression and target binding on the cell surface are simultaneously measured through the use of two different fluorophores. The region defined by the polygon indicates cells with maximum target binding at each protein expression level. Cells in this gated region would be collected in an affinity maturation sort to yield the top ~1% of binding clones within the total population.

Thus, one of the greatest advantages of FACS is quantitative control with single-cell resolution, as demonstrated in one example where fine affinity discrimination of clones with only a twofold difference in binding affinity was achieved (VanAntwerp and Wittrup 2000).

## MAGNETIC BEAD SORTING

If one does not have easy access to FACS instrumentation, cell surface display libraries can also be screened against a target that is immobilized on magnetic beads (Christmann et al. 1999; Yeung and Wittrup 2002; Miller et al. 2008). In a typical experiment, biotinylated targets are covalently coupled to magnetic beads that are coated with streptavidin. Cell surface display libraries are usually first incubated with uncoupled magnetic beads to deplete the library of members that recognize streptavidin or bind nonspecifically to the bead surface. The nonbound cells are then recovered and incubated with beads presenting the target ligand of interest. Significant library oversampling is often required to compensate for a high probability of losing binding clones, particularly in early rounds of sorting where individual clones may be represented in relatively small numbers. Following incubation and washes, a magnet is used to attract the beads, which are bound to cells displaying mutants that recognize the ligand. The supernatant, which contains the nonbinding clones, is discarded. Magnetic beads with bound cells are transferred to growth media to propagate clones for subsequent sort rounds.

Magnetic bead sorting may also play a complementary role in situations where FACS is available yet may not be practical for use as a sole screening method. For example, since magnetic bead screens can readily accommodate $10^9$–$10^{10}$ cells per sort, this can be used as a first step to reduce large libraries to a size more manageable for FACS, while simultaneously enriching for desired clones (Siegel et al. 2004). Also, naive libraries that are not based on a pre-existing binding interaction may have only a small subpopulation that recognizes the desired target, and it is likely that most of these clones will also have very weak binding affinity. In these situations, initial rounds of screening should be performed with relaxed stringency to collect as many clones as possible, including ones with weak binding ($K_D \sim \mu M$). However, FACS may not be the best method to use for screening naive libraries due to difficulty in discriminating between the low signal produced by a weak binding interaction and the background signal produced by clones that are devoid of binding. Furthermore, target can rapidly dissociate from weak-binding mutants during the unavoidable lag time between the final wash and loading onto the flow cytometer, and also during the time it takes to perform the sort. Since FACS often requires a much greater amount of target relative to magnetic bead sorting for isolating clones with weak binding affinity, magnetic bead sorting can also reduce the amount of target required for early sort rounds.

Despite the lack of quantitative control with magnetic bead sorting, sort stringency can be tuned by optimizing parameters such as cell suspension density and incubation time in order to enrich for desired clones (Yeung and Wittrup 2002). Magnetic bead sorting also carries the risk of enriching for clones that show apparent binding as a result of high expression levels and avidity effects rather than 1:1 mutant-target stoichiometry. Therefore, if possible, magnetic bead sorting should be interspersed with rounds of FACS to normalize binding with protein expression levels through immunofluorescent detection of an appropriate epitope tag.

## PANNING

Panning, a technique used extensively for screening phage display libraries, has been adapted for screening bacterial and yeast surface display libraries against mammalian cell targets (Wang and Shusta 2005; Zitzmann et al. 2005; Wang et al. 2007; Li et al. 2008; Yang et al. 2008). With this method, cell surface display libraries are incubated with living cell lines, and following wash steps, bound library members are recovered and propagated for future rounds of screening. Unlike FACS and magnetic bead sorting, cell panning does not require the production and purification of soluble target, which can be tedious or challenging if the target of interest is difficult to express or poorly behaved. In addition, cell panning allows libraries to be screened against targets in their native cell membrane environment. This screening method can also be used as a potent strategy for discovering binders that recognize previously unknown targets. For example, a yeast-displayed scFv library of $\sim 10^9$ clones was panned against brain endothelial cells to identify scFvs that bound tightly to cells, and in some cases, were internalized into cells (Wang and Shusta 2005; Wang et al. 2007). This finding was important because antibodies that engage endocytosis pathways have potential for delivering payloads, such as transporting therapeutic agents through the blood-brain barrier.

Density centrifugation is a common technique in immunological studies for separating mixtures of cell types into pure populations, and has recently been applied to screen cell surface display libraries against intact cells (Richman et al. 2006). In this method, a yeast-displayed scTCR library was incubated with a suspension of mammalian cells that expressed peptide-bound forms of the MHC protein on their surface. The yeast cell-mammalian cell conjugates were then separated from unbound yeast by centrifugation through a density gradient. Yeast-displaying scTCRs of interest were enriched 1000-fold after a single round of density centrifugation. This approach led to the isolation of scTCRs that were specific for either class I or II MHC proteins, and has important implications for protein engineering due to difficulties in obtaining soluble peptide-class II MHC ligands for library screening. Moreover, this screening methodology could potentially be applied to engineer other ligand-receptor interactions using a variety of cell surface display formats.

Despite the advantages just described, several important factors should be considered when panning cell surface display libraries against whole cells. As in magnetic sorting, cell panning lacks the rigorous quantitation provided by FACS, and does not allow discrimination of clones based on expression levels or avidity effects. Another concern is that mammalian cells express many more proteins besides the target on their surfaces; thus, there is a high likelihood of isolating clones that bind to other things. Therefore, negative screening steps should be incorporated whenever possible, such as depleting the library of clones that bind to similar mammalian cells that do not express the target. Moreover, this screening method is limited in that some membrane-bound targets may be inaccessible for binding to cell surface display library members, or may be expressed at levels that are too low for efficient recovery and propagation of binding clones.

Cell surface display libraries have also been panned against inorganic materials with the goal of discovering peptide motifs that are capable of binding to solid surfaces. These peptides could be used as novel affinity tags to immobilize proteins for biomaterials applications as an alternative to chemical conjugation, and would allow site-specific protein attachment for control of proper surface orientation. Early reports described the display of random peptide sequences within endogenous *E. coli* surface proteins for the identification of sequences that bind to materials such as iron oxide (Brown 1992), metallic gold and chromium (Brown 1997), and zinc oxide (Kjaergaard et al. 2000). More recently, the bacterial FliTrx cell surface display system has been used to identify disulfide-constrained dodecapeptides that bind semiconducting metal oxides (Thai et al. 2004). Due to their larger cell size compared to bacteria, yeast offer additional advantages for studying peptide-surface interactions through the ability to use conventional light microscopy in combination with techniques based on the application of mechanical force. Panning a yeast-displayed scFv library against CdS resulted in the identification of short scFv fragments, which were rare in the original library (Peelle et al. 2005a). Subsequent studies used yeast-displayed peptide libraries to pan against II-IV semiconductor and Au surfaces (Peelle et al. 2005b), as well as metal oxide surfaces (Krauland et al. 2007). Along with identifying peptides that bound to these materials, insight into the molecular interactions responsible for recognition was obtained through sequence-activity analysis of the surface-binding peptides. Uncovering the molecular details in these systems will aid nanobiotechnology efforts, and provide a better understanding of natural biomineralization mechanisms.

## CONSIDERATIONS FOR CHOOSING A CELL SURFACE DISPLAY PLATFORM FOR PROTEIN ENGINEERING

Researchers now have a variety of cell surface display systems to choose from when embarking on a protein engineering project. Understanding the strengths and limitations of each will help one choose a platform that maximizes the chances of success.

The first important factor to consider is the nature of the protein to be engineered. The proteins discussed throughout this chapter have an established track record in protein engineering projects using cell surface display, providing a point of reference for consultation when troubleshooting undesired or unanticipated behavior. If the protein of interest has not been previously used in cell surface display experiments, preliminary studies must be performed to test whether the wild-type protein can be displayed on the cell surface, and whether it is properly folded, as assayed by reactivity with conformational antibodies or cognate binding partners. The complexity of the protein fold suggests whether a eukaryotic host is needed or a prokaryotic host will suffice. Examples of important properties to consider when choosing a cell surface display format are the natural source of the protein of interest, the number of disulfide bonds or post-translational modifications it contains, and its oligomeric state. In addition, tethering the protein to the cell surface may also affect its fold and function. It should also be noted that protein engineering projects that involve screening polypeptide libraries displayed on cell surfaces always carry the risk that improved mutants will not retain the same function when decoupled from the cell surface (Weaver-Feldhaus et al. 2005). Enzyme engineering presents an additional challenge in maintaining the required physical link between genotype and phenotype, as most substrates form products that diffuse away from the cell surface. Recent reviews summarize new advances in enzyme engineering using cell surface display (Farinas 2006; Bershtein and Tawfik 2008).

For affinity maturation or the discovery of novel binders, the target plays a major role in deciding the best library screening method. Targets should be relatively accessible, easy to manipulate, and tolerant to chemical coupling for fluorescent labeling, biotinylation, or surface immobilization. Ideally, one should use a target that is most similar to the form that exists in the desired application. For example, if a receptor target only exists in a native conformation when presented within a cell membrane, an effective screening method would likely be panning a library against a cell line that expresses this receptor.

The choice of screening method will likely be influenced by the size of the library. While FACS allows high quantitative control over screening parameters, there is an upper limit to the library sizes that can be routinely screened with this method. This is in contrast to magnetic bead sorting or panning methods, which can process much larger libraries more efficiently. In all of these methods, negative screening steps may be required to avoid isolating undesired binders, such as mutants that bind to surfaces used for target immobilization or to reagents that are used in the screening process. In addition, if one wishes to engineer a protein for increased soluble expression and/or stability, mutants should be simultaneously screened for both increased cell surface display levels and preservation of wild-type protein function. Furthermore, while most cell surface display platforms allow mutants to be readily screened for improved binding affinity, stability, or expression, there is no guarantee that these properties will correlate with enhancements in biological function or activity (Jones et al. 2008). Therefore, one should have a clear idea of what specific biophysical or biological properties are required for the intended application and, if possible, design library screens appropriately. This also highlights a strong need for the development of innovative high-throughput screening strategies capable of discriminating mutants based on their biological function.

## CELL SURFACE DISPLAY OUTLOOK

Bacterial and yeast surface display are both well-established systems for protein engineering. Continued development of these methods will bring more goals within reach, such as engineering novel enzymes and integral membrane proteins. While currently less developed, insect cell and mammalian cell surface display have the potential to address the limitations of the more established platforms. Efforts that combine protein library screening with computational modeling and bioinformatics methods also hold much promise (Lippow et al. 2007; Armstrong and Tidor 2008). Cell surface display systems have already demonstrated success in a wide range of protein engineering applications, and will make valuable contributions in the future to medicine and biotechnology.

## ACKNOWLEDGMENTS

The authors would like to thank Douglas Jones and Jennifer Lahti for critical reading of the manuscript and assistance with figures, and Marco Mena for helpful discussions. SJM is supported by an NSF Graduate Research Fellowship and a Stanford Graduate Fellowship (Medtronic Fellow).

## REFERENCES

Akamatsu, Y., K. Pakabunto, Z. Xu, Y. Zhang and N. Tsurushita. 2007. "Whole IgG surface display on mammalian cells: Application to isolation of neutralizing chicken monoclonal anti-IL-12 antibodies." *J Immunol Methods* 327: 40–52.

Antipov, E., A. E. Cho, K. D. Wittrup and A. M. Klibanov. 2008. "Highly L and D enantioselective variants of horseradish peroxidase discovered by an ultrahigh-throughput selection method." *Proc Natl Acad Sci U S A* 105: 17694–9.

Armstrong, K. A. and B. Tidor. 2008. "Computationally mapping sequence space to understand evolutionary protein engineering." *Biotechnol Prog* 24: 62–73.

Bassi, A. S., D. N. Ding, G. B. Gloor and A. Margaritis. 2000. "Expression of single chain antibodies (ScFvs) for c-myc oncoprotein in recombinant *Escherichia coli* membranes by using the ice-nucleation protein of *Pseudomonas syringae.*" *Biotechnol Prog* 16: 557–63.

Becker, S., A. Michalczyk, S. Wilhelm, K. E. Jaeger and H. Kolmar. 2007. "Ultrahigh-throughput screening to identify *E. coli* cells expressing functionally active enzymes on their surface." *Chembiochem* 8: 943–9.

Becker, S., S. Theile, N. Heppeler, A. Michalczyk, A. Wentzel, et al. 2005. "A generic system for the *Escherichia coli* cell-surface display of lipolytic enzymes." *FEBS Lett* 579: 1177–82.

Beerli, R. R., M. Bauer, R. B. Buser, M. Gwerder, S. Muntwiler, et al. 2008. "Isolation of human monoclonal antibodies by mammalian cell display." *Proc Natl Acad Sci U S A* 105: 14336–41.

Bershtein, S. and D. S. Tawfik. 2008. "Advances in laboratory evolution of enzymes." *Curr Opin Chem Biol* 12: 151–8.

Bessette, P. H., J. J. Rice and P. S. Daugherty. 2004. "Rapid isolation of high-affinity protein binding peptides using bacterial display." *Protein Eng Des Sel* 17: 731–9.

Bhatia, S. K., J. S. Swers, R. T. Camphausen, K. D. Wittrup and D. A. Hammer. 2003. "Rolling adhesion kinematics of yeast engineered to express selectins." *Biotechnol Prog* 19: 1033–7.

Bidlingmaier, S. and B. Liu. 2006. "Construction and application of a yeast surface-displayed human cDNA library to identify post-translational modification-dependent protein–protein interactions." *Mol Cell Proteomics* 5: 533–40.

Boder, E. T., J. R. Bill, A. W. Nields, P. C. Marrack and J. W. Kappler. 2005. "Yeast surface display of a noncovalent MHC class II heterodimer complexed with antigenic peptide." *Biotechnol Bioeng* 92: 485–91.

Boder, E. T., K. S. Midelfort and K. D. Wittrup. 2000a. "Directed evolution of antibody fragments with monovalent femtomolar antigen-binding affinity." *Proc Natl Acad Sci U S A* 97: 10701–5.

Boder, E. T. and K. D. Wittrup. 1997. "Yeast surface display for screening combinatorial polypeptide libraries." *Nat Biotechnol* 15: 553–7.

Boder, E. T. and K. D. Wittrup. 1998. "Optimal screening of surface-displayed polypeptide libraries." *Biotechnol Prog* 14: 55–62.

Boder, E. T. and K. D. Wittrup. 2000b. "Yeast surface display for directed evolution of protein expression, affinity, and stability." *Methods Enzymol* 328: 430–44.

Boublik, Y., P. Di Bonito and I. M. Jones. 1995. "Eukaryotic virus display: Engineering the major surface glycoprotein of the Autographa californica nuclear polyhedrosis virus (AcNPV) for the presentation of foreign proteins on the virus surface." *Nat Biotechnol* 13: 1079–84.

Bowley, D. R., A. F. Labrijn, M. B. Zwick and D. R. Burton. 2007. "Antigen selection from an HIV-1 immune antibody library displayed on yeast yields many novel antibodies compared to selection from the same library displayed on phage." *Protein Eng Des Sel* 20: 81–90.

Brown, S. 1992. "Engineered iron oxide-adhesion mutants of the *Escherichia coli* phage lambda receptor." *Proc Natl Acad Sci U S A* 89: 8651–5.

Brown, S. 1997. "Metal-recognition by repeating polypeptides." *Nat Biotechnol* 15: 269–72.

Chao, G., J. R. Cochran and K. D. Wittrup. 2004. "Fine epitope mapping of anti-epidermal growth factor receptor antibodies through random mutagenesis and yeast surface display." *J Mol Biol* 342: 539–50.

Chao, G., W. L. Lau, B. J. Hackel, S. L. Sazinsky, S. M. Lippow, et al. 2006. "Isolating and engineering human antibodies using yeast surface display." *Nat Protoc* 1: 755–68.

Charbit, A., J. C. Boulain, A. Ryter and M. Hofnung. 1986. "Probing the topology of a bacterial membrane protein by genetic insertion of a foreign epitope; expression at the cell surface." *Embo J* 5: 3029–37.

Cho, C. M., A. Mulchandani and W. Chen. 2002. "Bacterial cell surface display of organophosphorus hydrolase for selective screening of improved hydrolysis of organophosphate nerve agents." *Appl Environ Microbiol* 68: 2026–30.

Christmann, A., K. Walter, A. Wentzel, R. Kratzner and H. Kolmar. 1999. "The cystine knot of a squash-type protease inhibitor as a structural scaffold for *Escherichia coli* cell surface display of conformationally constrained peptides." *Protein Eng* 12: 797-806.

Chung, K. M., G. E. Nybakken, B. S. Thompson, M. J. Engle, A. Marri, et al. 2006. "Antibodies against West Nile Virus nonstructural protein NS1 prevent lethal infection through Fc gamma receptor-dependent and -independent mechanisms." *J Virol* 80: 1340–51.

Cochran, J. R., Y. S. Kim, S. M. Lippow, B. Rao and K. D. Wittrup. 2006. "Improved mutants from directed evolution are biased to orthologous substitutions." *Protein Eng Des Sel* 19: 245–53.

Cochran, J. R., Y. S. Kim, M. J. Olsen, R. Bhandari and K. D. Wittrup. 2004. "Domain-level antibody epitope mapping through yeast surface display of epidermal growth factor receptor fragments." *J Immunol Methods* 287: 147–58.

Colby, D. W., Y. Chu, J. P. Cassady, M. Duennwald, H. Zazulak, et al. 2004a. "Potent inhibition of huntingtin aggregation and cytotoxicity by a disulfide bond-free single-domain intracellular antibody." *Proc Natl Acad Sci U S A* 101: 17616–21.

Colby, D. W., P. Garg, T. Holden, G. Chao, J. M. Webster, et al. 2004b. "Development of a human light chain variable domain (V(L)) intracellular antibody specific for the amino terminus of huntingtin via yeast surface display." *J Mol Biol* 342: 901–12.

Colby, D. W., B. A. Kellogg, C. P. Graff, Y. A. Yeung, J. S. Swers, et al. 2004c. "Engineering antibody affinity by yeast surface display." *Methods Enzymol* 388: 348–58.

Crawford, F., E. Huseby, J. White, P. Marrack and J. W. Kappler. 2004. "Mimotopes for alloreactive and conventional T cells in a peptide-MHC display library." *PLoS Biol* 2: E90.

Crawford, F., K. R. Jordan, B. Stadinski, Y. Wang, E. Huseby, et al. 2006. "Use of baculovirus MHC/peptide display libraries to characterize T-cell receptor ligands." *Immunol Rev* 210: 156–70.

Daugherty, P. S. 2007. "Protein engineering with bacterial display." *Curr Opin Struct Biol* 17: 474–80.

Daugherty, P. S., G. Chen, M. J. Olsen, B. L. Iverson and G. Georgiou. 1998. "Antibody affinity maturation using bacterial surface display." *Protein Eng* 11: 825–32.

Daugherty, P. S., M. J. Olsen, B. L. Iverson and G. Georgiou. 1999. "Development of an optimized expression system for the screening of antibody libraries displayed on the *Escherichia coli* surface." *Protein Eng* 12: 613–21.

Ernst, W., R. Grabherr, D. Wegner, N. Borth, A. Grassauer, et al. 1998. "Baculovirus surface display: Construction and screening of a eukaryotic epitope library." *Nucleic Acids Res* 26: 1718–23.

Esteban, O. and H. Zhao. 2004. "Directed evolution of soluble single-chain human class II MHC molecules." *J Mol Biol* 340: 81–95.

Farinas, E. T. 2006. "Fluorescence activated cell sorting for enzymatic activity." *Comb Chem High Throughput Screen* 9: 321–8.

Feldhaus, M. J. and R. W. Siegel. 2004. "Yeast display of antibody fragments: A discovery and characterization platform." *J Immunol Methods* 290: 69–80.

Feldhaus, M. J., R. W. Siegel, L. K. Opresko, J. R. Coleman, J. M. Feldhaus, et al. 2003. "Flow-cytometric isolation of human antibodies from a nonimmune *Saccharomyces cerevisiae* surface display library." *Nat Biotechnol* 21: 163–70.

Francisco, J. A., C. F. Earhart and G. Georgiou. 1992. "Transport and anchoring of beta-lactamase to the external surface of *Escherichia coli.*" *Proc Natl Acad Sci U S A* 89: 2713–7.

Furukawa, H., T. Tanino, H. Fukuda and A. Kondo. 2006. "Development of novel yeast cell surface display system for homo-oligomeric protein by coexpression of native and anchored subunits." *Biotechnol Prog* 22: 994–7.

Gai, S. A. and K. D. Wittrup. 2007. "Yeast surface display for protein engineering and characterization." *Curr Opin Struct Biol* 17: 467–73.

Garcia-Rodriguez, C., R. Levy, J. W. Arndt, C. M. Forsyth, A. Razai, et al. 2007. "Molecular evolution of antibody cross-reactivity for two subtypes of type A botulinum neurotoxin." *Nat Biotechnol* 25: 107–16.

Graff, C. P., K. Chester, R. Begent and K. D. Wittrup. 2004. "Directed evolution of an anti-carcinoembryonic antigen scFv with a 4-day monovalent dissociation half-time at 37 degrees C." *Protein Eng Des Sel* 17: 293–304.

Gunneriusson, E., P. Samuelson, J. Ringdahl, H. Gronlund, P. A. Nygren, et al. 1999. "Staphylococcal surface display of immunoglobulin A (IgA)- and IgE-specific in vitro-selected binding proteins (affibodies) based on *Staphylococcus aureus* protein A." *Appl Environ Microbiol* 65: 4134–40.

Gunneriusson, E., P. Samuelson, M. Uhlen, P. A. Nygren and S. Stahl. 1996. "Surface display of a functional single-chain Fv antibody on staphylococci." *J Bacteriol* 178: 1341–6.

Hackel, B. J., A. Kapila and K. D. Wittrup. 2008. "Picomolar affinity fibronectin domains engineered utilizing loop length diversity, recursive mutagenesis, and loop shuffling." *J Mol Biol* 381: 1238–52.

Hamilton, S. R., R. C. Davidson, N. Sethuraman, J. H. Nett, Y. Jiang, et al. 2006. "Humanization of yeast to produce complex terminally sialylated glycoproteins." *Science* 313: 1441–3.

Hansson, M., S. Stahl, T. N. Nguyen, T. Bachi, A. Robert, et al. 1992. "Expression of recombinant proteins on the surface of the coagulase-negative bacterium *Staphylococcus xylosus.*" *J Bacteriol* 174: 4239–45.

Harvey, B. R., G. Georgiou, A. Hayhurst, K. J. Jeong, B. L. Iverson, et al. 2004. "Anchored periplasmic expression, a versatile technology for the isolation of high-affinity antibodies from *Escherichia coli*-expressed libraries." *Proc Natl Acad Sci U S A* 101: 9193–8.

Harvey, B. R., A. B. Shanafelt, I. Baburina, R. Hui, S. Vitone, et al. 2006. "Engineering of recombinant antibody fragments to methamphetamine by anchored periplasmic expression." *J Immunol Methods* 308: 43–52.

Ho, M., S. Nagata and I. Pastan. 2006. "Isolation of anti-CD22 Fv with high affinity by Fv display on human cells." *Proc Natl Acad Sci U S A* 103: 9637–42.

Holler, P. D., P. O. Holman, E. V. Shusta, S. O'Herrin, K. D. Wittrup, et al. 2000. "In vitro evolution of a T cell receptor with high affinity for peptide/MHC." *Proc Natl Acad Sci U S A* 97: 5387–92.

Jeong, K. J., M. J. Seo, B. L. Iverson and G. Georgiou. 2007. "APEx 2-hybrid, a quantitative protein–protein interaction assay for antibody discovery and engineering." *Proc Natl Acad Sci U S A* 104: 8247–52.

Jones, D. S., A. P. Silverman, J. R. Cochran. 2008. "Developing therapeutic proteins by engineering ligand-receptor interactions." *Trends Biotechnol* 26: 498–505.

Jones, L. L., S. E. Brophy, A. J. Bankovich, L. A. Colf, N. A. Hanick, et al. 2006. "Engineering and characterization of a stabilized alpha1/alpha2 module of the class I major histocompatibility complex product Ld." *J Biol Chem* 281: 25734–44.

Jose, J. and T. F. Meyer. 2007. "The autodisplay story, from discovery to biotechnical and biomedical applications." *Microbiol Mol Biol Rev* 71: 600–19.

Jung, H. C., J. M. Lebeault and J. G. Pan. 1998. "Surface display of *Zymomonas mobilis levansucrase* by using the ice-nucleation protein of *Pseudomonas syringae.*" *Nat Biotechnol* 16: 576–80.

Jung, S. T., K. J. Jeong, B. L. Iverson and G. Georgiou. 2007. "Binding and enrichment of *Escherichia coli* spheroplasts expressing inner membrane tethered scFv antibodies on surface immobilized antigens." *Biotechnol Bioeng* 98: 39–47.

Kenrick, S., J. Rice and P. Daugherty. 2007. "Flow cytometric sorting of bacterial surface-displayed libraries." *Curr Protoc Cytom* Chapter 4: Unit4.6.

Kieke, M. C., E. V. Shusta, E. T. Boder, L. Teyton, K. D. Wittrup, et al. 1999. "Selection of functional T cell receptor mutants from a yeast surface-display library." *Proc Natl Acad Sci U S A* 96: 5651–6.

Kim, Y. S., R. Bhandari, J. R. Cochran, J. Kuriyan and K. D. Wittrup. 2006. "Directed evolution of the epidermal growth factor receptor extracellular domain for expression in yeast." *Proteins* 62: 1026–35.

Kimura, R. H., Z. Cheng, S. S. Gambhir and J. R. Cochran. 2009. "Engineered knottin peptides: A new class of agents for imaging integrin expression in living subjects." *Cancer Res* 69: 2435-42.

Kjaergaard, K., J. K. Sorensen, M. A. Schembri and P. Klemm. 2000. "Sequestration of zinc oxide by fimbrial designer chelators." *Appl Environ Microbiol* 66: 10–4.

Kondo, A. and M. Ueda. 2004. "Yeast cell-surface display—applications of molecular display." *Appl Microbiol Biotechnol* 64: 28–40.

Krauland, E. M., B. R. Peelle, K. D. Wittrup and A. M. Belcher. 2007. "Peptide tags for enhanced cellular and protein adhesion to single-crystalline sapphire." *Biotechnol Bioeng* 97: 1009–20.

Kronqvist, N., J. Lofblom, A. Jonsson, H. Wernerus and S. Stahl. 2008. "A novel affinity protein selection system based on staphylococcal cell surface display and flow cytometry." *Protein Eng Des Sel* 21: 247–55.

Kumara, M. T., N. Srividya, S. Muralidharan and B. C. Tripp. 2006. "Bioengineered flagella protein nanotubes with cysteine loops: Self-assembly and manipulation in an optical trap." *Nano Lett* 6: 2121–9.

Leclerc, C., A. Charbit, P. Martineau, E. Deriaud and M. Hofnung. 1991. "The cellular location of a foreign B cell epitope expressed by recombinant bacteria determines its T cell-independent or T cell-dependent characteristics." *J Immunol* 147: 3545–52.

Lee, S. Y., J. H. Choi and Z. Xu. 2003. "Microbial cell-surface display." *Trends Biotechnol* 21: 45–52.

Levy, R., C. M. Forsyth, S. L. LaPorte, I. N. Geren, L. A. Smith, et al. 2007. "Fine and domain-level epitope mapping of botulinum neurotoxin type A neutralizing antibodies by yeast surface display." *J Mol Biol* 365: 196–210.

Li, W., P. Lei, B. Yu, S. Wu, J. Peng, et al. 2008. "Screening and identification of a novel target specific for hepatoma cell line HepG2 from the FliTrx bacterial peptide library." *Acta Biochim Biophys Sin (Shanghai)* 40: 443–51.

Lin, Y., T. Tsumuraya, T. Wakabayashi, S. Shiraga, I. Fujii, et al. 2003. "Display of a functional hetero-oligomeric catalytic antibody on the yeast cell surface." *Appl Microbiol Biotechnol* 62: 226–32.

Lipovsek, D., E. Antipov, K. A. Armstrong, M. J. Olsen, A. M. Klibanov, et al. 2007a. "Selection of horseradish peroxidase variants with enhanced enantioselectivity by yeast surface display." *Chem Biol* 14: 1176–85.

Lipovsek, D., S. M. Lippow, B. J. Hackel, M. W. Gregson, P. Cheng, et al. 2007b. "Evolution of an interloop disulfide bond in high-affinity antibody mimics based on fibronectin type III domain and selected by yeast surface display: Molecular convergence with single-domain camelid and shark antibodies." *J Mol Biol* 368: 1024–41.

Lipovsek, D. and A. Pluckthun. 2004. "In-vitro protein evolution by ribosome display and mRNA display." *J Immunol Methods* 290: 51–67.

Lippow, S. M., K. D. Wittrup and B. Tidor. 2007. "Computational design of antibody-affinity improvement beyond in vivo maturation." *Nat Biotechnol* 25: 1171–6.

Lofblom, J., N. Kronqvist, M. Uhlen, S. Stahl and H. Wernerus. 2007a. "Optimization of electroporation-mediated transformation: *Staphylococcus carnosus* as model organism." *J Appl Microbiol* 102: 736–47.

Lofblom, J., J. Sandberg, H. Wernerus and S. Stahl. 2007b. "Evaluation of staphylococcal cell surface display and flow cytometry for postselectional characterization of affinity proteins in combinatorial protein engineering applications." *Appl Environ Microbiol* 73: 6714–21.

Lofblom, J., H. Wernerus and S. Stahl. 2005. "Fine affinity discrimination by normalized fluorescence activated cell sorting in staphylococcal surface display." *FEMS Microbiol Lett* 248: 189–98.

Lu, Z., E. R. LaVallie and J. M. McCoy. 2002. "Using bio-panning of FLITRX peptide libraries displayed on *E. coli* cell surface to study protein–protein interactions." *Methods Mol Biol* 205: 267–80.

Lu, Z., K. S. Murray, V. Van Cleave, E. R. LaVallie, M. L. Stahl, et al. 1995. "Expression of thioredoxin random peptide libraries on the Escherichia coli cell surface as functional fusions to flagellin: A system designed for exploring protein–protein interactions." *Nat Biotechnol (N Y)* 13: 366–72.

Lunder, M., T. Bratkovic, B. Doljak, S. Kreft, U. Urleb, et al. 2005. "Comparison of bacterial and phage display peptide libraries in search of target-binding motif." *Appl Biochem Biotechnol* 127: 125–31.

Makela, A. R. and C. Oker-Blom. 2006. "Baculovirus display: A multifunctional technology for gene delivery and eukaryotic library development." *Adv Virus Res* 68: 91–112.

Makela, A. R. and C. Oker-Blom. 2008. "The baculovirus display technology—an evolving instrument for molecular screening and drug delivery." *Comb Chem High Throughput Screen* 11: 86–98.

Maurer, J., J. Jose and T. F. Meyer. 1997. "Autodisplay: One-component system for efficient surface display and release of soluble recombinant proteins from *Escherichia coli*." *J Bacteriol* 179: 794–804.

Mazmanian, S. K., G. Liu, H. Ton-That and O. Schneewind. 1999. "*Staphylococcus aureus* sortase, an enzyme that anchors surface proteins to the cell wall." *Science* 285: 760–3.

Mazor, Y., T. Van Blarcom, B. L. Iverson and G. Georgiou. 2008. "E-clonal antibodies: Selection of full-length IgG antibodies using bacterial periplasmic display." *Nat Protoc* 3: 1766–77.

Mazor, Y., T. Van Blarcom, R. Mabry, B. L. Iverson and G. Georgiou. 2007. "Isolation of engineered, full-length antibodies from libraries expressed in *Escherichia coli*." *Nat Biotechnol* 25: 563–5.

Miller, K. D., N. B. Pefaur and C. L. Baird. 2008. "Construction and screening of antigen targeted immune yeast surface display antibody libraries." *Curr Protoc Cytom* Chapter 4: Unit4.7.

Mitra, N., S. Sinha, T. N. Ramya and A. Surolia. 2006. "N-linked oligosaccharides as outfitters for glycoprotein folding, form and function." *Trends Biochem Sci* 31: 156–63.

Narayanan, N. and C. P. Chou. 2008. "Physiological improvement to enhance *Escherichia coli* cell-surface display via reducing extracytoplasmic stress." *Biotechnol Prog* 24: 293–301.

Oliphant, T., M. Engle, G. E. Nybakken, C. Doane, S. Johnson, et al. 2005. "Development of a humanized monoclonal antibody with therapeutic potential against West Nile virus." *Nat Med* 11: 522–30.

Oliphant, T., G. E. Nybakken, M. Engle, Q. Xu, C. A. Nelson, et al. 2006. "Antibody recognition and neutralization determinants on domains I and II of West Nile Virus envelope protein." *J Virol* 80: 12149–59.

Olsen, M. J., D. Stephens, D. Griffiths, P. Daugherty, G. Georgiou, et al. 2000. "Function-based isolation of novel enzymes from a large library." *Nat Biotechnol* 18: 1071–4.

Orr, B. A., L. M. Carr, K. D. Wittrup, E. J. Roy and D. M. Kranz. 2003. "Rapid method for measuring ScFv thermal stability by yeast surface display." *Biotechnol Prog* 19: 631–8.

Ozhalici-Unal, H., C. L. Pow, S. A. Marks, L. D. Jesper, G. L. Silva, et al. 2008. "A rainbow of fluoromodules: A promiscuous scFv protein binds to and activates a diverse set of fluorogenic cyanine dyes." *J Am Chem Soc* 130: 12620–1.

Park, S., Y. Xu, X. F. Stowell, F. Gai, J. G. Saven, et al. 2006. "Limitations of yeast surface display in engineering proteins of high thermostability." *Protein Eng Des Sel* 19: 211–7.

Pazirandeh, M., B. M. Wells and R. L. Ryan. 1998. "Development of bacterium-based heavy metal biosorbents: Enhanced uptake of cadmium and mercury by *Escherichia coli* expressing a metal binding motif." *Appl Environ Microbiol* 64: 4068–72.

Peelle, B. R., E. M. Krauland, K. D. Wittrup and A. M. Belcher. 2005a. "Probing the interface between biomolecules and inorganic materials using yeast surface display and genetic engineering." *Acta Biomater* 1: 145–54.

Pepper, L. R., Y. K. Cho, E. T. Boder and E. V. Shusta. 2008. "A decade of yeast surface display technology: Where are we now?" *Comb Chem High Throughput Screen* 11:127–134.

Peelle, B. R., E. M. Krauland, K. D. Wittrup and A. M. Belcher. 2005b. "Design criteria for engineering inorganic material-specific peptides." *Langmuir* 21: 6929–33.

Piatesi, A., S. W. Howland, J. A. Rakestraw, C. Renner, N. Robson, et al. 2006. "Directed evolution for improved secretion of cancer-testis antigen NY-ESO-1 from yeast." *Protein Expr Purif* 48: 232–42.

Rakestraw, J. A., A. R. Baskaran and K. D. Wittrup. 2006. "A flow cytometric assay for screening improved heterologous protein secretion in yeast." *Biotechnol Prog* 22: 1200–8.

Rao, B. M., A. T. Girvin, T. Ciardelli, D. A. Lauffenburger and K. D. Wittrup. 2003. "Interleukin-2 mutants with enhanced alpha-receptor subunit binding affinity." *Protein Eng* 16: 1081–7.

Rice, J. J. and P. S. Daugherty. 2008. "Directed evolution of a biterminal bacterial display scaffold enhances the display of diverse peptides." *Protein Eng Des Sel* 21: 435–42.

Rice, J. J., A. Schohn, P. H. Bessette, K. T. Boulware and P. S. Daugherty. 2006. "Bacterial display using circularly permuted outer membrane protein OmpX yields high affinity peptide ligands." *Protein Sci* 15: 825–36.

Richman, S. A., S. J. Healan, K. S. Weber, D. L. Donermeyer, M. L. Dossett, et al. 2006. "Development of a novel strategy for engineering high-affinity proteins by yeast display." *Protein Eng Des Sel* 19: 255–64.

Richman, S. A. and D. M. Kranz. 2007. "Display, engineering, and applications of antigen-specific T cell receptors." *Biomol Eng* 24: 361–73.

Robert, A., P. Samuelson, C. Andreoni, T. Bachi, M. Uhlen, et al. 1996. "Surface display on staphylococci: A comparative study." *FEBS Lett* 390: 327–33.

Samuelson, P., H. Wernerus, M. Svedberg and S. Stahl. 2000. "Staphylococcal surface display of metal-binding polyhistidyl peptides." *Appl Environ Microbiol* 66: 1243–8.

Sarkar, C. A., I. Dodevski, M. Kenig, S. Dudli, A. Mohr, et al. 2008. "Directed evolution of a G protein-coupled receptor for expression, stability, and binding selectivity." *Proc Natl Acad Sci U S A* 105: 14808–13.

Schreuder, M. P., S. Brekelmans, H. van den Ende and F. M. Klis. 1993. "Targeting of a heterologous protein to the cell wall of *Saccharomyces cerevisiae*." *Yeast* 9: 399–409.

Schreuder, M. P., A. T. Mooren, H. Y. Toschka, C. T. Verrips and F. M. Klis. 1996. "Immobilizing proteins on the surface of yeast cells." *Trends Biotechnol* 14: 115–20.

Schwarzman, A. L., M. Tsiper, L. Gregori, D. Goldgaber, J. Frakowiak, et al. 2005. "Selection of peptides binding to the amyloid b-protein reveals potential inhibitors of amyloid formation." *Amyloid* 12: 199–209.

Schweickhardt, R. L., X. Jiang, L. M. Garone and W. H. Brondyk. 2003. "Structure-expression relationship of tumor necrosis factor receptor mutants that increase expression." *J Biol Chem* 278: 28961–7.

Shapiro, H. M. 2003. *Practical flow cytometry*, Wiley-Liss.

Shibasaki, S., M. Ueda, K. Ye, K. Shimizu, N. Kamasawa, et al. 2001. "Creation of cell surface-engineered yeast that display different fluorescent proteins in response to the glucose concentration." *Appl Microbiol Biotechnol* 57: 528–33.

Shiraga, S., M. Kawakami, M. Ishiguro and M. Ueda. 2005. "Enhanced reactivity of *Rhizopus oryzae* lipase displayed on yeast cell surfaces in organic solvents: Potential as a whole-cell biocatalyst in organic solvents." *Appl Environ Microbiol* 71: 4335–8.

Shusta, E. V., P. D. Holler, M. C. Kieke, D. M. Kranz and K. D. Wittrup. 2000. "Directed evolution of a stable scaffold for T-cell receptor engineering." *Nat Biotechnol* 18: 754–9.

Shusta, E. V., M. C. Kieke, E. Parke, D. M. Kranz and K. D. Wittrup. 1999. "Yeast polypeptide fusion surface display levels predict thermal stability and soluble secretion efficiency." *J Mol Biol* 292: 949–56.

Siegel, R. W., J. R. Coleman, K. D. Miller and M. J. Feldhaus. 2004. "High efficiency recovery and epitope-specific sorting of an scFv yeast display library." *J Immunol Methods* 286: 141–53.

Silverman, A. P., A. M. Levin, J. L. Lahti and J. R. Cochran. 2009. "Engineered cystine-knot peptides that bind alphav beta3 integrin with antibody-like affinities." *J Mol Biol* 385: 1064-75.

Stathopoulos, C., G. Georgiou and C. F. Earhart. 1996. "Characterization of *Escherichia coli* expressing an Lpp′OmpA(46-159)-PhoA fusion protein localized in the outer membrane." *Appl Microbiol Biotechnol* 45: 112–9.

Szent-Gyorgyi, C., B. F. Schmidt, Y. Creeger, G. W. Fisher, K. L. Zakel, et al. 2008. "Fluorogen-activating single-chain antibodies for imaging cell surface proteins." *Nat Biotechnol* 26: 235–40.

Tanino, T., H. Fukuda and A. Kondo. 2006. "Construction of a *Pichia pastoris* cell-surface display system using Flo1p anchor system." *Biotechnol Prog* 22: 989–93.

Thai, C. K., H. Dai, M. S. Sastry, M. Sarikaya, D. T. Schwartz, et al. 2004. "Identification and characterization of Cu(2)O- and ZnO-binding polypeptides by *Escherichia coli* cell surface display: Toward an understanding of metal oxide binding." *Biotechnol Bioeng* 87: 129–37.

Tripp, B. C., Z. Lu, K. Bourque, H. Sookdeo and J. M. McCoy. 2001. "Investigation of the 'switch-epitope' concept with random peptide libraries displayed as thioredoxin loop fusions." *Protein Eng* 14: 367–77.

van den Beucken, T., H. Pieters, M. Steukers, M. van der Vaart, R. C. Ladner, et al. 2003. "Affinity maturation of Fab antibody fragments by fluorescent-activated cell sorting of yeast-displayed libraries." *FEBS Lett* 546: 288–94.

VanAntwerp, J. J. and K. D. Wittrup. 2000. "Fine affinity discrimination by yeast surface display and flow cytometry." *Biotechnol Prog* 16: 31–7.

Varadarajan, N., J. Gam, M. J. Olsen, G. Georgiou and B. L. Iverson. 2005. "Engineering of protease variants exhibiting high catalytic activity and exquisite substrate selectivity." *Proc Natl Acad Sci U S A* 102: 6855–60.

Wadle, A., A. Mischo, J. Imig, B. Wullner, D. Hensel, et al. 2005. "Serological identification of breast cancer-related antigens from a *Saccharomyces cerevisiae* surface display library." *Int J Cancer* 117: 104–13.

Wang, X. X., Y. K. Cho and E. V. Shusta. 2007. "Mining a yeast library for brain endothelial cell-binding antibodies." *Nat Methods* 4: 143–5.

Wang, X. X. and E. V. Shusta. 2005. "The use of scFv-displaying yeast in mammalian cell surface selections." *J Immunol Methods* 304: 30–42.

Wang, Y., A. Rubtsov, R. Heiser, J. White, F. Crawford, et al. 2005a. "Using a baculovirus display library to identify MHC class I mimotopes." *Proc Natl Acad Sci U S A* 102: 2476–81.

Wang, Z., A. Mathias, S. Stavrou and D. Neville. 2005b. "A new yeast display vector permitting free scFv amino termini can augment ligand binding affinities." *Protein Eng Des Sel* 18: 337–343.

Weaver-Feldhaus, J. M., J. Lou, J. R. Coleman, R. W. Siegel, J. D. Marks, et al. 2004. "Yeast mating for combinatorial Fab library generation and surface display." *FEBS Lett* 564: 24–34.

Weaver-Feldhaus, J. M., K. D. Miller, M. J. Feldhaus and R. W. Siegel. 2005. "Directed evolution for the development of conformation-specific affinity reagents using yeast display." *Protein Eng Des Sel* 18: 527–36.

Wernerus, H. and S. Stahl. 2002. "Vector engineering to improve a staphylococcal surface display system." *FEMS Microbiol Lett* 212: 47–54.

Wernerus, H. and S. Stahl. 2004. "Biotechnological applications for surface-engineered bacteria." *Biotechnol Appl Biochem* 40: 209–28.

Wildt, S. and T. U. Gerngross. 2005. "The humanization of N-glycosylation pathways in yeast." *Nat Rev Microbiol* 3: 119–28.

Wolber, P. K. 1993. "Bacterial ice nucleation." *Adv Microb Physiol* 34: 203–37.

Wolkowicz, R., G. C. Jager and G. P. Nolan. 2005. "A random peptide library fused to CCR5 for selection of mimetopes expressed on the mammalian cell surface via retroviral vectors." *J Biol Chem* 280: 15195–201.

Yang, W., D. Luo, S. Wang, R. Wang, R. Chen, et al. 2008. "TMTP1, a novel tumor-homing peptide specifically targeting metastasis." *Clin Cancer Res* 14: 5494–502.

Yeung, Y. A. and K. D. Wittrup. 2002. "Quantitative screening of yeast surface-displayed polypeptide libraries by magnetic bead capture." *Biotechnol Prog* 18: 212–20.

Zitzmann, S., S. Kramer, W. Mier, M. Mahmut, J. Fleig, et al. 2005. "Identification of a new prostate-specific cyclic peptide with the bacterial FliTrx system." *J Nucleic Med* 46: 782–5.

# 3 Cell-Free Display Systems for Protein Engineering

*Pamela A. Barendt and Casim A. Sarkar*

## CONTENTS

Cell-free display methods offer an alternative to host-dependent methods (Chapters 1 and 2) for the directed evolution of proteins. The present chapter provides a brief overview of *in vitro* protein synthesis and describes how this method has been adapted for cell-free display. Several different cell-free platforms for protein engineering—ribosome display, mRNA display, covalent and noncovalent DNA display, and *in vitro* compartmentalization—are discussed in terms of their technical execution and their suitability for certain applications. The advantages and limitations of each of these methods, in comparison to cell-dependent or other cell-free methods, are examined.

## GENERAL APPROACHES TO CELL-FREE PROTEIN SYNTHESIS

### A TYPICAL *IN VITRO* PROTEIN SYNTHESIS REACTION

*In vitro* protein synthesis is generally conducted using three major components: an extract (minimally purified lysate) derived from cells engaged in a high rate of protein synthesis, a mixture of efficiency-enhancing additives, and an exogenous messenger RNA (mRNA) template. A DNA template may be used if transcription and translation can be coupled (see the section titled "Coupled Transcription and Translation" in this chapter). The components of the reaction are combined and incubated under conditions that produce a high yield of functional protein. The yield depends primarily upon the nature of the protein, the translation system, and the reaction conditions (see the section titled "Reaction Conditions" in this chapter). In general, *Escherichia coli*-based systems produce the largest amount of total protein, often milligrams per milliliter of reaction (Katzen et al. 2005). Wheat germ-based systems perform nearly as well, typically producing fractions of a milligram to milligrams per milliliter (Endo and Sawasaki 2006; Katzen et al. 2005). Rabbit reticulocyte-based systems have much lower yields, on the order of one microgram per milliliter (Endo and Sawasaki 2006). The functional yield of produced proteins is typically assessed based on measurable properties such as binding, catalysis, and fluorescence.

### Extract

Most commonly, *E. coli* extract is used when a prokaryotic system is desired, and rabbit reticulocyte or wheat germ extract is used when a eukaryotic system is preferred (Endo and Sawasaki 2006; Katzen et al. 2005). Other factors to consider in choosing an extract include cost, time, difficulty, and efficiency of synthesis. Codon limitations or particular template requirements (e.g., complex structure or disulfide bonding) might also influence the decision. In order to make an extract, whole cells must be lysed (i.e., the cell membrane must be ruptured). Mechanical methods are preferable to chemical methods for lysis, since certain additives may reduce the activity of the extract. Crude lysates are typically centrifuged twice at 30,000$g$ to remove large aggregates of cellular material. The resulting supernatant (often called "S30 extract" to indicate that it is the soluble fraction after centrifugation at 30,000$g$) contains all of the macromolecular components that are necessary for protein synthesis, including ribosomes; endogenous tRNAs; aminoacyl-tRNA synthetases; and initiation, elongation, and termination factors. Further processing of the S30 extract generally involves incubation at room temperature to allow for degradation of endogenous mRNAs and extensive dialysis against a suitable buffer at 4°C, followed by flash-freezing in liquid nitrogen and storage at −80°C (Amstutz et al. 2006). Successful protocols for making both prokaryotic (Nirenberg 1963; Pratt 1984; Zubay 1973) and eukaryotic (Anderson et al. 1983; Pelham and Jackson 1976) extracts have been in the literature for many years, although improvements are ongoing (Katzen et al. 2005).

## Additives

To increase the efficiency of synthesis, the reaction should also include a mixture of amino acids, energy sources (ATP and GTP), energy regenerating systems (pyruvate kinase and phosphoenolpyruvate for *E. coli* lysates, or creatine phosphokinase and creatine phosphate for rabbit reticulocyte and wheat germ lysates), key ions ($Mg^{2+}$ and $K^+$, in particular), and tRNAs. This mixture may also contain reagents specific to the synthesis at hand. To enhance correct disulfide bond formation and proper folding of eukaryote-derived proteins in *E. coli*-based translation systems, protein disulfide isomerase and chaperone systems, such as DnaK/DnaJ/GrpE and GroEL/GroES, may be introduced (Ryabova et al. 1997). One may also pretreat the extract with iodoacetamide, an alkylating agent that inactivates reductases in the extract and stabilizes the -SH/S-S redox potential, and then supplement the translation reaction with oxidized and reduced glutathione (in a ratio that provides a relatively oxidizing environment) and a disulfide isomerase (Kim and Swartz 2004; Yin and Swartz 2004). If core glycosylation is desired, microsomes may be added, depending on the system (Walter and Blobel 1983). Rabbit reticulocyte lysate is most commonly used for this purpose, although the production of glycoproteins *in vitro* remains a challenge, due to their natural complexity and dependence on intracellular transport (Katzen et al. 2005). In the case of transmembrane proteins, it is tremendously difficult to obtain high yields through *in vivo* expression, primarily due to aggregation, but detergents and lipids can be introduced during cell-free production in *E. coli* lysates to prevent aggregation and increase solubility (Klammt et al. 2004). In summary, the ability to modify and optimize the components of a translation reaction is a major advantage of *in vitro* synthesis, although much work remains to be done in the areas of glycosylation, membrane proteins, and large complex proteins.

## Template

The final component of the reaction, the exogenous mRNA or DNA template (see the section titled "Coupled Transcription and Translation" in this chapter), should be prepared according to the system in which it will be expressed. When using *E. coli* extract, for example, a Shine-Dalgarno sequence should be present at an appropriate distance upstream of the AUG start site. Analogously, in a eukaryotic system, a Kozak sequence should be used and the mRNA should also be modified with a 5′ cap and 3′ poly(A) tail, if possible. Codon usage should be optimized to reflect the relative abundances of codons in the organism from which the extract was derived.

## Reaction Conditions

In order to perform an *in vitro* protein synthesis reaction, the extract, additive mixture, and exogenous template should be combined and incubated at an optimal temperature (often 30–37°C) for an optimal time (often an hour or more) (Pelham and Jackson 1976; Zubay 1973). The goal is to maximize the yield of functional protein relative to the quantity of input extract, additives, and template. Higher temperatures result in faster translation kinetics (Endoh et al. 2006), but may cause misfolding of some proteins. Misfolding occurs when simple Brownian motion allows a nascent polypeptide to overcome an energy barrier and fall into a local, but not global, minimum on the energy landscape (Dill and Chan 1997). Additionally, longer incubation times are not necessarily beneficial. Energy sources become depleted over time, causing synthesis to slow, while protein degradation continuously reduces functional product. Systems for regenerating ATP in a "batch" reaction, in which the reagents are mixed and left to incubate, have been widely investigated and recently reviewed (Calhoun and Swartz 2007). For large-scale applications, many different configurations of reactors have been devised to replenish consumable reagents (Spirin 2004).

### COUPLED TRANSCRIPTION AND TRANSLATION

One notable variation on the basic synthesis protocol is the coupling of transcription and translation. In an "uncoupled" translation system, like the one described previously, transcription is performed in a separate reaction, the mRNA is purified, and translation is performed in a second step. In a "coupled" translation reaction, transcription and translation take place simultaneously in the same medium. The coupled format uses double-stranded DNA (dsDNA) as the template, and requires RNA polymerase and nucleoside triphosphates (NTPs). Translation generally starts before the transcription of an mRNA molecule is complete, and multiple ribosomes may translate a single message simultaneously, resulting in polysomes. Thus, the coupled reaction closely mimics what occurs naturally in prokaryotes. In contrast, "uncoupled" transcription and translation is the natural state in eukaryotic cells, as these processes are confined to the nucleus and cytoplasm, respectively. The advantage of using a decoupled system is that each step can be individually optimized in terms of buffer composition, incubation time, and temperature. However, the coupled system saves time and works well enough for many applications. Both prokaryotic and eukaryotic extracts have been used in coupled and uncoupled systems (He and Taussig 2002).

## THE PURE SYSTEM

One major advance in cell-free synthesis that warrants specific mention is the PURE (protein synthesis using recombinant elements) system, which uses purified, recombinantly expressed proteins from *E. coli* rather than extract from cells (Shimizu et al. 2001). All of the translation factors, aminoacyl-tRNA synthetases, and other enzymes in the PURE system are recombinant; only the ribosomes and tRNAs are purified from *E. coli* lysates. The main advantages of the PURE system, compared to lysate-based systems, are defined composition, lack of nucleases and proteases, and a simple procedure for isolating produced protein. Because all of the recombinant components are $(His)_6$-tagged, the actual protein product does not need any affinity tag for purification. The translation reaction can simply be subjected to ultrafiltration (to remove ribosomes) and affinity chromatography (to remove $(His)_6$-tagged components) to obtain highly pure, native protein product. One disadvantage of this reconstituted system is that it produces lower yields than an extract-based system does (Hillebrecht and Chong 2008). Additionally, the source of ribosomes, tRNAs, and recombinant components (*E. coli*) may restrict the types of proteins that can be successfully synthesized in the PURE system, although the functional yield of certain proteins can be improved by the introduction of molecular chaperones and oxidized glutathione in place of dithiothreitol (Shimizu et al. 2005). Typically, the PURE system is used as a coupled transcription/translation system, but it may be used as an uncoupled system as well.

## ADAPTATION OF CELL-FREE PROTEIN SYNTHESIS METHODS FOR DIRECTED EVOLUTION

### DIRECTED EVOLUTION AND GENOTYPE-PHENOTYPE LINKAGE

Cell-free protein synthesis methods can generally be adapted for use in directed evolution experiments. Briefly, directed evolution involves iterative cycles of diversification and selection for the purpose of creating proteins with certain desirable properties (Figure 3.1). Typically, one starts with a diverse library of nucleotide sequences and translates these into protein while maintaining some sort of physical or compartmental linkage between the nucleotide sequence (genotype) and corresponding protein (phenotype). Selections or screens can then be performed based on the properties of the displayed protein. [Note: The terms *selection* and *screen* are often confounded in the literature. For the purposes of this chapter, *selection* indicates a bulk process that allows only variants of interest to be recovered in aggregate, while *screen* indicates an active search in which each member of the library is assessed individually (Hilvert et al. 2002)]. After the selection or screen, the genetic information corresponding to the desired proteins is recovered and amplified, allowing one to pursue additional rounds of diversification followed by selection (often with increasing stringency). A combinatorial approach, in which selections are performed

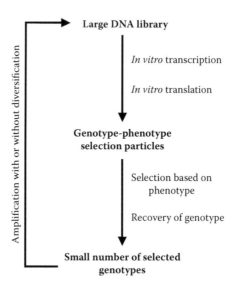

**FIGURE 3.1** Outline of *in vitro* directed evolution. All of the cell-free display methods discussed in this chapter—ribosome display, mRNA display, covalent and noncovalent DNA display, and *in vitro* compartmentalization—follow this basic procedure. First, a large, diverse library of DNA sequences is transcribed and translated to form genotype-phenotype selection particles. The link between genotype (nucleic acid sequence) and phenotype (protein) may be direct, as in ribosome display, mRNA display, and DNA display, or indirect, as in many forms of *in vitro* compartmentalization. Next, a selection pressure is applied to enrich the library for members having a desired property (e.g., high binding affinity for a target molecule, stability under certain conditions, ability to modify DNA). The desired genotypes are recovered and then amplified by polymerase chain reaction (PCR) for another round of selection. Error-prone PCR, DNA shuffling, or staggered extension process (StEP) may be used to introduce additional diversity during the amplification step.

without introducing additional mutations at intermediate steps, is also used in protein engineering, although the extent of sequence space that can be sampled is much smaller (Plückthun et al. 2000).

In both evolutionary and combinatorial approaches, each selection or screening round serves to enrich the library with members exhibiting the desired property. After several rounds, the level of enrichment may be high enough to sequence a subset of the library and thereby analyze the evolved (or selected) proteins. Significant enrichment is often observed after two to ten rounds, but this depends on a handful of variables, including the diversity of the initial library, the stringency of selection, and the extent of mutagenesis between rounds.

Chapters 1 and 2 describe phage display and cell surface display, respectively, which use a phage particle or whole cell to maintain genotype-phenotype linkage. While these methods are very powerful and have their own distinct advantages, there are a number of cell-free display technologies that have been adapted from cell-free protein synthesis methods and do not require whole cells at any step. Not using cells is an advantage in terms of obtainable library size, time, fewer restraints on translation and selection, reduced expression bias, and ease of introducing diversity between rounds (see the section titled "*In Vitro* versus *In Vivo* Display" in this chapter). The following cell-free methods are discussed in detail in this chapter and summarized in Table 3.1: ribosome display, mRNA display, covalent and noncovalent DNA display, and *in vitro* compartmentalization. Each uses a different method for linking genotype and phenotype, but their underlying principles are similar. The next section describes some general factors to consider in choosing a cell-free display platform, designing the nucleic acid template, and performing the selections or screens.

## CONSIDERATIONS IN CELL-FREE DISPLAY AND SELECTION/SCREENING

### Type of Selection or Screen

Many different properties of proteins may be engineered using *in vitro* techniques (Table 3.1). In general, ribosome display, mRNA display, and covalent and noncovalent display methods are equally well suited to improve protein equilibrium affinity, off-rate, stability, and folding. They all rely on a binding event between the displayed protein and an immobilized ligand to perform the selection. *In vitro* compartmentalization methods, in contrast, are primarily useful for the evolution of enzymes, which bind only transiently to their substrate. Water-in-oil-in-water emulsions, in particular, are notable for their ability to be screened by fluorescence-activated cell sorting (FACS), provided that a fluorescence-based assay is available for the detection of the desired activity. Clearly, the goal of the selection or screen will narrow down the choice of display technique.

Binding-based selections are very popular in directed evolution and warrant a brief explanation. First, the displayed library must be allowed to bind to an immobilized target molecule on a solid support (typically, an assay plate or beads). Unbound or weakly bound library members are washed away, so that only firmly bound library members remain. The method by which these members are recovered depends on the cell-free display technology used, but always involves PCR or reverse transcription followed by PCR (RT-PCR) to amplify enriched sequences. The washing times are relatively short in the early rounds to avoid losing valuable yet sparse clones due to stochastic effects. As the library becomes enriched, however, redundancy within the pool allows for more stringent washing. Both high and low affinity binders can be identified by examining the pool of binders before it is dominated by only a few sequences (Huang and Liu 2007).

An important variation on this basic affinity selection is off-rate selection. Selecting for slower off-rates leads to increased binding affinity, as on-rates are relatively constant for proteins (Northrup and Erickson 1992). After the washing steps, bound library members are allowed to dissociate from the immobilized target in the presence of a large excess of soluble target. Lower-affinity binders dissociate more rapidly from the immobilized target and are far more likely to bind to target molecules in solution, thus minimizing their rebinding to the immobilized target. After a given wait time, the solid support is washed, leaving only the highest affinity library members bound. These are amplified and carried forth to the next round. As usual, the selection pressure (time allowed for dissociation, in this case) is gradually increased.

**TABLE 3.1**

**Comparison of *In Vitro* Display Methods**

| Method | References | Library Size | Nature of Genotype-Phenotype Linkage | Properties Amenable to Selection/Screening | Selection/Screening Approaches | Size of Genotype-Phenotype Linkage (Excluding Genotype and Displayed Protein) |
|---|---|---|---|---|---|---|
| Ribosome display (Figure 3.2A) | Hanes and Plückthun 1997; Leemhuis et al. 2005; Lipovsek and Plückthun 2004 | $10^{12}$–$10^{14}$ | Noncovalent linkage of mRNA and protein through ribosome (C-terminal) | Equilibrium affinity, off-rate, stability, folding, single catalytic event | Binding to immobilized ligand | ~2.7 MDa (prokaryotic ribosome); ~4.2 MDa (eukaryotic ribosome) |
| mRNA display (Figure 3.2B) | Leemhuis et al. 2005; Lipovsek and Plückthun 2004; Roberts and Szostak 1997 | $10^{12}$–$10^{14}$ | Covalent linkage of mRNA and protein through DNA-puromycin linker (C-terminal) | " | " | ~10 kDa (DNA-puromycin linker) |
| **Covalent DNA Display Methods** | | | | | | |
| mRNA display-based DNA display (Figure 3.3A) | Kurz et al. 2001; Tabuchi et al. 2001 | $10^{12}$–$10^{14}$ | Covalent linkage of DNA and protein through various chemical linkers (C-terminal) | " | " | ~10 kDa (various low molecular weight linkers) |
| P2A fusion (Figure 3.3B) | Reiersen et al. 2005 | $10^{7}$ | Covalent linkage of DNA and P2A fusion protein (C-terminal) | " | " | 86 kDa (P2A endonuclease) |
| M.HaeIII fusion (Figure 3.3C) | Bertschinger and Neri 2004; Bertschinger et al. 2007 | $10^{10}$ | Covalent linkage of DNA and M.HaeIII fusion protein (N-terminal) | " | " | 38 kDa (M.HaeIII) |

| Method | Reference | Library size | Principle | Catalytic events | Selection | Protein/bead size |
|---|---|---|---|---|---|---|
| AGT fusion (Figure 3.3D) | Stein et al. 2007 | $10^{10}$ | Covalent linkage of BG-DNA and AGT fusion protein (N-terminal) | | " | 24 kDa (AGT) |
| **Noncovalent DNA Display Methods** | | | | | | |
| CIS display (Figure 3.4A) | Odegrip et al. 2004 | $10^{12}$ | Noncovalent linkage of DNA and RepA fusion protein (C-terminal) | | " | 33 kDa (RepA) |
| STABLE (Figure 3.4B) | Doi and Yanagawa 1999; Yonezawa et al. 2003; Yonezawa et al. 2004 | $10^{10}$ | Noncovalent linkage of biotinylated DNA and streptavidin fusion protein (N- or C-terminal) | | " | Up to 60 kDa (Streptavidin tetramer) |
| Microbead display (Figure 3.4C) | Sepp et al. 2002 | $10^{10}$ | Noncovalent linkage of DNA and protein through a microbead (N- or C-terminal tags possible) | | " Binding to immobilized ligand or binding to soluble ligand followed by FACS | ~1-μm bead |
| ***In Vitro* Compartmentalization Methods** | | | | | | |
| Water-in-oil (Figure 3.5A) | Tawfik and Griffiths 1998 | $10^{10}$ | Confinement of DNA and protein within aqueous compartment | Single catalytic event | Recovery of desired DNA by PCR | N/A (no direct genotype-phenotype linkage) |
| Water-in-oil with microbeads (Figure 3.5B) | Griffiths and Tawfik 2003 | $10^{10}$ | Confinement of DNA and protein within aqueous compartment and noncovalent linkage through a microbead (N- or C-terminal tags possible) | Multiple catalytic events | Binding to soluble anti-product Ab followed by affinity purification or FACS | ~1-μm bead |
| Water-in-oil-in-water (Figure 3.5C) | Bernath et al. 2004 | $10^{10}$ | Confinement of DNA and protein within aqueous compartment | Multiple catalytic events | FACS | N/A (no direct genotype-phenotype linkage) |

The types of selections and screens used in conjunction with *in vitro* compartmentalization are numerous and highly varied (see the section titled "*In Vitro* Compartmentalization Methods*" in this chapter). All involve some method for detecting enzymatic activity, whether it is the digestion of modified DNA with methylation-sensitive restriction enzymes followed by PCR (Tawfik and Griffiths 1998), labeling microbead-bound product with fluorescent antibodies (Griffiths and Tawfik 2003), or conversion of nonfluorescent substrate to fluorescent product (Bernath et al. 2004). In FACS-compatible methods, fluorescence data is collected from each particle, so that the distribution of fluorescent signal (corresponding to enzymatic activity) across the entire population is known. This knowledge greatly facilitates setting thresholds to determine which members of the library will be carried onto the next round.

### Types of Genotype–Phenotype Linkages

In order to adapt cell-free protein synthesis to cell-free protein display for directed evolution, the input template must be adjusted from a homogeneous complementary DNA (cDNA) template to a diverse cDNA library. Methods for creating such libraries are discussed in Chapter 4. Additionally, the DNA sequence may encode an N- or C-terminal fusion to the protein of interest for the purpose of establishing the genotype–phenotype linkage (see Table 3.1). Ribosome display, for example, requires a C-terminal tether that allows the nascent polypeptide to distance itself adequately from the ribosome and fold properly. mRNA display does not require such a tether, but it does require that the protein be fused to a DNA-puromycin linker at its C-terminus. Some DNA display methods may be compatible with either an N- or C-terminal fusion, while others specifically require one or the other. In contrast, *in vitro* compartmentalization methods do not require any fusion, unless the protein is linked to a microbead, in which case either N- or C-terminal tagging is possible. The relative importance of the N- and C-termini to the activity of the protein of interest determines which cell-free display techniques may be used. Such considerations must also be made when using phage- or cell-based display methods (see Chapters 1 and 2).

### Choice of Translation System

In choosing a translation system for the synthesis of display particles, two main considerations are the source of translation machinery and whether the system will be coupled or uncoupled. There is no reason, in principle, why any particular *in vitro* display platform would not be compatible with any of the major cell-free synthesis systems (*E. coli*, rabbit reticulocyte, or wheat germ), but there is a definite bias in the literature toward using *E. coli*-based extract. Of course, if the protein of interest cannot be functionally displayed in an *E. coli*-based system, other options must be pursued. Ribosome display is the platform most commonly used in conjunction with eukaryotic extracts. One must also consider whether or not transcription and translation should take place simultaneously, which is required for all of the DNA display and *in vitro* compartmentalization methods. In contrast, mRNA display requires separate transcription and translation reactions, as the mRNA template must be chemically modified in between these steps. Ribosome display has been used extensively with both coupled and uncoupled systems, but an uncoupled system may be preferable because the amount of input mRNA can be precisely regulated, leading to a higher number of functional ternary complexes (Plückthun et al. 2000).

## In Vitro versus In Vivo Display

One important advantage of cell-free methods is that a transformation step, in which foreign DNA is taken up by a host cell, is not required. In cell-based technologies, the relatively low efficiency of transformation typically limits library size to fewer than $10^8$ molecules (Leemhuis et al. 2005). Cell-free methods, in contrast, routinely handle library sizes in the range of $10^{12}$–$10^{13}$ displayed molecules, and sizes greater than $10^{14}$ have been achieved (Leemhuis et al. 2005). In principle, the library size is limited only by the translation volume one wishes to use and the number of display particles that can physically fit into a tube. A large library size is one of the most important parameters in directed evolution because it enhances the likelihood of isolating desirable sequences and increases the diversity of the enriched sequences (Huang and Liu 2007). Remarkably, entirely novel enzymes have been discovered within a naive protein library having very high diversity (greater than $10^{12}$ different sequences) (Seelig and Szostak 2007).

Another major advantage of cell-free methods is that translation and selections can be conducted without the constraint of having to maintain cell viability. Toxic proteins, or proteins that become toxic upon overexpression, may be investigated. The overexpression of transcription factors, for example, is often toxic to the host cells used in phage display or yeast one-hybrid systems (Tateyama et al. 2006). Also, cell-free translation facilitates the production of proteins containing unnatural amino acids (Noren et al. 1989), and selections can be carried out under relatively harsh chemical conditions, provided that the selection particle can maintain its integrity.

Along the same lines, cell-free display methods avoid selection bias due to variable sequence expression in the host cell or phage. Investigations involving calmodulin, for example, have met with limited success in phage, as many peptide binders of $Ca^{2+}$/calmodulin are positively charged amphipathic helices, which might influence expression, transport, and membrane insertion of the proteins, as well as viability of the host cells and infection efficiency of the phage (Huang and Liu 2007).

Finally, diversification between rounds of selection is straightforward in cell-free display methods. Recovered sequences may be subjected to error-prone PCR (Wong et al. 2006), staggered extension process (StEP) (Zhao et al. 1998), or DNA shuffling (Stemmer 1994), instead of traditional PCR, to introduce additional variation within the population (see Chapter 4). Cell-based methods, in contrast, require one of these more difficult approaches: (1) isolation of DNA from cells, randomization in vitro, and re-introduction into cells; or (2) use of a mutator strain, which can generate additional, undesired mutations in the plasmid and host genome.

One potential limitation of cell-free display is that proteins may not fold easily and the proper post-translational modifications may be hard to achieve. Often, however, the translation conditions may be manipulated to maximize the yield of functional protein (see the section titled "General Approaches to Cell-Free Protein Synthesis" in this chapter). Also, most in vitro selections are qualitative, as selection particles are too small to be screened by FACS. Two *in vitro* compartmentalization-dependent methods, microbead display (Sepp et al. 2002) and water-in-oil-in-water emulsions (Bernath et al. 2004), are exceptions to this general rule. In these cases, FACS allows for quantitative screening (i.e., the fluorescence of each *in vitro* compartment is precisely measured to determine if the corresponding DNA should be carried forth to the next round). FACS limits library size, however, as current instruments can sort only ~$10^7$–$10^8$ events per hour and the time lag between measuring the first and last particles may influence the quality of the screen. Additionally, FACS screens are typically used in cell-free systems for enzyme evolution, not evolution of binding affinity against a target of interest. By contrast, the evolution of better binders through cell surface display (Chapter 2) is highly compatible with FACS, and methods for determining the optimal ligand concentration for equilibrium screening and the optimal competition time for kinetic screening have been developed (Boder and Wittrup 1998).

## METHODS FOR CELL-FREE DIRECTED EVOLUTION

### RIBOSOME DISPLAY

### Technical Description

Ribosome display physically links genotype (mRNA) and phenotype (encoded protein) through a stalled ribosome. This selection particle is referred to as a ternary complex (Schaffitzel et al. 1999), a protein-ribosome-mRNA (PRM) complex (He and Taussig 2002), or, when applicable, an antibody-ribosome-mRNA (ARM) complex (He and Taussig 1997). A basic outline of the procedure for selecting protein binders from a diverse library is shown in Figure 3.2A. First, a library of dsDNA constructs is created by one of a variety of different means (see Chapter 4) and transcribed into mRNA. The mRNA does not contain a stop codon, but instead encodes a C-terminal polypeptide "tether," which allows the protein of interest to fully exit the ribosomal tunnel, fold, and remain associated with the ribosome for subsequent selection. After a short translation period (typically 10–15 minutes), the stalled ribosomal complexes (selection particles) are stabilized by dilution in ice-cold buffer that contains a high concentration of $Mg^{2+}$ (50 mM for *E. coli*-based extracts or 5 mM for rabbit reticulocyte-based extracts). $Mg^{2+}$ is used to "crosslink" the phosphate groups of the ribosomal RNA, thus preventing dissociation of the ribosomal complexes (Lipovsek and Plückthun 2004). It should be noted that the optimal translation time for maximizing ternary complex yield balances the increase in functional protein production with the increase in mRNA degradation. (In contrast, to maximize total protein yield, the *in vitro* reaction may be carried out for longer time periods since it is not important to have intact mRNA at the end of the reaction.)

Once translation has been stopped, the complexes are allowed to bind to a target, which is typically immobilized antigen on a plate or on beads. After a given wait time, the unbound complexes are removed through multiple washing steps, and the mRNA of the bound complexes is released by removing $Mg^{2+}$ from the buffer. The mRNA is purified and reverse transcribed to make cDNA, which is subsequently amplified by PCR with or without diversification. If beads are used, RT-PCR may conveniently be

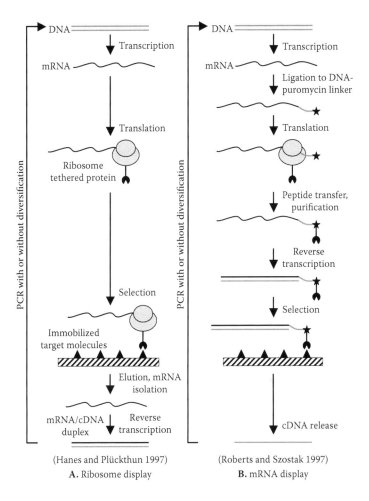

PCR with or without diversification

DNA ══════════
↓ Transcription
mRNA ∿∿∿∿∿
↓ Translation
Ribosome tethered protein
↓ Selection
Immobilized target molecules
↓ Elution, mRNA isolation
mRNA/cDNA duplex ↓ Reverse transcription

(Hanes and Plückthun 1997)
**A.** Ribosome display

DNA ══════════
↓ Transcription
mRNA ∿∿∿∿∿
↓ Ligation to DNA-puromycin linker
↓ Translation
↓ Peptide transfer, purification
↓ Reverse transcription
↓ Selection
↓ cDNA release

(Roberts and Szostak 1997)
**B.** mRNA display

**FIGURE 3.2** Ribosome display (A) and mRNA display (B). A simple affinity selection is pictured. (A) In ribosome display, a library of DNA constructs is transcribed *in vitro*. The unmodified mRNA, which contains no stop codon, is translated to form the selection particle, a stalled ribosomal complex that consists of mRNA, ribosome, and protein. These ternary complexes are allowed to bind to immobilized target molecules. Unbound or nonspecifically bound complexes are washed away, and mRNA from remaining complexes is eluted by removing $Mg^{2+}$ from the buffer. The released mRNA is isolated and reverse transcribed to form an mRNA/cDNA duplex. PCR is then used, with or without diversification, to amplify the selected members of the library. (B) mRNA display begins in the same manner as ribosome display, with *in vitro* transcription of a DNA library. The resulting mRNA must be modified, however, to include a DNA-puromycin linker at the 3' end. When this modified mRNA is translated, the ribosome stalls on the DNA, giving the conjugated puromycin an opportunity to enter the ribosomal A site and accept the nascent polypeptide. The mRNA-protein fusions are then purified, and the mRNA is reverse transcribed prior to selection. The selection step is conducted as in ribosome display, except that the biochemical conditions may be harsher, since the genotype-phenotype linkage is covalent in mRNA display. After washing, the cDNA is released by hydrolysis and amplified for another round.

performed *in situ*, without purifying the mRNA (He and Taussig 2007). When dsDNA has been recovered, one full round of ribosome display is complete. The entire process may be repeated as many times as necessary, and the stringency of the selection step may be increased with each round. Detailed protocols have been published using both prokaryotic (Zahnd et al. 2007a) and eukaryotic (He and Taussig 2007) extracts.

## History, Advances, and Applications

"Polysome display" using short peptides was first described by Mattheakis et al. (1994). Shortly thereafter, Hanes and Plückthun (1997) demonstrated a similar technique for proteins, which they called "ribosome display." Although polysomes can still form in the method that Hanes and Plückthun reported, only ribosomes near the 3' end of the transcript should contain functional, native protein. Additionally, when transcription and translation are decoupled, the ratio of purified mRNA to ribosomes in the extract can be more carefully controlled in order to minimize the likelihood of forming polysomes, which wastefully employ multiple ribosomes on the same transcript and may complicate affinity selections due to avidity effects.

Ribosome display was originally used to improve the binding properties of a single-chain variable fragment (scFv) of an antibody (Hanes and Plückthun 1997; Schaffitzel et al. 1999) and was then applied to select and evolve high affinity binders from synthetic naive scFv libraries (Hanes et al. 2000). Phage display, in a direct comparison with ribosome display, yielded less diverse, lower affinity scFvs (Groves et al. 2006). More recently, scaffolds other than scFvs have also been evolved successfully in ribosome display. Notably, designed ankyrin repeat protein (DARPin) libraries, which are stable, soluble, and well expressed in an S30 translation system, have been used in selections for high-affinity binders, initially to model ligands such as maltose binding protein (Binz et al. 2004) and subsequently to clinically relevant targets such as HER2 (Zahnd et al. 2007b), which is expressed on 20–30% of breast tumors.

Many properties other than binding affinity can also be evolved, provided that a selection pressure can be effectively applied. For example, scFvs that are stable and functional in the reducing cytoplasm (i.e., in the absence of disulfide bonds) have been evolved using decreasing redox potentials with each round (Jermutus et al. 2001). Ribosome display can also be used to evolve catalytic activity, provided that a mechanism-based inhibitor can be designed. Mechanism-based inhibitors, or "suicide inhibitors," bind to and inactivate an enzyme. This strategy was used in a proof-of-principle experiment to enrich a catalytically active β-lactamase over an inactive point mutant (Amstutz et al. 2002). Additionally, ribosome display has been used to evolve folding properties by selecting against proteins that are sensitive to proteases or have relatively large hydrophobic areas exposed (Matsuura and Plückthun 2003).

One particular aspect of the ribosome display protocol that investigators have experimented with is the mechanism for stalling. In the original protocol, mRNA lacking a stop codon was used, and the translation mix was supplemented with anti-ssrA oligonucleotides to prevent release of stalled ribosomes by ssrA in the extract (Hanes and Plückthun 1997). This method works relatively well at low temperatures and high $Mg^{2+}$ concentrations, as the complexes may be used as long as 20 days at 4°C (Plückthun et al. 2000), but the efficiency of ribosome display can still be improved. cDNA yields after a selection round can be increased by using the PURE system during translation, as ribonucleases (RNases), proteases, and ssrA are not naturally present (Villemagne et al. 2006). Introducing a pseudoknot into the transcript can also increase cDNA yield, as the ribosome is less likely to read to the end of the mRNA strand, where release can more readily occur (Kim et al. 2007). To stabilize the complexes at elevated temperatures, some have exploited peptide-mediated stalling by inserting the stalling sequence of *E. coli* secretion monitor (SecM) at the C-terminal end of the tether (Evans et al. 2005). This sequence interacts with the ribosome exit tunnel to cause elongation arrest (Nakatogawa and Ito 2002). This tactic has been used for *in vivo* ribosome display to evolve scFvs that fold and function efficiently in the cytoplasm of bacteria (Contreras-Martinez and DeLisa 2007). Yet another stalling technique, ribosome-inactivation display system (RIDS), employs the ricin A chain, which is capable of stabilizing eukaryotic ribosomal complexes even in the presence of a stop codon (Zhou et al. 2002). The template used in RIDS encodes a cDNA library, a short linker, the ricin A chain, and a C-terminal spacer. Once the ricin A chain has fully emerged from the ribosome, it catalyzes a single hydrolysis event in the large ribosomal subunit, which alters the binding site for elongation factors and stalls the complex.

## Advantages and Limitations among Cell-Free Methods

Ribosome display was the first fully *in vitro* display technique used for the directed evolution of proteins, and it is the most widely reported. It is straightforward to perform, as it involves relatively few processing steps, and it can handle large library sizes of ~$10^{12}$–$10^{14}$ (Leemhuis et al. 2005). Also, the ribosome may increase the solubility of proteins that, if displayed by other methods, might aggregate. For example, a protein that tends to form amyloid-like fibrils remains soluble when expressed on the ribosome, presumably because the ribosome sterically hinders aggregation (Matsuura and Plückthun 2003). This phenomenon has also been observed in the expression of the ligand-binding domain of the mammalian Nogo receptor (Schimmele et al. 2005).

One concern with ribosome display is that the ribosome or mRNA, rather than the expressed protein, might have some intrinsic affinity for the target molecule, which would complicate the selection process. Functional RNA molecules, or RNA aptamers, can actually be evolved through a similar evolutionary approach. In fact, this method, called SELEX (systematic evolution of ligands by exponential enrichment), predated ribosome display (Tuerk and Gold 1990). It is also possible that the ribosome or mRNA may sterically hinder binding of the displayed protein to the target.

Another potential concern is that RNase contamination can degrade the mRNA and reduce the number of clones that can be recovered. RNase-free pipette tips, reagents, and vessels should be used whenever possible, and gloves must be worn at all times. No matter what types of precautions are taken, however, there will always be a low level of RNase activity present in the extract, as all cells naturally produce nucleases. In order to minimize this problem in a prokaryotic system, the extract should be made using a strain of *E. coli* that is RNase-deficient, such as MRE600 (Wade and Robinson 1966). Additionally, stem-loop secondary structure at the 5' and 3' ends of the mRNA increases resistance to exonucleases (Hanes and Plückthun

1997). In a eukaryotic system, the rabbit reticulocyte lysate or wheat germ extract must be treated with RNase inhibitors. Regardless of the translation system used, however, all selection, washing, and elution steps should be performed in a cold room to minimize both RNase and protease activity. The only way to completely prevent RNase activity is to avoid using cell extracts.

The PURE system has been used to effectively eliminate the presence of RNases. Also, in the PURE system, release factors may be omitted from the translation mix to enhance the stability of the mRNA-ribosome-protein complex, enabling selections at temperatures up to 50°C (Matsuura et al. 2007). Traditional ribosome display has more limited uses, as it requires low temperatures (~4°C); however, all forms of ribosome display require high $Mg^{2+}$ concentrations and a relatively gentle biochemical environment free of denaturants. The stability of the ribosomal complex is always a concern, partially because the genotype-phenotype linkage is noncovalent.

## mRNA Display

### Technical Description

mRNA display links genotype (mRNA/cDNA) and phenotype (protein) covalently through a DNA-puromycin linker. The protocol for generating these mRNA-protein selection particles begins in much the same way as the ribosome display protocol, except that the 3′ end of the mRNA template must be covalently linked to a single-stranded DNA (ssDNA) linker bonded to puromycin (Figure 3.2B). The covalent linkage between the mRNA and ssDNA linker is generally accomplished through template-directed ligation, a process in which a DNA oligonucleotide complementary to the 3′ end of the mRNA and the 5′ end of the ssDNA linker is used as a splint to direct ligation between them. Puromycin is an antibiotic that structurally mimics the aminoacyl moiety of tRNA. It acts as a translation inhibitor by entering the ribosomal A site and forming a stable amide linkage to the polypeptide, causing it to release from the ribosome. In the context of mRNA display, puromycin allows the modified template mRNA to become covalently attached to the translated protein. The purpose of the DNA linker, which contains a stretch of ~21–27 adenine bases, is to pause translation long enough to allow puromycin to work (Keefe 2001). Without this linker, the yield of mRNA-protein fusions is drastically reduced (Nemoto et al. 1997).

After translation, a dilution buffer is added and the mRNA-protein particles are purified. A two-step purification process, which may not be necessary for all applications, is required to yield particles that contain both mRNA and a complete protein translated in-frame. First, the stopped translation is incubated with immobilized DNA oligonucleotides containing a stretch of thymine bases (T), which are complementary to the stretch of adenines (A) in the DNA linker. All mRNA present in the translation mixture is bound in this step: mRNA-protein fusions, free mRNA template, and any mRNA containing a poly(A) sequence that may have been present in the cell extract used for translation. Unbound and nonspecifically bound material is washed away, and the remaining species are eluted from the solid support. Next, mRNA-protein fusions are separated from free mRNA using a functionalized solid support that is able to bind to the expressed protein. The expressed protein may be engineered to contain an epitope tag [FLAG or $(His)_6$, for example] to facilitate this process. A C-terminal tag is preferable because only mRNA-protein fusions containing fully translated, in-frame proteins can bind. Free mRNA and mRNA-protein fusions containing frameshifts (deletions or insertions) are washed away. If for some reason it is detrimental to protein function to introduce a C-terminal tag, an N-terminal tag may be used, although most frame-shifted proteins will not be eliminated this way. In the absence of tags, disulfide bonding between immobilized sulfhydryl groups and cysteines present in the expressed protein can be used (Roberts and Szostak 1997).

Once the desired mRNA-protein fusions have been adequately purified, most protocols call for reverse transcription to form an mRNA/cDNA duplex. This eliminates any mRNA secondary structures that might interfere with selection and provides a cDNA template for PCR upon recovery of genotype. The mRNA-protein fusions are then ready to be used in selections. Selections are performed at 4°C, as in ribosome display, although the buffer composition is more flexible, for example, $Mg^{2+}$ need not be present, and strong denaturants are permissible. mRNA-protein fusions are allowed to bind to an immobilized target and, after washing, the bound mRNA-protein fusions are eluted by adding proteinase K or, if the affinity of the interaction is low enough, an excess of soluble target molecule. The cDNA of the mRNA/cDNA duplex is then purified and amplified by PCR. At this point, as in all *in vitro* techniques, the dsDNA may be used for further rounds of selection. Detailed protocols for mRNA display are available (Keefe 2001).

## History, Advances, and Applications

An mRNA-protein fusion (or "*in vitro* virus") was first reported by Nemoto et al. (1997). It was Roberts and Szostak, however, who carried out a proof-of-principle experiment for peptide evolution by mRNA display (1997). In this experiment, mRNA-peptide fusions containing a *myc* epitope were enriched 20- to 40-fold relative to mRNA-peptide fusions containing a random sequence.

The first protein scaffold to be used for directed evolution with mRNA display was the tenth fibronectin type III domain (10Fn3) (Xu et al. 2002). This scaffold has an immunoglobulin-like fold with three solvent-accessible loops, which are structurally analogous to the complementarity-determining regions of antibodies. It was chosen for its thermostability, solubility, high expression level in *E. coli*, and lack of cysteines (which can complicate folding). In the first reported instance of directed evolution on this scaffold using mRNA display, the loops were completely randomized, and variants binding to TNF-α with high affinity were isolated (Xu et al. 2002). More recently, 10Fn3 was used to evolve dual-specificity antagonists to human and mouse vascular endothelial growth factor (VEGF) receptor-2 (Getmanova et al. 2006). This latter study was the first to report the biological activity (inhibition of VEGF-dependent proliferation) of binding molecules developed on the 10Fn3 scaffold.

The second protein scaffold to be used in conjunction with mRNA display was the human retinoid-X-receptor (hRXRα) (Cho and Szostak 2006). This scaffold was chosen because it is small, folds stably, and has two recognition loops in close proximity. The loops were randomized, and mRNA display was used to isolate variants that specifically recognized adenosine triphosphate (ATP) (Cho and Szostak 2006). Additionally, scFvs have been affinity-matured successfully by mRNA display (Fukuda et al. 2006). In the first model system reported, error-prone PCR and StEP were used to diversify an anti-fluorescein scFv. Affinity was improved ~30-fold after four rounds of off-rate selection (Fukuda et al. 2006).

mRNA display has also been used to construct peptide libraries containing one or more unnatural amino acids (Li et al. 2002; Muranaka et al. 2006). In one report, four unnatural amino acids were introduced using three tRNAs containing different four-base anticodons and one tRNA containing an amber anticodon. From the mRNA-displayed library, a novel streptavidin-binding unnatural peptide was evolved, demonstrating the utility of this approach. Another method for generating peptides with unnatural amino acids is to reassign sense codons (Josephson et al. 2005). This requires more control over the tRNA species in the translation mix, so a well-defined chemical composition (e.g., using the PURE system) may be necessary.

Other applications of mRNA display include the study of protein–protein interactions (Huang and Liu 2007) and protein-DNA interactions (Tateyama et al. 2006). One study used a random peptide library and, in parallel, a natural library derived from human tissues to select peptides that bound to $Ca^{2+}$/calmodulin (Huang and Liu 2007). The purpose of using two different libraries in this study was to allow for comparisons between the selected peptides from each library. Indeed, it was found that the $Ca^{2+}$/calmodulin-binding peptides selected from the random peptide library correlated well with peptides selected from the natural library, which validated the results and demonstrated the usefulness of mRNA display for studying protein–protein interactions.

## Advantages and Limitations among Cell-Free Methods

The main advantage of mRNA display compared to ribosome display is that the covalent linkage between genotype and phenotype can withstand harsh biochemical treatments that would cause dissociation of the stalled ribosomal complex. For example, mRNA display has been used to evolve proteins that can bind ATP in the presence of 3 M guanidine hydrochloride, a strong denaturant (Chaput and Szostak 2004). Another notable advantage of mRNA display is that the size of the selection particle is minimized (i.e., there is no ribosome or bulky fusion protein that might have some unanticipated interaction with the target molecule). As far as library size, mRNA display is among the best of *in vitro* methods, similar to ribosome display and certain forms of covalent DNA display. The number of mRNA-protein fusions that can be generated is limited only by translation volume, translation efficiency, and the efficiency of puromycin bonding to the nascent polypeptide. As such, libraries containing greater than $10^{14}$ molecules have been made (Leemhuis et al. 2005).

As in ribosome display, care must be taken to avoid RNase contamination. While mRNA-protein fusions are generally stable at 4°C, they are not stable at room temperature in the presence of RNases. In contrast, ssDNA and dsDNA fusions are much more robust (Kurz et al. 2001). One disadvantage of mRNA display compared to ribosome display is that the "genotype" portion of the selection particle cannot be recovered simply by removing $Mg^{2+}$ from the buffer. This may be problematic when trying to recover the genotype of very tight binders by competitive elution.

In these cases, proteinase K digestion might be necessary, which is less specific. However, there is an alternative method that uses a photocleavable 2-nitrobenzyl linker between the mRNA and protein (Doi et al. 2007). In this system, the desired nucleic acid sequences may be recovered by brief (15 minute) UV irradiation.

## COVALENT DNA DISPLAY

As suggested by its name, covalent DNA display tethers a displayed protein to its encoding cDNA by a covalent linkage. There exist multiple strategies for accomplishing this feat, as described in this section and as shown in Figure 3.3. Because these strategies are all relatively new and not widespread, the technical descriptions, applications, and unique advantages and disadvantages are described separately for each technique. Finally, common advantages and disadvantages among all covalent DNA display methods are summarized.

### Covalent DNA Display Adapted from mRNA Display

One version of covalent DNA display is actually an adaptation of mRNA display (Tabuchi et al. 2001). In this protocol, the 3′ end of unmodified template mRNA is hybridized to a DNA primer/linker carrying puromycin at its 3′ end. During translation, the ribosome stalls at the mRNA/DNA duplex and the puromycin forms an amide linkage with the polypeptide chain. Reverse transcription then yields a stable cDNA-protein fusion (Figure 3.3A, *top*), which is used for selections. This strategy eliminates the need for modifying the mRNA template.

Alternatively, a photo-crosslinking method may be used (Kurz et al. 2001). In one version of this protocol, a modified mRNA template (mRNA-ssDNA-puromycin; see the section titled "mRNA Display, Technical Description" in this chapter) is translated to create an mRNA-protein fusion. Then, a DNA primer containing psoralen, a molecule that reacts to UV light, is hybridized over the junction between mRNA and ssDNA. UV irradiation allows the DNA primer to become covalently attached to the ssDNA portion of the modified mRNA template. Next, the mRNA is reverse transcribed to make cDNA, and RNase H treatment is used to degrade the mRNA. Finally, a DNA primer complementary to the 3′ end of the cDNA is hybridized and elongated to make dsDNA. The resulting DNA-protein fusion (Figure 3.3A, *middle*) is used for subsequent selections. In another version of the photo-crosslinking protocol, a DNA primer containing a branched phosphoramidite linked to puromycin is hybridized to the 3′ end of an unmodified mRNA template and covalently attached through psoralen. After *in vitro* translation and appropriate purification measures, reverse transcription, RNase H treatment, and second-strand synthesis are performed to create the final selection particle (Figure 3.3A, *bottom*).

Both ssDNA- and dsDNA-protein fusions, as described previously, offer valid alternatives to mRNA display, although their use has not been widely reported. It has been proposed that such fusions might be valuable for applications in which RNase activity cannot be effectively controlled (Kurz et al. 2001). Theoretically, any protein that can be displayed by mRNA display can be displayed by the previously mentioned methods. The two major disadvantages of mRNA display-based covalent

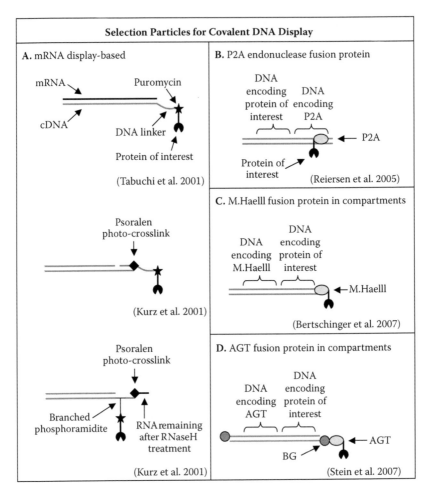

**FIGURE 3.3** Selection particles for covalent DNA display. (A) Multiple methods based on mRNA display have been developed. (B) The ability of the P2A endonuclease to nick and covalently bind to its own DNA has been exploited. The protein of interest is expressed as a P2A fusion. (C) The DNA methyltransferase M.HaeIII is expressed as a fusion with the protein of interest within *in vitro* compartments. M.HaeIII conjugates to a modified methylation target sequence, 5′-GGFC-3′ (F = 5-fluoro-2′-deoxycytidine). The selection particles are extracted from the emulsion prior to use. (D) A mutant of O⁶-alkylguanine-DNA alkyltransferase (AGT) is expressed as a fusion with the protein of interest within *in vitro* compartments. The AGT mutant covalently binds to O⁶-benzylguanine (BG), which is chemically conjugated to the DNA. The emulsion is broken to retrieve the selection particles.

DNA display methods are the requirement for reverse transcription and the number of handling steps.

## Covalent DNA Display Using P2A Endonuclease

A completely different covalent DNA display system exploits the natural propensity of the P2A endonuclease from the P2 bacteriophage to nick and covalently bind to its own DNA (Reiersen et al. 2005). In this method, a fusion protein consisting of the protein to be displayed and P2A is synthesized in a prokaryotic, coupled transcription and translation system. Shortly after the P2A endonuclease is synthesized, it creates a single-stranded site-specific nick at the viral origin of replication located within the *P2A* gene and becomes covalently attached to the 5′ phosphate at that site through its catalytic tyrosine residue (Y454). Following transcription/translation, these DNA-protein complexes (Figure 3.3B) are diluted in ice-cold buffer and allowed to bind to an immobilized target. Once unbound or weakly bound complexes are washed away, the DNA corresponding to the remaining binders is eluted using proteinase K. The recovered DNA is amplified by PCR.

This relatively new technology (called "covalent antibody display" or CAD) has been used to evolve scFv binding to tetanus toxoid (Reiersen et al. 2005). Although the library size was small by *in vitro* standards (~5 × 10$^7$), the authors noted that this number could be scaled up. A major advantage of CAD among *in vitro* methods is the speed with which selections can be performed. Within a few hours, unique sequences can be enriched, isolated, and amplified. Additionally, the authors of this study offered CAD as an alternative to a noncovalent DNA display method called CIS display, which involves cis binding of *E. coli* RepA to a specific DNA sequence (see section titled "Noncovalent DNA Display Using CIS Display" in this chapter).

One remaining challenge of CAD is that fusion proteins lacking a covalently bound DNA are readily formed during the transcription-translation process, and these can compete against fusion proteins that are covalently bound to DNA (Reiersen et al. 2005). To circumvent this problem, the authors proposed biotinylating the DNA and using streptavidin-coated beads to separate DNA-protein fusions from free protein prior to selection. They noted, however, that more complex, naive libraries would not encounter this problem until later rounds of selection because the amount of active protein in the initial library is typically very low. Additionally, P2A is relatively large (86 kDa), so it is possible that steric hindrance might inhibit selection.

### Covalent DNA Display Using *In Vitro* Compartmentalization

*In vitro* compartmentalization (see the section titled "*In Vitro* Compartmentalization Methods" in this chapter) has been used to create covalently bound DNA-protein selection particles (Bertschinger and Neri 2004; Stein et al. 2007). In one system (Bertschinger and Neri 2004), the protein of interest is expressed as a fusion to the DNA methyltransferase M.HaeIII, a 38-kDa enzyme that can covalently bond to the DNA sequence 5′-GGFC-3′ (F = 5-fluoro-2′-deoxycytidine). Library DNA encoding the fusion proteins is compartmentalized using water-in-oil emulsions. Each compartment contains, on average, one DNA molecule, so only one M.HaeIII-fused library member is expressed per compartment. Once the fusion protein has been transcribed and translated within its respective compartment, the M.HaeIII portion of the molecule reacts irreversibly with its DNA target sequence, forming a functional selection particle (Figure 3.3C). The emulsion is broken to recover these selection particles, and selections are performed against immobilized binding targets.

The researchers that developed this technique later applied an improved protocol toward the affinity maturation of a small protein domain (FynSH3, a Src homology 3 domain) that showed very low affinity for mouse serum albumin (Bertschinger et al. 2007). A focused library was generated by randomizing six amino acids in the n-Src-loop, and a number of improved binders were isolated after only two to four rounds of selection. This study also showed that the M.HaeIII system is able to generate covalent DNA-protein fusions quickly (in 2–3 h) and efficiently (with ~50% of input DNA molecules fused to a complete protein).

Another emulsion-based covalent DNA display method, which employs a fusion protein consisting of an $O^6$-alkylguanine-DNA alkyltransferase (AGT) mutant and the protein of interest, has been described (Stein et al. 2007). In this case, the DNA is derivatized with $O^6$-benzylguanine (BG), a substrate analog of the AGT mutant, for the purposes of forming a covalent genotype-phenotype linkage. Two advantages of this system over the previously reported M.HaeIII system are that valency can be varied and the AGT tag can be used to facilitate *in vitro* and *in vivo* functional assays.

One challenge of these methods is optimizing the compartmentalization protocol to maximize the number of correct DNA-protein fusions and, thereby, the enrichment efficiency. The authors of the first M.HaeIII display study (Bertschinger and Neri 2004) noted that the enrichment efficiency might be reduced if more than one DNA molecule were initially present in a compartment, or if some compartments were able to fuse. Also, they discussed the possibility that noncovalent DNA-protein interactions could occur upon breaking the emulsion, which would pair genotype and phenotype incorrectly. Moreover, a notable limitation of this method compared to other *in vitro* methods is the smaller library size imposed by *in vitro* compartmentalization; only ~$10^{10}$ compartments can fit into a typical one milliliter emulsion (see the section titled "*In Vitro* Compartmentalization Methods" in this chapter).

## Common Advantages and Limitations of Covalent DNA Display Strategies in Comparison to Other Cell-Free Methods

Stability is a common advantage shared by all covalent DNA display methods. The nucleic acid component of the selection particle is robust in the presence of RNases, and the genotype-phenotype linkage should be able to withstand harsh biochemical conditions (e.g., the presence of denaturing agents or organic solvents, high or low salt concentrations, extreme pH) and/or elevated temperatures (greater than 4°C) (Bertschinger and Neri 2004). Given these advantages, covalent DNA display may eventually become the *in vitro* method of choice for stability selections.

Because covalent DNA display techniques are newer than ribosome display and mRNA display, there is still a considerable amount of optimization and validation that needs to be done. In principle, there is no reason why compartment-independent versions of covalent DNA display cannot achieve the library sizes of ribosome display and mRNA display (~$10^{12}$–$10^{14}$), but this has not yet been reported.

## NONCOVALENT DNA DISPLAY

Noncovalent DNA display methods, as one would expect, involve the linkage of genotype (DNA) and phenotype (protein) in a noncovalent manner. Like covalent DNA display methods, noncovalent DNA display methods are relatively new and not commonly used. Only a few noncovalent DNA display platforms have been described in the literature, and these are summarized in the following sections and in Figure 3.4.

## Noncovalent DNA Display Using CIS Display

CIS display was the first droplet-independent DNA display technology to be described (Odegrip et al. 2004). It exploits the DNA replication initiator protein RepA, a cis-acting DNA binding protein that binds noncovalently to the DNA from which it is being expressed, thus permitting stable, noncovalent genotype-phenotype linkage of RepA-fused peptides and proteins. The DNA construct used in this method includes the coding sequence for the protein of interest, the coding sequence for RepA, the *CIS* element, and the R1 plasmid origin of replication (*ori*). In a standard, coupled *in vitro* reaction, RNA polymerase pauses at the *CIS* element, allowing the nascent RepA polypeptide to bind transiently to the *CIS* element, which in turn directs RepA to bind noncovalently to the adjacent *ori* (Figure 3.4A). RepA (33 kDa) might be more suitable for the display of polypeptide libraries than P2A (86 kDa; see the section titled "Covalent DNA Display Using P2A Endonuclease" in this chapter) because it is smaller, and steric hindrance that prevents the binding of the displayed protein to the target would be less likely.

## Noncovalent DNA Display Using *In Vitro* Compartmentalization

STABLE (streptavidin-biotin linkage in emulsions) was the first method reported for the production of protein-DNA selection particles (Doi and Yanagawa 1999). In this system, streptavidin-fused proteins are synthesized and allowed to bind their own biotinylated DNA within emulsion compartments (see the section titled "*In Vitro* Compartmentalization Methods" in this chapter). The emulsion is then broken and quenched to recover the particles for affinity selection (Figure 3.4B). When used in conjunction with a wheat germ translation system, the efficiency of fusion formation can be very high (greater than 95%) (Yonezawa et al. 2004). Also, a photocleavable linker between genotype and phenotype provides easy recovery of enriched genetic material (Doi et al. 2007). One important consideration in choosing this method is that streptavidin forms tetramers, such that up to four fusion proteins may be displayed on one DNA molecule. Avidity affects may aid in isolating low to moderate affinity binders, but high affinity binders may be difficult to isolate.

Another DNA-based display technology utilizing *in vitro* compartmentalization is called "microbead display" (Sepp et al. 2002). This particular method allows for the screening of binders using FACS. In the original protocol, a library of genes containing an epitope tag is bound to streptavidin-coated microbeads carrying biotinylated antibodies against the epitope tag. On average, one gene or fewer is attached per microbead. The microbeads are then compartmentalized such that the average number of beads per compartment is less than or equal to one. As transcription and translation are performed within this compartment, the proteins produced become attached to the bead via the affinity tag. Then, the emulsion is broken, the beads are recovered, and horseradish peroxidase (HRP)–conjugated ligand is allowed to bind to the displayed library. After removing unbound ligand and adding fluorescein tyramide, immobilized HRP converts this substrate into a free-radical intermediate that

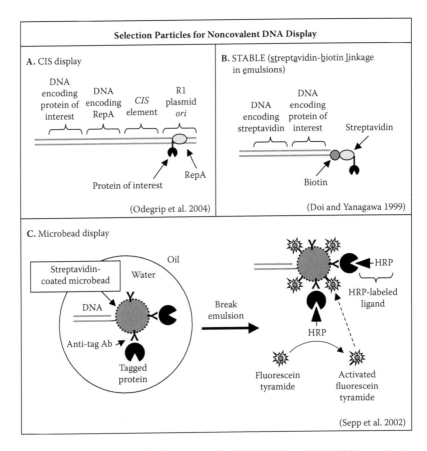

**FIGURE 3.4** Selection particles for noncovalent DNA display. (A) A DNA construct composed of the coding sequence for the protein of interest, the coding sequence for the DNA replication initiator protein RepA, the *CIS* element, and the R1 plasmid origin of replication (*ori*) is transcribed and translated in a standard, coupled *in vitro* reaction. RepA transiently binds to the *CIS* element, which directs it to bind noncovalently to the nearby *ori*. (B) Biotinylated DNA encoding a fusion protein of streptavidin and a protein of interest is transcribed and translated in the aqueous phase of a water-in-oil emulsion. The fusion protein is noncovalently linked to its own DNA through the high-affinity streptavidin-biotin interaction. The emulsion is broken, and the particles are used for selection. (C) Streptavidin-coated microbeads decorated with biotinylated anti-tag antibodies (Ab) and a single molecule of biotinylated DNA encoding a tagged protein are encapsulated in aqueous compartments (one bead per compartment), along with the reagents necessary for *in vitro* transcription and translation. As soon as the tagged protein is produced, it is able to bind to the anti-tag Ab on the bead. The emulsion is then broken, and ligand conjugated to horseradish peroxidase (HRP) is allowed to bind to the displayed library. Unbound ligand-HRP is washed away, and the beads with bound ligand-HRP are labeled fluorescently by the mechanism shown: HRP activates fluorescein tyramide to react rapidly with nearby proteins on the same bead. Beads carrying library members of interest can be recovered by FACS, which is a major advantage of this method if quantitative sorting metrics are desired. PCR can be performed on the beads to amplify the library for subsequent analysis or additional rounds of selection.

rapidly reacts with adjacent proteins, thus fluorescently labeling the corresponding microbeads for flow cytometry (Figure 3.4C).

## Common Advantages and Limitations of Noncovalent DNA Display Strategies in Comparison to Other Cell-Free Methods

Although the genotype-phenotype linkages just described are not technically as strong as a covalent bond, the lifetime of the noncovalent bond used is often orders of magnitude longer than the time needed for selection and therefore not a limitation. Additionally, as in covalent DNA display, the genotype component of the selection particle (dsDNA) is highly stable, allowing for selections in the presences of RNases. The main limitation of noncovalent DNA display methods is that very few studies have been published, so their general usefulness is still uncertain. As more literature accumulates, these platforms will become increasingly attractive for future protein engineering efforts.

## In Vitro Compartmentalization Methods

### Technical Description

Another set of cell-free display methods relies on *in vitro* compartmentalization, in which individual DNA molecules and the proteins they encode are trapped together in the aqueous phase of a water-in-oil emulsion (Tawfik and Griffiths 1998). These "cell-like compartments" have been used as reaction vessels for the selection process or simply for forming genotype-phenotype selection particles that are then subjected to binding selections outside of the emulsion (see the sections titled "Covalent DNA Display Using *In Vitro* Compartmentalization" and "Noncovalent DNA Display Using *In Vitro* Compartmentalization" in this chapter). In order to form these cell-like compartments, an ice-cold coupled transcription and translation reaction (aqueous) is stirred into an ice-cold mixture containing mineral oil and surfactants (e.g., Span 80 and Tween 80). Under optimal emulsification conditions, each resulting aqueous droplet will ideally contain at most one DNA molecule, which is then transcribed and translated within that compartment.

In a typical protocol, 50 µL of the aqueous reaction might be dispersed into 950 µL of oil-surfactant mixture to yield $\sim10^{10}$ droplets having an average volume of $\sim5$ fL and an average diameter of $\sim2$ µm. However, there will be a distribution of sizes, and not all compartments will necessarily contain a DNA molecule. Once the droplets are formed, the emulsion is warmed to 23–30°C, and transcription/translation is allowed to proceed for 1–4 h. In most *in vitro* compartmentalization protocols, the protein being evolved is an enzyme, and the substrate of that enzyme is either the DNA molecule itself or a protein or small molecule also present in the compartment. If the enzyme in a given compartment is capable of converting the substrate to product, the DNA itself or the non-DNA substrate will be modified. In the former case, the emulsion can be broken by centrifugation to separate the water and oil phases, and the desired DNA molecules can be recovered from the aqueous phase. In the latter case, FACS may be employed prior to breaking the emulsion. As in all *in vitro* display techniques, the recovered DNA is amplified with or without diversification for the next round of selection. Many variations on this basic protocol exist, most notably in the selection process (Miller et al. 2006). Some important examples are discussed in the following sections (see also Figure 3.5).

## History, Advances, and Applications

Proof of principle for this technique was first demonstrated by selecting genes encoding the DNA methyltransferase M.HaeIII from a mixture containing a $10^7$-fold excess of genes encoding dihydrofolate reductase (DHFR), an unrelated enzyme (Tawfik and Griffiths 1998). M.HaeIII methylates the restriction/methylation sites attached to M.HaeIII genes (Figure 3.5A), making them resistant to digestion by HaeIII endonuclease. Transcription and translation of M.HaeIII or DHFR was carried out in droplets, the emulsion was broken, and the aqueous phase was extracted. The DNA (biotinylated) was captured on streptavidin beads and incubated with HaeIII endonuclease. Only genes encoding M.HaeIII remained intact after digestion and could therefore be amplified by subsequent PCR. A similar strategy was later used to evolve variants of M.HaeIII having altered specificity (Cohen et al. 2004). *In vitro* compartmentalization has also been used to evolve restriction endonucleases, DNA polymerases, phosphotriesterases, β-galactosidases, thiolactonases, and transcription factors (Fen et al. 2007; Miller et al. 2006).

An adaptation of microbead display (see the section titled "Noncovalent DNA Display Using *In Vitro* Compartmentalization" and Figure 3.4C) enables screening for enzymatic activity (Griffiths and Tawfik 2003). Microbeads displaying a single gene and multiple corresponding proteins are recompartmentalized, one bead to a compartment, along with a soluble protein substrate conjugated to caged biotin (Figure 3.5B). Active enzymes cause turnover of the substrate within their droplet, and the biotin is uncaged (activated) by irradiation, allowing both substrate and product to associate with the bead. This emulsion is broken, and antibodies to the product are allowed to associate. The beads carrying active enzyme may now be purified based on the presence of these antibodies, either by affinity purification or by reaction with a fluorescently labeled secondary antibody and sorting by FACS. Using this latter method, the catalytic rate of a phosphodiesterase was dramatically increased (Griffiths and Tawfik 2003). This report was novel in multiple ways. It was the first instance of *in vitro* compartments being used to evolve enzymes having protein (rather than DNA) substrates. Also, it uncoupled translation from the enzymatic reaction and selected for multiple turnovers. Moreover, the authors proposed that this system might be employed to evolve thermophilic enzymes, as the emulsions are stable up to 99°C.

Another application of *in vitro* compartmentalization uses FACS without microbeads (Bernath et al. 2004). The traditional water-in-oil emulsion cannot be used in FACS because the viscosity of the oil phase is too high; however, water-in-oil droplets may be re-emulsified in an aqueous phase containing Tween 20 to create a water-in-oil-in-water emulsion. In this system, the substrate is designed such that the desired enzymatic activity will generate a fluorescent signal. Droplets containing a high fluorescent signal (indicating multiple catalytic turnover events) can be recovered, and the encapsulated genes may be subjected to further rounds of screening.

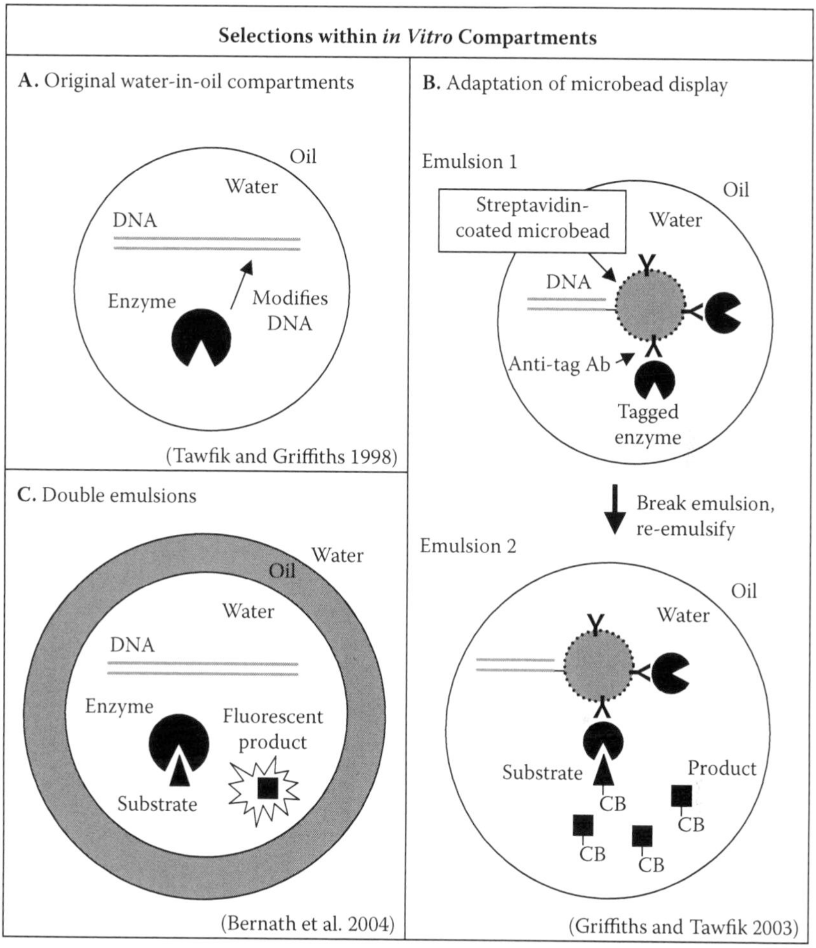

**FIGURE 3.5** Selections within *in vitro* compartments. (A) In the original protocol, DNA encoding an enzyme that acts on DNA is encapsulated. The enzyme is synthesized in the compartment, and it is allowed to act on its DNA. The emulsion is broken and only those DNA molecules that have been modified are isolated and amplified. (B) Microbead display (see Figure 3.4C) has been adapted for use with enzymes. First, *in vitro* compartmentalization is used to create beads carrying one gene and multiple copies of the enzyme that it encodes (Emulsion 1). Then, the emulsion is broken, the decorated microbeads are recompartmentalized, soluble substrate attached to caged biotin (CB) is added, and the enzyme is allowed to act (Emulsion 2). Next, the biotin is uncaged, which allows all substrate and product to associate with the bead. The emulsion is broken, the beads are incubated with anti-product antibody, and the library is enriched by affinity purification or by FACS, if labeling with fluorescent antibody. (C) Water-in-oil droplets can be re-emulsified to form water-in-oil-in-water emulsions. The viscosity of these latter emulsions makes them amenable to screening by FACS, provided that the system is designed such that enzymatic turnover yields a fluorescent product. Note that selections by FACS are technically called *screens* since each particle is individually analyzed to determine whether it meets the defined sorting criteria.

## Advantages and Limitations among Cell-Free Methods

One major advantage of *in vitro* compartmentalization is that the expressed protein, usually an enzyme, need not be directly bound (by covalent or other means) to a nucleic acid. This is useful in any situation where modifying the enzyme might alter its activity. Additionally, *in vitro* compartmentalization has been adapted for quantitative screening by FACS. Ribosome display, mRNA display, and covalent and noncovalent DNA display all involve qualitative selections hinging upon a binding event. When these binding-dependent methods are used to select enzymes, the substrate must be a mechanism-based inhibitor, meaning that only one catalytic event can be monitored. Traditional water-in-oil emulsions in which the protein of interest is a DNA-modifying enzyme (e.g., a DNA methyltransferase or restriction endonuclease) are also restricted to one catalytic event. In contrast, the FACS-compatible compartmentalization methods using water-in-oil-in-water emulsions and microbeads allow repeated substrate turnover, which is useful for the selection of highly active enzymes.

As far as limitations, the typical library size (~$10^{10}$) of *in vitro* compartmentalization methods tends to be rather low among cell-free methods because only ~$10^{10}$ compartments can fit into a convenient one milliliter volume. Scaling up the reaction volumes by a factor of 10 or more, at least for the first round of selection, could partially mitigate this disadvantage; however, library sizes of $10^{12}$–$10^{14}$, which are obtainable through ribosome display and mRNA display (Leemhuis et al. 2005), are impractical for routine use because they would require much larger volumes. If particles are being screened by FACS, the library size that can be assayed is even smaller. FACS machines can generally process ~20,000–50,000 events per second, so only $10^7$–$10^8$ events can be handled this way. Additionally, the requirement for a fluorescent readout can be problematic, as a fluorescent assay for enzyme function is not always readily available. Finally, both water-in-oil and water-in-oil-in-water emulsions may require significant optimization of the emulsion process, despite the availability of detailed protocols.

## CONCLUDING REMARKS

A variety of cell-free display methods have been developed and validated, especially in the last decade. While all of the described approaches offer larger library sizes and less intrinsic bias than cell-based methods, the *in vitro* method of choice depends largely upon the specific protein of interest and the type of function to be evolved, as each approach has its own advantages and limitations. If the goal is to quickly select binders to a target molecule from a large library, and the biochemical conditions will be relatively mild, ribosome display may be the best option. On the other hand, if enhanced protein stability under denaturing conditions is desired, mRNA display or covalent DNA display may be more appropriate. Finally, multiple forms of *in vitro* compartmentalization are well suited for the evolution of enzymes and other applications in which a physical linkage between genotype and phenotype is not easily achieved. As these methods are further refined, our ability to evolve proteins with interesting and beneficial properties should improve as well.

## REFERENCES

Amstutz, P., H. K. Binz, C. Zahnd and A. Plückthun. 2006. "Ribosome display: *In vitro* selection of protein–protein interactions." In *Cell biology, A laboratory handbook*, ed. J. E. Celis. 3rd ed. Vol. 1, 497–503. San Diego, CA: Elsevier Academic Press.

Amstutz, P., J. N. Pelletier, A. Guggisberg, L. Jermutus, S. Cesaro-Tadic, et al. 2002. "*In vitro* selection for catalytic activity with ribosome display." *J Am Chem Soc* 124: 9396–403.

Anderson, C. W., J. W. Straus and B. S. Dudock. 1983. "Preparation of a cell-free protein-synthesizing system from wheat germ." *Methods Enzymol* 101: 635–44.

Bernath, K., M. Hai, E. Mastrobattista, A. D. Griffiths, S. Magdassi, et al. 2004. "*In vitro* compartmentalization by double emulsions: Sorting and gene enrichment by fluorescence activated cell sorting." *Anal Biochem* 325: 151–7.

Bertschinger, J., D. Grabulovski and D. Neri. 2007. "Selection of single domain binding proteins by covalent DNA display." *Protein Eng Des Sel* 20: 57–68.

Bertschinger, J. and D. Neri. 2004. "Covalent DNA display as a novel tool for directed evolution of proteins *in vitro*." *Protein Eng Des Sel* 17: 699–707.

Binz, H. K., P. Amstutz, A. Kohl, M. T. Stumpp, C. Briand, et al. 2004. "High-affinity binders selected from designed ankyrin repeat protein libraries." *Nat Biotechnol* 22: 575–82.

Boder, E. T. and K. D. Wittrup. 1998. "Optimal screening of surface-displayed polypeptide libraries." *Biotechnol Prog* 14: 55–62.

Calhoun, K. A. and J. R. Swartz. 2007. "Energy systems for ATP regeneration in cell-free protein synthesis reactions." *Methods Mol Biol* 375: 3–17.

Chaput, J. C. and J. W. Szostak. 2004. "Evolutionary optimization of a nonbiological ATP binding protein for improved folding stability." *Chem Biol* 11: 865–74.

Cho, G. S. and J. W. Szostak. 2006. "Directed evolution of ATP binding proteins from a zinc finger domain by using mRNA display." *Chem Biol* 13: 139–47.

Cohen, H. M., D. S. Tawfik and A. D. Griffiths. 2004. "Altering the sequence specificity of HaeIII methyltransferase by directed evolution using *in vitro* compartmentalization." *Protein Eng Des Sel* 17: 3–11.

Contreras-Martinez, L. M. and M. P. DeLisa. 2007. "Intracellular ribosome display via SecM translation arrest as a selection for antibodies with enhanced cytosolic stability." *J Mol Biol* 372: 513–24.

Dill, K. A. and H. S. Chan. 1997. "From Levinthal to pathways to funnels." *Nat Struct Biol* 4: 10–19.

Doi, N., H. Takashima, A. Wada, Y. Oishi, T. Nagano, et al. 2007. "Photocleavable linkage between genotype and phenotype for rapid and efficient recovery of nucleic acids encoding affinity-selected proteins." *J Biotechnol* 131: 231–9.

Doi, N. and H. Yanagawa. 1999. "STABLE: Protein-DNA fusion system for screening of combinatorial protein libraries *in vitro*." *FEBS Lett* 457: 227–30.

Endo, Y. and T. Sawasaki. 2006. "Cell-free expression systems for eukaryotic protein production." *Curr Opin Biotechnol* 17: 373–80.

Endoh, T., T. Kanai, Y. T. Sato, D. V. Liu, K. Yoshikawa, et al. 2006. "Cell-free protein synthesis at high temperatures using the lysate of a hyperthermophile." *J Biotechnol* 126: 186–95.

Evans, M. S., K. G. Ugrinov, M. A. Frese and P. L. Clark. 2005. "Homogeneous stalled ribosome nascent chain complexes produced *in vivo* or *in vitro*." *Nat Methods* 2: 757–62.

Fen, C. X., D. W. Coomber, D. P. Lane and F. J. Ghadessy. 2007. "Directed evolution of p53 variants with altered DNA-binding specificities by *in vitro* compartmentalization." *J Mol Biol* 371: 1238–48.

Fukuda, I., K. Kojoh, N. Tabata, N. Doi, H. Takashima, et al. 2006. "*In vitro* evolution of single-chain antibodies using mRNA display." *Nucleic Acids Res* 34: e127.

Getmanova, E. V., Y. Chen, L. Bloom, J. Gokemeijer, S. Shamah, et al. 2006. "Antagonists to human and mouse vascular endothelial growth factor receptor 2 generated by directed protein evolution *in vitro*." *Chem Biol* 13: 549–56.

Griffiths, A. D. and D. S. Tawfik. 2003. "Directed evolution of an extremely fast phosphotriesterase by *in vitro* compartmentalization." *EMBO J* 22: 24–35.

Groves, M., S. Lane, J. Douthwaite, D. Lowne, D. G. Rees, et al. 2006. "Affinity maturation of phage display antibody populations using ribosome display." *J Immunol Methods* 313: 129–39.

Hanes, J. and A. Plückthun. 1997. "*In vitro* selection and evolution of functional proteins by using ribosome display." *Proc Natl Acad Sci U S A* 94: 4937–42.

Hanes, J., C. Schaffitzel, A. Knappik and A. Plückthun. 2000. "Picomolar affinity antibodies from a fully synthetic naive library selected and evolved by ribosome display." *Nat Biotechnol* 18: 1287–92.

He, M. and M. J. Taussig. 2007. "Eukaryotic ribosome display with *in situ* DNA recovery." *Nat Methods* 4: 281–8.

He, M. and M. J. Taussig. 2002. "Ribosome display: Cell-free protein display technology." *Brief Funct Genomic Proteomic* 1: 204–12.

He, M. and M. J. Taussig. 1997. "Antibody-ribosome-mRNA (ARM) complexes as efficient selection particles for *in vitro* display and evolution of antibody combining sites." *Nucleic Acids Res* 25: 5132–4.

Hillebrecht, J. R. and S. Chong. 2008. "A comparative study of protein synthesis in *in vitro* systems: From the prokaryotic reconstituted to the eukaryotic extract-based." *BMC Biotechnol* 8: 58.

Hilvert, D., S. V. Taylor and P. Kast. 2002. "Using evolutionary strategies to investigate the structure and function of chorismate mutases." In *Directed molecular evolution of proteins or how to improve enzymes for biocatalysis*, eds. S. Brakmann and K. Johnsson, 29–62. Weinheim, Germany: Wiley-VCH.

Huang, B. C. and R. Liu. 2007. "Comparison of mRNA-display-based selections using synthetic peptide and natural protein libraries." *Biochemistry* 46: 10102–12.

Jermutus, L., A. Honegger, F. Schwesinger, J. Hanes and A. Plückthun. 2001. "Tailoring *in vitro* evolution for protein affinity or stability." *Proc Natl Acad Sci U S A* 98: 75–80.

Josephson, K., M. C. Hartman and J. W. Szostak. 2005. "Ribosomal synthesis of unnatural peptides." *J Amer Chem Soc* 127: 11727–35.

Katzen, F., G. Chang and W. Kudlicki. 2005. "The past, present and future of cell-free protein synthesis." *Trends Biotechnol* 23: 150–6.

Keefe, A. D. 2001. "Protein selection using mRNA display." In *Current protocols in molecular biology*, ed. F. M. Ausubel et al., 24.5.1–34. Hoboken, NJ: Wiley.

Kim, D. M. and J. R. Swartz. 2004. "Efficient production of a bioactive, multiple disulfide-bonded protein using modified extracts of *Escherichia coli*." *Biotechnol Bioeng* 85: 122–9.

Kim, J. M., H. J. Shin, K. Kim and M. S. Lee. 2007. "A pseudoknot improves selection efficiency in ribosome display." *Mol Biotechnol* 36: 32–7.

Klammt, C., F. Lohr, B. Schafer, W. Haase, V. Dotsch, et al. 2004. "High level cell-free expression and specific labeling of integral membrane proteins." *Eur J Biochem* 271: 568–80.

Kurz, M., K. Gu, A. Al-Gawari and P. A. Lohse. 2001. "cDNA-protein fusions: Covalent protein-gene conjugates for the *in vitro* selection of peptides and proteins." *Chembiochem* 2: 666–72.

Leemhuis, H., V. Stein, A. D. Griffiths and F. Hollfelder. 2005. "New genotype-phenotype linkages for directed evolution of functional proteins." *Curr Opin Struct Biol* 15: 472–8.

Li, S., S. Millward and R. Roberts. 2002. "*In vitro* selection of mRNA display libraries containing an unnatural amino acid." *J Amer Chem Soc* 124: 9972–3.

Lipovsek, D. and A. Plückthun. 2004. "*In-vitro* protein evolution by ribosome display and mRNA display." *J Immunol Methods* 290: 51–67.

Matsuura, T. and A. Plückthun. 2003. "Selection based on the folding properties of proteins with ribosome display." *FEBS Lett* 539: 24–8.

Matsuura, T., H. Yanagida, J. Ushioda, I. Urabe and T. Yomo. 2007. "Nascent chain, mRNA, and ribosome complexes generated by a pure translation system." *Biochem Biophys Res Commun* 352: 372–7.

Mattheakis, L. C., R. R. Bhatt and W. J. Dower. 1994. "An *in vitro* polysome display system for identifying ligands from very large peptide libraries." *Proc Natl Acad Sci U S A* 91: 9022–6.

Miller, O. J., K. Bernath, J. J. Agresti, G. Amitai, B. T. Kelly, et al. 2006. "Directed evolution by *in vitro* compartmentalization." *Nat Methods* 3: 561–70.

Muranaka, N., T. Hohsaka and M. Sisido. 2006. "Four-base codon mediated mRNA display to construct peptide libraries that contain multiple nonnatural amino acids." *Nucleic Acids Res* 34: e7.

Nakatogawa, H. and K. Ito. 2002. "The ribosomal exit tunnel functions as a discriminating gate." *Cell* 108: 629–36.

Nemoto, N., E. Miyamoto-Sato, Y. Husimi and H. Yanagawa. 1997. "*In vitro* virus: Bonding of mRNA bearing puromycin at the 3′-terminal end to the C-terminal end of its encoded protein on the ribosome *in vitro*." *FEBS Lett* 414: 405–8.

Nirenberg, M. W. 1963. "Cell-free protein synthesis directed by messenger RNA." *Methods Enzymol* 6: 17–23.

Noren, C. J., S. J. Anthony-Cahill, M. C. Griffith and P. G. Schultz. 1989. "A general method for site-specific incorporation of unnatural amino acids into proteins." *Science* 244: 182–8.

Northrup, S. H. and H. P. Erickson. 1992. "Kinetics of protein–protein association explained by Brownian dynamics computer simulation." *Proc Natl Acad Sci U S A* 89: 3338–42.

Odegrip, R., D. Coomber, B. Eldridge, R. Hederer, P. A. Kuhlman, et al. 2004. "CIS display: *In vitro* selection of peptides from libraries of protein-DNA complexes." *Proc Natl Acad Sci U S A* 101: 2806–10.

Pelham, H. R. and R. J. Jackson. 1976. "An efficient mRNA-dependent translation system from reticulocyte lysates." *Eur J Biochem* 67: 247–56.

Plückthun, A., C. Schaffitzel, J. Hanes and L. Jermutus. 2000. "*In vitro* selection and evolution of proteins." *Adv Protein Chem* 55: 367–403.

Pratt, J. M. 1984. "Coupled transcription-translation in prokaryotic cell-free systems." In *Transcription and translation: A practical approach*, eds. B. D. Hames and S. J. Higgins, 179–209. Oxford, U.K.: IRL Press.

Reiersen, H., I. Lobersli, G. A. Loset, E. Hvattum, B. Simonsen, et al. 2005. "Covalent antibody display—an *in vitro* antibody-DNA library selection system." *Nucleic Acids Res* 33: e10.

Roberts, R. W. and J. W. Szostak. 1997. "RNA-peptide fusions for the *in vitro* selection of peptides and proteins." *Proc Natl Acad Sci U S A* 94: 12297–302.

Ryabova, L. A., D. Desplancq, A. S. Spirin and A. Plückthun. 1997. "Functional antibody production using cell-free translation: Effects of protein disulfide isomerase and chaperones." *Nat Biotechnol* 15: 79–84.

Schaffitzel, C., J. Hanes, L. Jermutus and A. Plückthun. 1999. "Ribosome display: An *in vitro* method for selection and evolution of antibodies from libraries." *J Immunol Methods* 231: 119–35.

Schimmele, B., N. Grafe and A. Plückthun. 2005. "Ribosome display of mammalian receptor domains." *Protein Eng Des Sel* 18: 285–94.

Seelig, B. and J. W. Szostak. 2007. "Selection and evolution of enzymes from a partially randomized non-catalytic scaffold." *Nature* 448: 828–31.

Sepp, A., D. S. Tawfik and A. D. Griffiths. 2002. "Microbead display by *in vitro* compartmentalisation: Selection for binding using flow cytometry." *FEBS Lett* 532: 455–8.

Shimizu, Y., A. Inoue, Y. Tomari, T. Suzuki, T. Yokogawa, et al. 2001. "Cell-free translation reconstituted with purified components." *Nat Biotechnol* 19: 751–5.

Shimizu, Y., T. Kanamori and T. Ueda. 2005. "Protein synthesis by pure translation systems." *Methods* 36: 299–304.

Spirin, A. S. 2004. "High-throughput cell-free systems for synthesis of functionally active proteins." *Trends Biotechnol* 22: 538–45.

Stein, V., I. Sielaff, K. Johnsson and F. Hollfelder. 2007. "A covalent chemical genotype-phenotype linkage for *in vitro* protein evolution." *Chembiochem* 8: 2191–4.

Stemmer, W. P. 1994. "Rapid evolution of a protein *in vitro* by DNA shuffling." *Nature* 370: 389–91.

Tabuchi, I., S. Soramoto, N. Nemoto and Y. Husimi. 2001. "An *in vitro* DNA virus for *in vitro* protein evolution." *FEBS Lett* 508: 309–12.

Tateyama, S., K. Horisawa, H. Takashima, E. Miyamoto-Sato, N. Doi, et al. 2006. "Affinity selection of DNA-binding protein complexes using mRNA display." *Nucleic Acids Res* 34: e27.

Tawfik, D. S. and A. D. Griffiths. 1998. "Man-made cell-like compartments for molecular evolution." *Nat Biotechnol* 16: 652–6.

Tuerk, C. and L. Gold. 1990. "Systematic evolution of ligands by exponential enrichment: RNA ligands to bacteriophage T4 DNA polymerase." *Science* 249: 505–10.

Villemagne, D., R. Jackson and J. A. Douthwaite. 2006. "Highly efficient ribosome display selection by use of purified components for *in vitro* translation." *J Immunol Methods* 313: 140–8.

Wade, H. E. and H. K. Robinson. 1966. "Magnesium ion-independent ribonucleic acid depolymerases in bacteria." *Biochem J* 101: 467–79.

Walter, P. and G. Blobel. 1983. "Preparation of microsomal membranes for cotranslational protein translocation." *Methods in Enzymol* 96: 84–93.

Wong, T. S., D. Roccatano, M. Zacharias and U. Schwaneberg. 2006. "A statistical analysis of random mutagenesis methods used for directed protein evolution." *J Mol Biol* 355: 858–71.

Xu, L., P. Aha, K. Gu, R. G. Kuimelis, M. Kurz, et al. 2002. "Directed evolution of high-affinity antibody mimics using mRNA display." *Chem Biol* 9: 933–42.

Yin, G. and J. R. Swartz. 2004. "Enhancing multiple disulfide bonded protein folding in a cell-free system." *Biotechnol Bioeng* 86: 188–95.

Yonezawa, M., N. Doi, T. Higashinakagawa and H. Yanagawa. 2004. "DNA display of biologically active proteins for *in vitro* protein selection." *J Biochem* 135: 285–8.

Yonezawa, M., N. Doi, Y. Kawahashi, T. Higashinakagawa and H. Yanagawa. 2003. "DNA display for *in vitro* selection of diverse peptide libraries." *Nucleic Acids Res* 31: e118.

Zahnd, C., P. Amstutz and A. Plückthun. 2007a. "Ribosome display: Selecting and evolving proteins *in vitro* that specifically bind to a target." *Nat Methods* 4: 269–79.

Zahnd, C., E. Wyler, J. M. Schwenk, D. Steiner, M. C. Lawrence, et al. 2007b. "A designed ankyrin repeat protein evolved to picomolar affinity to Her2." *J Mol Biol* 369: 1015–28.

Zhao, H., L. Giver, Z. Shao, J. A. Affholter and F. H. Arnold. 1998. "Molecular evolution by staggered extension process (StEP) *in vitro* recombination." *Nat Biotechnol* 16: 258–61.

Zhou, J. M., S. Fujita, M. Warashina, T. Baba and K. Taira. 2002. "A novel strategy by the action of ricin that connects phenotype and genotype without loss of the diversity of libraries." *J Amer Chem Soc* 124: 538–43.

Zubay, G. 1973. "*In vitro* synthesis of protein in microbial systems." *Ann Rev Genet* 7: 267–87.

# 4 Library Construction for Protein Engineering

*Daša Lipovšek, Marco Mena, Shaun M. Lippow, Subhayu Basu, and Brian M. Baynes*

## CONTENTS

Diversity-oriented protein engineering, also known as directed molecular evolution or directed protein evolution, relies on the construction of large libraries of variant genes, followed by high-throughput screening or selection to identify those members of each library that encode proteins with desired properties.

Typically, a protein-engineering library is based on the sequence of one or a small number of starting proteins, which already have properties similar to those required. For example, a diversity-based engineering project may start with a well-characterized antibody that binds to a specific antigen, with the goal of identifying a related antibody that will bind the same antigen with a higher affinity or specificity. Similarly, an enzyme may be re-engineered to increase its activity or thermostability, or to modify its substrate specificity. More recently, natural proteins with favorable biophysical properties have been used as scaffolds to design families of stable proteins selected for their ability to bind target macromolecules.

In order to modify or optimize an existing protein, a library of variant genes is designed and constructed with two, often conflicting, goals in mind:

First, library members need to be sufficiently similar in sequence to the starting protein to share a similar structure and function. For example, in the case of antibody affinity maturation, as many variants as possible should be similar enough to the starting antibody to also fold into functional antibodies that recognize the same antigen. Similarly, in the case of engineering an enzyme for change in substrate specificity, the variant enzymes should be similar enough in structure to the starting enzyme to catalyze the same chemical reaction.

Second, library members need to be sufficiently different in sequence from the starting protein to be slightly different in structure and thus in the functional property of interest. For example, in the case of antibody affinity maturation, the variant proteins should bind to the antigen of interest with different affinities. Similarly, in the case of engineering an enzyme for change in substrate specificity, the variant enzymes should differ enough in structure to bind a range of substrates other than the natural substrate of the starting enzyme.

In practice, striking the optimal balance between conservation and diversification of the sequence, and thus of properties to be modified, is one of the biggest challenges in library design. Ideally, all available information on the relationship between the sequence, structure, and function of the starting protein and its relatives should be used in library design, as it would have been used in computational design of a small number of improved, site-directed mutants. The amount of design incorporated into library construction is limited by the level of structural and functional understanding of the protein being engineered, as well as by technical constraints of the library-construction method being used.

Other factors that contribute to library design and construction are: (1) the disparity between accessible library size and the theoretical sequence space of interest, (2) the limited number of approaches that can incorporate knowledge and design rules into a library, (3) the use and control of randomization, (4) the natural versus synthetic origin of the diverse population, and (5) library quality.

This chapter opens with an overview of the library-construction methods most commonly used today. It then discusses these methods in light of the considerations listed in the previous paragraph, including the implications for their use in conjunction with different screening and selection methods, and for their application to different types of protein-engineering problems. We conclude with an emphasis on the complementarity and synergy between different library-construction methods, as well as between the use of diverse libraries and computational protein design.

## ESTABLISHED METHODS FOR LIBRARY CONSTRUCTION

The three major classes of library-construction methods applied to protein engineering are based on random mutagenesis (Figure 4.1A), recombination (Figure 4.1B), and site-directed diversification (Figure 4.1C). In their pure form, random mutagenesis and recombination mimic the random generation of variants during natural evolution, albeit at a much accelerated and tunable rate. In contrast, even in its most naive form, site-directed diversification requires judgment on which sites in a protein are most likely to yield beneficial mutations when diversified; thus, site-diversification methods are informed by a higher level of statistical and structure-based protein design.

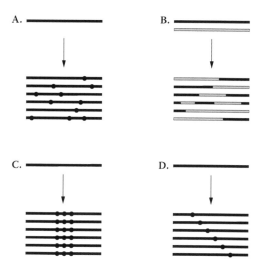

FIGURE 4.1   Major classes of library-construction methods used in protein engineering. The black and gray lines represent polypeptide sequences of the starting clone or clones, and the bullets represent mutations. (A) Random mutagenesis. (B) Recombination. (C) Site-directed diversification. (D) Scanning mutagenesis.

Scanning mutagenesis (Figure 4.1D), formally a type of site-directed diversification, can be used either to generate improved protein variants or to collect sequence-function correlations that will inform the construction of more complex and diverse libraries. Finally, libraries in which diversity is derived from naturally diverse gene sets present in an organism, tissue, or a complex environmental sample make possible the identification of naturally occurring proteins with desired phenotype.

Whereas the simplest examples of the major classes of library-construction methods are easy to distinguish, more sophisticated versions of these methods share underlying concepts and methodology with each other. Some of the most successful diversity-oriented applications of protein engineering have exploited several different types of libraries. All of these library-construction methods will be discussed in detail below.

## RANDOM MUTAGENESIS

In this family of methods, the gene that encodes a starting protein is modified by introducing relatively random mutations—substitutions, deletions, and insertions—at random positions in the gene. This method mimics the introduction of such errors over the millions of years of natural evolution, but vastly increases the rate of mutagenesis by artificially increasing the error rate of DNA replication.

The earliest random-mutagenesis methods obtained the increase in error rate by exposing dividing cells to mutagenic conditions such as UV light, x-ray radiation, or chemical mutagens (Doudney and Haas 1959; Ong and De Serres 1972; Myers et al. 1985), or by propagating the gene of interest in mutator cell strains deficient in DNA repair (Cox 1976; Low et al. 1996; Nguyen and Daugherty 2003). In the last

20 years, these methods were surpassed by polymerase chain reaction (PCR) performed under conditions of reduced replication fidelity (Leung et al. 1989; Cadwell and Joyce 1994), which has the advantage of speed, technical simplicity, and specificity to the gene of interest.

Any gene that can be amplified by standard PCR can also be randomized using error-prone PCR by changing buffer composition, manipulating ratios of free nucleotides, adding unnatural nucleotide analogs, or using polymerase mutants with a high propensity for incorporating errors (Cadwell and Joyce 1994; Zaccolo et al. 1996; Cirino et al. 2003). The mutagenesis rate can be manipulated by fine-tuning PCR conditions and the number of mutagenic PCR cycles, and can reach the rate of one mutation per five base pairs (Zaccolo and Gherardi 1999). Preassembled error-prone PCR kits that reliably mutagenize DNA at a set rate are available commercially from Clontech and Stratagene.

Since error-prone PCR introduces mutations throughout the DNA sequence being amplified with little or no positional bias, it is particularly well suited to applications where no information is available to direct diversification to particular "hot spots" in the sequence, positions where mutations would be the most likely to affect the phenotype of interest. Error-prone PCR tuned for a low rate of mutagenesis is commonly used to discover such hot spots (Takase et al. 2003; Hamamatsu et al. 2006). It is important to keep in mind that, while error-prone PCR is an excellent way to scan the length of the gene for promising positions to mutagenize, the mutations that it introduces are biased, especially at the protein level. This is because many amino-acid mutations require two or even three nucleotide mutations per codon, whereas error-prone PCR is most likely to introduce one nucleotide mutation per codon. Depending on the wild-type codon, only between three and seven amino-acid substitutions per amino acid are typically achieved by error-prone PCR (Wong et al. 2006), and the other twelve to sixteen substitutions are not sampled at all. Similarly, beneficial double or multiple mutations are sampled very rarely. In theory, the rate of occurrence of beneficial multiple mutations could be increased by increasing the overall mutagenesis rate, but in practice that is risky. Given that a random mutation is much more likely to be detrimental than beneficial, increasing the mutagenesis rate increases the risk of any beneficial multiple mutations being obscured by deleterious mutations occurring in the same clone. As a consequence, high mutagenesis rates are best combined with a recombination approach, described in the next section, that redistributes mutations in a wild-type background.

### RECOMBINATION

This family of methods mimics a second mechanism of natural evolution: the exchange of pieces between related genes using homologous recombination. In recombination-based library construction, two or more related starting genes are recombined, resulting in a library of variant genes with new combinations of sequences that were present in the starting gene. The starting genes can be either naturally occurring, closely related members of the same gene family or a mixture of naturally occurring genes and mutants generated *in vitro*, often by error-prone PCR.

The first *in vitro* recombination method described, DNA shuffling (Stemmer 1994a; Stemmer 1994b), is simple, powerful, and still widely used. Here the starting genes are randomly fragmented using DNAse I, and the fragments are annealed and reassembled using PCR. Highly homologous but nonidentical fragments from different parental genes cross-anneal and are extended into longer, chimeric fragments, eventually leading to the construction of full-length novel genes that contain sequences from two or more different sources.

When compared to library-generation methods that involve random mutagenesis (described in the preceding section) or saturation mutagenesis of defined parts of the starting gene (described in the next section), standard DNA shuffling from related, naturally occurring genes (Crameri et al. 1998) is a relatively conservative method, as it combines pieces of related functional proteins, generating new sequences that have a relatively high probability of being compatible with the desired protein structure and function. As with any method that relies on PCR-like DNA replication, DNA shuffling does incorporate a low level of random mutagenesis due to imperfect fidelity of even high-fidelity PCR. In addition, error-prone PCR can be employed during fragment assembly to add random point mutagenesis to recombination (Stemmer 1994a). Finally, DNA shuffling and other recombination-based methods are an excellent way of counteracting one of the weaknesses of randomization by error-prone PCR: A library with a high density of random mutations introduced by error-prone PCR can be back-crossed by shuffling with excess of the original wild-type gene, thus separating beneficial mutations from deleterious and neutral ones (Stemmer 1994a).

Numerous alternative DNA-recombination methods have been reported:

In staggered extension process (StEP) (Zhao et al. 1998), recombination of genetic information between several starting genes occurs when extension of the growing DNA strand from the first template is interrupted before a full-length gene is copied. The mixture of DNA template and product is denatured, re-annealed, and re-extended, allowing the growing strand to anneal to a different, homologous template, thus combining sequences from two or more templates. Extension time in each cycle, rather than the size of parental fragments in DNA shuffling, controls the frequency of crossover events. Another related method, random chimeragenesis on transient templates (RACHITT) (Coco et al. 2001), uses a full-length, uracil-containing template to assemble complementary, short, single-stranded fragments copied from other templates; the uracil-containing template is eventually degraded. RACHITT has the advantage of allowing high frequency of recombination between genes with low homology.

In contrast to the previously mentioned methods, incremental truncation for the creation of hybrid enzymes (ITCHY) (Lutz et al. 2001a; Ostermeier and Lutz 2003) achieves recombination without relying on hybridization of homologous DNA, and thus can recombine genes with little or no sequence homology. In this method, fragments of two or more gene templates are generated by digesting the templates from the 5' or 3' end, and then ligating the resulting fragments. Whereas this method allows recombination of any genes, it does not guide the recombination between analogous parts of two genes and is thus less likely to generate functional proteins than are homology-guided recombination methods. Elements of ITCHY and DNA shuffling are combined in a method named SCRATCHY (Lutz et al. 2001b).

Whereas DNA shuffling, StEP, RACHITT, and ITCHY generate recombination sites at random homologous sites, several recombination methods have been developed that use synthetic oligonucleotides to guide recombination to a specific site or sites in the starting genes. Degenerate oligonucleotide gene shuffling (DOGS) (Gibbs et al. 2001) achieves this aim by using amplification primers that contain two regions that recognize two different templates. Direct amplification of one template in the first round of PCR results in a fragment that anneals to a fragment amplified from the second template in the second round of PCR. The ultimate control of cross-over frequency and location is achieved when a set of synthetic oligos itself encodes amino-acid residues from different parental genes (Ness et al. 2002; Zha et al. 2003). In such a case, no physical template is required, and the library is generated synthetically using a design based on a number of parental sequences. In addition to allowing a complete control of recombination events, the use of synthetic oligonucleotides allows the introduction of any additional desired point mutations or randomized regions, thus including elements of site-directed diversification (described in the next section). Alternatively, in a method named site-directed chimeragenesis (SISDC) (Hiraga and Arnold 2003), synthetic oligonucleotides can be used to introduce into the starting genes tags containing restriction endonuclease sites, which then direct the fragmentation of the starting genes and their reassembly by ligation.

## SITE-DIRECTED DIVERSIFICATION

In this collection of methods, diversification is directed to a specific position or set of positions, and the remaining protein sequence is fixed as wild-type. In its classical form, known as site-directed randomization, an oligonucleotide that spans the codon or codons of interest is synthesized *in vitro*, and each wild-type codon of interest is replaced by a mixture of codons (Georgescu et al. 2003; Steffens and Williams 2007). If only a few clustered codons are being diversified, the oligonucleotide or oligonucleotides containing the desired mutations can be incorporated into the starting gene by site-directed mutagenesis, such as the PCR-based method of the Stratagene QuikChange kit (Miyazaki 2003; Zheng et al. 2004). If multiple codons are being diversified, the desired diversified gene product can be assembled from a mixture of constant and diversified oligonucleotides using PCR-based gene assembly (Ho et al. 1989; Stemmer et al. 1995; Bessette et al. 2003), ligation (Hughes et al. 2003), or a combination of the two methods.

The most widely used method of obtaining a site-directed mixture of codons is to synthesize a set of oligonucleotides where the wild-type bases in each codon of interest are replaced by mixtures of nucleotides described as NNN, NNS, or NNK, where N represents an equimolar mixture of A, G, C, and T; S represents an equimolar mixture of G and C; and K represents an equimolar mixture of G and T. Each of these three codon patterns encodes a mixture of all 20 naturally occurring amino-acid residues as well as translational stop codons, but the NNS and NNK pattern have the advantage of encoding stop codons at a frequency of 1/32 rather than 3/64 for NNN, thus reducing the proportion of truncated proteins encoded by the library.

Due to the redundancy and coding imbalance of the genetic code, site-saturation libraries using NNN, NNS, or NNK encode amino acids with different probabilities (Figure 4.2), but codon bias of NNS- or NNK-encoded site-directed diversity is much lower than that of error-prone PCR.

The major limitation of NNS- and NNK-encoded diversification is their demand on physical library size. Diversification of $n$ positions can generate $32^n$ possible combinations of codons, and the exponential increase of possible sequences with the number of positions diversified means that the number of possible unique codon combinations defined by a particular design quickly overwhelms the capacity of physical library construction methods or the screening and selection methods to test them (Figure 4.3). An additional pressure on physical library size comes from statistical considerations in sampling: A library must be oversampled threefold to ensure a 95% probability that any one of its unique clones will be included at least once, and oversampled 10- to 25-fold to ensure a greater than 99% probability of capturing the entire library (Bosley and Ostermeier 2005). As a consequence, only libraries with relatively few NNS- or NNK-diversified positions can be sampled thoroughly. As

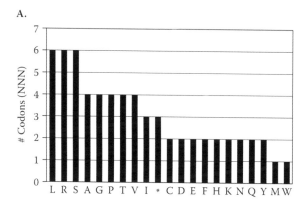

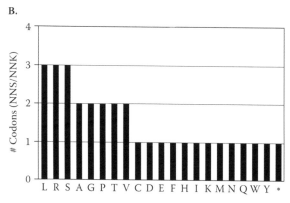

**FIGURE 4.2** Codon bias in diversification using mixtures of nucleotides. The bar graphs indicate the number of codons encoding each amino-acid residue and the stop codon (*) when the diversified region is encoded by NNN (A) vs. NNS or NNK (B).

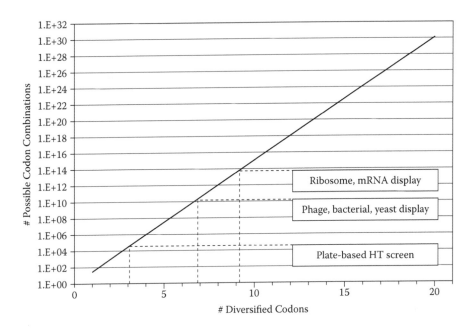

**FIGURE 4.3** Number of possible codon combinations encoded by NNS or NNK increases exponentially with the number of codons that are diversified. The boxes show typical ranges in library size that can be sampled by plate-based high-throughput (HT) screening, microbial display, and *in vitro* display.

illustrated in Figure 4.3, a typical plate-based, high-throughput screening assay can thoroughly sample a library with no more than two to three NNS- or NNK-diversified positions; phage, bacterial, or yeast display, a library with no more than five to seven diversified positions; and mRNA and ribosome display, a library with no more than eight to nine diversified positions. Once these thresholds are exceeded, only a small proportion of each possible combination of residues allowed by the diversification scheme is sampled, and it becomes highly unlikely that the globally best combination of residues in the diversified positions will be tested and identified.

The sampling problem associated with site-saturation libraries made with mixtures of nucleotides can be mitigated in several different ways. The simplest approach is to limit the number of residues diversified in any single library, and to assume that many mutations discovered in separate libraries with different diversified residues will be additive. An extreme example of this approach are libraries created by scanning-mutagenesis (described in the next section), which diversify only one position at a time. A more aggressive approach, combinatorial active-site saturation test (CAST) (Reetz et al. 2005), was first applied to amino-acid residues found in enzyme active sites. CASTing uses available structural information on the starting enzyme to identify pairs of residues that are close in space, and are thus presumed to interact or to have synergistic effects on enzyme function. A separate site-saturation library is made for each pair of residues, yielding $32^2 = 1024$ unique sequences per library, which are then oversampled by screening 3000 randomly picked transformants. This concept can be extrapolated to display-based selection, for example, by constructing sets of libraries with six diversified positions per library for phage display, or with eight diversified positions per library for mRNA display. An example of this approach is walk-through mutagenesis, which was used to affinity-mature antibodies by constructing site-saturation libraries of one complementarity determining region (CDR) at a time (Barbas et al. 1994). Another major problem of site-saturation libraries that use degenerate oligonucleotides with NNS or NNK codons is the introduction of unwanted stop codons into the diversified region due to the 1 in 32 translational stop encoded by each diversified codon. This problem, too, can be limited by limiting the number of diversified residues.

An alternative to limiting the number of diversified positions is to limit the depth of diversification at each position to fewer than 20 amino acids. The subset of amino-acid residues chosen in this approach depends on information available on the system being engineered, and on the limitations of the method used to encode that subset. An elegant example of such an approach is the use of restricted-alphabet libraries in antibody engineering (Fellouse et al. 2004; Fellouse et al. 2005), which takes advantage of the fact that the two amino-acid residues found most commonly at the interface between antibodies and antigens, tyrosine and serine, are encoded by a single degenerate codon. The theoretical complexity of binary tyrosine/serine antibody libraries, where $n$ diversified positions encode four different amino acids, grows as $2^n$ rather than $32^n$, allowing the efficient sampling by phage display of libraries with more than 20 diversified positions. Restricted-alphabet libraries can also use a limited codon set that encodes chemically diverse amino-acid residues (Reetz et al. 2008), or they can be informed by structure- or homology-based protein design. The relative advantages of thorough sampling versus a highly diverse sequence space depend on the specific system and design used, with published examples where limiting the alphabet size yielded selected antibodies with lower affinity (Munoz and Deem 2008). Due to the restrictions of genetic code, design-based restricted-alphabet libraries that use nucleotide mixtures require constant compromise between including extra residues that are not part of the design and excluding some of the desired residues (Mena and Daugherty 2005).

Despite their suboptimal sampling of sequence space, the preceding methods, which are based on oligonucleotides synthesized using mixtures of nucleotides, remain the most popular approaches to constructing libraries with site-directed diversity, primarily due to their technical simplicity and relatively low reagent cost. Most commercial suppliers of synthetic oligonucleotides also sell affordable oligonucleotides diversified in this manner. However, the last decade has seen the development of several new methods that allow fine control over the exact sequence of diversified regions, including the identity and proportion of specific codons allowed at each diversified position in synthetic oligonucleotides, the exclusion of translational stops, and error reduction. Whereas these new methods allow tighter control over library composition and quality, they are technically demanding, less readily available commercially, and more expensive. These methods are described in detail in the next few paragraphs.

Oligonucleotides that contain defined mixtures of codons unrestricted by the genetic code can be synthesized using defined mixtures of trinucleotide phosphoramidite codons (Virnekas et al. 1994; Kayushin et al. 1996; Yanez et al. 2004), split-and-mix strategies (Glaser et al. 1992; Lahr et al. 1999), or enzymatic ligation of defined trinucleotides (Van den Brulle et al. 2008; Xiong et al. 2008). These methods can be used to introduce into the diversified position a mixture of codons for all 20 amino-acid residues or for a smaller set that follows specific rules, such as those derived by computational protein design. Amino-acid residues inconsistent with library design are avoided, and library sampling is greatly improved.

A further improvement in user control over library sequences has been made possible by recent advances in parallel oligonucleotide synthesis (Singh-Gasson et al. 1999; Pirrung 2002; Cleary et al. 2004; Zhou et al. 2004), which allow the simultaneous small-scale synthesis of 1000 to 100,000 oligonucleotides, each with a defined sequence. Libraries built using such complex pools of defined-sequence oligonucleotides as the source of diversity (Cleary et al. 2004; Richmond et al. 2004; Tian et al. 2004) allow the control of not only specific amino-acid residues allowed at each diversified position, but also which residues are found close to each other in primary sequence. This in turn makes possible the control over many protein properties that are defined by primary oligopeptide sequence, such as net charge, average hydrophobicity, and the presence of protease cleavage sites, deamination sites, N-glycosylation sites, and predicted T-cell epitopes. Many structure- and homology-based design constraints can also be incorporated into libraries using this method. Library diversity restricted to oligopeptides compatible with protein design yields a higher density of clones with predicted favorable properties, leading to a vast improvement in physical sampling of the theoretical sequence space of interest. Table 4.1 illustrates this improvement for the case of an enzyme-engineering problem, by comparing the physical library size required to represent a particular protein design using different site-directed library-construction methods (S. M. Lippow, S. Basu, K. Prather, and T. S. Moon, unpublished data).

An additional advantage of libraries assembled from complex mixtures of defined-sequence oligonucleotides is that they are the only source of library diversity compatible with error correction during the assembly process, reducing the impact of mutations introduced during oligonucleotide synthesis. The method used, error correction by consensus filtering (Figure 4.4) (Carr et al. 2004), requires that every oligonucleotide used to assemble the diversified gene be synthesized in both its forward- and reverse-complementary form, then allowed to anneal. Given the random nature of errors introduced during synthesis, it is highly unlikely that the same error will occur in both the forward and the reverse strand encoding a particular variant; thus, a double-stranded fragment containing an oligonucleotide with an error is almost certain to contain a mismatch between the forward and the reverse strand. MutS, a protein that binds preferentially to mismatches, small insertions and small deletions in double-stranded DNA, is then used to remove such mismatched fragments from the mixture, greatly improving the content of wild-type and designed-diversity sequences. For oligonucleotides encoding constant regions in the library or diversified positions with a small number of changes, each oligonucleotide pair can be error-corrected in a separate reaction. For oligonucleotides encoding regions of high diversity, which require the use of a mixture of hundreds or thousands of oligonucleotides, the forward- and the reverse-complementary forms of each sequence

## TABLE 4.1A
## Top 48 Predicted Sequences Spanning Residues
## 324–334 in Galactose Oxidase

| Amino-Acid Position | 324 | 326 | 329 | 330 | 333 | 334 |
|---|---|---|---|---|---|---|
| Wild-type GaOx | **Asp** | **Gln** | **Tyr** | **Arg** | **Asn** | **His** |
| Variant #1 | **Asp** | Arg | Arg | Asn | **Asn** | **His** |
| Variant #2 | **Asp** | Arg | Arg | Asp | **Asn** | **His** |
| Variant #3 | **Asp** | Arg | Arg | Gln | **Asn** | **His** |
| Variant #4 | **Asp** | Arg | Arg | His | **Asn** | **His** |
| Variant #5 | **Asp** | Arg | Arg | Lys | **Asn** | **His** |
| Variant #6 | **Asp** | Arg | Arg | Met | **Asn** | **His** |
| Variant #7 | **Asp** | Arg | Arg | Ser | **Asn** | **His** |
| Variant #8 | **Asp** | Arg | Asn | Lys | **Asn** | **His** |
| Variant #9 | **Asp** | Arg | Glu | Lys | **Asn** | **His** |
| Variant #10 | **Asp** | Arg | Glu | Ser | **Asn** | **His** |
| Variant #11 | **Asp** | Arg | His | Asp | **Asn** | **His** |
| Variant #12 | **Asp** | Arg | Lys | His | **Asn** | **His** |
| Variant #13 | **Asp** | Arg | Lys | Tyr | **Asn** | **His** |
| Variant #14 | **Asp** | Arg | **Tyr** | Asp | **Asn** | **His** |
| Variant #15 | **Asp** | Arg | **Tyr** | Gln | **Asn** | **His** |
| Variant #16 | **Asp** | Arg | **Tyr** | His | **Asn** | **His** |
| Variant #17 | **Asp** | Arg | **Tyr** | Lys | **Asn** | **His** |
| Variant #18 | **Asp** | Arg | **Tyr** | Met | **Asn** | **His** |
| Variant #19 | **Asp** | Arg | **Tyr** | Ser | **Asn** | **His** |
| Variant #20 | **Asp** | Asn | Arg | Asn | **Asn** | **His** |
| Variant #21 | **Asp** | Asn | Arg | Asp | **Asn** | **His** |
| Variant #22 | **Asp** | Asn | Arg | His | **Asn** | **His** |
| Variant #23 | **Asp** | Asn | Arg | Lys | **Asn** | **His** |
| Variant #24 | **Asp** | Asn | Lys | His | **Asn** | **His** |
| Variant #25 | **Asp** | Asn | **Tyr** | His | **Asn** | **His** |
| Variant #26 | **Asp** | Asp | Arg | Asn | **Asn** | **His** |
| Variant #27 | **Asp** | **Gln** | Arg | Gln | **Asn** | **His** |
| Variant #28 | **Asp** | **Gln** | Arg | Lys | **Asn** | **His** |
| Variant #29 | **Asp** | **Gln** | Arg | Ser | **Arg** | **His** |
| Variant #30 | **Asp** | **Gln** | Asn | Gln | **Asn** | **His** |
| Variant #31 | **Asp** | **Gln** | Lys | Gln | **Asn** | **His** |
| Variant #32 | **Asp** | **Gln** | Lys | Lys | **Asn** | **His** |
| Variant #33 | **Asp** | **Gln** | **Tyr** | **Arg** | **Asn** | **His** |
| Variant #34 | **Asp** | **Gln** | **Tyr** | Lys | **Asn** | **His** |
| Variant #35 | **Asp** | Glu | Arg | Gln | **Asn** | **His** |
| Variant #36 | **Asp** | Glu | Arg | Lys | **Asn** | **His** |
| Variant #37 | **Asp** | Glu | Arg | Ser | **Asn** | **His** |
| Variant 38 | **Asp** | Glu | Lys | Lys | **Asn** | **His** |

**TABLE 4.1A (continued)**

**Top 48 Predicted Sequences Spanning Residues 324–334 in Galactose Oxidase**

| | | | | | | |
|---|---|---|---|---|---|---|
| Variant #39 | **Asp** | Glu | **Tyr** | Gln | **Asn** | **His** |
| Variant #40 | **Asp** | Lys | Arg | Gln | **Asn** | **His** |
| Variant #41 | **Asp** | Lys | Arg | Glu | **Asn** | **His** |
| Variant #42 | **Asp** | Lys | Lys | Glu | **Asn** | **His** |
| Variant #43 | **Asp** | Lys | **Tyr** | Gln | **Asn** | **His** |
| Variant #44 | **Asp** | Lys | **Tyr** | Glu | **Asn** | **His** |
| Variant #45 | **Asp** | Ser | Arg | Lys | **Asn** | **His** |
| Variant #46 | Gly | Arg | **Tyr** | Ile | **Asn** | **His** |
| Variant #47 | Ser | Arg | Arg | Ser | **Asn** | **His** |
| Variant #48 | Ser | **Gln** | Arg | Asp | Arg | Glu |

*Note:* Program Rosetta (Kuhlman and Baker 2000; Meiler and Baker 2006) was used to predict combinations of amino-acid substitutions near the active site of galactose oxidase (Figure 4.6) to change enzyme specificity from galactose to glucose. This table lists the substitutions between D324 and H334 in the 48 variants predicted to be the most consistent with glucose binding (S. Lippow, unpublished results). Table 4.1B compares different library-construction methods that can be used to construct libraries that contain these 48 variants and the minimal number of library clones required to sample those 48 variants.

are synthesized in the same physical oligonucleotide. The complex mixture of self-annealed hairpins that results is then exposed to MutS, and hairpins containing mismatches are removed as in the case of single oligonucleotide pairs (Figure 4.4).

## SCANNING MUTAGENESIS

The scanning-mutagenesis family of methods generates libraries that exhaustively sample a very simple library design: a collection of single point mutations at every position in the protein, or at least at every position in the region of interest. The earliest form of this method, alanine scanning (Cunningham and Wells 1989; Clackson and Wells 1995; Weiss et al. 2000), mutates each residue in turn to alanine, thus interrogating mostly the effect of a loss of chemical functionality at different positions in the sequence. In order to capture the information inherent in the loss-of-function phenotype of mutants in an alanine-scanning library, a high-throughput screen that obtains information on each individual mutant is required. The results from screening an alanine-scanning library are typically used to focus the design of a more thorough library synthesized by in-depth diversification (see the section titled "Site-Directed Diversification" in this chapter), focusing on the residues found to be sensitive to alanine scanning.

More complex scanning libraries can be generated by mutating each codon position of interest to a number of different codons. Typically, degenerate synthetic oligonucleotides NNN, NNS, or NNK (defined in the preceding section) are used to replace each

**TABLE 4.1B**

**Minimal Library Size Required to Encode Top 48 Predicted Sequences Spanning Residues 324–334 in Galactose Oxidase**

| Library Type | 324 | 326 | 329 | 330 | 333 | 334 | Minimal Library Size |
|---|---|---|---|---|---|---|---|
| Mix of defined-sequence oligos | Oligonucleotides encoding variants #1–48 | | | | | | 48 |
| Mix of codons | Asp | Arg | Arg | Arg | Arg | Glu | $3\times7\times6\times11\times2\times2 =$ |
| | Gly | Asn | Asn | Asn | Asn | His | 5,544 |
| | Ser | Asp | Glu | Asp | | | |
| | | Gln | His | Gln | | | |
| | | Glu | Lys | Glu | | | |
| | | Lys | Tyr | His | | | |
| | | Ser | | Ile | | | |
| | | | | Lys | | | |
| | | | | Met | | | |
| | | | | Ser | | | |
| | | | | Tyr | | | |
| Degenerate oligos | RRC: | VRS: | NRS: | NDS: | MRC: | SAS: | $4\times12\times16\times24\times4\times4 =$ |
| | Asn | Arg(3) | Arg(3) | Arg(3) | Arg | Asp | $2.9\times10^5$ |
| | Asp | Asn | Asn | Asn | Asn | Gln | |
| | Gly | Asp | Asp | Asp | His | Glu | |
| | Ser | Gln | Cys | Cys | Ser | His | |
| | | Glu | Gln | Gln | | | |
| | | Gly(2) | Glu | Glu | | | |
| | | His | Gly(2) | Gly(2) | | | |
| | | Lys | His | His | | | |
| | | Ser | Lys | Ile | | | |
| | | | Trp | Leu(3) | | | |
| | | | Tyr | Lys | | | |
| | | | Ser | Met | | | |
| | | | Stop | Phe | | | |
| | | | | Trp | | | |
| | | | | Tyr | | | |
| | | | | Ser | | | |
| | | | | Val(2) | | | |
| | | | | Stop | | | |
| NNS | 32 | 32 | 32 | 32 | 32 | 32 | $32^6 = 1.1\times10^9$ |
| | | | | | | | $(20^6 = 6.4\times10^7)$ |

*S. Lippow, S. Basu, K. Prather, and T.S. Moon, unpublished results.*

*Note:* Comparison of the different library-construction methods that can be used to construct libraries containing the 48 variants listed in Table 4.1A, and the minimal number of total library clones required to sample those 48 variants. No oversampling was assumed. Since the variants in this method will be screened using a 96-well–based assay, methods other than library construction from a mixture of defined-sequence oligonucleotides will clearly lead to severe undersampling of the designed variables.

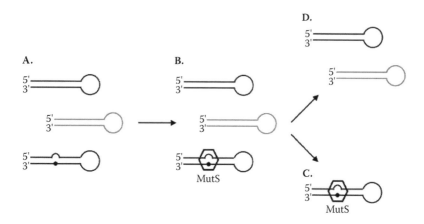

**FIGURE 4.4** Error correction of pools of defined-sequence oligonucleotides by consensus filtering. (A) A large pool of oligonucleotides encoding a diversified region in a library is synthesized, with each oligonucleotide containing both the forward and the reverse-complementary strand encoding a specific variant, and is allowed to form a hairpin structure. An error in one of the strands leads to a mismatch between the two strands. (B) MutS protein, which recognizes mismatched double-stranded DNA, is added to the pool, and binds to error-containing hairpins. (C) Error-containing hairpins bound to MutS are removed by affinity purification, and the remaining oligonucleotide pool (D) is enriched for error-free sequences.

codon of interest in turn with the mixture of codons for all 20 naturally occurring amino-acid residues (Weiner et al. 1994; Braman et al. 1996; Kretz et al. 2004). If a defined mixture of oligonucleotides is used to drive the mutagenesis reactions, any subset of amino-acid substitutions can be made. The possibilities include the 18 possible noncysteine amino-acid substitutions; a substitution with a smaller set of amino-acid residues that represent the broad classes of chemical functionalities present in the full set (look-through mutagenesis) (Rajpal et al. 2005); substitutions based on residues found at a specific position in naturally occurring, homologous proteins; or substitutions computationally predicted to be favorable at each position. Scanning libraries of even higher complexity can be constructed using mixtures of defined-sequence oligonucleotides to generate all possible double as well as single mutations in the region of interest, for example, within a specific complementarity-determining region of an antibody undergoing affinity maturation. Since such libraries are expected to contain some members with improved properties of interest, the mutants can be combined into a single pool and selected for the desired property in a high-throughput manner using an *in vitro*, display-based method. Typically, low-stringency selection is used to identify a large number of beneficial mutations, which are then either combined and tested as multiple mutants or used to design a second-generation, site-directed library (see the section titled "Site-Directed Diversification" in this chapter).

## DIVERSITY FROM NATURAL SOURCES

Libraries that exploit protein diversity found in nature, with no design-driven manipulation of diversity, may not be considered protein-engineering libraries in the strictest definition of the term because of the lack of the engineering component. However, they are mentioned here because of the technical parallels between their construction and use and that of the other library-generation methods described previously. Typically, libraries derived from natural sources are made by PCR-amplifying genetic information from a diverse set of genes found in a particular organism or tissue, for example, the antibody variable domains found in human peripheral blood cells (Marks and Bradbury 2004; Hoogenboom 2005), or all polypeptides transcribed in a particular organism or tissue under particular conditions (Cujec et al. 2002). Alternatively, microbial genes homologous to a gene encoding a known enzyme of interest can be PCR-amplified from DNA present in complex environmental samples, eliminating the need to identify and culture the source microorganisms (Lafferty and Dycaico 2004; Mathur et al. 2005).

Antibody libraries amplified from germline sequences (Marks et al. 1991; Vaughan et al. 1996; Sheets et al. 1998) have been used with great success to discover high affinity, target-binding antibodies using *in vitro* selection methods such as phage display (Bradbury and Marks 2004), bacterial display (Jeong et al. 2007), yeast display (Feldhaus et al. 2003), and ribosome display (Hanes et al. 1998). These protein-engineering technologies are discussed in detail in Chapters 1–3. Given that antibody variable heavy and light domains are amplified separately and then paired randomly into antibody fragments that contain both domains, an element of recombination or shuffling is present in all antibody libraries derived from natural sources. Similarly, the use of PCR to amplify the genes invariably introduces some random mutations. Further protein-engineering, knowledge-based approaches commonly applied to antibody libraries from natural sources are the use of specific framework classes with favorable biophysical properties (Pini et al. 1998; Hoet et al. 2005), recombination between complementarity-determining regions present in the natural repertoire (Soderlind et al. 2000), and the incorporation of random (Barbas et al. 1992; Hoogenboom and Winter 1992) or designed (de Kruif et al. 1995) synthetic diversity into complementarity-determining regions (Knappik et al. 2000; Rothe et al. 2008). Finally, target-binding antibodies selected from natural-diversity libraries are often affinity-matured using engineered libraries based on the sequence of the initial hits (Roskos et al. 2007).

## CRITICAL FACTORS IN EVALUATION OF LIBRARY-CONSTRUCTION METHODS

Library design and construction involve numerous compromises between (1) imposing a design and maximizing unbiased randomization, (2) wanting to test every possible permutation of every possible mutation yet having dramatic physical constraints on library size, (3) taking conservative and aggressive approaches, (4) investing resources in library quality and in quality control, and (5) obtaining information on each clone in a smaller library using a high-throughput screen and identifying only the best clones in a larger library using a display-based selection method.

## Extent and Nature of Diversification

All library construction methods allow at least some control over the balance between conserving and diversifying the sequence, structure, and function of the starting protein or proteins. For example, the average error rate in error-prone PCR is controlled by adjusting PCR conditions, the number and location of crossovers in recombination-based techniques is controlled by selecting the appropriate method and the number of starting variants of a gene, and the extent of site-directed diversification is controlled by selecting the number and location of positions diversified and the number of substitutions in each position.

Methods that produce mutants close in sequence to the starting protein (such as scanning mutagenesis and error-prone PCR tuned to generate primarily single and double mutants) are most useful when only a small change in properties is desired, when there is reason to believe that effect of individual mutations will be additive, or when not enough is known about the protein being engineered to direct more aggressive mutagenesis to a particular region of the protein. Recombination methods that produce variants close in sequence to the starting proteins, such as recombination between closely related homologous proteins or between a protein and its mutants, are similarly well suited to fine-tuning existing properties. Once the most improved members of lightly mutagenized libraries are identified, they can be further evolved by subjecting them to another round of light diversification and screening or selection, or by directing more intensive mutagenesis to amino-acid positions identified in the first round.

More aggressive methods such as highly mutagenic error-prone PCR, extensive saturation mutagenesis, and recombination between less closely related proteins are often required to effect a large change in protein properties (Xu et al. 2002; Otey et al. 2004; Seelig and Szostak 2007). Since only a small fraction of all possible protein sequences can fold into a stable structure, and only a small proportion of folded proteins can carry out a desired function, the vast majority of random mutations are fatal or deleterious. Thus any mutagenic method of library construction that combines a large number of random mutations is guaranteed to produce a high proportion of unfolded and nonfunctional proteins. As a consequence, aggressive protein-engineering approaches require particularly careful library design based on an understanding of the protein being engineered and of the statistical implications of the library-construction method being used.

Ideally, an x-ray or NMR structure of the protein being engineered is used to direct diversification to solvent-exposed positions, where the diversification is less likely to interfere with protein stability, and to positions most directly involved with protein function, such as residues at or close to a ligand binding site or an active site. Even when no experimentally determined three-dimensional structure is available, it is often possible to build a low-resolution three-dimensional model of the protein of interest based on known structures of homologous proteins. Alternatively, some information on tolerance of amino-acid substitution and on sequence-function relationship can be obtained by comparing the sequence of the starting protein with those of naturally occurring related proteins, or with sequences of both favorable and deleterious mutations identified in a preliminary screen or selection (typically from an error-prone PCR or site-directed scanning library). In addition, the library construction strategy should ensure that a sufficient proportion of wild-type residues and conservative substitutions are present to allow the identification of favorable mutations, rather than have them masked by deleterious mutations that happen to occur in the same clone. When using site-directed approaches, this goal can be achieved by limiting the theoretical sequence space to ensure thorough sampling, such as by reducing the number of diversified positions or the number of substitutions allowed at each position. Alternatively, when using either aggressive error-prone PCR or extensive site-directed diversification, a highly mutagenized library can be backcrossed with the starting wild-type gene using a recombination method, generating a "diluted" version of the library, where clones containing a smaller number of favorable mutations can be identified.

## SAMPLING PROBLEM

How much sequence diversity can be sampled thoroughly is limited by the number of clones that can be tested experimentally. With the exception of traditional site-directed scanning mutagenesis, where the scope of a screen can be limited by the effort and expense of constructing a large number of defined mutants, all library-construction methods described in this chapter can easily produce very large populations of gene variants. Thus the number of individual clones sampled is limited by either the system for transcribing and translating those genes into proteins or by the throughput of the screening method used. For example, the physical size of a library used in a microbial display system such as phage display, bacterial display, or yeast display is typically limited to $10^6$–$10^{10}$ clones by the transformation efficiency for each microorganism, and the size of a library used in an *in vitro* display system such as ribosome or mRNA display is limited to $10^{11}$–$10^{13}$ clones by the availability of *in vitro* transcription/translation reagents (Figure 4.3). In the case of high-throughput, plate-based screening, where each clone is tested individually, the amount of automation available and the cost per assay typically limit the number of clones that can be tested to $10^2$–$10^4$. Selection methods that rely on fluorescence-activated cell sorting (FACS) to examine each cell in a mixture are limited by the throughput of the flow cytometer to testing approximately $10^8$ clones per experiment. This throughput can be increased to approximately $10^{10}$–$10^{11}$ clones by prefiltering the cell-displayed library using magnetic beads (MACS).

It is important to take into account the throughput of the selection or screening method when designing and constructing a library. Whereas there are many examples of successful selections from vastly undersampled libraries constructed by diversifying positions highly tolerant of substitution (Xu et al. 2002; Silverman et al. 2005), undersampling is likely to be a more serious problem with library designs that diversify positions critical for protein stability and function, such as positions in an enzyme active site, in the hydrophobic core, or at the interface between subunits of a multimeric protein.

## LIBRARY QUALITY

A challenge closely related to library size and sampling is that of library quality. At the DNA level, library quality refers to the proportion of gene sequences in the library that conform to the library design. For example, in the case of a typical site-saturation library, a stringent measure of library quality is the proportion of clones in the unselected library that contain mutations only in the positions chosen to be diversified, that contain only the types of mutations specified, and that contain no substitutions, insertions, or deletions in the proposed constant regions. A less stringent measure of library quality is the proportion of clones encoding full-length proteins, with or without mutations in constant regions.

Whereas the understanding of library quality is critical for an accurate estimate of effective library size, published data on library quality is scarce, probably because it requires extensive sequencing of unselected clones, which are of no phenotypic interest. Sources of poor library quality are generally believed to be out-of-frame or out-of-alignment recombination, errors in oligonucleotide synthesis, and errors in PCR amplification (except as desired during construction of libraries by error-prone PCR). In our personal experience with site-directed diversification, library quality has been limited by the quality of commercially available synthetic oligonucleotides, whose error rates range between 0.2% and 1% per base for standard-scale oligonucleotide synthesis (with higher error rates for oligonucleotides longer than 80 base pairs), and between 0.5% and 2% per base for microscale synthesis (S. Basu, unpublished data). Mistakes in DNA replication by high-fidelity polymerases, on the order of $10^{-6}$ per base per extension (Hengen 1995), is a distant second source of errors, and only becomes significant in methods that use repeated PCR amplification between rounds of selection, such as ribosome display and mRNA display. Thus the easiest way to maximize library quality is to minimize the number and length of synthetic oligonucleotides used in library construction. This is straightforward when the sites to be diversified are close enough in sequence to be encoded by a single degenerate oligonucleotide; on the other hand, when diversified positions are far apart in sequence, multiple diversified oligonucleotides are required to make the library, leading to a higher error rate. One approach to limit the errors even in a library constructed from multiple diversified oligonucleotides is to use mixtures of defined-sequence oligonucleotides, and subject them to consensus-based error correction (see the section titled "Site-Directed Diversification" in this chapter). We find that this method reduces the error rate in starting oligos by between 10- and 20-fold (S. Basu, unpublished results), which is consistent with previously published results (Carr et al. 2004). The ability to remove errors from synthetic oligonucleotide pools becomes critical for libraries with long diversified (or otherwise oligonucleotide-derived) regions, since the probability of an error-free clone decreases exponentially with length (Figure 4.5), and for screens using very expensive or low-throughput assays.

## DESIGN VERSUS RANDOMIZATION

Whereas it is common to contrast *in vitro* molecular evolution with protein design due to the reliance of molecular evolution on randomization, a minimal amount

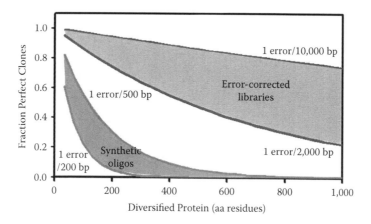

**FIGURE 4.5 (see color insert following page 178)** Comparison in library quality with and without error correction. The fraction of clones in a library that conform to the intended library design drop exponentially with the length of the fragment being diversified. Orange: error rates in standard commercially available synthetic oligonucleotides, and fraction of perfect clones when a library is assembled from standard oligonucleotides. Green: error rates in synthetic oligonucleotides that have been error-corrected by consensus filtering (see Figure 4.4, and the section titled "Library Quality" earlier in this chapter), and fraction of perfect clones when a library is assembled from error-corrected oligonucleotides.

of design is at least implicit even in the library-construction approaches that aim to avoid all sequence bias. For example, site-saturation mutagenesis requires the choice of the positions to be randomized, recombination methods require the choice of starting sequences, and error-prone PCR requires the choice of the average mutagenesis rate. On the other hand, recent methods of library construction that allow explicit control over sequences present in a library, such as oligonucleotide-directed recombination (see the section titled "Recombination" in this chapter), site-directed diversification using a defined mixture of codons, and site-directed diversification using mixtures of defined-sequence oligonucleotides (see the section titled "Site-Directed Diversification" in this chapter), can be used to build libraries that closely follow sequence-level protein design. For example, the design of ankyrin-based scaffold libraries has taken into account which amino-acid residues are compatible with the secondary structure at diversified positions (Binz et al. 2004), and synthetic antibody libraries followed the distribution of amino-acid residues found naturally at specific positions of complementarity-determining regions (Rothe et al. 2008).

In parallel with the advances in library-construction methods, which allow library sequences to be guided by protein design, computational protein modeling has become capable of increasingly reliable predictions of which sequences are likely to improve protein function (Lippow and Tidor 2007; Wong et al. 2007). We anticipate that, as both computational protein modeling and sophisticated library-construction methods become more widely understood and available, an increasing proportion of diversity-based protein-engineering projects will combine these two components to exploit large libraries guided by protein design. Such a combination of computational

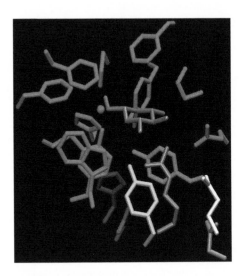

**FIGURE 4.6 (see color insert following page 178)** Active site of galactose oxidase. Galactose (in gray) was modeled into the experimentally determined crystal structure of the enzyme [PDB ID: 1GOF (Ito et al. 1994)] using the program Schrödinger (Glide, version 4.5, and Prime, version 1.6, Schrödinger, LLC, New York, NY, 2007) (Friesner et al. 2004) and displayed using VMD (Humphrey et al. 1996); for clarity, none of the hydrogens are shown. Active site residues 194, 227, 228, 272, 290, 405, 406, 463, 464, 495, 496, 581, and the copper ion are shown in green. Side-chains of six residues that were mutated in a library designed to change the specificity of galactose oxidase to glucose (Table 4.1) are shown in color: Asp 324 (cyan), Gln 326 (white), Tyr 329 (yellow), Arg 330 (orange), Asn 333 (red), and His 334 (blue).

and diversity-based protein engineering will enable a more efficient selection of proteins with new and improved properties. Even more importantly, we expect that it will facilitate the evaluation and further development of protein-modeling methods, which in turn will lead to a deeper understanding of the relationship between protein sequence, structure, and function.

## CONCLUSIONS

We are entering the golden age of directed molecular evolution, with a large choice of ever-improving methods of library construction that are either modeled on natural evolutionary processes or designed to take full advantage of newly available tools in molecular and synthetic biology. Ideally the library-construction method applied to a particular problem should be matched to the known properties of the protein being engineered and to the strengths and weaknesses of the screening or selection method that will be used to identify improved protein variants in the library. The location, extent, and nature of diversification in particular will be driven by the nature of the starting protein and the desired properties of the engineered variants, whereas library size relative to the theoretical sequence space, library quality, and associated thoroughness of sampling will be driven by the throughput and cost of the screening assay or display technology.

Whereas many protein-engineering problems can be solved by a number of alternative combinations of library-construction and screening or selection methods, in other cases the most effective approach is to combine or iterate different approaches. More information is needed on the effect of different library-construction and selection methods on the outcome. With the decrease in cost and effort required to construct and screen or select libraries, more such comparative studies are likely to be performed and published in the near future. A particularly interesting comparison will be between the effectiveness of library designs that rely on extensive randomization and library designs that are guided by high-resolution, structure- or homology-based protein design, which bridge the divide between diversity-based protein engineering and computational protein design.

## ACKNOWLEDGMENTS

We thank Kristala Prather and Tae Seok Moon for allowing us to disclose information on an ongoing collaboration, and Andy Rakestraw for critical reading of the manuscript.

## REFERENCES

Barbas, C. F., D. Hu, N. Dunlop, L. Sawyer, D. Cababa, et al. 1994. "In vitro evolution of a neutralizing human antibody to human immunodeficiency virus type 1 to enhance affinity and broaden strain cross-reactivity." *Proc Natl Acad Sci U S A* 91: 3809–13.

Barbas, C. F., J. D. Bain, D. M. Hoekstra and R. A. Lerner. 1992. "Semisynthetic combinatorial antibody libraries: A chemical solution to the diversity problem." *Proc Natl Acad Sci U S A* 89: 4457–61.

Bessette, P. H., M. A. Mena, A. W. Nguyen and P. S. Daugherty. 2003. "Construction of designed protein libraries using gene assembly mutagenesis." *Methods Mol Biol* 231: 29–37.

Binz, H. K., P. Amstutz, A. Kohl, M. T. Stumpp, C. Briand, et al. 2004. "High-affinity binders selected from designed ankyrin repeat protein libraries." *Nat Biotechnol* 22: 575–82.

Bosley, A. D. and M. Ostermeier. 2005. "Mathematical expressions useful in the construction, description and evaluation of protein libraries." *Biomol Eng* 22: 57–61.

Bradbury, A. R. and J. D. Marks. 2004. "Antibodies from phage antibody libraries." *J Immunol Methods* 290: 29–49.

Braman, J., C. Papworth and A. Greener. 1996. "Site-directed mutagenesis using double-stranded plasmid DNA templates." *Methods Mol Biol* 57: 31–44.

Cadwell, R. C. and G. F. Joyce. 1994. "Mutagenic PCR." *PCR Methods Appl* 3: S136–40.

Carr, P. A., J. S. Park, Y. J. Lee, T. Yu, S. Zhang, et al. 2004. "Protein-mediated error correction for de novo DNA synthesis." *Nucleic Acids Res* 32: e162.

Cirino, P. C., K. M. Mayer and D. Umeno. 2003. "Generating mutant libraries using error-prone PCR." *Methods Mol Biol* 231: 3–9.

Clackson, T. and J. A. Wells. 1995. "A hot spot of binding energy in a hormone-receptor interface." *Science* 267: 383–6.

Cleary, M. A., K. Kilian, Y. Wang, J. Bradshaw, G. Cavet, et al. "Production of complex nucleic acid libraries using highly parallel in situ oligonucleotide synthesis." *Nat Methods* 1: 241–8.

Coco, W. M., W. E. Levinson, M. J. Crist, H. J. Hektor, A. Darzins, et al. 2001. "DNA shuffling method for generating highly recombined genes and evolved enzymes." *Nat Biotechnol* 19: 354–9.

Cox, E. C. 1976. "Bacterial mutator genes and the control of spontaneous mutation." *Annu Rev Genet* 10: 135–56.

Crameri, A., S. A. Raillard, E. Bermudez and W. P. Stemmer. 1998. "DNA shuffling of a family of genes from diverse species accelerates directed evolution." *Nature* 391: 288–91.

Cujec, T. P., P. F. Medeiros, P. Hammond, C. Rise and B. L. Kreider. 2002. "Selection of v-abl tyrosine kinase substrate sequences from randomized peptide and cellular proteomic libraries using mRNA display." *Chem Biol* 9: 253–64.

Cunningham, B. C. and J. A. Wells. 1989. "High-resolution epitope mapping of hGH-receptor interactions by alanine-scanning mutagenesis." *Science* 244: 1081–5.

de Kruif, J., E. Boel and T. Logtenberg. 1995. "Selection and application of human single chain Fv antibody fragments from a semi-synthetic phage antibody display library with designed CDR3 regions." *J Mol Biol* 248: 97–105.

Doudney, C. O. and F. L. Haas. 1959. "Mutation induction and macromolecular synthesis in bacteria." *Proc Natl Acad Sci U S A* 45: 709–22.

Feldhaus, M. J., R. W. Siegel, L. K. Opresko, J. R. Coleman, J. M. Feldhaus, et al. 2003. "Flow-cytometric isolation of human antibodies from a nonimmune *Saccharomyces cerevisiae* surface display library." *Nat Biotechnol* 21: 163–70.

Fellouse, F. A., B. Li, D. M. Compaan, A. A. Peden, S. G. Hymowitz, et al. 2005. "Molecular recognition by a binary code." *J Mol Biol* 348: 1153–62.

Fellouse, F. A., C. Wiesmann and S. S. Sidhu. 2004. "Synthetic antibodies from a four-amino-acid code: A dominant role for tyrosine in antigen recognition." *Proc Natl Acad Sci U S A* 101: 12467–72.

Friesner, R. A., J. L. Banks, R. B. Murphy, T. A. Halgren, J. J. Klicic, et al. 2004. "Glide: A new approach for rapid, accurate docking and scoring. 1. Method and assessment of docking accuracy." *J Med Chem* 47: 1739–49.

Georgescu, R., G. Bandara and L. Sun. 2003. "Saturation mutagenesis." *Methods Mol Biol* 231: 75–83.

Gibbs, M. D., K. M. Nevalainen and P. L. Bergquist. 2001. "Degenerate oligonucleotide gene shuffling (DOGS): A method for enhancing the frequency of recombination with family shuffling." *Gene* 271: 13–20.

Glaser, S. M., D. E. Yelton and W. D. Huse. 1992. "Antibody engineering by codon-based mutagenesis in a filamentous phage vector system." *J Immunol* 149: 3903–13.

Hamamatsu, N., Y. Nomiya, T. Aita, M. Nakajima, Y. Husimi, et al. 2006. "Directed evolution by accumulating tailored mutations: Thermostabilization of lactate oxidase with less trade-off with catalytic activity." *Protein Eng Des Sel* 19: 483–9.

Hanes, J., L. Jermutus, S. Weber-Bornhauser, H. R. Bosshard and A. Pluckthun. 1998. "Ribosome display efficiently selects and evolves high-affinity antibodies in vitro from immune libraries." *Proc Natl Acad Sci U S A* 95: 14130–5.

Hengen, P. N. 1995. "Fidelity of DNA polymerases for PCR." *Trends Biochem Sci* 20: 324–5.

Hiraga, K. and F. H. Arnold. 2003. "General method for sequence-independent site-directed chimeragenesis." *J Mol Biol* 330: 287–96.

Ho, S. N., H. D. Hunt, R. M. Horton, J. K. Pullen and L. R. Pease. 1989. "Site-directed mutagenesis by overlap extension using the polymerase chain reaction." *Gene* 77: 51–9.

Hoet, R. M., E. H. Cohen, R. B. Kent, K. Rookey, S. Schoonbroodt, et al. 2005. "Generation of high-affinity human antibodies by combining donor-derived and synthetic complementarity-determining-region diversity." *Nat Biotechnol* 23: 344–8.

Hoogenboom, H. R. 2005. "Selecting and screening recombinant antibody libraries." *Nat Biotechnol* 23: 1105–16.

Hoogenboom, H. R. and G. Winter. 1992. "By-passing immunisation. Human antibodies from synthetic repertoires of germline VH gene segments rearranged in vitro." *J Mol Biol* 227: 381–8.

Hughes, M. D., D. A. Nagel, A. F. Santos, A. J. Sutherland and A. V. Hine. 2003. "Removing the redundancy from randomised gene libraries." *J Mol Biol* 331: 973–9.

Humphrey, W., A. Dalke and K. Schulten. 1996. "VMD: Visual molecular dynamics." *J Mol Graph* 14: 33–8, 27–8.

Ito, N., S. E. Phillips, K. D. Yadav and P. F. Knowles. 1994. "Crystal structure of a free radical enzyme, galactose oxidase." *J Mol Biol* 238: 794–814.

Jeong, K. J., M. J. Seo, B. L. Iverson and G. Georgiou. 2007. "APEx 2-hybrid, a quantitative protein–protein interaction assay for antibody discovery and engineering." *Proc Natl Acad Sci U S A* 104: 8247–52.

Kayushin, A. L., M. D. Korosteleva, A. I. Miroshnikov, W. Kosch, D. Zubov, et al. 1996. "A convenient approach to the synthesis of trinucleotide phosphoramidites—synthons for the generation of oligonucleotide/peptide libraries." *Nucleic Acids Res* 24: 3748–55.

Knappik, A., L. Ge, A. Honegger, P. Pack, M. Fischer, et al. 2000. "Fully synthetic human combinatorial antibody libraries (HuCAL) based on modular consensus frameworks and CDRs randomized with trinucleotides." *J Mol Biol* 296: 57–86.

Kretz, K. A., T. H. Richardson, K. A. Gray, D. E. Robertson, X. Tan, et al. 2004. "Gene site saturation mutagenesis: A comprehensive mutagenesis approach." *Methods Enzymol* 388: 3–11.

Kuhlman, B. and D. Baker. 2000. "Native protein sequences are close to optimal for their structures." *Proc Natl Acad Sci U S A* 97: 10383–8.

Lafferty, M. and M. J. Dycaico. 2004. "GigaMatrix: A novel ultrahigh throughput protein optimization and discovery platform." *Methods Enzymol* 388: 119–34.

Lahr, S. J., A. Broadwater, C. W. Carter, Jr., M. L. Collier, L. Hensley, et al. 1999. "Patterned library analysis: A method for the quantitative assessment of hypotheses concerning the determinants of protein structure." *Proc Natl Acad Sci U S A* 96: 14860–5.

Leung, D. W., E. Chen and D. V. Goeddel. 1989. "A method for random mutagenesis of a defined DNA segment using a modified polymerase chain reaction." *Techniques* 1: 11–15.

Lippow, S. M. and B. Tidor. 2007. "Progress in computational protein design." *Curr Opin Biotechnol* 18: 305–11.

Low, N. M., P. H. Holliger and G. Winter. 1996. "Mimicking somatic hypermutation: Affinity maturation of antibodies displayed on bacteriophage using a bacterial mutator strain." *J Mol Biol* 260: 359–68.

Lutz, S., M. Ostermeier and S. J. Benkovic. 2001a. "Rapid generation of incremental truncation libraries for protein engineering using alpha-phosphothioate nucleotides." *Nucleic Acids Res* 29: E16.

Lutz, S., M. Ostermeier, G. L. Moore, C. D. Maranas and S. J. Benkovic. 2001b. "Creating multiple-crossover DNA libraries independent of sequence identity." *Proc Natl Acad Sci U S A* 98: 11248–53.

Marks, J. D. and A. Bradbury. 2004. "PCR cloning of human immunoglobulin genes." *Methods Mol Biol* 248: 117–34.

Marks, J. D., H. R. Hoogenboom, T. P. Bonnert, J. McCafferty, A. D. Griffiths, et al. 1991. "By-passing immunization. Human antibodies from V-gene libraries displayed on phage." *J Mol Biol* 222: 581–97.

Mathur, E. J., G. Toledo, B. D. Green, M. Podar, T. H. Richardson, et al. 2005. "A biodiversity-based approach to development of performance enzymes." *Industrial Biotechnology* 1: 283–287.

Meiler, J. and D. Baker. 2006. "ROSETTALIGAND: Protein-small molecule docking with full side-chain flexibility." *Proteins* 65: 538–48.

Mena, M. A. and P. S. Daugherty. 2005. "Automated design of degenerate codon libraries." *Protein Eng Des Sel* 18: 559–61.

Miyazaki, K. 2003. "Creating random mutagenesis libraries by megaprimer PCR of whole plasmid (MEGAWHOP)." *Methods Mol Biol* 231: 23–8.

Munoz, E. and M. W. Deem. 2008. "Amino acid alphabet size in protein evolution experiments: Better to search a small library thoroughly or a large library sparsely?" *Protein Eng Des Sel* 21: 311–7.

Myers, R. M., L. S. Lerman and T. Maniatis. 1985. "A general method for saturation mutagenesis of cloned DNA fragments." *Science* 229: 242–7.

Ness, J. E., S. Kim, A. Gottman, R. Pak, A. Krebber, et al. 2002. "Synthetic shuffling expands functional protein diversity by allowing amino acids to recombine independently." *Nat Biotechnol* 20: 1251–5.

Nguyen, A. W. and P. S. Daugherty. 2003. "Production of randomly mutated plasmid libraries using mutator strains." *Methods Mol Biol* 231: 39–44.

Ong, T. M. and F. J. De Serres. 1972. "Mutagenicity of chemical carcinogens in *Neurospora crassa*." *Cancer Res* 32: 1890–3.

Ostermeier, M. and S. Lutz. 2003. "The creation of ITCHY hybrid protein libraries." *Methods Mol Biol* 231: 129–41.

Otey, C. R., J. J. Silberg, C. A. Voigt, J. B. Endelman, G. Bandara, et al. 2004. "Functional evolution and structural conservation in chimeric cytochromes p450: Calibrating a structure-guided approach." *Chem Biol* 11: 309–18.

Pini, A., F. Viti, A. Santucci, B. Carnemolla, L. Zardi, et al. 1998. "Design and use of a phage display library. Human antibodies with subnanomolar affinity against a marker of angiogenesis eluted from a two-dimensional gel." *J Biol Chem* 273: 21769–76.

Pirrung, M. C. 2002. "How to make a DNA chip." *Angew Chem Int Ed* 41: 1276–1289.

Rajpal, A., N. Beyaz, L. Haber, G. Cappuccilli, H. Yee, et al. 2005. "A general method for greatly improving the affinity of antibodies by using combinatorial libraries." *Proc Natl Acad Sci U S A* 102: 8466–71.

Reetz, M. T., M. Bocola, J. D. Carballeira, D. Zha and A. Vogel. 2005. "Expanding the range of substrate acceptance of enzymes: Combinatorial active-site saturation test." *Angew Chem Int Ed Engl* 44: 4192–6.

Reetz, M. T., D. Kahakeaw and R. Lohmer. 2008. "Addressing the numbers problem in directed evolution." *Chembiochem* 9: 1797–804.

Richmond, K. E., M. H. Li, M. J. Rodesch, M. Patel, A. M. Lowe, et al. 2004. "Amplification and assembly of chip-eluted DNA (AACED): A method for high-throughput gene synthesis." *Nucleic Acids Res* 32: 5011–8.

Roskos, L., S. Klakamp, M. Lilang, R. Arends and L. Green. 2007. "Molecular engineering II: Antibody affinity." In: *Handbook of therapeutic antibodies.* S. Duebel. Weinheim, Wiley-VCH: 145–69.

Rothe, C., S. Urlinger, C. Lohning, J. Prassler, Y. Stark, et al. 2008. "The human combinatorial antibody library HuCAL GOLD combines diversification of all six CDRs according to the natural immune system with a novel display method for efficient selection of high-affinity antibodies." *J Mol Biol* 376: 1182–200.

Seelig, B. and J. W. Szostak. 2007. "Selection and evolution of enzymes from a partially randomized non-catalytic scaffold." *Nature* 448: 828–31.

Sheets, M. D., P. Amersdorfer, R. Finnern, P. Sargent, E. Lindquist, et al. 1998. "Efficient construction of a large nonimmune phage antibody library: The production of high-affinity human single-chain antibodies to protein antigens." *Proc Natl Acad Sci U S A* 95: 6157–62.

Silverman, J., Q. Liu, A. Bakker, W. To, A. Duguay, et al. 2005. "Multivalent avimer proteins evolved by exon shuffling of a family of human receptor domains." *Nat Biotechnol* 23: 1556–61.

Singh-Gasson, S., R. D. Green, Y. Yue, C. Nelson, F. Blattner, et al. 1999. "Maskless fabrication of light-directed oligonucleotide microarrays using a digital micromirror array." *Nat Biotechnol* 17: 974–8.

Soderlind, E., L. Strandberg, P. Jirholt, N. Kobayashi, V. Alexeiva, et al. 2000. "Recombining germline-derived CDR sequences for creating diverse single-framework antibody libraries." *Nat Biotechnol* 18: 852–6.

Steffens, D. L. and J. G. Williams. 2007. "Efficient site-directed saturation mutagenesis using degenerate oligonucleotides." *J Biomol Tech* 18: 147–9.

Stemmer, W. P. 1994a. "DNA shuffling by random fragmentation and reassembly: In vitro recombination for molecular evolution." *Proc Natl Acad Sci U S A* 91: 10747–51.

Stemmer, W. P. 1994b. "Rapid evolution of a protein in vitro by DNA shuffling." *Nature* 370: 389–91.

Stemmer, W. P., A. Crameri, K. D. Ha, T. M. Brennan and H. L. Heyneker. 1995. "Single-step assembly of a gene and entire plasmid from large numbers of oligodeoxyribonucleotides." *Gene* 164: 49–53.

Takase, K., S. Taguchi and Y. Doi. 2003. "Enhanced synthesis of poly(3-hydroxybutyrate) in recombinant *Escherichia coli* by means of error-prone PCR mutagenesis, saturation mutagenesis, and in vitro recombination of the type II polyhydroxyalkanoate synthase gene." *J Biochem* 133: 139–45.

Tian, J., H. Gong, N. Sheng, X. Zhou, E. Gulari, et al. 2004. "Accurate multiplex gene synthesis from programmable DNA microchips." *Nature* 432: 1050–4.

Van den Brulle, J., M. Fischer, T. Langmann, G. Horn, T. Waldmann, et al. 2008. "A novel solid phase technology for high-throughput gene synthesis." *Biotechniques* 45: 340–3.

Vaughan, T. J., A. J. Williams, K. Pritchard, J. K. Osbourn, A. R. Pope, et al. 1996. "Human antibodies with sub-nanomolar affinities isolated from a large non-immunized phage display library." *Nat Biotechnol* 14: 309–14.

Virnekas, B., L. Ge, A. Pluckthun, K. C. Schneider, G. Wellnhofer, et al. 1994. "Trinucleotide phosphoramidites: Ideal reagents for the synthesis of mixed oligonucleotides for random mutagenesis." *Nucleic Acids Res* 22: 5600–7.

Weiner, M. P., G. L. Costa, W. Schoettlin, J. Cline, E. Mathur, et al. 1994. "Site-directed mutagenesis of double-stranded DNA by the polymerase chain reaction." *Gene* 151: 119–23.

Weiss, G. A., C. K. Watanabe, A. Zhong, A. Goddard and S. S. Sidhu. 2000. "Rapid mapping of protein functional epitopes by combinatorial alanine scanning." *Proc Natl Acad Sci U S A* 97: 8950–4.

Wong, T. S., D. Roccatano and U. Schwaneberg. 2007. "Steering directed protein evolution: Strategies to manage combinatorial complexity of mutant libraries." *Environ Microbiol* 9: 2645–59.

Wong, T. S., D. Zhurina and U. Schwaneberg. 2006. "The diversity challenge in directed protein evolution." *Comb Chem High Throughput Screen* 9: 271–88.

Xiong, A. S., R. H. Peng, J. Zhuang, J. G. Liu, F. Gao, et al. 2008. "Non-polymerase-cycling-assembly-based chemical gene synthesis: strategies, methods, and progress." *Biotechnol Adv* 26: 121–34.

Xu, L., P. Aha, K. Gu, R. G. Kuimelis, M. Kurz, et al. 2002. "Directed evolution of high-affinity antibody mimics using mRNA display." *Chem Biol* 9: 933–42.

Yanez, J., M. Arguello, J. Osuna, X. Soberon and P. Gaytan. 2004. "Combinatorial codon-based amino acid substitutions." *Nucleic Acids Res* 32: e158.

Zaccolo, M. and E. Gherardi. 1999. "The effect of high-frequency random mutagenesis on in vitro protein evolution: A study on TEM-1 beta-lactamase." *J Mol Biol* 285: 775–83.

Zaccolo, M., D. M. Williams, D. M. Brown and E. Gherardi. 1996. "An approach to random mutagenesis of DNA using mixtures of triphosphate derivatives of nucleoside analogues." *J Mol Biol* 255: 589–603.

Zha, D., A. Eipper and M. T. Reetz. 2003. "Assembly of designed oligonucleotides as an efficient method for gene recombination: A new tool in directed evolution." *Chembiochem* 4: 34–9.

Zhao, H., L. Giver, Z. Shao, J. A. Affholter and F. H. Arnold. 1998. "Molecular evolution by staggered extension process (StEP) in vitro recombination." *Nat Biotechnol* 16: 258–61.

Zheng, L., U. Baumann and J. L. Reymond. 2004. "An efficient one-step site-directed and site-saturation mutagenesis protocol." *Nucleic Acids Res* 32: e115.

Zhou, X., S. Cai, A. Hong, Q. You, P. Yu, et al. 2004. "Microfluidic PicoArray synthesis of oligodeoxynucleotides and simultaneous assembling of multiple DNA sequences." *Nucleic Acids Res* 32: 5409–17.

# 5 Design and Engineering of Synthetic Binding Proteins Using Nonantibody Scaffolds

*Shohei Koide*

## CONTENTS

Highly specific molecular recognition is one of the fundamental functions of proteins. In an effort to explore and exploit this functionality, intense efforts have been concentrated on the development of molecular systems that consistently produce novel proteins for molecular recognition. It is important to note that the goal of these efforts is to design molecular systems for diverse applications, and thus they should be clearly distinguished from traditional research and development efforts for a single, specific function. The use of the term *design* in the title of this chapter is deliberate with a purpose to emphasize this viewpoint. While traditional approaches have relied on natural systems, such as molecules involved in adaptive immunity as the source of interface diversity, recent advances in protein engineering technologies and also in our understanding of the principles governing molecular recognition by proteins have made it possible to create novel molecular recognition sites on other proteins.

## THE *MOLECULAR SCAFFOLD* CONCEPT

The idea of constructing a new molecular recognition function through the engineering of a section of protein surface (i.e., a patch; Figure 5.1A) originates from the molecular architecture of immunoglobulins (Figure 5.1B). The immune system can produce antibodies against virtually any type of antigen by, to a first-order approximation, simply tuning the amino-acid sequences of a total of six short segments termed complementarity determining regions (CDRs). CDRs are surface-exposed loops located at one end of the immunoglobulin structure (Figure 5.1B). They collectively form a contiguous patch for molecular recognition. Sequence analysis revealed the presence of hypervariable regions that are primarily responsible for antigen binding (Wu et al. 1993). From these observations, the concept of *molecular scaffold* has been developed. This term usually refers to a protein framework that is essentially invariant and provides positions that can accommodate extensive sequence variations (substitution, insertion, and deletion) for the purpose of generating desired functions (Figure 5.1A). In the antibody example, the protein outside the CDRs can be considered as molecular scaffold that supports the CDRs, although it is well known that the surfaces outside the antigen-binding site are involved in many critical biological functions.

While the manifestation of the molecular scaffold concept is particularly evident in the immunoglobulins, it should be emphasized that this concept is perhaps a unifying theme in protein architecture. There are a number of protein folds, such as the triosephosphateisomerase (TIM) barrel and oligonucleotide/oligosaccharide binding (OB) fold (Nagano et al. 2002; Agrawal and Kishan 2003), that are used for recognition of a wide variety of ligands. These examples have established the "one fold-many functions" paradigm (Nagano et al. 2002). Small interaction domains are present in many eukaryotic proteins, and they are engaged in diverse molecular recognition functions. Comparisons of these domains uncover a set of positions that play major roles in target recognition (Figure 5.1C). While these proteins are not involved in adaptive immunity, their folds are clearly capable of supporting distinct amino-acid motifs for diverse functions, and thus conceptually they are molecular scaffolds

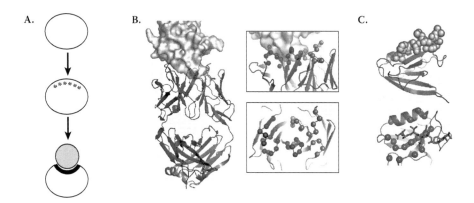

**FIGURE 5.1** Examples of natural protein–ligand interactions that illustrate the molecular scaffold concept. (A) A schematic representation of synthetic interface engineering. The oval at the top indicates an inert scaffold. The asterisks denote sequence diversity. The bottom is a binder-target complex with the optimized interface shown in black. (B) Antibody-antigen interaction (PDB: 1TZI) (Fellouse et al. 2004). The Fab is shown as a cartoon model with the heavy chain shown in darker gray. Only a portion of the antigen is shown as a surface model. The antigen-antibody interaction interface is shown expanded on the right, where the Cα positions of the CDR residues are marked with spheres. (C) PDZ domain-peptide interaction (PDB: 1MFG) (Birrane et al. 2003). The PDZ domain is shown as a cartoon model and the peptide is shown as spheres. The bottom figure shows the same complex in an approximately perpendicular view, with the spheres denoting the Cα positions of residues that are found to be important for ligand specificity (Chen et al. 2008). The molecular graphics were made with PyMOL (www.pymol.org).

just like immunoglobulins. The only difference is that, in these cases, new functions emerge through natural selection over a long time period, whereas the immune system has the capacity to perform directed evolution within days. Both of these examples demonstrate that natural proteins have the potential to evolve surfaces into distinct molecular recognition interfaces. Therefore, with suitable technologies, one should be able to create a series of new functions by evolving appropriate positions on a molecular scaffold.

Traditionally, molecular scaffolds have been classified into antibody fragments and nonantibody, "alternative" scaffolds. This distinction is based on the premise that only the antibody scaffolds (including T-cell receptors that have essentially the same binding site architecture) are used in nature to build a system that consistently produces highly functional recognition interfaces. Recent studies do not support such a narrow view. Perhaps the most illustrative case in point is the development of repeat proteins (e.g., leucine-rich repeats and ankyrin repeat proteins; see the next section) as "alternative" scaffolds by Plückthun and colleagues (Forrer et al. 2003), which was followed by the discovery that leucine-rich repeats are used as the scaffold for adaptive immunity in the jawless fish (Pancer et al. 2004). Clearly, natural evolution is opportunistic and it is possible that other classes of protein could be capable scaffolds for supporting adaptive immunity.

## MOTIVATIONS FOR THE DEVELOPMENT
## OF NONANTIBODY SCAFFOLDS

The development of systems to produce novel recognition functions using nonantibody scaffolds has been driven by several motivations. First, the physicochemical properties of natural antibodies are not ideal in many applications in research, biotechnology, and therapy. Immunoglobulins are large molecules (~150 kDa) made of two copies each of the heavy and light chains. Their stability is dependent on intradomain and interchain disulfide bonds and thus they do not fold under reducing conditions. Their production is cumbersome and expensive. The heterodimeric structure makes reformatting for downstream applications difficult. Smaller fragments of immunoglobulins (e.g., Fab and single-chain Fv) have been used for antibody-engineering applications, but they inherit many of these limitations (Worn and Plückthun 2001). The hope is that these limitations can be overcome by switching to synthetic binders built on a simpler scaffold. Second, from a basic science point of view, the exercise of engineering synthetic interfaces critically tests and enhances our understanding of the principles governing molecular recognition and explores the capacity of protein engineering technology. Third, the intellectual property landscape associated with recombinant antibody engineering is so complex that it has become practically impossible to sort out the situation. This cause is not intellectually stimulating, but nevertheless critically important in the industry. From a commercial point of view, an "alternative" scaffold system with little prior intellectual property would be highly attractive.

Although the concept of creating a synthetic recognition interface on a scaffold is simple, it is technically daunting, because one essentially needs to recapitulate a directed evolution system that rivals that used by the adaptive immunity. Such a system minimally consists of a "library" containing large sequence diversity and a method to isolate members with a desired property. Fundamental technological breakthroughs that have established the field of synthetic interface engineering are collectively known as *molecular display technologies*. These include phage display, mRNA display, yeast surface display, and yeast two-hybrid, which all establish unambiguous linkage between protein phenotype and genotype. Such phenotype-genotype linkage is critically important for library selection and directed evolution, because one ultimately needs to isolate (i.e., clone) and determine the identity of proteins that have been selected. Equally important, they enable the generation of large combinatorial libraries whose size rivals or exceeds natural immune diversity. Molecular display technologies and library construction methods are described in detail in Chapters 1–4.

## FACTORS GOVERNING PROTEIN–PROTEIN INTERACTIONS

Because synthetic proteins produced using these molecular scaffolds are still proteins made of natural amino acids and they obey the same set of rules that govern natural protein–ligand interactions, it is useful to briefly review the current understanding of recognition interfaces in natural protein systems. Structural studies of many protein–ligand complexes have repeatedly shown that proteins use all types of structural motifs (i.e., loop, turn, helix, and sheet) to interact with other molecules. The underlying principle in this apparent chaos is that proteins accomplish specific recognition of their cognate targets by two types of complementarity: shape and electrostatics (i.e., charge-charge interactions and hydrogen bonds). In the case of high-affinity protein–protein interactions, the interface is contiguous and quite large with an average buried surface of ~1600 $\text{Å}^2$ (Lo Conte et al. 1999). Such interfaces are also generally tightly packed (Lawrence and Colman 1993). Systematic mutagenesis studies of high affinity interfaces have revealed two general trends. Binding interfaces tend to have "hot-spots" and outside the hot-spots the interfaces are remarkably plastic to amino-acid substitution (DeLano 2002; Reichmann et al. 2007). This trend results in an "O-ring" architecture in which a few hot-spot residues are surrounded by energetically inert residues that were proposed to act as an O-ring to shield hot-spot residues from water molecules (Bogan and Thorn 1998). Overall, studies of natural proteins suggest that in order to produce highly functional recognition interfaces, one needs to achieve (1) shape complementarity, (2) chemical complementarity, and (3) the O-ring architecture in the interface. This is a rather intimidating set of requirements for a protein engineer to satisfy.

While the first two requirements remain valid, recent studies have proven that the O-ring architecture is not a fundamental prerequisite. An "Ala-shave" analysis of the human growth hormone and its receptor showed that over half of the interface residues could be *simultaneously* changed to alanine while still retaining high affinity as long as most of the hot-spot residues are maintained (Kouadio et al. 2005). Complementary to this study, affinity maturation of a small interface (~600 $\text{Å}^2$ per molecule) in a camelid single-domain antibody ($V_HH$)/antigen complex by exhaustive sequence sampling resulted in hot-spot residues constituting 76% of the surface and complete elimination of the O-ring architecture (Koide et al. 2007b). Thus, these recent "semi-synthetic" studies strongly suggest that a highly functional interface can be minimized essentially to a cluster of hot-spot residues. This notion in turn suggests that a scaffold system that can produce sufficient shape diversity and chemical diversity should be able to recognize diverse target molecules.

## DESIGN CONSIDERATIONS FOR NEW SCAFFOLDS

To date, all scaffolds have been derived from natural proteins. In principle, any protein that can tolerate extensive surface mutations can potentially be used as a molecular scaffold. However, as structural classifications of proteins have shown that a small number of protein folds are used in many proteins with diverse functions, not all protein architecture classes are equally suited as molecular scaffolds.

What are the requirements for a highly functional molecular scaffold? As outlined in the previous section, one of the main motivations in the development of new scaffolds is to avoid the issues associated with undesirable physical properties of the immunoglobulin scaffolds. These include the presence of multiple disulfide bonds and a heterodimeric architecture. Clearly, a scaffold must also have multiple surface positions that can be mutated to construct a contiguous interface. Extensive mutations frequently reduce the conformational stability of a protein. Decreased stability due to introduced mutations can also promote transient exposure of secondary structure elements (e.g., helix or strand) and the hydrophobic interior, which can lead to protein aggregation and domain swapping (Liu and Eisenberg 2002). For these reasons, it is advantageous to start with a highly stable protein as a molecular scaffold. Depending on the intended application, different types of stability, such as thermal stability and resistance to extreme pH or organic solvent, might be important. It is often possible to improve the conformational stability of a protein using rational design, library selection, or computational design (Malakauskas and Mayo 1998; Sieber et al. 1998; Perl et al. 2000; Koide et al. 2001). Therefore, a high degree of stability in a starting scaffold is desirable but not absolutely required.

A scaffold should be small (less than ~100 residues) for the ease of manipulation, production, and potential access by chemical synthesis, and simple in terms of its functional structure, meaning it should not contain prosthetic groups or disulfide bonds. The absence of disulfide bonds is particularly important if intended uses include intracellular expression. The scaffold should also be easily produced in bacteria and highly soluble. Solubility in particular is an important point to consider for downstream applications. Not unexpectedly, introducing amino-acid residues that promote new molecular recognition often reduces the solubility of the resulting binders (Bianchi et al. 1994). These issues often need to be dealt with empirically because it is usually difficult to pinpoint the cause of protein precipitation and aggregation. Furthermore, a scaffold should be compatible with common molecular display techniques, and it should have minimal spectroscopic signatures because strong inherent signals limit flexibility in derivatization (e.g., fluorescence labels). Several systems that meet these stringent requirements have been successfully developed and are reviewed in the next section.

## SUCCESSFUL SCAFFOLDS

Over the years, attempts have been made to use virtually all classes of protein architecture as scaffolds for constructing synthetic interfaces. While loops are the most common location into which amino-acid diversity is introduced, surface-exposed residues on α-helices or β-strands have also successfully been used. The examples presented in this section demonstrate that a synthetic interface can be engineered as long as a contiguous surface (patch) of a substantial size (≥ ~15 residues) is available. However, this does not mean that all scaffolds possess equal potential. During the "scaffold bubble" early in this decade, many new scaffolds were reported and then disappeared without making a significant impact.

Representative examples of alternative scaffolds are described in this section. Because a number of recent reviews have provided extensive coverage of molecular scaffolds (Binz et al. 2004; Binz et al. 2005; Binz and Plückthun 2005; Hosse et al. 2006; Skerra 2007), only select examples of each architectural class are discussed. The reader is referred to a comprehensive table in Binz and Plückthun (2005) that summarizes properties of alternative molecular scaffolds. An important factor in choosing these examples was the availability of three-dimensional (3D) structures of the synthetic interfaces formed with a particular scaffold. The ability to determine the 3D structure of the resulting synthetic binder is a rigorous test for good biophysical properties, providing ultimate validation of the design. This does not imply, however, that these are the best molecules in each class; they may be superceded in the future.

## IMMUNOGLOBULIN-LIKE β-SANDWICH

Immunoglobulin domains have a β-sandwich architecture consisting of two β-sheets, and their CDRs structurally correspond to loops connecting β-strands at one end of the molecule. The antigen-binding site of standard immunoglobulins is constructed from surfaces of two immunoglobulin domains ($V_H$ and $V_L$). In contrast, the antigen-binding domain of antibodies from the camelid and shark consists of a single immunoglobulin heavy-chain domain, whose high functionality indicates the potential of a single β-sandwich domain as a scaffold (Hamers-Casterman et al. 1993; Stanfield et al. 2004). Thus, it is not surprising that β-sandwich proteins were among the first nonantibody scaffolds used (Pessi et al. 1993). In this class, the fibronectin type III domain (FN3; Figure 5.2A) is the most widely used scaffold today (Koide et al. 1998; Xu et al. 2002; Richards et al. 2003; Karatan et al.

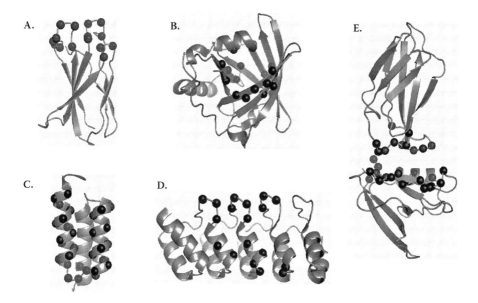

**FIGURE 5.2** The three-dimensional structures of representative scaffold architectures. Only those scaffolds for which crystal structures of synthetic binding interfaces in complex with a target have been determined are shown. Their respective target molecules are not shown. The Cα positions for residues within 5 Å of the target are shown as spheres. (A) FN3 binding to maltose-binding protein (MBP) (PBD: 2OBG) (Koide et al. 2007a). (B) Lipocalin binding to a small organic molecule, digoxigenin (PDB: 1LNM) (Korndorfer et al. 2003). (C) Affibody binding to another affibody (PDB: 2B87) (Lendel et al. 2006). (D) Ankyrin repeat protein binding to MBP (PDB: 1SVX) (Binz et al. 2004). (E) Affinity clamp constructed with PDZ and FN3 domains binding to a short peptide (PDB: 2QBW) (Huang et al. 2008). The affibody structure was determined by NMR spectroscopy, and the others by x-ray crystallography.

2004; Lipovsek et al. 2007). FN3 has a clear structural homology to the Ig domain, and FN3 family members are involved in many types of molecular recognition (Campbell and Spitzfaden 1994). However, unlike immunoglobulin domains, FN3 does not rely on an intradomain disulfide, and thus it combines the demonstrated potency of Ig scaffolds with simple production and the possibility of intracellular applications (Koide et al. 2002).

## β-Barrel

The β-barrel architecture is similar to the immunoglobulin-like β-sandwich in that it presents multiple loops at one end of the molecule. The anticalin scaffold has been developed from lipocalins, a ubiquitous protein family that is usually involved in binding to small hydrophobic compounds (Figure 5.2B). High affinity synthetic interfaces have successfully been developed against small molecules as well as for protein targets from libraries in which four anticalin loops are diversified (Beste et al. 1999; Schlehuber and Skerra 2005).

## β-Sheet Surface

Residues on the β-sheet surface, as opposed to loops connecting β-strands in the aforementioned classes, have also been used to generate synthetic recognition interfaces. A four-stranded β-sheet in a 56-residue hyperthermophilic IgG-binding protein was used to present diversified positions and binding proteins to beta-amyloid fibrils were generated (Smith et al. 2006).

## Single Loop

This class represents the simplest form of a scaffold and involves the presentation of a single contiguous peptide segment. Historically, this class has been derived from two approaches. Disulfide-constrained peptides, which were among the earliest types of phage-displayed peptide libraries (Sidhu et al. 2000), represent the first approach. Numerous functional peptides have been generated from this type of peptide library. Interestingly, functional peptides generated from these libraries are often highly structured. Data on this class of functional peptides suggest that although they are only short segments, a portion of these peptides appears to serve as a scaffold to present the remaining functional groups in a defined geometry.

The second approach has been derived from the grafting of peptide segments onto full protein scaffolds. The Kunitz domain, for example, is a small (~60 residues), disulfide-stabilized protein that is common among natural protease inhibitors. This scaffold has been successfully used to develop novel inhibitors for proteases (Roberts et al. 1992; Dennis and Lazarus 1994). The amylase inhibitor tendamistat, another disulfide-stabilized scaffold, was used to present integrin-binding segments (Li et al. 2003). The knottin family of proteins are ~30–50 residues long and contain a topological cystine-knot motif, a specific pattern of three disulfide bonds. Knottins have also been used as a scaffold for grafted peptide segments (Krause et al. 2007;

Silverman et al. 2009). In another example, thioredoxin (108 residues) has been used as a scaffold for grafted peptide segments, using an *in vivo* yeast two-hybrid approach to produce molecules termed *peptide aptamers* (Colas et al. 1996). Peptide aptamers have been used primarily for perturbing target function in cellular contexts (Geyer et al. 1999). However, little structural information has been obtained for this class of synthetic binding proteins.

## α-HELICAL SCAFFOLDS

The previous examples utilize the plasticity of loops connecting β-strands for a recognition interface. In contrast, the recognition interface of affibodies is constructed from residues that reside on two α-helices of an antibody-binding domain of *Staphylococcus aureus* protein A (Figure 5.2C) (Nord et al. 1997). Affibodies that bind to many targets have been developed (Todd et al. 2002), demonstrating the feasibility of employing residues held on a comparatively rigid scaffold.

## REPEAT PROTEINS

Repeat proteins are characterized by the presence of copies of a short motif (repeats) that together form a rigid structure. Repeat proteins such as ankyrin- and leucine-rich repeats, are ubiquitous in nature and are usually involved in molecular recognition. Designed ankyrin repeats (DARPins; Figure 5.2D) have been developed using a consensus repeat sequence that has been optimized for stability (Binz et al. 2004). Libraries are constructed where residues in a loop and those on a helix are diversified in each repeat and multiple repeats are concatenated. An interesting feature of DARPins is that the number of repeats can be adjusted according to the size of a target molecule to be recognized, although most work to date has been done with three repeats (Binz et al. 2004; Kawe et al. 2006; Schweizer et al. 2007; Zahnd et al. 2007).

## MULTIDOMAIN SCAFFOLDS

The scaffolds described thus far are a single entity presenting one contiguous interface. Several scaffolds utilize multiple copies of discrete domains that allow for additive effects (avidity) from multiple interfaces recognizing distinct epitopes within a single target molecule. This mode of interaction is common in natural DNA-binding proteins. For example, the zinc finger domain, a small zinc-stabilized globular domain that presents a recognition helix, is often found in multiple copies in tandem, with each copy recognizing different surfaces on the target DNA. Tandem zinc fingers have been used to engineer proteins with novel DNA-binding specificity (Choo and Klug 1994; Jamieson et al. 1994; Rebar and Pabo 1994; Beerli and Barbas 2002). Avimers are tandem repeats of small (35 amino acids), disulfide-stabilized A-domains derived from the low-density lipoprotein receptor. Avimers consisting of up to seven A-domain units have been constructed, and high affinity binders were obtained for several targets using two- and three-repeat avimers (Silverman et al. 2005). While the units used in these examples function as small building blocks of a multidomain scaffold without making the total scaffold size large and unwieldy, the concept of expanding interface areas by concatenation could potentially be applied to any scaffold.

### RECOGNITION INTERFACES AT THE JUNCTION OF TWO DOMAINS

All the scaffolds described in the previous section present a highly exposed recognition interface, that is, the type of interaction interfaces seen in antibodies. However, many natural proteins, particularly enzymes, have a cleft-like recognition site located at the interface between two domains. Inspired by this type of interface topography (which is distinctly different from that of an antibody), we have recently developed a new class of scaffolds by connecting two domains and constructing a recognition interface between these two domains (Huang et al. 2008). In our proof-of-concept experiments, we used a PDZ domain that binds to the C-terminal extreme of proteins as one-half of the scaffold and the FN3 scaffold described previously as the other half. In order to produce the intended orientation between the two domains, the termini of the PDZ domain were relocated by circular permutation. The recognition loops of the FN3 scaffold were then diversified to generate a phage-display library and clones were selected that had >1000-fold higher affinity and specificity than the original PDZ domain. These "affinity clamps" have a clamshell architecture (Figure 5.2E) that enables extremely high affinity and specificity toward short, flexible peptides, a class of antigens that has been difficult to target using the traditional rigid scaffolds.

## ROLES OF INTERFACE TOPOGRAPHY

Studies of natural interfaces have suggested that the interface topography is a major determinant of protein–protein interaction. The topography of the intended paratope (the interface on the antibody/binder side that interacts with the epitope) is strongly influenced by the choice of scaffold surfaces to be used, and it may be a greater determinant of scaffold capacity than the secondary structure class. The overall shape of a paratope can be convex, flat, or concave. Furthermore, the convex and concave surfaces can have different levels of protrusion and depth, respectively. Clearly, different scaffolds are predisposed to different paratope shapes.

A comparison of two synthetic binders to a common target protein has provided direct experimental support for this view. An FN3-based binder targeted a deep cleft of maltose-binding protein (MBP) (Figure 5.3A,B), while a DARPin targeted MBP's slightly convex surface (Figure 5.3C) (Binz et al. 2004; Koide et al. 2007a; Gilbreth et al. 2008). The paratope of the former has a convex surface, and the latter concave. Thus, although the role of interface amino-acid composition may also contribute to this difference in epitope, these results strongly suggest the importance of paratope shape on epitope selection and the utility of distinct scaffolds in producing diverse binding modes.

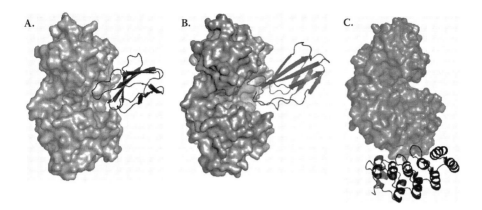

**FIGURE 5.3** Paratope topography influences epitope selection. The three-dimensional structures of synthetic binders made with distinct scaffolds, FN3 (A and B) and DARPin (C) bound to a common target, MBP, are shown. Two FN3-based binders bind to the concave surface of MBP's active site (Koide et al. 2007a; Gilbreth et al. 2008), while the DARPin binds to a convex surface (Binz et al. 2004).

## ADVANCES IN DISPLAY TECHNOLOGIES

An important aspect of successful interface design and engineering is the ability to generate and test a large number of amino-acid sequence diversities. Thus, molecular display technologies are an indispensable tool in directed evolution-oriented interface engineering. Phage display (Chapter 1) has been the most commonly used display technology in this field. However, other methods such as yeast display (Chapter 2), and ribosome display and mRNA display (Chapter 3) have been successfully used. In general, nonantibody scaffolds developed for synthetic binders are small, stable, and devoid of disulfide bonds, and thus they are compatible with a broader range of display technologies than are antibody fragments. For example, FN3-based binders have been generated using virtually all display technologies available (Koide et al. 1998; Koide et al. 2002; Xu et al. 2002; Koide et al. 2007a; Lipovsek et al. 2007; Garcia-Ibilcieta et al. 2008).

A larger starting library can sample a larger sequence space and should in principle provide a greater chance of producing a binder if the same selection or screening procedures are used. In practice, however, most experiments are performed with the largest library size that an investigator can produce with a particular display technology, and it is difficult to determine whether success or failure of an attempt can be attributed solely to the library size. For example, although it is difficult to produce a library greater than $10^8$ using yeast surface display (a level that is potentially five orders of magnitude smaller than those constructed with ribosome display or mRNA display), highly functional interfaces have been engineered using this technique (Lipovsek et al. 2007). The quality of a library is equally important for successful generation of functional molecules. A larger library containing a large fraction of "junk" (e.g., stop codons and frameshifts) may be outperformed by a much smaller one in which most library members are functional molecules. It can also be useful to combine different display methods, for example, phage display and yeast display, to exploit their complementary strengths (Koide et al. 2007a; Koide et al. 2007b).

In the conventional filamentous phage display systems, proteins are fused to phage coat proteins that are secreted into the periplasm through the SecB-mediated posttranslational secretion pathway. In this secretion mechanism, fully translated proteins are threaded through the membrane in an unfolded state (Sidhu and Koide 2007). While this mechanism is suitable for proteins that are marginally stable in the bacterial cytoplasm, such as antibody fragments, it presents a major bottleneck for the display of stable proteins that fold rapidly in the cytoplasm (O'Neil et al. 1995). Because nonantibody scaffolds often fall in the latter category, this bottleneck is potentially a serious problem.

This secretion bottleneck has been overcome by use of the cotranslational secretion pathway mediated by the signal recognition particle (SRP) (Steiner et al. 2006). Use of the SRP-dependent pathway increased display levels of highly stable proteins, such as DARPins, as much as 700-fold relative to the Sec pathway. The SRP-dependent phage display system was also effective in increasing the display level of another highly stable scaffold, FN3 (Koide et al. 2007a). Thus, the use of SRP-dependent systems considerably broadens the applicability of phage display. An interesting possibility is that a highly promising scaffold may fail because of a low level of phage display caused by its high stability and rapid folding. This introduces a caveat to high scaffold stability, which, as indicated earlier, is an otherwise favorable property of a molecular scaffold.

Compared with phage display, relatively small numbers of scaffold systems have been tested with other display technologies such as yeast surface display and mRNA/ribosome display. Thus, it is difficult to determine pros and cons of these display technologies in engineering synthetic binders based on nonantibody scaffolds. Potential problems with yeast surface display include smaller library sizes and non-natural glycosylation. In addition, although early work suggested that yeast surface display could be used to improve protein stability, recent reports demonstrated that this capacity is rather limited, particularly when applied to a small protein (Park et al. 2006; Dutta et al. 2008). Potential problems with mRNA and ribosome display include the difficulty to produce protein homo- and hetero-oligomers. Together, it is extremely important to consider the biology and potential constraints of a molecular display technology when developing a new scaffold.

## ADVANCES IN SEQUENCE DIVERSITY DESIGN

Library design is critically important for the success of scaffold development. One first needs to decide what part of a scaffold should be diversified in a library. An approximate location of the paratope is usually defined at the time of scaffold selection (for example, surface loops constitute the intended paratope of a β-sandwich scaffold). However, it should be noted that refining the diversified positions requires an iterative and systematic process of constructing a library, selecting binders, and identifying positions that contribute to binding.

Even for a small interface consisting of 15–20 amino-acid residues, the number of total possible sequences ($20^{15} = {\sim}3 \times 10^{19}$) far exceeds the size of the largest experimental libraries (~$10^{13}$). Most libraries made in the early days of synthetic interface engineering were based on total randomization, typically using the NNK codon where N is the equal mixture of the A/T/G/C bases and K is the equal mixture of G/T (Nord et al. 1997; Koide et al. 1998). In these attempts, typically only low affinity binders with $K_D$ values in the low μM range were selected. Because such a completely random library includes a very small fraction of all possible sequences, it is perhaps not surprising that this approach often exhibits low levels of success in finding a highly optimized interface required for high affinity.

Analyses of natural protein–protein interfaces, in particular those of antibody-antigen complexes, have revealed the presence of strong bias in amino-acid composition. Aromatic amino acids with hydrogen-bonding capability (Tyr and Trp) are highly enriched, while those with long side chains (Met, Lys and Glu) are disfavored (Mian et al. 1991; Bogan and Thorn 1998; Lo Conte et al. 1999). These findings clearly indicate that a small subset of genetically encoded amino acids is particularly effective in forming a recognition interface and another subset is effective in disrupting an interface. Clearly, certain amino acids are capable of forming a larger number of productive interactions for molecular recognition such as tight packing, hydrophobic interaction, hydrogen bonding, and salt bridges. It is also useful to look at this from the opposite side. It may be particularly costly to bury some amino acids in an interface, for example, due to the large desolvation cost of a charge and high entropic cost of immobilizing a flexible side chain (Derewenda 2004; Doye et al. 2004).

The effectiveness of tailoring the amino-acid composition in the library has convincingly been established by a series of studies pioneered by Sidhu and colleagues (Fellouse et al. 2004; Fellouse et al. 2005; Fellouse et al. 2007; Birtalan et al. 2008). These researchers showed that a library containing only a few amino-acid types (e.g., Tyr, Ser, Ala, and Asp) is highly effective in producing high affinity and high specificity binders. In the most extreme case, a library consisting of the binary code of Tyr and Ser produced high affinity binders. Tyrosine is the most prevalent amino acid in the germline diversity of the immunoglobulins, as well as in the antigen-binding interfaces of matured antibodies (Mian et al. 1991; Zemlin et al. 2003), indicating that the highly functional nature of tyrosine has been exploited in nature.

The effectiveness of the Tyr/Ser binary code was also demonstrated using the FN3 scaffold, the most widely used alternative scaffold (Koide et al. 2007a). In addition, it has been shown that supplementing the Tyr/Ser binary code with lower concentrations of other amino acids further enhances the capacity of antibody and FN3 libraries to produce high affinity binding proteins to diverse target molecules (Fellouse et al. 2007; Gilbreth et al. 2008).

Structural analyses of these minimalist synthetic interfaces have shown that Tyr dominates the contacting surfaces and that other amino-acid types play an important role in providing conformational diversity to the paratope structure but a less important role in composing the interaction surface itself, demonstrating that chemically, tyrosine alone still proves quite effective for recognition (Fellouse et al. 2005; Fellouse et al. 2006; Koide et al. 2007a; Gilbreth et al. 2008). This conclusion rationalizes why these minimalist libraries function so well when diversifying CDRs and CDR-like loops. It remains to be seen whether the same approach is as effective in different classes of scaffolds.

The high efficacy of such minimalist designs of amino-acid diversity greatly diminishes the fundamental problem of constructing synthetic interface libraries by reducing the number of encoded sequences, often to a level that makes complete sequence sampling possible even under the practical limitations of commonly used display and selection methods. Importantly, the high functionality of minimalist libraries suggests that minimalist design dramatically increases the fraction of functional sequences within a library relative to a traditional, fully random library, and thus this type of library can produce functional molecules with fewer iterations of library sorting.

## APPLICATIONS

### ANTIBODY ALTERNATIVES

Synthetic binding proteins built with nonantibody scaffolds can obviously be used as alternatives to antibodies in immunochemical applications. A clear advantage of nonantibody recombinant scaffolds is the ease of reformatting them into a variety of fusion proteins. It is usually trivial to fuse a synthetic binder to a reporter enzyme or a fluorescent protein or to introduce a chemical group such as biotin and fluorescent dye (de Graaf et al. 2002). FN3-based binders to the Src SH2 domain as well as the PDZ affinity clamps described previously were successfully used in immunoprecipitation ("pull down") and Western blotting (Karatan et al. 2004; Huang et al. 2008). Development of an economical and high-throughput pipeline for synthetic binder generation will be essential for widespread use of synthetic binders as true alternatives to antibodies.

### THERAPEUTICS

The emphasis of synthetic binders in the biotechnology and pharmaceutical industries has, understandably, been on the development of therapeutic molecules. The successes of antibody-based therapeutics clearly support the notion that drugs based on synthetic binders can be developed using available technologies. Given their excellent physical properties and lower production costs afforded by microbial expression systems, such synthetic binders could end up being superior to antibody-based therapeutics. A simpler intellectual property landscape is another important advantage in commercial applications of synthetic binders. In addition, different modes of biological action can be envisioned, such as directed receptor agonism or antagonism. Furthermore, synthetic binders can be an "addressing" reagent to deliver cytotoxic molecules to cells and tissues. A number of synthetic binders are already in clinical trials including a VEGF receptor–targeted FN3 (Getmanova et al. 2006) and an IL-6–targeted avimer (Silverman et al. 2005). The reader is referred to recent review articles on commercial activities in this area (Binz et al. 2005; Sheridan 2007; Skerra 2007). Importantly, these pioneering clinical trials have so far detected no significant immunological responses, a widespread concern for introducing engineered proteins into humans. It is important to note that these small synthetic binding proteins are no more foreign than the antigen-binding site of human antibodies. A series of recent acquisitions of biotech ventures by biopharmaceutical companies indicates that the concept of drug development using nonantibody scaffolds has been generally accepted in the pharmaceutical industry (Sheridan 2007).

## AFFINITY PURIFICATION

Synthetic binders may be more effective alternatives to antibodies in immunoaffinity chromatography, a technique where an antibody is immobilized onto a solid support and used to purify a binding partner from a mixture of proteins. Common problems in immunoaffinity chromatography are the high cost of antibody production and the sensitivity of antibodies to experimental conditions such as changes in pH and salt concentration. The high stability and lower cost of production of nonantibody scaffolds make them well suited for replacing antibodies in affinity purification applications. Furthermore, it is likely that small and simple molecular scaffolds can withstand harsh treatments for column cleaning and regeneration better than complex molecular scaffolds. Protein A is widely used for affinity purification of immunoglobulins; thus it is self-evident that affibodies, which are built on the protein A scaffold, can be effectively used as affinity chromatography reagents (Nord et al. 2000). Also, FN3-based affinity chromatography was able to specifically purify one active form of the estrogen receptor from *E. coli* crude lysate also containing misfolded forms of the receptor (Huang et al. 2006).

## CRYSTALLIZATION CHAPERONES

A highly successful strategy to determine the structures of elusive structural biology systems such as membrane proteins has been the use of "crystallization chaperones" (Hunte and Michel 2002). Fab-based chaperones have been the enabling factor for determining a number of paradigm-shifting structures, such as the KcsA potassium channel (Zhou et al. 2001) and the $\beta_2$-adrenergic receptor (Rasmussen et al. 2007). Chaperones promote crystallization by reducing conformational heterogeneity (i.e., reducing flexibility), by masking hydrophobic surfaces and thus increasing solubility, and by providing primary contact points between molecules in the crystal lattice. A number of crystal structures have recently been determined for complexes of a synthetic protein with its binding partner (Binz et al. 2004; Fellouse et al. 2004; Fellouse et al. 2005; Fellouse et al. 2007; Koide et al. 2007a; Schweizer et al. 2007; Bandeiras et al. 2008), including membrane proteins (Sennhauser et al. 2007) and structured RNA (Ye et al. 2008). In all cases, crystallization chaperones were responsible for creating a majority of contacts in the lattice, either through the binding interface or through crystal contacts. Two examples, FN3/ER-LBD (PDB ID 2OCF) and DARPin/AcrB (Sennhauser et al. 2007), suggest that a chaperone does not have to be similar in size to its target to be effective.

INTRACELLULAR APPLICATIONS

Synthetic binders can be powerful tools for perturbing the functions of proteins inside cells. Such perturbation information is useful for dissecting complex signaling networks and for validating potential drug targets. A clear advantage of many nonantibody scaffolds is their lack of disulfide bonds, which generally do not form in the cytoplasm, making it straightforward to express them inside cells and to perform binder selection using *in vivo* methods. Peptide aptamers have been extensively used in this type of application. They are selected from yeast two-hybrid libraries, and they are then expressed to perturb cellular functions (Brent and Finley 1997; Fabbrizio et al. 1999). In addition, FN3-based binders were used to discriminate different states of the human estrogen receptor in the nucleus and inhibit them (Koide et al. 2002). DARPins that inhibit a bacterial kinase in cells have also been generated (Kawe et al. 2006). Clearly, this is an area of applications that is uniquely suited for synthetic binders and it is likely that this area will see significant growth in the near future. It is important to note, however, that there is still a significant technological hurdle in delivering such binders and antibodies into cells except for expression-based methods.

## CONCLUSIONS

Examples described here clearly illustrate that engineering of synthetic molecular interaction interfaces using nonantibody scaffolds is a mature branch of protein engineering. The breadth of available scaffolds and their demonstrated effectiveness in producing highly functional interfaces suggest that establishing another molecular scaffold will be an exercise of diminishing returns, unless such a scaffold offers a clearly distinct advantage over the current ones. Although these scaffold systems have been developed primarily for overcoming limitations of natural antibodies, structural and biophysical analyses of binders produced from these systems have expanded the types of protein recognition motifs beyond those that are present in natural proteins and have provided fundamental insights into the molecular mechanisms underlying protein–ligand interactions.

The common goal of scaffold development has been to establish a single, highly versatile system that consistently produces high affinity and high specificity binders to all types of targets. However, this "one size fits all" goal should probably be reexamined. As described previously, the interface topography is a major determinant of protein–protein interaction interfaces. Therefore, scaffolds presenting different paratope shapes are expected to have different binding preferences. Consequently, the most successful strategy might be to use a few distinct and complementary scaffolds, which should increase the probability of successfully generating high-performance binders to diverse targets. It is interesting that recent studies in the synthetic antibody field show that this type of scaffold diversity is highly effective (Persson et al. 2006; Cobaugh et al. 2008). Given the diverse and complementary nature of highly functional scaffold systems that are already available, it is likely that we, as a community, will be able to achieve this goal of establishing a small number of complementary scaffolds sufficient for most applications in the near future.

Applications of synthetic binders are still in their infancy. Synthetic binders can be in many instances superior alternatives to natural antibodies as affinity reagents. Particularly, the invariant scaffolds facilitate the construction of highly efficient generation pipelines that can be easily scaled up. This is because, unlike natural antibodies that contain many variations in the scaffold outside the CDRs, a standardized manipulation of all members from a synthetic binder library can be easily established. However, the true value of synthetic binders as novel tools lies in their applications in areas that are deemed impossible or extremely difficult with natural and synthetic antibodies. These include intracellular applications, such as intrabodies and protein tracking in live cells, and implementation of higher-order functionalities such as sensors. In the future, it is highly likely that synthetic binders will prove to be an enabling factor in many unforeseen applications.

## ACKNOWLEDGMENTS

I thank Ryan Gilbreth and Dr. Akiko Koide for critical reading of this manuscript. This work was supported by NIH grants R01-GM72688, R21-CA132700, R21-DA025725, and U54 GM74946.

## REFERENCES

Agrawal, V., and Kishan, K.V. 2003. OB-fold: Growing bigger with functional consistency. *Curr Protein Pept Sci* 4: 195–206.

Bandeiras, T.M., Hillig, R.C., Matias, P.M., Eberspaecher, U., Fanghanel, J., Thomaz, M., Miranda, S., Crusius, K., Putter, V., Amstutz, P., et al. 2008. Structure of wild-type Plk-1 kinase domain in complex with a selective DARPin. *Acta Crystallogr D Biol Crystallogr* 64: 339–53.

Beerli, R.R., and Barbas, C.F., 3rd. 2002. Engineering polydactyl zinc-finger transcription factors. *Nat Biotechnol* 20: 135–141.

Beste, G., Schmidt, F.S., Stibora, T., and Skerra, A. 1999. Small antibody-like proteins with prescribed ligand specificities derived from the lipocalin fold. *Proc Natl Acad Sci U S A* 96: 1898–1903.

Bianchi, E., Venturini, S., Pessi, A., Tramontano, A., and Sollazzo, M. 1994. High level expression and rational mutagenesis of a designed protein, the minibody. From an insoluble to a soluble molecule. *J. Mol. Biol.* 236: 649–59.

Binz, H.K., Amstutz, P., Kohl, A., Stumpp, M.T., Briand, C., Forrer, P., Grutter, M.G., and Plückthun, A. 2004. High-affinity binders selected from designed ankyrin repeat protein libraries. *Nat Biotechnol* 22: 575–82.

Binz, H.K., Amstutz, P., and Plückthun, A. 2005. Engineering novel binding proteins from nonimmunoglobulin domains. *Nat Biotechnol* 23: 1257–68.

Binz, H.K., and Plückthun, A. 2005. Engineered proteins as specific binding reagents. *Curr Opin Biotechnol* 16: 459–69.

Birrane, G., Chung, J., and Ladias, J.A. 2003. Novel mode of ligand recognition by the Erbin PDZ domain. *J Biol Chem* 278: 1399–1402.

Birtalan, S., Zhang, Y., Fellouse, F.A., Shao, L., Schaefer, G., and Sidhu, S.S. 2008. The intrinsic contributions of tyrosine, serine, glycine and arginine to the affinity and specificity of antibodies. *J Mol Biol* 377: 1518–28.

Bogan, A.A., and Thorn, K.S. 1998. Anatomy of hot spots in protein interfaces. *J Mol Biol* 280: 1–9.

Brent, R., and Finley, R.L., Jr. 1997. Understanding gene and allele function with two-hybrid methods. *Annu Rev Genet* 31: 663–704.

Campbell, I.D., and Spitzfaden, C. 1994. Building proteins with fibronectin type III modules. *Structure* 2: 233–337.

Chen, J.R., Chang, B.H., Allen, J.E., Stiffler, M.A., and MacBeath, G. 2008. Predicting PDZ domain-peptide interactions from primary sequences. *Nat Biotechnol* 26: 1041–45.

Choo, Y., and Klug, A. 1994. Toward a code for the interactions of zinc fingers with DNA: Selection of randomized fingers displayed on phage. *Proc Natl Acad Sci U S A* 91: 11163–67.

Cobaugh, C.W., Almagro, J.C., Pogson, M., Iverson, B., and Georgiou, G. 2008. Synthetic antibody libraries focused towards peptide ligands. *J Mol Biol* 378: 622–33.

Colas, P., Cohen, B., Jessen, T., Grishina, I., McCoy, J., and Brent, R. 1996. Genetic selection of peptide aptamers that recognize and inhibit cyclin-dependent kinase 2. *Nature* 380: 548–50.

de Graaf, M., van der Meulen-Muileman, I.H., Pinedo, H.M., and Haisma, H.J. 2002. Expression of scFvs and scFv fusion proteins in eukaryotic cells. *Methods Mol Biol* 178: 379–87.

DeLano, W.L. 2002. Unraveling hot spots in binding interfaces: Progress and challenges. *Curr Opin Struct Biol* 12: 14–20.

Dennis, M.S., and Lazarus, R.A. 1994. Kunitz domain inhibitors of tissue factor-factor VIIa. I. Potent inhibitors selected from libraries by phage display. *J Biol Chem* 269: 22129–36.

Derewenda, Z.S. 2004. Rational protein crystallization by mutational surface engineering. *Structure (Camb)* 12: 529–35.

Doye, J.P., Louis, A.A., and Vendruscolo, M. 2004. Inhibition of protein crystallization by evolutionary negative design. *Phys Biol* 1: P9–13.

Dutta, S., Koide, A., and Koide, S. 2008. High-throughput analysis of the protein sequence-stability landscape using a quantitative yeast surface two-hybrid system and fragment reconstitution. *J Mol Biol* 382: 721–33.

Fabbrizio, E., Le Cam, L., Polanowska, J., Kaczorek, M., Lamb, N., Brent, R., and Sardet, C. 1999. Inhibition of mammalian cell proliferation by genetically selected peptide aptamers that functionally antagonize E2F activity. *Oncogene* 18: 4357–63.

Fellouse, F.A., Barthelemy, P.A., Kelley, R.F., and Sidhu, S.S. 2006. Tyrosine plays a dominant functional role in the paratope of a synthetic antibody derived from a four amino acid code. *J Mol Biol* 357: 100–114.

Fellouse, F.A., Esaki, K., Birtalan, S., Raptis, D., Cancasci, V.J., Koide, A., Jhurani, P., Vasser, M., Wiesmann, C., Kossiakoff, A.A., et al. 2007. High-throughput generation of synthetic antibodies from highly functional minimalist phage-displayed libraries. *J Mol Biol* 373: 924–40.

Fellouse, F.A., Li, B., Compaan, D.M., Peden, A.A., Hymowitz, S.G., and Sidhu, S.S. 2005. Molecular recognition by a binary code. *J Mol Biol* 348: 1153–62.

Fellouse, F.A., Wiesmann, C., and Sidhu, S.S. 2004. Synthetic antibodies from a four-amino-acid code: A dominant role for tyrosine in antigen recognition. *Proc Natl Acad Sci U S A* 101: 12467–72.

Forrer, P., Stumpp, M.T., Binz, H.K., and Plückthun, A. 2003. A novel strategy to design binding molecules harnessing the modular nature of repeat proteins. *FEBS Lett* 539: 2–6.

Garcia-Ibilcieta, D., Bokov, M., Cherkasov, V., Sveshnikov, P., and Hanson, S.F. 2008. Simple method for production of randomized human tenth fibronectin domain III libraries for use in combinatorial screening procedures. *Biotechniques* 44: 559–62.

Getmanova, E.V., Chen, Y., Bloom, L., Gokemeijer, J., Shamah, S., Warikoo, V., Wang, J., Ling, V., and Sun, L. 2006. Antagonists to human and mouse vascular endothelial growth factor receptor 2 generated by directed protein evolution in vitro. *Chem Biol* 13: 549–56.

Geyer, C.R., Colman-Lerner, A., and Brent, R. 1999. "Mutagenesis" by peptide aptamers identifies genetic network members and pathway connections. *Proc Natl Acad Sci U S A* 96: 8567–72.

Gilbreth, R.N., Esaki, K., Koide, A., Sidhu, S.S., and Koide, S. 2008. A dominant conformational role for amino acid diversity in minimalist protein–protein interfaces. *J Mol Biol* 381: 407–418.

Hamers-Casterman, C., Atarhouch, T., Muyldermans, S., Robinson, G., Hamers, C., Songa, E.B., Bendahman, N., and Hamers, R. 1993. Naturally occurring antibodies devoid of light chains. *Nature* 363: 446–48.

Hosse, R.J., Rothe, A., and Power, B.E. 2006. A new generation of protein display scaffolds for molecular recognition. *Protein Sci* 15: 14–27.

Huang, J., Koide, A., Makabe, K., and Koide, S. 2008. Design of protein function leaps by directed domain interface evolution. *Proc Natl Acad Sci U S A* 105: 6578–83.

Huang, J., Koide, A., Nettle, K.W., Greene, G.L., and Koide, S. 2006. Conformation-specific affinity purification of proteins using engineered binding proteins: Application to the estrogen receptor. *Protein Expr Purif* 47: 348–54.

Hunte, C., and Michel, H. 2002. Crystallisation of membrane proteins mediated by antibody fragments. *Curr Opin Struct Biol* 12: 503–508.

Jamieson, A.C., Kim, S.-H., and Wells, J.A. 1994. In vitro selection of zinc fingers with altered DNA-binding specificity. *Biochemistry* 33: 5689–95.

Karatan, E., Merguerian, M., Han, Z., Scholle, M.D., Koide, S., and Kay, B.K. 2004. Molecular recognition properties of FN3 monobodies that bind the Src SH3 domain. *Chem Biol* 11: 835–44.

Kawe, M., Forrer, P., Amstutz, P., and Plückthun, A. 2006. Isolation of intracellular proteinase inhibitors derived from designed ankyrin repeat proteins by genetic screening. *J Biol Chem* 281: 40252–63.

Koide, A., Abbatiello, S., Rothgery, L., and Koide, S. 2002. Probing protein conformational changes by using designer binding proteins: Application to the estrogen receptor. *Proc Natl Acad Sci USA* 99: 1253–58.

Koide, A., Bailey, C.W., Huang, X., and Koide, S. 1998. The fibronectin type III domain as a scaffold for novel binding proteins. *J Mol Biol* 284: 1141–51.

Koide, A., Gilbreth, R.N., Esaki, K., Tereshko, V., and Koide, S. 2007a. High-affinity single-domain binding proteins with a binary-code interface. *Proc Natl Acad Sci U S A* 104: 6632–37.

Koide, A., Jordan, M.R., Horner, S.R., Batori, V., and Koide, S. 2001. Stabilization of a fibronectin type III domain by the removal of unfavorable electrostatic interactions on the protein surface. *Biochemistry* 40: 10326–33.

Koide, A., Tereshko, V., Uysal, S., Margalef, K., Kossiakoff, A.A., and Koide, S. 2007b. Exploring the capacity of minimalist protein interfaces: Interface energetics and affinity maturation to picomolar KD of a single-domain antibody with a flat paratope. *J Mol Biol* 373: 941–53.

Korndorfer, I.P., Schlehuber, S., and Skerra, A. 2003. Structural mechanism of specific ligand recognition by a lipocalin tailored for the complexation of digoxigenin. *J Mol Biol* 330: 385–96.

Kouadio, J.L., Horn, J.R., Pal, G., and Kossiakoff, A.A. 2005. Shotgun alanine scanning shows that growth hormone can bind productively to its receptor through a drastically minimized interface. *J Biol Chem* 280: 25524–32.

Krause, S., Schmoldt, H.U., Wentzel, A., Ballmaier, M., Friedrich, K., and Kolmar, H. 2007. Grafting of thrombopoietin-mimetic peptides into cystine knot miniproteins yields high-affinity thrombopoietin antagonists and agonists. *Febs J* 274: 86–95.

Lawrence, M.C., and Colman, P.M. 1993. Shape complementarity at protein/protein interfaces. *J Mol Biol* 234: 946–50.

Lendel, C., Dogan, J., and Hard, T. 2006. Structural basis for molecular recognition in an affibody:affibody complex. *J Mol Biol* 359: 1293–1304.

Li, R., Hoess, R.H., Bennett, J.S., and DeGrado, W.F. 2003. Use of phage display to probe the evolution of binding specificity and affinity in integrins. *Protein Eng* 16: 65–72.

Lipovsek, D., Lippow, S.M., Hackel, B.J., Gregson, M.W., Cheng, P., Kapila, A., and Wittrup, K.D. 2007. Evolution of an interloop disulfide bond in high-affinity antibody mimics based on fibronectin type III domain and selected by yeast surface display: Molecular convergence with single-domain camelid and shark antibodies. *J Mol Biol* 368: 1024–41.

Liu, Y., and Eisenberg, D. 2002. 3D domain swapping: As domains continue to swap. *Protein Sci* 11: 1285–99.

Lo Conte, L., Chothia, C., and Janin, J. 1999. The atomic structure of protein–protein recognition sites. *J Mol Biol* 285: 2177–98.

Malakauskas, S.M., and Mayo, S.L. 1998. Design, structure and stability of a hyperthermophilic protein variant. *Nat Struct Biol* 5: 470–75.

Mian, I.S., Bradwell, A.R., and Olson, A.J. 1991. Structure, function and properties of antibody binding sites. *J Mol Biol* 217: 133–51.

Nagano, N., Orengo, C.A., and Thornton, J.M. 2002. One fold with many functions: The evolutionary relationships between TIM barrel families based on their sequences, structures and functions. *J Mol Biol* 321: 741–65.

Nord, K., Gunneriusson, E., Ringdahl, J., Stahl, S., Uhlen, M., and Nygren, P.A. 1997. Binding proteins selected from combinatorial libraries of an alpha-helical bacterial receptor domain. *Nat Biotechnol* 15: 772–77.

Nord, K., Gunneriusson, E., Uhlen, M., and Nygren, P.A. 2000. Ligands selected from combinatorial libraries of protein A for use in affinity capture of apolipoprotein A-1M and taq DNA polymerase. *J Biotechnol* 80: 45–54.

O'Neil, K.T., Hoess, R.H., Raleigh, D.P., and DeGrado, W.F. 1995. Thermodynamic genetics of the folding of the B1 immunoglobulin-binding domain from streptococcal protein G. *Proteins* 21: 11–21.

Pancer, Z., Amemiya, C.T., Ehrhardt, G.R., Ceitlin, J., Gartland, G.L., and Cooper, M.D. 2004. Somatic diversification of variable lymphocyte receptors in the agnathan sea lamprey. *Nature* 430: 174–80.

Park, S., Xu, Y., Stowell, X.F., Gai, F., Saven, J.G., and Boder, E.T. 2006. Limitations of yeast surface display in engineering proteins of high thermostability. *Protein Eng Des Sel* 19: 211–17.

Perl, D., Mueller, U., Heinemann, U., and Schmid, F.X. 2000. Two exposed amino acid residues confer thermostability on a cold shock protein [see comments]. *Nat Struct Biol* 7: 380–83.

Persson, H., Lantto, J., and Ohlin, M. 2006. A focused antibody library for improved hapten recognition. *J Mol Biol* 357: 607–20.

Pessi, A., Bianchi, E., Crameri, A., Venturini, S., Tramontano, A., and Sollazzo, M. 1993. A designed metal-binding protein with a novel fold. *Nature* 362: 367–69.

Rasmussen, S.G., Choi, H.J., Rosenbaum, D.M., Kobilka, T.S., Thian, F.S., Edwards, P.C., Burghammer, M., Ratnala, V.R., Sanishvili, R., Fischetti, R.F., et al. 2007. Crystal structure of the human beta2 adrenergic G-protein-coupled receptor. *Nature* 450: 383–87.

Rebar, E.J., and Pabo, C.O. 1994. Zinc finger phage: Affinity selection of fingers with new DNA-binding specificities. *Science* 263: 671–73.

Reichmann, D., Rahat, O., Cohen, M., Neuvirth, H., and Schreiber, G. 2007. The molecular architecture of protein–protein binding sites. *Curr Opin Struct Biol* 17: 67–76.

Richards, J., Miller, M., Abend, J., Koide, A., Koide, S., and Dewhurst, S. 2003. Engineered fibronectin type III domain with a RGDWXE sequence binds with enhanced affinity and specificity to human alphavbeta3 integrin. *J Mol Biol* 326: 1475–88.

Roberts, B.L., Markland, W., Ley, A.C., Kent, R.B., White, D.W., Guterman, S.K., and Ladner, R.C. 1992. Directed evolution of a protein: Selection of potent neutrophil elastase inhibitors displayed on M13 fusion phage. *Proc Natl Acad Sci U S A* 89: 2429–33.

Schlehuber, S., and Skerra, A. 2005. Lipocalins in drug discovery: From natural ligand-binding proteins to "anticalins." *Drug Discov Today* 10: 23–33.

Schweizer, A., Roschitzki-Voser, H., Amstutz, P., Briand, C., Gulotti-Georgieva, M., Prenosil, E., Binz, H.K., Capitani, G., Baici, A., Plückthun, A., et al. 2007. Inhibition of caspase-2 by a designed ankyrin repeat protein: Specificity, structure, and inhibition mechanism. *Structure* 15: 625–36.

Sennhauser, G., Amstutz, P., Briand, C., Storchenegger, O., and Grutter, M.G. 2007. Drug export pathway of multidrug exporter AcrB revealed by DARPin inhibitors. *PLoS Biol* 5: e7.

Sheridan, C. 2007. Pharma consolidates its grip on post-antibody landscape. *Nat Biotechnol* 25: 365–66.

Sidhu, S.S., and Koide, S. 2007. Phage display for engineering and analyzing protein interaction interfaces. *Curr Opin Struct Biol* 17: 481–87.

Sidhu, S.S., Lowman, H.B., Cunningham, B.C., and Wells, J.A. 2000. Phage display for selection of novel binding peptides. *Methods Enzymol* 328: 333–63.

Sieber, V., Plückthun, A., and Schmid, F.X. 1998. Selecting proteins with improved stability by a phage-based method. *Nat Biotechnol* 16: 955–60.

Silverman, A.P., Levin, A.M., Lahti, J.L., and Cochran, J.R. 2009. Engineered cystine-knot peptides that bind alpha(v)beta(3) integrin with antibody-like affinities. *J Mol Biol* 385:1064–1075.

Silverman, J., Liu, Q., Bakker, A., To, W., Duguay, A., Alba, B.M., Smith, R., Rivas, A., Li, P., Le, H., et al. 2005. Multivalent avimer proteins evolved by exon shuffling of a family of human receptor domains. *Nat Biotechnol* 23: 1556–61.

Skerra, A. 2007. Alternative non-antibody scaffolds for molecular recognition. *Curr Opin Biotechnol* 18: 295–304.

Smith, T.J., Stains, C.I., Meyer, S.C., and Ghosh, I. 2006. Inhibition of beta-amyloid fibrillization by directed evolution of a beta-sheet presenting miniature protein. *J Am Chem Soc* 128: 14456–57.

Stanfield, R.L., Dooley, H., Flajnik, M.F., and Wilson, I.A. 2004. Crystal structure of a shark single-domain antibody V region in complex with lysozyme. *Science* 305: 1770–73.

Steiner, D., Forrer, P., Stumpp, M.T., and Plückthun, A. 2006. Signal sequences directing cotranslational translocation expand the range of proteins amenable to phage display. *Nat Biotechnol* 24: 823–31.

Todd, A.E., Orengo, C.A., and Thornton, J.M. 2002. Sequence and structural differences between enzyme and nonenzyme homologs. *Structure* 10: 1435–51.

Worn, A., and Plückthun, A. 2001. Stability engineering of antibody single-chain Fv fragments. *J Mol Biol* 305: 989–1010.

Wu, T.T., Johnson, G., and Kabat, E.A. 1993. Length distribution of CDRH3 in antibodies. *Proteins: Struct Funct Genet* 16: 1–7.

Xu, L., Aha, P., Gu, K., Kuimelis, R., Kurz, M., Lam, T., Lim, A., Liu, H., Lohse, P., Sun, L., et al. 2002. Directed evolution of high-affinity antibody mimics using mRNA display. *Chem Biol* 9: 933–42.

Ye, J.D., Tereshko, V., Frederiksen, J.K., Koide, A., Fellouse, F.A., Sidhu, S.S., Koide, S., Kossiakoff, A.A., and Piccirilli, J.A. 2008. Synthetic antibodies for specific recognition and crystallization of structured RNA. *Proc Natl Acad Sci U S A* 105: 82–87.

Zahnd, C., Wyler, E., Schwenk, J.M., Steiner, D., Lawrence, M.C., McKern, N.M., Pecorari, F., Ward, C.W., Joos, T.O., and Plückthun, A. 2007. A designed ankyrin repeat protein evolved to picomolar affinity to her2. *J Mol Biol* 369: 1015–28.

Zemlin, M., Klinger, M., Link, J., Zemlin, C., Bauer, K., Engler, J.A., Schroeder, H.W., Jr., and Kirkham, P.M. 2003. Expressed murine and human CDR-H3 intervals of equal length exhibit distinct repertoires that differ in their amino acid composition and predicted range of structures. *J Mol Biol* 334: 733–49.

Zhou, Y., Morais-Cabral, J.H., Kaufman, A., and MacKinnon, R. 2001. Chemistry of ion coordination and hydration revealed by a K+ channel-Fab complex at 2.0 A resolution. *Nature* 414: 43–48.

# 6 Combinatorial Enzyme Engineering

*Patrick C. Cirino and Christopher S. Frei*

## CONTENTS

While our understanding of biological systems may one day be sufficiently comprehensive to engineer proteins in a direct and targeted fashion with predictable outcomes, biological complexity continues to stifle even seemingly simple "rational" engineering plans, often leaving combinatorial approaches as the most likely means of achieving a desired phenotype. A key step in biocatalyst design is the development of high-throughput screening procedures to identify a specific, improved cellular phenotype (e.g., productivity) or enzyme property of interest. Although an abundance of recombinant DNA and molecular cloning methods provide the protein engineering community with a myriad of approaches to genetic library construction (refer to Chapter 4 in this volume), we are often constrained by the current capabilities of high-throughput screening or selection technologies.

The use of combinatorial methods to engineer enzymes is similar to their applications in engineering noncatalytic proteins. High-throughput screening involves isolating and analyzing individual library members, often with no prior enrichment of function. A large fraction of the mutants screened is likely to have poorer function than the parent protein, depending on the library's diversity. In contrast, selection typically refers to an ability to isolate only those variants possessing a minimum level of function from a pool of all mutants (or cells harboring mutants). This commonly means isolating only those variants having sufficient activity to confer host cell viability, with all other mutants never having to be "observed." Flow cytometry and binding-based display technologies are also often considered forms of selection. The requirement that enzymatic activity (i.e., catalytic conversion of substrate) be retained or possibly improved through mutagenesis poses important challenges that differentiate screening and selection systems for enzyme engineering from those for engineering protein binding. Bearing in mind the popular directed evolution adage "you get what you screen for" and the fact that accumulation of mutations leads to loss of functions not intentionally retained (e.g., substrate specificity, stability, activity) (Hibbert and Dalby 2005; Schmidt-Dannert and Arnold 1999), it is important to incorporate activity determination into a high-throughput assay for enzyme evolution, even when the function to be evolved is not activity *per se* (e.g., thermostability).

Combinatorial methods applied to enzyme engineering therefore require careful modification of these same techniques applied to engineering properties of noncatalytic proteins. Regardless of the protein engineering objective, a requirement common to all high-throughput screening methods is maintenance of a genotype-phenotype linkage. Diffusion of products away from the enzyme is a central challenge of linking genotype to phenotype for the case of enzymes. As discussed in other chapters, screening methods such as phage and cell surface display are well-suited for isolating improvements in binding properties. Similarly, selection systems such as the yeast two-hybrid or other genetic reporter-based assays are readily designed to identify protein-based binding events. It is more challenging to develop screens that allow one to monitor catalysis or generalized selections in which cell survival requires a threshold level of enzyme activity.

This chapter highlights technologies that have been developed to enable combinatorial enzyme engineering, with particular attention to features that differentiate them from similar techniques used in engineering noncatalytic proteins. Specific examples in which each technology has been applied are provided. The application of evolutionary design principles to engineer enzymes is widespread, and the number of screening techniques is vast and ever-growing. While not comprehensive, this chapter describes representative examples of the most common, innovative, or effective strategies for successfully identifying improved enzymes from combinatorial libraries. For more examples or more detailed descriptions of experimental procedures, the reader is encouraged to refer to the technical manuscripts cited in this chapter, a number of excellent review articles (Aharoni et al. 2005b; Boersma et al. 2007a; Lin and Cornish 2002; Link et al. 2007; Olsen et al. 2000a; Olsen et al. 2000b; Taylor et al. 2001), and an entire book volume dedicated to enzyme selection and screening methods (Arnold and Georgiou 2003).

We first describe the many ways in which the ultra high-throughput capabilities of flow cytometry and fluorescence-activated cell sorting (FACS) have been adapted to enzyme evolution. This includes screening whole-cells in which the intracellular reaction results in a fluorescent signal (including via activation of a genetic reporter), whole-cells displaying the enzyme of interest on their surface, and cell-free compartments containing enzyme variants along with their unique genetic information.

We next describe examples of enzymes engineered by genetic selection or high-throughput microbial colony-based screens. This includes examples in which the enzyme catalyzes a reaction that directly affects cell viability (e.g., auxotroph complementation) as well as applications of genetic reporter systems for enzyme engineering. Then the adaptation of phage and ribosome display for enzyme engineering is addressed. Finally, we provide several examples of enzyme assays developed and implemented in high-throughput, liquid handling–dependent screening systems requiring spatial address to maintain genotype-phenotype linkage.

## FLOW CYTOMETRY

Flow cytometry allows for the rapid measurement of multiple optical properties of individual cells, particles, or compartmentalized droplets. Fluorescence-activated cell sorting (FACS) involves the flow cytometric separation of members of the cell population with desired fluorescent properties (and is not limited only to cells). Modern FACS instruments are capable of sorting up to $10^9$ cells in one or a few days. Note that these screening capabilities approach the limits of library cloning and transformation using bacterial expression hosts (Georgiou 2000). The distinction between screening and selection is somewhat blurred at the level of flow cytometric sorting. In one sense this method is a screen, since improved variants are isolated only by measuring the phenotype of every library member. However, this may also be considered a selection method since the entire pool of variants is acted on at one time, and the improved variants are subsequently pooled together, as opposed to spatially separated, and typically enriched through multiple rounds. The high-throughput capabilities combined with the increasing availability of FACS instrumentation have led to the use of flow cytometry as one of the most enabling tools in enzyme engineering. The use of FACS requires the generation of a fluorescence signal that reflects the phenotype of interest, and a major challenge in implementing FACS for enzyme evolution is therefore correlating enzyme activity with cell or particle fluorescence. This has been accomplished in a variety of ways, which are summarized in many reviews (Becker et al. 2004; Farinas 2006; Georgiou 2000; Link et al. 2007). Here we describe some of the important applications of FACS for enzyme engineering.

### INTRACELLULAR REACTIONS

In the simplest case, the enzyme of interest is expressed and maintained intracellularly, and action on a substrate triggers an intracellular fluorescent signal. An example of this kind is an assay developed by Withers and coworkers for FACS-based sorting of *E. coli* cells expressing a library of sialyltransferase (ST; belonging to the glycosyltransferase superfamily) (Aharoni et al. 2006). In this case, fluorescently labeled lactose-containing acceptor sugars are freely transported into and out of the cell. Upon intracellular ST-catalyzed addition of sialic acid to the fluorescent sugar, the larger and charged fluorescent sialylated lactoside product is trapped inside the cell. Unreacted fluorescent sugar is washed away and cells retaining fluorescence are sorted by FACS, with increased cellular fluorescence correlating with increased intracellular enzyme activity. Using this approach, a library of >10$^6$ ST mutants was screened and yielded a variant with up to 400-fold improved catalytic efficiency (Aharoni et al. 2006).

In a similar example, Shim et al. reported the development of a FACS-based method for isolation of aldolase antibodies (Shim et al. 2004). They used a fluorescein-linked aldol derivative as a substrate, which upon reaction with an intracellular aldolase releases chloromethylfluorescein, which in turn reacts with nucleophiles causing cellular retention of fluorescence (Shim et al. 2004). The authors demonstrated the ability to apply this screen to isolate catalytic antibodies from a plasmid-encoded library introduced to and expressed in mammalian cells.

Glutathione transferases (GSTs) represent a family of detoxification isoenzymes that catalyze the conjugation of reactive electrophilic compounds to glutathione (GSH). Combinatorial libraries of GST were expressed in *E. coli* and screening was accomplished by monitoring conjugation of GSH with the fluorogenic substrate monochlorobimane (MCB), which gives rise to a unique fluorescent signal (Eklund et al. 2002). Using a similar FACS-based assay, Georgiou and coworkers evolved highly active GSTs using homology-independent recombination methods for library creation (Griswold et al. 2005). Mutant activity was screened using the substrate 7-amino-4-chloromethyl-coumarin (CMAC), which fluoresces upon conjugation with GSH (Griswold et al. 2005). Another example of the use of FACS to isolate enzymes with improved intracellular activity was reported by Kwon et al. (Kwon et al. 2004), who developed a ferrochelatase screen to improve metalloporphyrin production by *E. coli*. A library of 2.4 × 10$^6$ mutants was expressed in protoporphyrin-overproducing recombinant *E. coli* cells. The screen was based on the decrease in fluorescence of protoporphyrin IX upon ferrochelatase-catalyzed insertion of Fe(II), resulting in production of heme. In a final example, pH-sensitive mutants of GFP (pHluorins) can act as fluorescent sensors for pH shifts and offer great potential for reporting intracellular activity of hydrolytic enzymes (Schuster et al. 2005).

In order to achieve incorporation of unnatural amino acids at specific sites in a protein *in vivo*, Schultz and coworkers combined selection with FACS-based screening to evolve aminoacyl-tRNA synthetase specificity toward desired tyrosine analogs (Santoro et al. 2002). To accomplish this, a plasmid was constructed that carries the GFPuv gene under control of a T7 promoter, along with the gene encoding T7 RNA polymerase containing amber codons inserted at positions previously determined to "tolerate" amino-acid substitutions. The T7 polymerase should therefore be functional (and able to express GFPuv) as long as amino acids are incorporated at the amber codons during translation. The same plasmid additionally carries the gene encoding a chloramphenicol resistance reporter (Cm$^R$), also containing an amber codon, and the *Mj*YtRNA$_{CUA}$ gene encoding an orthogonal amber suppressor tRNA$^{Tyr}$ derived from *Methanococcus jannaschii*. This tRNA enables amino-acid incorporation at amber codons and can be aminoacylated by the *M. jannaschii* tyrosyl-tRNA synthetase (*Mj*YRS), and not by an *E. coli* tRNA synthetase. A second plasmid expresses a mutant library of 10$^9$ variants of *Mj*YRS, which was to be evolved to accept selectively chemical groups other than the phenolic side chain of tyrosine. Using *E. coli*, a round of selection on chloramphenicol in the presence of the desired tyrosine analog isolated clones expressing *Mj*YRS mutants that allow for amber codon incorporation leading to functional expression of Cm$^R$. This "positive" selection, however, does not eliminate clones expressing nonspecific $Mj$YRS mutants that load the $Mj$tRNA$^{Tyr}$ with amino acids other than the desired tyrosine analog. Amber suppression during T7 RNA polymerase expression and in the absence of the desired Tyr analog is the result of nonspecific amino-acid incorporation and enables GFPuv expression from the T7 promoter. "Negative" sorting by FACS was therefore next employed to eliminate fluorescent clones in the absence of the specific Tyr analog used in the positive selection. Recovery of nonfluorescent clones therefore yields $Mj$YRS mutants specific to only the unnatural amino acid that was used in the positive selection. The authors isolated variants that selectively accommodated one of a variety of unnatural amino acids (e.g., $p$-isopropyl-phenylalanine) and no natural amino acid.

The Schultz lab applied another FACS-based evolution approach to engineer Cre recombinase from bacteriophage P1 to mediate recombination at new DNA recognition sites (Santoro and Schultz 2002). In this system, a reporter plasmid contained the genes for two GFP variants having different excitation and fluorescence emission properties, EYFP and GFPuv. Only EYFP was placed downstream of a promoter. The plasmid also contained two natural or mutated Cre recognition sequences (*loxP*). A second plasmid contained a library of Cre variants (~10$^8$ library clones). Without Cre-mediated recombination, *E. coli* cells harboring the reporter plasmid express only EYFP. Cells expressing Cre mutants capable of recombination at the mutated recognition sequence reversibly rearrange the reporter plasmid so that GFPuv and EYFP exchange positions. This in turn causes those cells to express GFPuv from some plasmids, and EYFP from others. Positive sorting by FACS for GFPuv expression using mutated *loxP* sequences yielded Cre variants with activity toward the new recognition sequence. Subsequent negative sorting for clones only having EYFP fluorescence (and no GFPuv) using native *loxP* sequences in the reporter plasmid allowed for isolation of Cre mutants with selectivity toward the new *loxP* sequence and not the native one.

## CELL SURFACE DISPLAY

One of the key limitations to intracellular expression and screening of enzyme libraries is the requirement that substrate be transported across cell membranes. Other potential limitations include the possibility that the intracellular environment (e.g., pH, reduction potential) may not be suitable for supporting a properly folded and functional enzyme of interest (e.g., bacterial cytoplasms do not readily support protein disulfide bonds). Displaying the enzyme on the cell surface enables retention of a genotype-phenotype linkage and provides a potential solution to these intracellular screening problems. Surface display technologies coupled to binding-based assays are superb for selecting protein variants with improved ligand binding properties, but have proven difficult to implement for enzyme evolution since reaction products often diffuse away from the enzyme. Therefore, correlating a screenable cellular phenotype (typically fluorescence) with catalytic properties of the displayed enzyme poses a challenge.

In a pioneering application of these methods, Georgiou and coworkers displayed a library of the serine protease OmpT on the *E. coli* surface and used FACS to isolate variants with improved activity and substrate specificity (Olsen et al. 2000b). Their strategy is depicted in Figure 6.1. A cleavable fluorescence resonance energy transfer (FRET) peptide substrate containing a fluorophore and quenching partner separated by the target scissile bond was embedded on the cell surface via electrostatic interaction (Olsen et al. 2000b). Cleavage at the target peptide sequence results in fluorescence due to disruption of the FRET interaction, allowing for a correlation between activity and fluorescence. Using this approach protease activity on the sequence Arg-Val was improved 60-fold.

Applying directed evolution to improve enzyme activity typically results in mutants with relaxed substrate specificity, unless specificity is included in the screening criteria during evolution. Georgiou and coworkers improved their protease display system by incorporating a means of screening for substrate specificity (Varadarajan et al. 2005). In this system, the substrate with the desired scissile bond sequence was essentially the same as the FRET substrate described previously, in which cleavage results in cell fluorescence. Also included in the screen was a fluorescent "counter-selection" substrate containing an undesired protease cleavage site. The charge on the counter-selection substrate is such that it will not adhere to the cell surface unless cleaved. Positive selection and counter-selection substrates are different colors, allowing simultaneous activity and specificity screening using multicolor

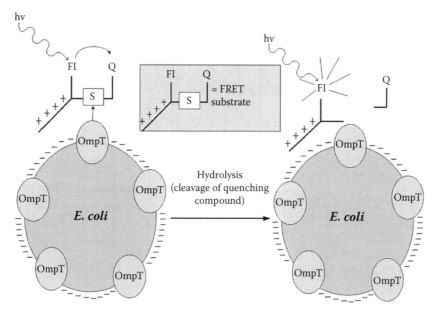

**FIGURE 6.1** FACS-based screen for protease mutants (Olsen et al. 2000b; Varadarajan et al. 2005). A FRET substrate containing fluorophore (Fl) and quenching compound (Q) binds to the surface of *E. coli* through electrostatic attraction. The quenching compound is released through hydrolysis of the scissile bond (S) by a surface-displayed OmpT mutant, resulting in a fluorescent signal on the cell surface.

flow cytometry to sort cells with the desired fluorescent properties (cells having fluorescence indicative of FRET substrate cleavage and not counter-selection substrate fluorescence). This approach resulted in an OmpT variant with both high activity on a non-native substrate as well as exceptional substrate specificity (Varadarajan et al. 2005). This approach has the potential to be generally applicable, but it requires substrates derivatized with FRET reagents and will be significantly more difficult to implement for enzyme reactions where substrates are not cleaved (e.g., oxidations or reductions). Similar reaction-specific methods of anchoring fluorescently labeled enzyme reaction products to the cell surface as a means of monitoring catalysis are now being developed for FACS screening (Wilhelm et al. 2007).

The use of bacteria for displaying and engineering eukaryotic proteins requiring disulfide bonds or posttranslational modifications has inherent limitations. Yeast surface display of proteins is an established technology and has recently been implemented for isolating variants of horseradish peroxidase (HRP) with enhanced enantioselectivity (Lipovsek et al. 2007). HRP contains four disulfide bonds and a heme prosthetic group, and is not expressed in a soluble form in bacteria. Two different mutant libraries of the HRP gene (one random mutation library generated by error-prone PCR and one library generated by active-site-directed saturation mutagenesis) were transformed into an *S. cerevisiae* strain engineered to display the expressed mutant proteins on the cell surface. HRP-catalyzed radical polymerization of cell surface tyrosine residues with fluorescently labeled L-tyrosinol or D-tyrosinol results in cell surface incorporation of the fluorescent label. To select for mutants with enantioselectivity for D-tyrosinol over L-tyrosinol, FACS was used in alternating rounds of positive and negative selection, by isolating clone populations with high fluorescence following exposure to labeled D-tyrosinol (positive selection) and low fluorescence following exposure to labeled L-tyrosinol (negative selection). The opposite selection strategy instead enriches mutants selective for L-tyrosinol. Screening the saturation mutagenesis library using this strategy yielded mutants with 3.8-fold and 7.5-fold improved selectivity for D- over L-tyrosinol and L- over D-tyrosinol, respectively (compared to wild-type HRP). In contrast, the most highly represented clones selected from the random mutagenesis library were not improved relative to wild-type.

## COMPARTMENTALIZATION OF WHOLE CELLS

Often a FACS-based whole-cell screen for enzyme activity is not possible, even when enzymatic activity can be linked to fluorescence. For example, the fluorescent molecule may rapidly diffuse out of the cell, or if the enzyme is displayed on the cell surface, the fluorescent product may not remain associated with the cell. In these situations compartmentalization of the cells in water-in-oil (w/o) emulsions can be used to keep the phenotype of interest (fluorescence production) associated with the genotype (the cell that produced the fluorescence) within a microdroplet (Taly et al. 2007). The w/o emulsion can be further converted into a water-in-oil-in-water (w/o/w) emulsion, which can be readily analyzed or sorted via flow cytometry. This was demonstrated by Tawfik and coworkers (Aharoni et al. 2005). A library of mutants of the mammalian enzyme serum paraoxonase (PON1) was expressed in *E.*

*coli*, and single cells were encapsulated in w/o/w droplets (<10 fL per drop). Droplets containing mutants with improved thiolactonase activity were isolated by FACS, using thiobutryolactone as substrate combined with a fluorogenic thiol-detecting dye that is retained in the droplets. More than $10^7$ mutants were screened, resulting in isolation of clones with 100-fold improved catalytic efficiency (Aharoni et al. 2005).

## In Vitro Compartmentalization

The compartmentalization techniques described in the previous section were first developed for *in vitro* genotype-phenotype linkage and screening applications, in a process termed *in vitro* compartmentalization (IVC) (Aharoni et al. 2005b; Boersma et al. 2007b; Farinas 2006; Taly et al. 2007). In this system, aqueous w/o or w/o/w emulsion microcompartments or microdroplets can act as artificial cells in which enzyme variants are expressed by *in vitro* transcription and translation. Following *in vitro* protein expression (IVE), the microdroplets are subject to a fluorescence-based enzyme activity assay. Microdroplets can be the size of a small bacterium, and $>10^{10}$ individual microdroplets per mL can be readily produced and screened by FACS. The IVC technique offers several potential benefits over the use of cells to express enzyme libraries. Substrates, products, and reaction conditions that are incompatible with *in vivo* systems are more likely to be compatible with emulsions. In addition, problems related to substrate transport across cell membranes are also eliminated, and the simplified environment of microdroplets as compared to cells reduces enzyme inhibition and cross-reactivity during screening. As an example, IVC was used to evolve the β-galactosidase activity of the *E. coli* protein Ebg, whose natural function is unknown (Mastrobattista et al. 2005). In this example the aqueous compartments of w/o/w emulsions contained a library of mutated *ebg* genes (on average one mutant gene per droplet), all necessary IVE reagents, and a fluorogenic substrate for screening β-galactosidase activity (fluorescein-di-β-D-galactopyranoside) by FACS. Following IVE the microdroplets were sorted by FACS and variants with more than 300-fold enhanced catalytic efficiency toward the fluorogenic substrate were isolated.

Another benefit of IVC for enzyme engineering is that the microdroplet environment during screening can vary significantly from that required for gene expression. If conditions are to be altered following IVE, the initial emulsions containing the genetic information and corresponding mutant enzyme may need to be broken so that new emulsions can be formed under the desired screening conditions. In this case it is necessary to physically link the genetic information to the expressed enzyme. This has been accomplished via microbead display, whereby a streptavidin-coated bead acts as scaffold to support a single copy of a biotinylated mutant gene. Following compartmentalized IVE, multiple copies of the expressed enzyme variant are captured onto the bead via affinity tags. Griffith and Tawfik used this method to screen for phosphotriesterase (PTE) mutants with improved activity (Griffiths and Tawfik 2003). Following IVE and protein capture in one emulsion, a second emulsion containing phosphortriester substrate was used to initiate the PTE reaction. In this example, product was also coupled on the beads, and after the PTE reaction the emulsion was broken and the beads were treated with fluorescently labeled antiproduct antibodies and screened by FACS. This method resulted in identification of a PTE mutant with 63-fold higher activity than wild-type enzyme (Griffiths and Tawfik 2003).

## *IN VIVO* SELECTIONS

### AUXOTROPH COMPLEMENTATION

The implementation of genetic selections in biocatalyst optimization has proven difficult. The most common examples involve auxotroph complementation, in which the enzymatic activity of interest (substrate depletion or product formation) is required for cell viability. Selection conditions often involve host genetic modifications (e.g., deleting genes that naturally serve the function of interest), as well as the use of carefully selected growth media. While many examples exist in which auxotroph complementation has been successfully implemented for enzyme engineering, this approach is not generally applicable and it is often not possible to couple enzymatic activity to cellular metabolism such that it is required for growth. This is particularly true for enzymes whose substrates are not native metabolites, or enzymes that are involved in secondary metabolism or other nonessential pathways (i.e., the majority of industrially relevant enzymes). Here we provide notable examples of enzyme engineering via auxotroph complementation.

To recover catalytic antibodies capable of decarboxylating orotate, Smiley and Benkovic screened a library of recombinant antibody fragments (Fabs) by complementation of a pyrimidine auxotroph *pyrF E. coli* strain [lacking orotidine-5'-phosphate (OMP) decarboxylase activity] grown on pyrimidine-free medium (Smiley and Benkovic 1994). The *pyrF* deletion blocks the native pathway for UMP production from orotate, but direct orotate decarboxylation to uracil results in a new pathway to UMP via uracil phosphoribosyltransferase. The combinatorial antibody library was isolated from mice immunized with a hapten designed to resemble the orotate-binding and catalytic portion of the active site of OMP decarboxylase. The library was then used to produce phage for the auxotroph complementation assay. This approach yielded a catalytic orotate decarboxylase antibody with specific activity estimated at ~$10^8$-times the background rate.

In another example, Boersma and coworkers demonstrated the ability to select for enantioselectivity of a lipase via auxotroph complementation (Boersma et al. 2008). In this example, *E. coli* deficient in aspartate synthesis were used to isolate mutants of *Bacillus subtilis* lipase A capable of hydrolyzing the aspartate ester of the desired enantiomer (S)-(+)-1,2-O-isopropylidene-sn-glycerol (IPG). A phosphonate ester of the undesired (R)-(-) enantiomer was included in the selection medium to inhibit growth of less enantioselective variants. A mutant was isolated with a switch in enantioselectivity: enantiomeric excess (ee) shifted from −29.6% to +73.1% toward the (S)-(+)-enantiomer.

HisA and TrpF catalyze the irreversible isomerization of an aminoaldose to an aminoketose in the biosynthetic pathways of histidine and tryptophan, respectively (Jurgens et al. 2000). Using random mutagenesis and auxotroph complementation of an *E. coli* strain deficient in TrpF, Jürgens and coworkers were able to isolate a variant of HisA from *Thermotoga maritima* possessing TrpF activity (Jurgens et al. 2000). As a final example, Nair and coworkers developed an *E. coli*-based selection system to isolate xylose reductase (XR) mutants with increased substrate specificity for D-xylose over L-arabinose (Nair and Zhao 2008). The *E. coli* enzyme XylA converts D-xylose to D-xylulose, while in yeasts xylose is reduced to xylitol via XR, and xylitol is then oxidized to xylulose via xylitol dehydrogenase (XDH). *E. coli xylA* was deleted so that growth on xylose required heterologous expression of functional XR and XDH. Selection on D-xylose thus ensured activity of XR mutants on D-xylose. The selection strain was also deficient in L-arabinose metabolism and over-expressed L-ribulokinase, capable of phosphorylating L-arabitol to a toxic metabolite. Thus, XR mutants able to reduce L-arabinose to L-arabitol would result in poor or no growth on medium containing D-xylose and L-arabinose, while mutants with improved specificity for D-xylose over L-arabinose would grow faster. This selection strain allowed for the evolution of an XR with nearly eightfold improved specificity for D-xylose over L-arabinose (Nair and Zhao 2008).

## CHEMICAL COMPLEMENTATION

An alternative to relying on enzymatic activity to affect metabolism and cell viability is to monitor enzyme activity indirectly via a genetic reporter system that controls cell viability (e.g., antibiotic resistance). This requires that the enzyme activity of interest communicate with the reporter system via substrate depletion or product formation. Hence we will call these *chemical reporter*, or *chemical complementation*, systems. A recent surge in the design of biomolecules as sensors and switches (Bayer and Smolke 2005; Buskirk et al. 2004; Chockalingam et al. 2005; de Lorimier et al. 2002; Derr et al. 2006; Doi and Yanagawa 1999; Dueber et al. 2003; Gryczynski and Schleif 2004; Guntas et al. 2005; Looger et al. 2003; Ostermeier 2005; Tucker and Fields 2001; Vuyisich and Beal 2002) has addressed a general biotechnological need for sensitive and specific molecular recognition and reporting. Although a logical extension of these technologies is in biocatalysis screening applications, a relatively small number of molecular recognition research efforts have focused on this important area.

Benkovic and coworkers demonstrated the use of a chemical inducer of dimerization (CID) in a three-hybrid system to correlate *in vivo* enzyme activity with expression of a reporter gene (Firestine et al. 2000). In this example, the CID bridges DNA-binding and transcriptional activation domains to activate transcription from the $P_{BAD}$ promoter in *E. coli*, and the bridging interaction is inhibited by the substrate of the desired enzyme reaction (in this case, scytalone dehydratase). Depletion of substrate via catalytic turnover resulted in increased reporter gene expression via assembly of the three-hybrid system (Firestine et al. 2000). Cornish and coworkers extended this concept of chemical complementation to a variety of yeast three-hybrid systems in which enzyme-mediated disruption or formation of the CID enables use of the reporter for enzyme library screening, as depicted in Figure 6.2 (Baker et al. 2002; Lin et al. 2004). A methotrexate-dexamethasone heterodimer with a cephem linker (Mtx-Cephem-Dex) was used as a CID for a LexA DNA-binding domain (DBD) fused to dihydrofolate reductase (DHFR, which binds to Mtx) and a B42 activation domain

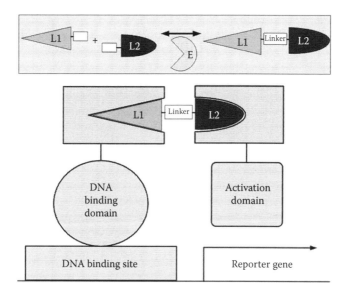

**FIGURE 6.2** Screening enzymatic activity by chemical complementation via the yeast three-hybrid system (Lin and Cornish 2002). A chemical inducer of dimerization (CID) contains ligands L1 and L2, which interact with a DNA binding domain and an activation domain, respectively. Enzyme (E) catalyzes a reaction that either forms or breaks a bond that links L1 to L2, resulting in formation or disruption of the CID and three-component transcriptional activating complex. The subsequent change in reporter gene expression is used to indicate enzyme activity.

(which binds to dexamethasone) fused to the glucocorticoid receptor (GR) (Baker et al. 2002). A LacZ reporter coupled to this system was used to screen for cephalosporinase activity via hydrolysis of the β-lactam bond in Mtx-Cephem-Dex. The same system was later developed to select for glycosynthase activity via formation of the CID (Lin et al. 2004). In this case, LEU2 served as the selection marker and Mtx and Dex were each derivatized with a disaccharide. Glycosynthase-catalyzed formation of a glycosidic bond between Mtx and Dex results in assembly of the CID and LEU2 expression. Applying this selection to screen a small library of the endoglucanase Cel7B yielded a mutant having five-fold increased glycosynthase activity.

## COLONY SCREENS

In many cases the same or very similar genetic constructs used in selections can be adapted to colony screening. That is, auxotroph complementation or chemical reporter systems can be used either to identify enzymes conferring sufficient activity to enable viability or to differentiate between more and less active variants in a library by screening colonies having more or less of the resulting phenotype. The former case allows for much higher throughput library analysis since the vast majority of variants having insufficient activity to sustain viability are never observed. This approach primarily provides "yes or no" information about mutants, although the stringency of selection conditions can often be adjusted (e.g., by changing plasmid copy number or enzyme expression levels, or increasing antibiotic concentration). The latter case is lower throughput as it involves screening all library clones based on colony properties such as size, fluorescence, color, and local clearing zone (visible depletion of a substrate).

There are many examples of colony-based screens, in which the enzymatic reaction of interest is correlated with a colony phenotype. Typically this phenotype can be identified by the naked eye under proper illumination conditions (e.g., color or colony size), although it is common to use digital imaging equipment so that the range of colony phenotypes can be more readily quantified and library statistics can be compiled. Here we describe a few interesting examples of colony-based screens for improved enzyme activity.

Bornscheuer and coworkers screened an esterase library expressed in *E. coli* by plating onto minimal medium agar plates containing the pH indicator substances neutral red and crystal violet (Bornscheuer et al. 1998). Esterase-catalyzed hydrolysis of a sterically hindered 3-hydroxy ester was detected visually by the formation of red color due to a decrease in pH. In another example, screening by colony color was used to identify novel carotenoid compounds synthesized by chimeric phytoene desaturase variants expressed in *E. coli*, in the context of a heterologously expressed carotenoid biosynthetic pathway (Schmidt-Dannert et al. 2000). One desaturase chimera introduced six rather than four double bonds into phytoene, to favor production of the fully conjugated carotenoid 3,4,3′,4′-tetradehydrolycopene. This new pathway was extended with a second library of shuffled lycopene cyclases to produce a variety of colored products including the cyclic compound torulene (Schmidt-Dannert et al. 2000).

Joern et al. describe a solid-phase, high-throughput screen for dioxygenase activity in which the cis-dihydrodiol product of dioxygenase bioconversion is converted to a phenol or catechol, which is subsequently reacted with Gibb's reagent to yield colored products easily detected by eye, spectrophotometer, or digital imaging (Joern et al. 2001). Finally, a colony size-based screen for ribulose 1,5-bisphosphate carboxylase/oxygenase (RuBisCO) variants with increased activity was developed (Parikh et al. 2006). The Calvin Cycle was partially reconstructed in *E. coli* such that the engineered strain required RuBisCO for growth in minimal media supplemented with a pentose. Random mutagenesis of a RuBisCO subunit gene and transformation into this engineered strain for screening resulted in colonies with different growth rates, the fastest of which corresponded to RuBisCO variants with enhanced specific activity.

## *IN VITRO* SELECTIONS

*In vitro* selection techniques allow for unhindered accessibility of substrate to enzyme and better control over reaction conditions compared to *in vivo* methods. Here we describe the use of phage and ribosome display for selecting improved enzymes. While these methods have been widely used as protein engineering tools, they rely on the ability of the expressed mutant protein to bind to a ligand and have primarily found applications in engineering protein affinity. Most demonstrations of the use of these methods for catalyst design have focused on method development, where the enzyme and reaction have been chosen primarily for proof of concept.

## PHAGE DISPLAY

Phage display is described in detail in Chapter 1 of this volume. Many enzymes and catalytic antibodies have been successfully displayed on phage coat proteins (Aharoni et al. 2005b; Boersma et al. 2007; Lin and Cornish 2002; Olsen et al. 2000a; Taylor et al. 2001). Enzymes lower reaction activation energy by stabilizing the high-energy, substrate-to-product transition state via binding energy. An active site with increased affinity toward a compound resembling a reaction transition state (yet unable to react with that compound) may then also have increased activity toward a substrate that is converted to product through the same transition state. In the spirit of traditional immunization protocols for affinity maturation, most examples of using phage display to design catalytic proteins have relied on active site affinity toward immobilized substrate, product, or transition state analog (TSA) ligands (Fernandez-Gacio et al. 2003). A suicide inhibitor is a compound that resembles a true substrate or transition state, but becomes an irreversible inhibitor of the enzyme of interest (by covalent attachment or irreversible binding) only after being reacted upon by that enzyme, and immobilized suicide inhibitors have been used to select for phage-displayed catalytic proteins (Hansson et al. 1997; Soumillion et al. 1994). Examples of these types include the use of a variety of phosphonate TSA compounds for isolation of phage-displayed catalytic antibodies with different esterase activities (Arkin and Wells 1998; Baca et al. 1997), and the use of immobilized ligands resembling transition states to isolate novel glutathione transferase activities from a library of glutathione transferase mutants expressed as fusions with pIII (Widersten and Mannervik 1995). In addition, the selection of *B. subtilis* lipase A variants with improved and inverted enantioselectivity toward a chiral substrate (IPG) was achieved using phosphonate esters of each substrate enantiomer (Droge et al. 2006). In this example, the use of both enantiomers in two selection stages allowed for isolation of mutants that could bind only one of the substrates.

In general, the indirect approaches for selecting catalysts are limited. As in the case of cell surface display, it is preferable to directly select on the basis of catalytic turnover whenever possible. Since reaction products readily diffuse away from the reaction site (in this case the phage surface), these methods rely either on a physical link between phage and substrate, or on reaction chemistry that forms a link upon catalysis. Schultz and coworkers demonstrated the ability to isolate phage-displayed nucleases based on their ability to catalyze "intraphage" hydrolysis of an oligonucleotide tether that releases the phage particle from a support and allows for its capture (Pedersen et al. 1998). Using a similar approach, Demartis et al. developed a system in which the enzymes to be screened were fused to calmodulin and displayed on the surface of phage (Demartis et al. 1999). This allowed for the use of high affinity calmodulin-binding peptides to target substrate to the phage surface via noncovalent interactions. Active enzymes displayed on the phage surface converted neighboring substrates to products that remain tethered to the particle. Antibodies with affinity for the anchored product were then used to capture phage particles.

In another example, Ponsard et al. used a catalytic elution technique to select for increased β-lactamase activity toward benzylpenicillin from a library of phage-displayed, zinc-dependent metallo-β-lactamase from *Bacillus cereus* (Ponsard et

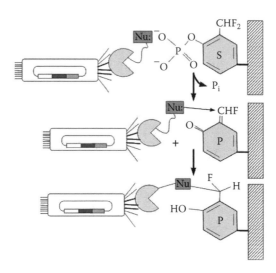

**FIGURE 6.3** Use of phage display for turnover-dependent *in vitro* selection (Cesaro-Tadic et al. 2003). Hydrolysis of phosphate monoester substrate (S) catalyzed by a phage-displayed antibody creates an electrophilic product (P). A nucleophile (Nu:) bound to the antibody covalently attaches to the product, allowing for capture of phage displaying catalytically active antibodies.

al. 2001). Here, removal of the essential cofactor zinc from the library of phage-displayed enzymes allowed for their binding to the immobilized benzylpenicillin substrate, but not catalysis. Nonbinding phage were washed away, and addition of zinc reactivated catalytic mutants, resulting in their elution and recovery.

Atwell and Wells selected phage-displayed subtiligases with increased ligase activity based on their ability to catalyze ligation of their own extended N-termini to a biotin-labeled peptide, which was subsequently captured (Atwell and Wells 1999). In a final example, Plückthun and coworkers employed a turnover-based *in vitro* selection method to isolate novel catalytic phosphatase antibodies from a naive library of antibodies displayed on the phage coat. As depicted in Figure 6.3, phosphatase activity was selected for using an immobilized aryl phosphate substrate, which upon hydrolysis covalently attaches to (and thus immobilizes) the phage particle displaying the corresponding active antibody (Cesaro-Tadic et al. 2003).

## Ribosome Display

With ribosome display, a genotype-phenotype linkage is achieved by physically linking the mRNA transcript of a mutant gene to its encoded protein via the ribosome (Lin and Cornish 2002). Transcription and translation are performed *in vitro*, and the ribosome remains bound to the mRNA due to the lack of a stop codon and/or the addition of translation inhibitors. The mRNA-ribosome-protein complex can be directly used for selection. The recovered mRNA-protein hybrids are then reverse transcribed, and the corresponding cDNA is amplified to identify the genotype of the selected proteins. A detailed description of the ribosome display platform is described in Chapter

3 of this volume. Since cell transformations are not necessary with ribosome display, this technique offers the potential to screen significantly larger gene libraries than what can be achieved with cloning-dependent procedures. While ribosome display has been a popular and successful technique for evolving functional binding proteins (Yan and Xu 2006), it has not been widely used for enzyme evolution, largely owing to difficulties similar to those associated with phage and cell surface display (i.e., linking binding to catalytic activity). Amstutz et al. reported the first use of ribosome display for selection of enzyme activity (Amstutz et al. 2002). Their approach involved the use of a biotinylated suicide inhibitor of β-lactamase to select for catalytic turnover of ribosome-displayed β-lactamase. After translation, displayed proteins were incubated with the inhibitor and rescued with streptavidin-coated beads. From a mixture of ribosome-displayed wild-type β-lactamase and an inactive β-lactamase mutant, the active mutant was enriched >100-fold per round of selection.

## HIGH-THROUGHPUT SCREENING IN MICROTITER PLATES

Perhaps the most traditional and straightforward high-throughput screening approaches involve monitoring the activities of individual enzyme variants expressed in clones separated in the wells of microtiter plates. These assays typically involve a spectrophotometric or fluorimetric readout and may be adapted to measure *in vivo* activity from whole or permeabilized cells, or more commonly *in vitro* activity from cell lysates, in which case a high-throughput and reproducible cell lysis procedure must be implemented. The phenotype measured in a well is directly linked to the clone (genotype) contained in that same well, or to the corresponding clone used for inoculation, located at a spatial address either in a master liquid culture well plate or as a colony on an agar plate. These screens often require robotics for colony picking or liquid handling and in general are significantly lower throughput than the other approaches described in this chapter (no more than ~$10^5$ variants can be screened). However, the basic enzyme directed evolution approach of iteratively screening small libraries of variants containing few mutations for modest improvements in each round of screening has repeatedly proven to be an effective strategy. These lower throughput screens also typically offer the benefit of providing detailed kinetic or biochemical information about library members, since each variant's kinetics or product profile can be monitored (e.g., via microtiter plates or mass spectrometry, respectively). And whereas it may be difficult or impossible to develop selection, surface display, or *in vivo* screening systems for an enzymatic reaction of interest, it is often possible to develop a microtiter plate-based, liquid-phase, high-throughput screening assay.

A common liquid-based high-throughput enzyme screen might proceed as follows. The gene library of interest is cloned into an expression vector and transformed into the microbial screening host. Individual colonies representing individual mutant clones are picked and transferred to agar plate grids or into microtiter plate wells. These serve as the "master" plates in which each mutant has a defined spatial address. Each clone is then used to inoculate a microtiter plate culture (e.g., 96-well or 384-well; 20 μL, 200 μL, or 1 mL culture volumes) for expression of the mutant enzymes. If the enzyme is not secreted and the reaction cannot be monitored *in vivo*, the cells must be chemically lysed prior to screening. A typical high-throughput lysis procedure may involve centrifugation of the microtiter plates to pellet the cells within each well, removal of culture broth, addition of a lysis buffer (e.g., containing lysozyme and/or detergents), high-throughput cell resuspension and lysis, centrifugation of the plates to pellet lysed cell debris, and transfer of clarified lysates to microtiter plates containing the reagents required for spectrophotometrically assaying enzyme activity.

The requirement for the generation of a catalysis-correlated signal that is rapidly detected in a high-throughput manner (e.g., via change in color or fluorescence) often requires that the screening conditions or chemistry be modified from the intended reaction. A common example is the use of a "surrogate" substrate that resembles the substrate of interest but with readily measured spectral properties that change upon reaction. This in turn requires that improved variants be carefully rescreened under the conditions in which the enzyme is intended to be used, and with the intended substrate. This is becoming less of a restriction with the increasing availability of high-throughput chromatography and mass spectrometry technologies, since these do not require modification of the enzyme reaction of interest for accommodation of spectroscopic analysis. The many examples of high-throughput, liquid-based enzyme assays are each unique to the desired conditions and reactions catalyzed. Many popular screening assays are described in a volume entitled *Directed Enzyme Evolution: Screening and Selection Methods* (Arnold and Georgiou 2003). Here we briefly provide representative examples.

Enzymatic redox reactions that consume or produce the reduced cofactor NADH or NADPH are readily monitored spectrophotometrically due to the characteristic absorbance peak of these compounds at 340 nm (e.g., Glieder et al. 2002). However, high-throughput screening of NAD(P)H depletion or production from crude cell lysates is often complicated by background activities and high background absorbance by the lysate and plastic screening plates. The 340 nm signal also may not be sufficiently strong for detection, particularly if activity is low on the substrate used in the screen. In these cases, it may be preferable to indirectly monitor NAD(P)H by reduction of a tetrazolium salt such as nitroblue tetrazolium (NBT) to a formazan dye, allowing for sensitive colorimetric analysis at 580 nm, as depicted in Figure 6.4. The redox mediator phenazine methosulfate (PMS) is commonly used to enhance electron transfer from NAD(P)H to NBT. Whereas direct spectrophotometric monitoring of NAD(P)H is amenable to kinetic analysis, the NBT/PMS assay is typically used as an end-point assay. Mayer and Arnold describe a version of this assay suitable for high-throughput screening of the kinetics of 6-phosphogluconate dehydrogenase in crude cell lysates (Mayer and Arnold 2002).

Drawbacks to screening catalytic activity via monitoring cofactor utilization rather than substrate consumption or product formation include the fact that the screen is not sensitive to substrate or reaction specificity, and the possibility of selecting for variants with reduced coupling efficiency (i.e., more NAD(P)(H) is consumed but without concomitant substrate reduction or oxidation). Cytochrome P450 BM-3 is a soluble, NADPH-dependent fatty acid hydroxylase. Schwaneberg and coworkers developed a continuous spectrophotometric assay for regiospecific P450 BM-3-catalyzed hydroxylation using fatty acid surrogate substrates (Schwaneberg et al. 1999). Here, ω-hydroxylation of *p*-nitrophenoxycarboxylic acids (the fatty

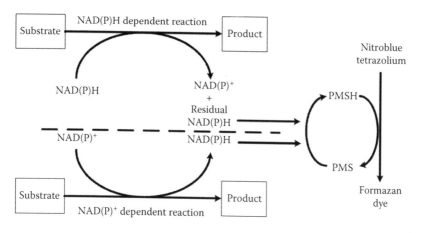

**FIGURE 6.4** Colorimetric screens for NAD(P)H depletion (top) or formation (bottom). The cofactor NAD(P)H is indirectly used to screen for catalytic activity in NAD(P)H- and NAD(P)⁺-dependent reactions. Residual or generated NAD(P)H reduces PMS to PMSH, which mediates reduction of nitroblue tetrazolium to formazan dye, readily detected at 580 nm.

acid surrogate) results in formation of an unstable hemiacetal that dissociates to the ω-oxycarboxylate and the readily monitored chromophore *p*-nitrophenolate. While the use of surrogate substrates enable high-throughput screening of reactions that are otherwise difficult to monitor (e.g., hydroxylation), an important disadvantage is that the selected mutants may have poorer activity on the actual substrate of interest.

Reetz and coworkers have described a variety of medium- to high-throughput liquid handling assays designed to screen for enzyme enantioselectivity, and these have been thoroughly reviewed (Reetz 2003; Reetz 2006). Assays are based on a variety of detection methods including UV/Vis spectroscopy, fluorescence, NMR spectroscopy, gas chromatography, and mass spectrometry. As an example, in the directed evolution of enantioselective hydrolytic desymmetrization of meso-1,4-diacetoxy-cyclopentene by a lipase from *Bacillus subtilis*, Reetz synthesized the pseudomeso-substrate in which one acetyl group is deuterated. To determine enantioselectivity of enzyme library variants, a mass spectrometry-based high throughput screen was used to distinguish between hydrolysis products, as they differ in absolute configuration and mass (Reetz 1999). Similar pseudomeso-¹³C-labeled compounds were synthesized for NMR-based high-throughput enzyme screening for enantioselectivity (Reetz 2002).

## CONCLUSIONS

In this chapter we have summarized common and innovative screening and selection approaches to identify enzymes with improved catalytic properties from large combinatorial libraries. While many of the methods are similar to their counterparts used for screening nonenzymatic protein properties, important modifications must be made to ensure the enzyme variants are catalytically active. As in the case of engineering noncatalytic proteins, the screening or selection approach implemented will depend on a variety of factors. Important criteria include the cloning capabilities of the chosen expression host; whether the enzyme can be functionally expressed in the cytoplasm, on the cell surface, or *in vitro*; and whether substrate can be transported into the cell without lysis or permeabilization. The reaction-specific link between catalytic turnover and an observable phenotype or affinity capture presents unique enzyme-dependent challenges for protein library screening.

Library sizes that can be analyzed using genetic selections and ultra-high-throughput sorting- and binding-based screens (i.e., phage and cell surface display and microcompartmentalization) are limited mainly by the efficiency of library construction and transformation. In contrast, the throughputs of colony and microtiter plate-based screens limit library size, although these methods generally provide more detailed biochemical information about each library member screened. Iterative rounds of screening relatively small libraries using lower throughput, but information rich, methods is an effective approach for improving upon an existing catalytic property by accumulating random point mutations throughout the protein. However, if more diverse regions of sequence space must be explored to identify extremely rare mutants possessing a unique property of interest, a higher throughput method is necessary. An example is the use of saturation mutagenesis to randomize multiple residue positions within an active site so that a novel substrate can be accepted.

While a great variety of combinatorial enzyme library screening methods have been developed, the success of enzyme directed evolution is still ultimately limited by screening capabilities. Enzyme engineers are continuously seeking ways to increase the fitness density of libraries so fewer variants need be searched and to improve the throughput of screening methods without compromising accuracy or information garnered. Steady advances in the speed, cost, and technologies associated with robotics, microfluidics, FACS, chemical analytics, and *in vitro* and *in vivo* molecular recognition continuously guide improvements in enzyme library screening.

The authors thank the National Science Foundation for support under Grant No. 0644678.

## REFERENCES

Aharoni, A., G. Amitai, K. Bernath, S. Magdassi and D. S. Tawfik. 2005a. "High-throughput screening of enzyme libraries: Thiolactonases evolved by fluorescence-activated sorting of single cells in emulsion compartments." *Chem Biol* 12: 1281–9.

Aharoni, A., A. D. Griffiths and D. S. Tawfik. 2005b. "High-throughput screens and selections of enzyme-encoding genes." *Curr Opin Chem Biol* 9: 210–6.

Aharoni, A., K. Thieme, C. P. Chiu, S. Buchini, L. L. Lairson, et al. 2006. "High-throughput screening methodology for the directed evolution of glycosyltransferases." *Nat Methods* 3: 609–14.

Amstutz, P., J. N. Pelletier, A. Guggisberg, L. Jermutus, S. Cesaro-Tadic, et al. 2002. "In vitro selection for catalytic activity with ribosome display." *J Am Chem Soc* 124: 9396–403.

Arkin, M. R. and J. A. Wells. 1998. "Probing the importance of second sphere residues in an esterolytic antibody by phage display." *J Mol Biol* 284: 1083–94.

Arnold, F. H. and G. Georgiou. 2003. *Directed enzyme evolution: Screening and selection methods.* Humana Press, Totowa, N.J.

Atwell, S. and J. A. Wells. 1999. "Selection for improved subtiligases by phage display." *Proc Natl Acad Sci U S A* 96: 9497–502.

Baca, M., T. S. Scanlan, R. C. Stephenson and J. A. Wells. 1997. "Phage display of a catalytic antibody to optimize affinity for transition-state analog binding." *Proc Natl Acad Sci U S A* 94: 10063–8.

Baker, K., C. Bleczinski, H. Lin, G. Salazar-Jimenez, D. Sengupta, et al. 2002. "Chemical complementation: A reaction-independent genetic assay for enzyme catalysis." *Proc Natl Acad Sci U S A* 99: 16537–42.

Bayer, T. S. and C. D. Smolke. 2005. "Programmable ligand-controlled riboregulators of eukaryotic gene expression." *Nat Biotechnol* 23: 337–43.

Becker, S., H. U. Schmoldt, T. M. Adams, S. Wilhelm and H. Kolmar. 2004. "Ultra-high-throughput screening based on cell-surface display and fluorescence-activated cell sorting for the identification of novel biocatalysts." *Curr Opin Biotechnol* 15: 323–9.

Boersma, Y. L., M. J. Droge and W. J. Quax. 2007. "Selection strategies for improved biocatalysts." *FEBS J* 274: 2181–95.

Boersma, Y. L., M. J. Droge, A. M. van der Sloot, T. Pijning, R. H. Cool, B. W. Dijkstra, et al. 2008. "A novel genetic selection system for improved enantioselectivity of *Bacillus subtilis* lipase A." *Chembiochem* 9: 1110–5.

Bornscheuer, U. T., J. Altenbuchner and H. H. Meyer. 1998. "Directed evolution of an esterase for the stereoselective resolution of a key intermediate in the synthesis of epothilones." *Biotechnol Bioeng* 58: 554–9.

Buskirk, A. R., Y. C. Ong, Z. J. Gartner and D. R. Liu. 2004. "Directed evolution of ligand dependence: Small-molecule-activated protein splicing." *Proc Natl Acad Sci U S A* 101: 10505–10.

Cesaro-Tadic, S., D. Lagos, A. Honegger, J. H. Rickard, L. J. Partridge, et al. 2003. "Turnover-based in vitro selection and evolution of biocatalysts from a fully synthetic antibody library." *Nat Biotechnol* 21: 679–85.

Chockalingam, K., Z. Chen, J. A. Katzenellenbogen and H. Zhao. 2005. "Directed evolution of specific receptor-ligand pairs for use in the creation of gene switches." *Proc Natl Acad Sci U S A* 102: 5691–6.

de Lorimier, R. M., J. J. Smith, M. A. Dwyer, L. L. Looger, K. M. Sali, et al. 2002. "Construction of a fluorescent biosensor family." *Protein Sci* 11: 2655–75.

Demartis, S., A. Huber, F. Viti, L. Lozzi, L. Giovannoni, et al. 1999. "A strategy for the isolation of catalytic activities from repertoires of enzymes displayed on phage." *J Mol Biol* 286: 617–33.

Derr, P., E. Boder and M. Goulian. 2006. "Changing the specificity of a bacterial chemoreceptor." *J Mol Biol* 355: 923–32.

Doi, N. and H. Yanagawa. 1999. "Design of generic biosensors based on green fluorescent proteins with allosteric sites by directed evolution." *FEBS Lett* 453: 305–7.

Droge, M. J., Y. L. Boersma, G. van Pouderoyen, T. E. Vrenken, C. J. Ruggeberg, et al. 2006. "Directed evolution of *Bacillus subtilis* lipase A by use of enantiomeric phosphonate inhibitors: Crystal structures and phage display selection." *Chembiochem* 7: 149–57.

Dueber, J. E., B. J. Yeh, K. Chak and W. A. Lim. 2003. "Reprogramming control of an allosteric signaling switch through modular recombination." *Science* 301: 1904–8.

Eklund, B. I., M. Edalat, G. Stenberg and B. Mannervik. 2002. "Screening for recombinant glutathione transferases active with monochlorobimane." *Anal Biochem* 309: 102–8.

Farinas, E. T. 2006. "Fluorescence activated cell sorting for enzymatic activity." *Comb Chem High Throughput Screen* 9: 321–8.

Fernandez-Gacio, A., M. Uguen and J. Fastrez. 2003. "Phage display as a tool for the directed evolution of enzymes." *Trends Biotechnol* 21: 408–14.

Firestine, S. M., F. Salinas, A. E. Nixon, S. J. Baker and S. J. Benkovic. 2000. "Using an AraC-based three-hybrid system to detect biocatalysts in vivo." *Nat Biotechnol* 18: 544–7.

Georgiou, G. 2000. "Analysis of large libraries of protein mutants using flow cytometry." *Adv Protein Chem* 55: 293–315.

Glieder, A., E. T. Farinas and F. H. Arnold. 2002. "Laboratory evolution of a soluble, self-sufficient, highly active alkane hydroxylase." *Nat Biotechnol* 20: 1135–9.

Griffiths, A. D. and D. S. Tawfik. 2003. "Directed evolution of an extremely fast phosphotri-esterase by in vitro compartmentalization." *EMBO J* 22: 24–35.

Griswold, K. E., Y. Kawarasaki, N. Ghoneim, S. J. Benkovic, B. L. Iverson, et al. 2005. "Evolution of highly active enzymes by homology-independent recombination." *Proc Natl Acad Sci U S A* 102: 10082–7.

Gryczynski, U. and R. Schleif. 2004. "A portable allosteric mechanism." *Proteins* 57: 9–11.

Guntas, G., T. J. Mansell, J. R. Kim and M. Ostermeier. 2005. "Directed evolution of protein switches and their application to the creation of ligand-binding proteins." *Proc Natl Acad Sci U S A* 102: 11224–9.

Hansson, L. O., M. Widersten and B. Mannervik. 1997. "Mechanism-based phage display selection of active-site mutants of human glutathione transferase A1-1 catalyzing SNAr reactions." *Biochemistry* 36: 11252–60.

Hibbert, E. G. and P. A. Dalby. 2005. "Directed evolution strategies for improved enzymatic performance." *Microb Cell Fact* 4: 29.

Joern, J. M., T. Sakamoto, A. Arisawa and F. H. Arnold. 2001. "A versatile high throughput screen for dioxygenase activity using solid-phase digital imaging." *J Biomol Screen* 6: 219–23.

Jurgens, C., A. Strom, D. Wegener, S. Hettwer, M. Wilmanns, et al. 2000. "Directed evolution of a (beta alpha)8-barrel enzyme to catalyze related reactions in two different metabolic pathways." *Proc Natl Acad Sci U S A* 97: 9925–30.

Kwon, S. J., R. Petri, A. L. DeBoer and C. Schmidt-Dannert. 2004. "A high-throughput screen for porphyrin metal chelatases: Application to the directed evolution of ferrochelatases for metalloporphyrin biosynthesis." *Chembiochem* 5: 1069–74.

Lin, H. and V. W. Cornish. 2002. "Screening and selection methods for large-scale analysis of protein function." *Angew Chem Int Ed Engl* 41: 4402–25.

Lin, H., H. Tao and V. W. Cornish. 2004. "Directed evolution of a glycosynthase via chemical complementation." *J Am Chem Soc* 126: 15051–9.

Link, A. J., K. J. Jeong and G. Georgiou. 2007. "Beyond toothpicks: New methods for isolating mutant bacteria." *Nat Rev Microbiol* 5: 680–8.

Looger, L. L., M. A. Dwyer, J. J. Smith and H. W. Hellinga. 2003. "Computational design of receptor and sensor proteins with novel functions." *Nature* 423: 185–90.

Lipovsek, D., E. Antipov, K. A. Armstrong, M. J. Olsen, A. M. Klibanov, et al. 2007. "Selection of horseradish peroxidase variants with enhanced enantioselectivity by yeast surface display." *Chem Biol* 14: 1176–85.

Mastrobattista, E., V. Taly, E. Chanudet, P. Treacy, B. T. Kelly, et al. 2005. "High-throughput screening of enzyme libraries: In vitro evolution of a beta-galactosidase by fluorescence-activated sorting of double emulsions." *Chem Biol* 12: 1291–300.

May, O., P. T. Nguyen and F. H. Arnold. 2000. "Inverting enantioselectivity by directed evolution of hydantoinase for improved production of L-methionine." *Nat Biotechnol* 18: 317–20.

Mayer, K. M. and F. H. Arnold. 2002. "A colorimetric assay to quantify dehydrogenase activity in crude cell lysates." *J Biomol Screen* 7: 135–40.

Nair, N. U. and H. Zhao. 2008. "Evolution in reverse: Engineering a D-xylose-specific xylose reductase." *Chembiochem* 9: 1213–5.

Olsen, M., B. Iverson and G. Georgiou. 2000a. "High-throughput screening of enzyme libraries." *Curr Opin Biotechnol* 11: 331–7.

Olsen, M. J., D. Stephens, D. Griffiths, P. Daugherty, G. Georgiou, et al. 2000b. "Function-based isolation of novel enzymes from a large library." *Nat Biotechnol* 18: 1071–4.

Ostermeier, M. 2005. "Engineering allosteric protein switches by domain insertion." *Protein Eng Des Sel* 18: 359–64.

Parikh, M. R., D. N. Greene, K. K. Woods and I. Matsumura. 2006. "Directed evolution of RuBisCO hypermorphs through genetic selection in engineered *E. coli*." *Protein Eng Des Sel* 19: 113–9.

Pedersen, H., S. Holder, D. P. Sutherlin, U. Schwitter, D. S. King, et al. 1998. "A method for directed evolution and functional cloning of enzymes." *Proc Natl Acad Sci U S A* 95: 10523–8.

Ponsard, I., M. Galleni, P. Soumillion and J. Fastrez. 2001. "Selection of metalloenzymes by catalytic activity using phage display and catalytic elution." *Chembiochem* 2: 253–9.

Reetz, M. T., M. H. Becker, H.W. Klein, and D. Stockigt 1999. "A method for high-throughput screening of enantioselective catalysts." *Angew Chem Int Ed* 38: 1758–1761.

Reetz, M. T., A. Eipper, P. Tielmann, and R. Mynott 2002. "A practical NMR-based high-throughput assay for screening enantioselective catalysts and biocatalysts." *Advanced Synthesis & Catalysis* 344: 1008–1016.

Reetz, M. T. 2003. "An overview of high-throughput screening systems for enantioselective enzymatic transformations." *Methods Mol Biol* 230: 259–82.

Reetz, M. T. 2004. "Controlling the enantioselectivity of enzymes by directed evolution: Practical and theoretical ramifications." *Proc Natl Acad Sci U S A* 101: 5716–22.

Reetz, M. T. 2006. in *Advances in Catalysis*.

Reetz, M. T. and L. W. Wang. 2006. "High-throughput selection system for assessing the activity of epoxide hydrolases." *Comb Chem High Throughput Screen* 9: 295–9.

Reetz, M. T., L. W. Wang and M. Bocola. 2006. "Directed evolution of enantioselective enzymes: Iterative cycles of CASTing for probing protein-sequence space." *Angew Chem Int Ed Engl* 45: 1236–41.

Santoro, S. W. and P. G. Schultz. 2002. "Directed evolution of the site specificity of Cre recombinase." *Proc Natl Acad Sci U S A* 99: 4185–90.

Santoro, S. W., L. Wang, B. Herberich, D. S. King and P. G. Schultz. 2002. "An efficient system for the evolution of aminoacyl-tRNA synthetase specificity." *Nat Biotechnol* 20: 1044–8.

Schmidt-Dannert, C. and F. H. Arnold. 1999. "Directed evolution of industrial enzymes." *Trends Biotechnol* 17: 135–6.

Schmidt-Dannert, C., D. Umeno and F. H. Arnold. 2000. "Molecular breeding of carotenoid biosynthetic pathways." *Nat Biotechnol* 18: 750–3.

Schuster, S., M. Enzelberger, H. Trauthwein, R. D. Schmid and V. B. Urlacher. 2005. "pHluorin-based in vivo assay for hydrolase screening." *Anal Chem* 77: 2727–32.

Schwaneberg, U., C. Otey, P. C. Cirino, E. Farinas and F. H. Arnold. 2001. "Cost-effective whole-cell assay for laboratory evolution of hydroxylases in *Escherichia coli*." *J Biomol Screen* 6: 111–7.

Schwaneberg, U., C. Schmidt-Dannert, J. Schmitt and R. D. Schmid. 1999. "A continuous spectrophotometric assay for P450 BM-3, a fatty acid hydroxylating enzyme, and its mutant F87A." *Anal Biochem* 269: 359–66.

Shim, H., A. Karlstrom, S. M. Touami, R. P. Fuller and C. F. Barbas, 3rd. 2004. "Flow cytometric screening of aldolase catalytic antibodies." *Bioorg Med Chem Lett* 14: 4065–8.

Smiley, J. A. and S. J. Benkovic. 1994. "Selection of catalytic antibodies for a biosynthetic reaction from a combinatorial cDNA library by complementation of an auxotrophic *Escherichia coli*: Antibodies for orotate decarboxylation." *Proc Natl Acad Sci U S A* 91: 8319–23.

Soumillion, P., L. Jespers, M. Bouchet, J. Marchand-Brynaert, P. Sartiaux, et al. 1994. "Phage display of enzymes and in vitro selection for catalytic activity." *Appl Biochem Biotechnol* 47: 175–89; discussion 189–90.

Taly, V., B. T. Kelly and A. D. Griffiths. 2007. "Droplets as microreactors for high-throughput biology." *Chembiochem* 8: 263–72.

Taylor, S. V., P. Kast and D. Hilvert. 2001. "Investigating and engineering enzymes by genetic selection." *Angew Chem Int Ed Engl* 40: 3310–35.

Tucker, C. L. and S. Fields. 2001. "A yeast sensor of ligand binding." *Nat Biotechnol* 19: 1042–6.

Varadarajan, N., J. Gam, M. J. Olsen, G. Georgiou and B. L. Iverson. 2005. "Engineering of protease variants exhibiting high catalytic activity and exquisite substrate selectivity." *Proc Natl Acad Sci U S A* 102: 6855–60.

Vuyisich, M. and P. A. Beal. 2002. "Controlling protein activity with ligand-regulated RNA aptamers." *Chem Biol* 9: 907–13.

Widersten, M. and B. Mannervik. 1995. "Glutathione transferases with novel active sites isolated by phage display from a library of random mutants." *J Mol Biol* 250: 115–22.

Wilhelm, S., F. Rosenau, S. Becker, S. Buest, S. Hausmann, et al. 2007. "Functional cell-surface display of a lipase-specific chaperone." *Chembiochem* 8: 55–60.

Yan, X. and Z. Xu. 2006. "Ribosome-display technology: Applications for directed evolution of functional proteins." *Drug Discov Today* 11: 911–6.

# 7 Engineering of Therapeutic Proteins

*Fei Wen, Sheryl B. Rubin-Pitel, and Huimin Zhao*

## CONTENTS

We live in a pharmaceutical age with access to more medicines than any other time in human history. Among these, there are two major classes of drugs in the market: small molecule drugs and protein therapeutics. Although the former is currently the principal therapeutic agent, the impact of protein therapeutics is increasing thanks to advances in recombinant DNA technology and improved understanding of disease pathology. Currently, the U.S. Food and Drug Administration (FDA) has approved over 150 different protein drugs including monoclonal antibodies (mAbs), insulin, erythropoietin, interferons, and somatropin (human growth hormone) (Wishart et al. 2008). The protein therapeutics market is continuously growing and has more than doubled in the past few years, jumping from $25 billion in 2001 to $57 billion in 2006 with strong sales in insulin, erythropoietin, and interferon segments (KaloramaInformation 2006; RNCOS 2007). In addition to small molecule drugs and protein therapeutics, there are also other approaches being researched, including gene therapy, RNA interference (RNAi), stem cells, and nanotechnology-based solutions. These technologies hold great promise as future pharmaceuticals, but they are still a long way from routine use. This chapter will mainly focus on therapeutic proteins that are either in clinical use or under development. After a brief overview of the molecular basis of protein therapeutics, we will discuss some recent advances in protein engineering and design technologies, with an emphasis on diversity-oriented methods and their applications to protein therapeutics.

## PROTEIN THERAPEUTICS VERSUS SMALL MOLECULE DRUGS

Small molecule drugs have several advantages, including oral bioavailability, ability to reach intracellular targets, ease of manufacturing, and generally a long shelf life. These characteristics make them favorable over protein drugs in the pharmaceutical industry (see the section titled "Challenges in Pharmaceutical Translation of New Therapeutic Proteins," in this chapter). However, small molecule drugs, which typically have a molecular weight less than 1000 Daltons, have limited surface area available to contact a target protein. Furthermore, forming a favorable interaction requires the presence of a deep hydrophobic pocket in the target protein, limiting the number of potential druggable targets (Hopkins and Groom 2002; Hopkins and Groom 2003). In contrast, protein drugs are usually large in size and do not have this limitation, making them indispensable therapeutic tools for human disease treatment. By using protein therapeutics, some debilitating diseases that were previously untreatable, such as chronic renal failure, dwarfism, and infertility, are now successfully managed (Johnson-Leger et al. 2006).

Protein therapeutics have higher binding selectivity and specificity compared to small molecule drugs; therefore they can target specific steps in disease pathology. For example, before the advent of protein therapeutics, drugs used to suppress the immune system in chronic inflammatory disorders were limited to small molecule drugs, such as corticosteroids and cyclosporine A. These drugs act broadly and inhibit both protective and harmful immune responses indiscriminately, thus having serious side effects. In contrast, mAbs such as Infliximab (Remicade®, Centocor Inc.) are considered immune-modulating. Infliximab targets tumor necrosis factor-α, a key proinflammatory cytokine in the pathogenesis of chronic immune disorders, and leaves the protective immune response intact (Rutgeerts et al. 2006). In the past decades, the development of new protein therapeutics has revised the treatment paradigm of certain diseases (Flamant and Bourreille 2007; Gergely and Fekete 2007), and they are gradually replacing or supplementing small molecule drug therapies (Johnson-Leger et al. 2006; Eng 2007; Flamant and Bourreille 2007; Gergely and Fekete 2007).

## SOURCES OF PROTEIN THERAPEUTICS

The human body has evolved an elegant immune system that helps combat and control diseases. Insufficient, deficient, or improper action of any molecular component of the immune system results in disorders to various extents. Therefore, extrinsic modulation of the immune system using natural human immune regulators represents an appealing strategy to cure diseases, and the human genome, completely sequenced in 2003, provides a huge source for drug target mining. In fact, before the introduction of recombinant DNA technology, therapeutic proteins such as growth hormone and follicle stimulating hormone were isolated directly from the human body. With the advance of recombinant DNA technology, therapeutic proteins in the market now include recombinant antibodies, hormones, cytokines, interferons, and enzymes of human origin produced industrially in bacterial, yeast, or mammalian expression systems (Johnson-Leger et al. 2006).

In addition to human immunoregulatory proteins, many viruses are also "experts" at manipulating the human immune system to facilitate their propagation in the host. Viruses evade or subvert host immune detection and destruction by encoding and expressing a diverse array of immunomodulatory proteins, which target pathways of antibody response, cytokine-mediated signaling, and major histocompatibility complex (MHC)–restricted antigen presentation (Tortorella et al. 2000; Alcami 2003). After eons of coevolution, these virus-engineered immunomodulatory proteins have exquisite potency and specificity unrivalled by commercial pharmaceuticals, providing a powerful platform of protein therapeutics development (Lucas and McFadden 2004). Therefore, similar to the concept of human genome mining described previously, human "virome" mining has been proposed to uncover more drug candidates (Anderson et al. 2003; DeFilippis et al. 2003).

## TARGETS OF PROTEIN THERAPEUTICS AND MODES OF ACTION

Any molecule that has implications in the pathogenesis of a disease is a potential target for protein therapeutics. In contrast to small molecule drugs that are able to diffuse across cell membranes, protein therapeutics typically cannot traverse this cellular barrier due to their large size. Therefore, they almost exclusively target cell surface receptors or extracellular molecules. In recent years, researchers have also explored the possibility of directing protein therapeutics to intracellular targets (Stocks 2004; Bernal et al. 2007).

Broadly speaking, protein therapeutics have three different modes of action based on the pathology of a disease. First, if the disease is caused by unwanted extracellular molecules such as cell metabolites or cell lysate, enzyme therapeutics can degrade these targets. Second, if the disease is caused by a deficiency in certain proteins, such as enzymes, protein therapeutics can be used to replace them and restore an individual's health. Third, if the disease involves improper immune responses or dysregulated signaling pathways, such as chronic inflammatory diseases, autoimmune diseases, infectious diseases, and cancers, protein therapeutics act as inhibitors or activators of cell surface receptors. Among these three categories, the last one has attracted the most attention of researchers and represents an active area of research (see the sections titled "Examples of Protein Therapeutics," and "New Classes of Therapeutic Proteins under Development" in this chapter).

## ENGINEERING EFFECTIVE PROTEIN THERAPEUTICS

Protein therapeutics clearly have indisputable importance among modern pharmaceuticals. However, they remain at an early stage of development and application, and substantial improvement must be made in almost all aspects, including drug target identification, protein engineering and design, protein expression and purification, drug delivery, and marketing. Here we will focus on recent advances in protein engineering and design technologies.

### CHALLENGES IN PHARMACEUTICAL TRANSLATION OF NEW THERAPEUTIC PROTEINS

Although the human genome represents a rich source of candidate proteins for therapeutics, these proteins were not evolved for therapeutic purposes, and thus do not have optimal affinity, specificity, activity, and/or stability for disease treatment. Protein instability and immunogenicity are among the key challenges affecting the success of protein therapeutics. Proteins often have limited physical and chemical stability, which has several implications: (1) They have short half-life in the body, which in turn leads to limited efficacy and frequent dosage; (2) they are difficult to produce and have short shelf life, both of which are responsible for high cost of pharmaceutical commercialization; (3) they need to be administered through injection because they are quickly digested in the intestines if taken orally, which affects patient compliance and therapeutic outcomes (Kefalides 1998). Therefore, development of protein therapeutics with improved stability, efficacy, pharmacokinetics, pharmacodynamics, and expression productivity is required.

Safety is the priority criterion of all drugs. Immunogenicity is a unique issue associated with protein therapeutics. The human immune system responds to pathogens through recognition of their proteins or processed protein products. Similarly, patients who receive protein therapeutics can potentially develop immune responses against the protein drug, producing antidrug antibodies (Barbosa and Celis 2007; De Groot and Scott 2007). Such immune responses can reduce the efficacy of protein drugs, and in rare cases, they can lead to life-threatening situations (Barbosa and Celis 2007; De Groot and Scott 2007). Therefore, an immunogenicity assessment is essential to protein therapeutics development.

## STRATEGIES FOR DESIGNING EFFECTIVE PROTEIN THERAPEUTICS

To address the issues in protein therapeutics development described previously, various protein engineering and design strategies have been developed, including protein post-translational modification, protein fusions, and genetic engineering, as illustrated in Figure 7.1.

### Strategies for Improving Pharmacokinetics

The efficacy of a therapeutic protein in the human body can be improved by a number of strategies, including fusions, glycosylation, and chemical modification. Recombinant DNA technology enables the fusion of a protein of interest to an endogenous human protein (for example, human serum albumin, an antibody Fc fragment, and transferrin), thereby increasing the effective size of the protein and reducing clearance in the kidneys, which occurs for proteins below a molecular weight of approximately 70,000 Daltons (Caliceti and Veronese 2003; Beals and Shanafelt 2006). Fusing a protein to an antibody Fc fragment affords a second benefit: This takes advantage of the interaction between Fc and the FcRn receptor, which protects IgG antibodies in the body by allowing them to be released into the plasma rather than degraded in the lysosome, a natural "recycling" system (Lobo et al. 2004).

*Glycosylation*, the term for decoration of a protein's surface with carbohydrates, also increases protein size, thus reducing renal clearance. In addition, glycosylation can further help enhance protein solubility, stabilize against damage from heat and free radicals, and shield a protein from proteolysis and immune surveillance, ultimately resulting in enhanced serum half-life (Sinclair and Elliott 2005). Targeted introduction of glycosylation motifs into the sequence of a therapeutic protein is termed *glycoengineering*. Production of recombinant protein by mammalian cell

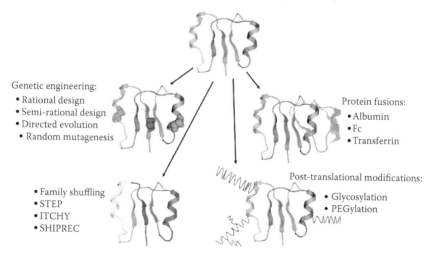

Genetic engineering:
• Rational design
• Semi-rational design
• Directed evolution
• Random mutagenesis

• Family shuffling
• STEP
• ITCHY
• SHIPREC

Protein fusions:
• Albumin
• Fc
• Transferrin

Post-translational modifications:
• Glycosylation
• PEGylation

**FIGURE 7.1 (see color insert following page 178)** Strategies for designing effective protein therapeutics. StEP = staggered extension process, ITCHY = incremental truncation for the creation of hybrid enzymes, SHIPREC = sequence homology-independent protein recombination (see Chapter 4 for a detailed description of each method).

culture, particularly Chinese hamster ovary (CHO) cells, is used industrially to synthesize glycosylated therapeutic proteins. The use of CHO cell culture for protein production does have drawbacks, namely significantly higher cost versus culture of bacteria or yeast. It is also possible that CHO cell cultures can harbor viral or prion contamination, and the glycosylation by CHO cells can be heterogeneous, leading to therapeutic proteins with a range of efficacies (Sethuraman and Stadheim 2006).

Finally, chemical modification by PEGylation, the conjugation of polyethylene glycol (PEG), is a common strategy to enhance the serum half-life of protein therapeutics (Beals and Shanafelt 2006). The increased size of the protein-PEG conjugate reduces clearance from the kidneys, and the bulky PEG molecule also protects therapeutic proteins from degradation by proteases via steric hindrance (Veronese and Pasut 2005). However, PEGylation also suffers from several limitations. Low molecular weight proteins are especially susceptible to partial or complete inactivation after conjugation of a PEG molecule (Shechter et al. 2008). Further, interference of protein–protein binding is beneficial vis-à-vis proteases, but reduces the effectiveness of antibody-based protein therapeutics or those acting through a receptor (Kubetzko et al. 2005; Shechter et al. 2008). This has led to the development of "reversible PEGylation," where the conjugated PEG molecule can be considered a prodrug, and undergoes spontaneous hydrolysis under physiological conditions to release the active therapeutic protein (Shechter et al. 2008).

## Strategies for Reducing Immunogenicity

In general, reduction of immunogenicity involves altering protein therapeutics such that they can avoid immune surveillance. This includes avoidance of antibodies, binding to antigen presenting cell (APC) surface receptors leading to receptor-mediated endocytosis, subsequent proteolysis to peptide fragments that bind to MHC class II molecules, and, finally, avoidance of binding by B and T cell receptors (Chirino et al. 2004). Many strategies are available to reduce immunogenicity and are similar to techniques used to improve pharmacokinetics. PEGylation, which possesses other useful properties as discussed previously, is nontoxic and reduces immunogenicity and antigenicity (Caliceti and Veronese 2003). Specifically, the PEG molecule shields immunoreactive sites on recombinant proteins from recognition by antibodies or surface receptors (Caliceti and Veronese 2003). PEG also enhances solubility, which prevents the accumulation of highly immunogenic protein aggregates (Chirino et al. 2004). Conjugated PEG also deters proteolysis, which may help PEGylated therapeutics to avoid cleavage into peptides capable of display on MHC class II molecules.

Glycosylation, which helps to enhance the serum half-life of therapeutic proteins, is also thought to interfere with antibody binding (De Groot and Scott 2007). Ideally, glycosylation should be of a human pattern for greatest effectiveness (Brooks 2006). This has led to considerable interest in humanizing the glycosylation pathways of the organisms currently used to express recombinant proteins, and fully humanized, sialylated glycoproteins can be produced from the yeast *Pichia pastoris* (Hamilton et al. 2003). Changes to the primary sequence of a therapeutic protein can also help reduce immunogenicity. The humanization of murine and mouse-human chimeric antibodies (i.e., the removal of as much nonhuman content from the constant and variable regions as possible) helps to reduce the formation of human antimouse antibodies (De Groot and Scott 2007). In addition, identifying and eliminating antibody and T-cell epitopes and class II MHC agretopes are strategies currently employed to reduce the immunogenicity of the next generation of protein therapeutics (Chirino et al. 2004).

## Genetic Engineering

Genetic engineering strategies, consisting of three broad categories—rational design, directed evolution, and semirational design—have long been valuable tools in engineering proteins with altered physical and chemical properties and/or creating novel functions. In the context of protein therapeutics, these well-established methods are facing new challenges due to consideration of additional engineering parameters including pharmacokinetics, pharmacodynamics, and immunogenicity as described in the section in this chapter titled "Challenges in Pharmaceutical Translation of New Therapeutic Proteins." Numerous rational and semirational design strategies have been developed (detailed in other chapters in this volume) and applied to the engineering of protein therapeutics. Rational or computation design methods have been applied to improve stability and solubility, or to predict and to reduce immunogenicity of protein therapeutics (reviewed in Marshall et al. 2003; Rosenberg and Goldblum 2006; De Groot and Moise 2007). The primary drawback of rational design is the requirement for knowledge of protein structure, mechanisms, and protein structure-function relationships to a certain extent. In contrast, directed evolution does not have this limitation because it creates molecular diversity at the DNA level in a stochastic manner (see Chapter 4). This also leads to the key challenge of directed evolution: how to find the variant with desired property in a library of up to a billion variants. Therefore, high-throughput selection or screening methods are highly desirable for directed evolution. A variety of library selection and screening methods have been developed for different applications (reviewed in Arnold and Georgiou 2003). For each directed evolution experiment, the selection or screening method must be prudently chosen or developed, because the first principle of directed evolution is "you get what you select (screen) for."

Among the library selection and screening approaches, display technologies have been increasingly used in therapeutic protein engineering, and have proved to be especially powerful for engineering protein drugs for improved affinity and specificity. The shared principle of different display technologies is to create a physical linkage between the genotype and the protein displayed on the platform, so that a library of target protein variants is directly accessible to binding analysis and thus selectable and recoverable for further engineering. By using surface display, *in vitro* affinity maturation of an antibody yielded variants with the highest affinity reported (femtomolar range, which is orders of magnitude beyond natural antibodies) (Boder et al. 2000), and new classes of therapeutic proteins are being developed (see the section titled "New Classes of Therapeutic Proteins under Development," in this chapter). Over the past decades, a number of display platforms have been developed, including phage display, cell surface display, and cell-free display (see Chapters 1–3). These different platforms have advantages and disadvantages that make them more conducive to certain protein engineering applications.

Phage display was the earliest developed platform and has since been utilized most often for protein engineering (Sergeeva et al. 2006). Recent advances have enabled selection of phage libraries in more complex biological systems, such as cultured cells and *in vivo* (Sergeeva et al. 2006). Although phage display has been successfully used for engineering of peptides, antibodies, and for epitope mapping, it has achieved limited success with more complex human membrane proteins such as MHC and T cell receptors (TCRs) (see the section titled "New Classes of Therapeutic Proteins under Development," in this chapter). This is because the bacterial host required for phage propagation has limited ability in terms of protein folding and post-translational modifications that are important for mammalian protein functions.

A cell surface display library is usually generated by transforming cells with DNA variants and screening for mutants with a desired phenotype by fluorescence activated cell sorting (FACS). FACS enables high-throughput enrichment of positive clones in a quantitative manner, but is not applicable for phage display libraries due to the small size of phage particles (Georgiou et al. 1997; Boder and Wittrup 1998). Several different cell types have been explored for their ability to display protein libraries, including bacteria, yeast, insect, and mammalian cells. Among these platforms, yeast display has attracted the most attention. Yeast display has the advantage of possessing post-translational processing pathways, which enable folding and glycosylation of complex human proteins (Kondo and Ueda 2004). Starting with the same library, yeast display was shown to sample the immune antibody repertoire considerably more fully than phage display, selecting twice as many novel antibodies as phage display (Bowley et al. 2007). Studies have also shown that the surface display level of a protein on yeast cell surface is strongly correlated with both thermal stability and soluble expression level (Shusta et al. 1999).

Cell-free display (also known as *in vitro* display) represents an emerging technology that has proven useful for discovery and engineering of therapeutic proteins with high affinity (FitzGerald 2000; Rothe et al. 2006). For example, ribosome (polysome) display, which was the first cell-free system developed for complete *in vitro* protein engineering, has been shown to generate antibodies with higher affinities (picomolar range) than those obtained from a phage display library (nanomolar range) (Groves et al. 2006). The biggest advantage of cell-free display methods is that the transcription and/or translation steps completely take place *in vitro*, abolishing the need of introducing DNA into host cells, which often limits the library size accessible to other display approaches. Library sizes created by *in vitro* display platforms are usually several orders of magnitude higher than that obtained with other display methods (up to $10^{14}$–$10^{15}$) (FitzGerald 2000). In addition, the cell-free feature of *in vitro* display methods might make them more amenable to automation, potentially allowing ultra high-throughput identification of new drug targets on a genomic level (FitzGerald 2000).

## EXAMPLES OF PROTEIN THERAPEUTICS

The previously described protein engineering and design strategies have been applied to engineer a wide variety of protein therapeutics for enhanced activity, stability, affinity, specificity, pharmacokinetics, pharmacodynamics, reduced immunogenicity, and improved productivity.

## Monoclonal Antibody Therapeutics

Extensive review articles and book volumes have been devoted to the subject of monoclonal antibodies and antibody engineering; therefore, a detailed discussion will not be provided here. However, it is worth mentioning that their exquisite specificity, ease of engineering fragments by display technologies, and chimerization or humanization to enhance stability have propelled antibodies to become a swiftly expanding class of therapeutics for treating a wide variety of human conditions [46]. Future directions of research include the development of multiantibody cocktails for synergistic effects (Logtenberg 2007) and the conjugation of antibodies to immunotoxic drugs to enhance tumor killing (Zafir-Lavie et al. 2007).

## Enzyme Therapeutics Acting on Extracellular Targets

In addition to binding proteins targeting soluble molecules and membrane-bound receptors, enzyme therapeutics can also target other molecules in the extracellular environment. In the clinical setting, the primary example is the use of amino acid–degrading enzymes as an anticancer strategy. Unlike healthy cells, rapidly growing tumor cells may be auxotrophic for certain metabolites, whose depletion in the plasma can selectively inhibit tumor growth. For example, lymphoid tumors are auxotrophic for asparagine because they lack asparagine synthetase activity, and recombinant, PEGylated L-asparaginase (Oncaspar®, Enzon) is an FDA-approved leukemia treatment (Pasut et al. 2008). A second enzyme, PEG-arginine deiminase (ADI-PEG 20, Pheonix Pharmacologics), is currently undergoing clinical trials for the treatment of the arginine-auxotrophic tumors melanoma (clinical trial phase I/II completed) and hepatocellular carcinoma (clinical trial phase II/III in progress) (Ni et al. 2008). Recent work indicates that renal cell carcinoma, a cancer that often metastasizes and is notoriously difficult to treat by conventional therapies, may also be susceptible to arginine deprivation [51]. A diverse group of solid tumor types including lung, prostate, and bladder tumors are methionine-dependent (Mecham et al. 1983), and PEGylated recombinant methioninase has been studied as a cancer therapy in animal models (Pasut et al. 2008). Notably, the aforementioned enzymes are immunogenic, likely because they originate from bacterial sources, and as a result PEGylation was needed to reduce immunogenicity and prolong serum half-life (Pasut et al. 2008). Finally, PEGylated recombinant human arginase has been proposed as a treatment of arginine-dependent hepatocellular carcinoma resistant to PEG-arginine deiminase and is currently undergoing preclinical study [53].

Enzyme therapeutics have also found use in the treatment of cystic fibrosis. Frequent bacterial infections lead to accumulation and eventual lysis of neutrophils in the lungs, releasing extracellular DNA to form abnormally viscous mucus. Dornase alfa, or recombinant human DNase I (Pulmozyme®, Genentech), is delivered to the lungs as an aerosol and degrades extracellular DNA to improve lung function, quality of life, and prevent exacerbations of disease (Thomson 1995). Dornase alfa is produced by CHO cell culture, resulting in a glycosylated protein with minimized immunogenicity, but this also contributes to the high cost of the drug.

PROTEIN THERAPEUTICS AS REPLACEMENTS FOR DEFECTIVE OR DEFICIENT PROTEINS

## Protein Hormones

Some medical conditions characterized by deficiency or complete loss of an endogenous protein can be treated by replacement therapy. The most widely recognized protein therapeutics in this area are the protein hormones, including insulin and human growth hormone. Tens of millions of individuals worldwide suffer from Type I and Type II diabetes, and many are dependent on injection of insulin, a peptide consisting of one 21-amino-acid chain (A chain) and a separate 31-amino-acid chain (B chain) linked by two disulfide bonds. Extensive molecular engineering of this simple molecule has resulted in numerous short- and long-acting insulin analogs that are used to control mealtime glucose spikes and meet the body's basal insulin need, respectively. These analogs can differ in terms of primary amino-acid sequence or may be chemically modified. For example, long-acting insulin glargine (Lantus®, Aventis) differs from human insulin by a substitution of asparagine for glycine at position 21 on the A chain and by the addition of two arginine residues to the C-terminus of the B chain (Sadrzadeh et al. 2007). Diarginyl insulin, possessing two extra arginines on the B chain, occurs naturally at low concentrations. A shifted isoelectric point makes this molecule soluble at mildly acidic pH, but it precipitates under physiological conditions; when injected, most diarginyl insulin degrades in subcutaneous tissue before it can be absorbed (Home and Ashwell 2002). The amino-acid substitution at position 21 enhances stability and increases bioavailability, providing a steady release of active insulin with only once-a-day injection (Bahr et al. 1997; Home and Ashwell 2002). Other long-acting forms of insulin were engineered by conjugation with fatty acids (Levemir®, Novo Nordisk) and PEGylation (InsuLAR, PR Pharmaceuticals) (Sadrzadeh et al. 2007). Alternatively, in fast-acting insulin lispro (Humalog®, Eli Lilly), the lysine at position 28 and the proline at position 29 of the B chain are inverted (Sadrzadeh et al. 2007). These mutations originated from the related protein hormone insulin-like growth factor I. Due to variation in the C-terminal ends of their respective B chains, natural insulin forms dimers and hexamers that slow absorption after subcutaneous injection, while insulin-like growth factor I does not self-associate (Holleman and Hoekstra 1997). By reversing residues 28 and 29, the self-association of insulin is abolished due to steric hindrance (Holleman and Hoekstra 1997). Injected insulin monomers absorb rapidly with a short duration of activity to control the spike in glucose associated with meals (Eckardt and Eckel 2008).

The second well-known protein hormone, recombinant human growth hormone (rhGH), has been available for several decades for the treatment of growth hormone deficiency in children and adults. rhGH also finds use in treating growth failure in children caused by other disorders (Bajpai and Menon 2005), as well as in treatment of HIV-associated wasting and lipodystrophy [61]. Current manufacturing processes generate two pharmacologically equivalent products: a 191-amino-acid protein identical to the natural human growth hormone or a 192-amino-acid protein possessing an additional N-terminal methionine [62]. The short plasma half-life of rhGH forces daily injection to maximize therapeutic benefit, making compliance an issue, particularly for pediatric patients. A sustained-release injection was developed via the collaboration of Genentech, Inc. and Akermes,

Inc., but commercialization of this product, Nutropin Depot®, was discontinued in 2004. Recently, conjugation of human growth hormone to human serum albumin [63] and PEG (Cox et al. 2007) have been studied as a means to create more stable, longer-acting formulations for less frequent injections. Further, computational design of human growth hormone was carried out by Dahiyat and colleagues to create thermostabilized rhGH mutants (Filikov et al. 2002). To lower the potential for immunogenicity, analysis was focused on the core of the protein rather than its surface, and resulted in mutants with improved van der Waals interaction, hydrophobic substitutions at polar serine and threonine residues, and better burial of hydrophobic groups (Filikov et al. 2002). Higher thermostability may improve the stability of rhGH in the body, thereby increasing its half-life.

## Coagulation Factors

Management of acquired or genetic loss of coagulation factors is another long-standing field of protein replacement therapy, and transfusion of human plasma–derived coagulation factors is the traditional treatment of bleeding disorders. Unfortunately, the use of human-derived products exposes transfusion recipients to risk of infection, and in the 1970s and 1980s, contaminated clotting factor concentrates led to HIV infection in nearly half of the United States' population of hemophiliacs (Key and Negrier 2007). The sequencing of genes encoding human coagulation factors, as well as advances in molecular biology techniques and mammalian cell culture, has made recombinant production a reality (Pipe 2005). The availability of safe, pathogen-free coagulation factors has also created the opportunity to explore these therapies in "off-label" uses in nonhemophiliac patients. For example, while plasma-derived protein C is used to treat inherited protein C deficiency, recombinant activated protein C has been investigated in treatment of sepsis, an often fatal condition that leads to a rapid depletion of protein C and limited protein C activation (Key and Negrier 2007). Reasoning that the ability of activated protein C to reduce mortality in septic patients is related to its anti-inflammatory and anti-apoptotic activity but unrelated to its anticoagulant activity, researchers used site-directed mutagenesis to engineer a mutant protein C with reduced coagulant activity but normal anti-apoptotic activity (Mosnier et al. 2004). In the future, this mutant may prove effective at treating activated protein C depletion during sepsis without leading to severe bleeding complications (Mosnier et al. 2007).

## Enzyme Replacement Therapy

The third category is enzyme replacement therapy (ERT) to treat the acquired or hereditary loss of an enzyme. ERT has revolutionized the treatment of several rare genetic diseases, including lysosomal storage disorders. In this case, successful therapy is contingent upon delivering exogenously supplied replacement enzyme to the intracellular lysosome and targeting specific cell types for maximum effectiveness. Gaucher disease ERT (Cerezyme®, Genzyme Corp.) is accomplished by post-translational modification of the replacement enzyme β-Glucocerebrosidase (GCase). Glycosylated GCase is produced in CHO cells, and sugar residues are subsequently removed from the carbohydrate chains of the glycosylated protein to generate a terminal mannose (Beck 2007). These mannose-terminated glycosylation residues interact with the mannose receptor, which is restricted to the macrophage plasma membrane, to deliver GCase specifically to macrophage lysosomes (Grabowski and Hopkin 2003).

## CYTOKINES AND THEIR RECEPTORS AS PROTEIN THERAPEUTICS

Cytokines, regulatory proteins secreted by white blood cells and a number of other cell types, and their receptors are important both as targets and tools in designing protein therapeutics. The term *cytokine* encompasses a variety of endogenous proteins, including colony stimulating factors (CSFs), epidermal growth factors (EGFs), interleukins, interferons, tumor necrosis factors, and others. It is well understood that cytokines can both protect from and contribute to disease, and accordingly, both recombinant cytokines and anticytokine antibodies or soluble receptors find clinical use (Vilcek and Feldmann 2004).

### Colony Stimulating Factors

The most widely recognized cytokine therapeutic, owing to its misuse as a performance-enhancing drug among professional athletes, is likely erythropoietin, a CSF that increases proliferation of red blood cells (Segura et al. 2007). In the clinical setting, recombinant human erythropoietin (Epogen®, Amgen Inc.; Procrit, Ortho Biotech Products, LP) treats anemia in patients undergoing chemotherapy or suffering from renal failure. Erythropoietin can also reduce the requirement for blood transfusion during surgery. AraNESP® (Amgen, Inc.), or Darbepoetin alfa, was glycoengineered to contain five N-linked, sialic acid–containing carbohydrate chains, two more than recombinant human erythropoietin, which provided approximately threefold higher serum half-life, and therefore can be administered less often (Egrie and Browne 2001). Two other recombinant CSFs, granulocyte CSF (Neupogen and Neulasta, PEGylated, Amgen Inc.) and granulocyte-macrophage CSF (Leucomax, Novartis, and others) are approved for use in patients undergoing chemotherapy or following a bone marrow transplant (Vilcek and Feldmann 2004). Recent *in vivo* pharmacokinetic studies of a mono-PEGylated recombinant human granulocyte CSF in rodents showed up to 40-fold enhanced serum half-life (Lee et al. 2008). Granulocyte CSF was also engineered via computational design to incorporate beneficial histidines on the binding surface of the protein. Sites were chosen such that under physiological conditions, a histidine of neutral charge does not adversely affect binding to the receptor, but in the slightly acid endosome, protonation results in a positive charge that disrupts receptor binding (Sarkar et al. 2002). In the endosome, weaker binding of granulocyte CSF to its receptor favors release of the cytokine and its receptor on the surface (recycling) over trafficking to the lysosome and subsequent degradation; as a result, improved potency and half-life were observed (Sarkar et al. 2002).

### Interleukins and Their Receptors

Two recombinant human interleukins (ILs) are currently approved for therapeutic use. The first is a modified recombinant human IL-2 produced in *E. coli*. This IL-2 analog, Aldesleukin (Proleukin®, Chiron Corp.), is not glycosylated, lacks an

N-terminal alanine, and possesses a cysteine-to-serine mutation at residue number 125. Proleukin® is FDA-approved for treatment of both metastatic renal cell carcinoma and metastatic melanoma and is employed as a standard treatment either alone or in combination with chemotherapy or interferon-α (Dillman 1999). However, the high systemic doses of IL-2 used in treatment of these cancers also leads to significant toxicity due to activation of natural killer (NK) cells (Atkins 2002). To reduce toxicity and enhance tumor killing, fusions of IL-2 with single chain T-cell receptor [79] or antibodies (Gillies et al. 1992; Lee et al. 2006) have been tested as a means to target IL-2 to tumor cells. Additionally, genetic engineering was used to enhance the affinity of IL-2 to a subunit of its receptor (IL-2Rα) found on human T cells but not NK cells (Rao et al. 2003). Higher affinity IL-2 mutants were isolated by yeast surface display and were shown to activate human T cells more potently than were wild-type IL-2 (Rao et al. 2005). Improved activation of T cells may allow for lower dosage of recombinant IL-2 in the clinic, thereby reducing toxicity. As an added benefit, tight binding to the IL-2 receptors on activated human T cells essentially sequesters these IL-2 mutants so that they cannot interact with IL-2 receptors on NK cells (Rao et al. 2004).

The IL-2 receptor is also a useful clinical target. Fusion of IL-2 to toxins has been used to achieve targeted killing of IL-2 receptor-expressing cells. Specifically, a fusion of IL-2 and diphtheria toxin, Denileukin diftitox (Ontak®, Seragen Inc.), is approved for treatment of persistent or recurrent cutaneous T-cell lymphoma in patients whose malignant cells express the CD25 component of the IL-2 receptor (Wong et al. 2007). Recently, a fusion of modified aerolysin and IL-2 was reported. Aerolysin is a bacterial toxin that binds to glycosylphosphatidylinositol-anchored proteins on the cell membrane and oligomerizes to create channels that trigger cell death. A variant of aerolysin with reduced binding to its native target was fused to IL-2 to create a chimera active only against cells expressing the IL-2 receptor [86]. This IL-2 fusion is a promising candidate for therapeutic use, and a fusion of modified aerolysin to prostate-specific antigen for the treatment of localized prostate cancer is already undergoing phase I clinical trials [86, 87]. Further, two monoclonal antibodies against the IL-2 receptor, basiliximab (Simulect®, Novartis Pharmaceuticals) and daclizumab (Zenapax®, Hoffmann-La Roche Inc.) are used to prevent rejection following organ transplantation (Van Gelder et al. 2004).

The second interleukin approved for clinical application is Oprelvekin (Neumega®, Genetics Institute Inc.), a nonglycosylated recombinant IL-11 produced in *E. coli*, which is also missing an N-terminal proline (Dorner et al. 1997). Oprelvekin prevents severe thrombocytopenia and reduces platelet transfusion requirements following myelosuppressive chemotherapy and bone marrow failure [90, 91]. Recently, PEGylation was reported to prolong serum half-life and enhance pharmacological efficacy of recombinant human IL-11, which may allow for less frequent injection [92].

## Interferons

The interferons are a family of secreted proteins that regulate resistance to viral infections, enhance innate and acquired immune responses, and modulate normal and tumor cell survival (Borden et al. 2007). A number of interferon therapeutics are approved for human use. Interferon-α (IFN-α) is approved both as the purified natural human protein and as the nonglycosylated recombinant proteins IFN-α-2a and IFN-α-2b, which differ by a substitution at amino acid position 23 (Roferon A®, Hoffmann-La Roche Inc.; Intron A®, Schering Corp.). The advantages of PEGylation led to the availability of both of these as PEG conjugates (Pegasys®, Hoffman-La Roche Inc.; PEG-Intron®, Schering Corp.). IFN-α and PEGylated IFN-α are standard treatments for chronic Hepatitis B and C and have been applied in the treatment of a number of different viral infections (Borden et al. 2007).

Variants of IFN-α have also been created using genetic engineering techniques. A novel IFN-α protein, termed *consensus interferon* (CIFN), was created by assigning the most commonly encountered amino acids at each position of several IFN-α nonallelic subtypes to create a consensus sequence (Keeffe and Hollinger 1997). Also known as *interferon alfacon-1* (Infergen®, InterMune Inc.), CIFN was shown to be more potent than natural IFN-α *in vitro* and is currently approved for clinical use (Alberti 1999). In 2007, DNA shuffling was used to create new IFN-α variants with higher antiviral activity than IFN-α-2a, IFN-α-2b, and CIFN, and importantly, decreased antiproliferative activity [96]. The antiproliferative activity of IFN-α is helpful in the treatment of cancers, but in the context of chronic viral illness, it may lead to bone marrow suppression and therefore limits the dosage used. The IFN-α mutants obtained in this study are thus "tailored" for treatment of viral infections due to their higher ration of antiviral to antiproliferative activity. The genetic engineering strategies applied to IFN-α are summarized in Figure 7.2. Additionally, to improve the serum half-life of IFN-α, glycoengineering was employed to create new, heavily glycosylated IFN-α analogs (Ceaglio et al. 2008). A fusion of IFN-α to human albumin was created for the same purpose, and this recombinant fusion protein is undergoing phase III clinical trials (Subramanian et al. 2007).

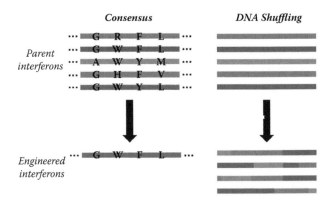

**FIGURE 7.2    (see color insert following page 178)** Comparison of genetic engineering strategies used to create improved interferon-α mutants. The consensus method (left) was used to create highly active consensus interferon (CIFN), while DNA shuffling (right) was used to create interferon-α mutants tailored for treatment of chronic viral illness.

# NEW CLASSES OF THERAPEUTIC PROTEINS UNDER DEVELOPMENT

T cells play a central role in cell-mediated immunity; however, the molecular mechanism underlying TCR recognition was not well understood until the late 1980s. Unlike antibodies that can recognize pathogens or their toxins directly, T cells recognize only short peptides derived from pathogens in complex with MHC molecules on the surface of antigen presenting cells (APCs) through their TCRs. The nature of the interaction between TCRs and peptide-MHC (pMHC) complexes determines the function of the induced cellular immune responses. Therefore, both TCRs and MHC molecules can be potentially used in protein therapeutics (see the section titled "Targets of Protein Therapeutics and Modes of Action," in this chapter).

## MAJOR HISTOCOMPATIBILITY COMPLEX (MHC) PROTEINS

Malfunction of the MHC system, consisting of class I (MHCI) and class II (MHCII) MHC proteins, has been implicated in many diseases, such as malaria, rheumatoid arthritis, type-I diabetes, and graft rejection. This has spurred great interest in developing MHC-based immunotherapeutics and immunodiagnostics methods. The development of pMHC tetramer, a multimeric form of peptide-MHC complexes, has revolutionized the field of T cell research. It enables direct detection and identification of antigen-specific T cells, modulation of T cell responses *in vivo* to treat graft rejection and autoimmune diseases, and detailed monitoring of cellular immune responses induced by immunotherapy, which is critical for a better understanding of tumor immunology and improved immune-based therapies. However, there are some limitations of pMHC tetramers that stem from their difficult recombinant production and the low affinity of the pMHC monomer. Therefore, it is highly desirable to engineer MHC molecules with improved solubility or higher TCR-binding affinity.

### Phage Display

Phage display of MHCs is challenging, because MHC molecules are large, heterodimeric membrane proteins with high glycosylation and multiple disulfide bonds. In addition, in the absence of their transmembrane domains, the α and β polypeptide chains are unable to assemble properly and tend to aggregate. To date, there are only three successful examples of MHCI phage display, with the first one reported in 2000 (Le Doussal et al. 2000; Vest Hansen et al. 2001; Kurokawa et al. 2002). Although the displayed pMHC complexes were correctly folded and capable of binding specific antigenic peptides, no significant interaction with relevant T cells was detected. Therefore, it is necessary to further optimize and develop novel design of the pMHC phage display system for efficient and stable T cell recognition.

### Cell Surface Display

Compared to phage display, yeast surface display is more effective for the display of MHC proteins and has been used to express both MHCI (Brophy et al. 2003; Jones et al. 2006) and MHCII (Starwalt et al. 2003; Esteban and Zhao 2004; Boder et al. 2005; Wen et al. 2008) proteins. Functional display of a mutant single-chain murine

MHCI protein was evidenced not only by recognition of conformation-specific antibodies but also by direct binding of a specific TCR that has been engineered to have high affinity (see the section titled "T Cell Receptors (TCRs)," in this chapter) (Brophy et al. 2003). More significantly, yeast cells displaying pMHC complexes upregulated the surface expression level of an early activation marker on naive T cells isolated from mice. Although the authors did not rule out the possibility of T cell autostimulation, this study clearly suggested that yeast display could be used for directed evolution of pMHC complexes. Indeed, the same group later successfully isolated stabilized mutants of a single-chain murine MHCII protein (which is known to be unstable and difficult to work with) from either a focused library created by site-directed mutagenesis or a library created by random mutagenesis (Starwalt et al. 2003). In a similar study, mutants of a single-chain human MHCII protein without a covalently attached peptide were stably displayed on yeast cell surface (Esteban and Zhao 2004). These MHCII proteins exhibited specific and fast peptide-binding kinetics, as well as high thermostability. Interestingly, although the single-chain gene construct did not include any peptide, the peptide-binding groove of the displayed MHCII mutants was not empty, but occupied presumably by yeast endogenous peptides, indicating the importance of binding peptides in stabilizing MHCII expression. By incorporating an MHCII-binding peptide in the single-chain construct, the wild-type pMHC complex was functionally displayed on yeast cell surface without introduction of any mutations and capable of activating immobilized hybridoma T cells (Wen et al. 2008). More importantly, the authors demonstrated that yeast display could be used in combination with expression cloning to identify T cell epitopes from a pathogen-derived peptide library.

Recently, baculovirus-infected insect cells have also been used for MHC display (Crawford et al. 2004; Wang et al. 2005; Crawford et al. 2006). As higher-order eukaryotic cells, insect cells have fewer problems with MHC expression compared to yeast. So far, baculovirus display of MHC has been used only to identify T cell epitopes/mimotopes from peptide libraries. Nevertheless, the ability of insect cell–displayed pMHC complexes to directly activate relevant T cells (Crawford et al. 2004) makes this system a promising platform for pMHC engineering.

## T Cell Receptors (TCRs)

TCRs are a pivotal element in almost every aspect of T lymphocytes, including their development, proliferation, differentiation, activity, and specificity. Therefore, researchers are exploring the potential of soluble TCRs to be used as immunotherapeutic or immunodiagnostic reagents to specifically target the pMHC complex (Miles et al. 2006), just as antibodies are used to neutralize or opsonize their antigens. However, there are several obstacles impeding therapeutic applications of TCRs, including difficult recombinant production, instability, and low pMHC binding affinity. Therefore, efforts have been devoted to the engineering and design of soluble, stable, and high-affinity TCRs.

## Protein Fusions and Rational Design

Until recently, there was no generally applicable method of producing soluble TCRs (Molloy et al. 2005). Many of the initial strategies, such as removing exposed hydrophobic residues, or fusion to antibody constant regions or thioredoxin (Andrews et al. 1996; Shusta et al. 1999), worked for a very limited number of TCRs. Later, Jun/Fos leucine zipper domains were introduced as fusions to the C-termini of the α/β TCR extracellular domains, respectively (Willcox et al. 1999). The incorporation of leucine zipper domains significantly stabilized the TCR, while maintaining its ligand specificity; however, it raised the potential of immunogenicity. Another generally applicable method involved introducing a non-native interchain disulfide bond, predicted by molecular modeling based on the TCR crystal structure, in the TCR invariant region (Boulter et al. 2003). The resulting disulfide-stabilized TCR (dsTCR) was highly stable, and the sequence/structural change was minimal compared to wild-type TCR, reducing its possibility of being immunogenic. More importantly, the dsTCR construct enabled phage display (Figure 7.3A), which represented a powerful directed evolution platform for TCR engineering (Li et al. 2005). By using the dsTCR format, ten different human class I– and class II–restricted TCRs were successfully displayed on phage particles (Li et al. 2005).

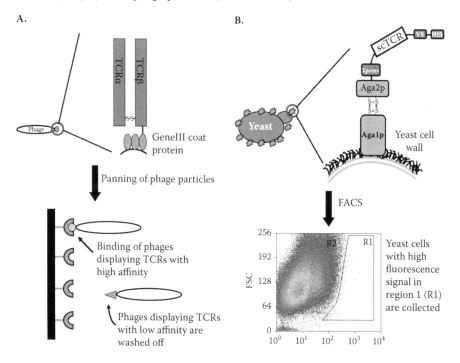

**FIGURE 7.3** **(see color insert following page 178)** TCR engineering using (A) phage display and (B) yeast display, adapted from (Boder and Wittrup 1997). A phage display library of TCRs is usually screened by panning phage particles on a surface/matrix coated with purified pMHC complexes, while a yeast display library can be rapidly screened by FACS using fluorophore-labeled (such as phycoerythrin, PE) pMHC complexes.

## Phage Display

The highest affinity TCRs yet reported (with picomolar affinities for their specific pMHC ligand) were achieved using phage display and directed evolution (Li et al. 2005) (Figure 7.3A). By using random mutagenesis, human TCR mutants with pMHC affinities up to 26 pM were isolated, representing more than a million-fold improvement over the wild-type TCR. In addition, the half-life of pMHC binding was also improved by ~8000-fold to ~1000 minutes at 25°C. More significantly, these high affinity TCRs showed a high degree of antigen specificity and no cross-reactivity with endogenous pMHC complexes, and they enabled direct visualization of specific pMHC complexes on tumor cells for the first time (Purbhoo et al. 2006). When transfected into human T cells, the mutant TCR with highest pMHC affinity completely lost its antigen specificity, while those expressing TCRs with lower pMHC affinity (with $K_D$ values of 450 nM and 4 μM) responded in an antigen-specific manner (Zhao et al. 2007). This indicates that genetically engineered T cells with midrange pMHC affinity might be useful in immunotherapeutics.

## Yeast Surface Display

As with MHC proteins, directed evolution and yeast surface display have also been applied to engineer soluble TCRs (Figure 7.3B). By using an *E. coli* mutator strain XL1-Red, FACS screening, and combining mutations, TCR mutants with increased thermal stability and secretion efficiency were identified (Kieke et al. 1999; Shusta et al. 1999). Selected mutations were combined and resulted in a TCR mutant that was stable for >1 hour at 65°C, had a solubility of over 4 mg/mL, and had a shake-flask expression level of 7.5 mg/L (Shusta et al. 2000). More importantly, although mutations were introduced, the resulting TCRs retained their ligand-binding specificity, making yeast display an attractive engineering platform for engineering TCRs with high affinity to pMHC (Kieke et al. 1999; Shusta et al. 2000). The first example of *in vitro* affinity maturation of a TCR was reported in 2000 using yeast surface display (Holler et al. 2000). A focused library of MHCI-restricted single-chain TCR was constructed by mutating the CDR3 (complementarity-determining region three) of the α-chain, and mutants with greater than 100-fold higher pMHC binding affinity were identified. Unlike the wild-type TCR, the soluble monomeric form of the high-affinity TCR was capable of directly detecting specific pMHC complexes on APCs. *In vivo* studies showed that a mouse T cell hybridoma transfected with the high affinity TCR responded to a significantly lower concentration of antigenic peptide (Holler et al. 2001). In several follow-up studies, it was shown that mutations in the CDR1 and CDR2 regions could also contribute to improving the pMHC binding affinity of TCRs (Chlewicki et al. 2005; Weber et al. 2005).

## TCR-LIKE ANTIBODIES

Researchers have used phage display to create TCR-like antibodies, which recognize specific peptides in complex with MHC molecules (Andersen et al. 1996; Denkberg and Reiter 2006). There have been ~100 different TCR-like antibodies generated to date, recognizing antigens involved in several infectious diseases and cancer.

However, the *in vivo* targeting capability of these TCR-like antibodies has not yet been demonstrated. With further development, this new class of antibodies might lead to novel therapeutic and diagnostic solutions.

## CONCLUSION AND FUTURE PROSPECTS

Protein therapeutics are a very important part of modern medicine, and in certain situations they are the only effective therapies. Expanding with an ever-increasing speed, the protein therapeutics market is projected to reach $87 billion by year 2010 (Kalorama Information 2006). To realize such great potential, continuous efforts are required to optimize their efficacy, while simultaneously discovering novel protein drugs. As exemplified in this chapter, protein engineering and design have long been valuable tools in developing effective protein therapeutics with improved affinity, specificity, activity, stability, pharmacokinetics, pharmacodynamics, and productivity, as well as reduced immunogenicity. In the future, with better understanding of protein structure-function relationships and rapid development of *in silico* bioinformatics and systems biology techniques, we are more likely to see an increased synergy between protein engineering and design strategies. In terms of discovering novel protein therapeutics, it is necessary to incorporate strengths from multiple disciplines, such as molecular biology, pathology, immunology, and nanotechnology. With orchestrated efforts, the "Golden Age" of protein therapeutics can be realized.

## REFERENCES

Alberti, A. 1999. "Interferon alfacon-1: A novel interferon for the treatment of chronic hepatitis C." *BioDrugs* 12: 343–57.

Alcami, A. 2003. "Viral mimicry of cytokines, chemokines and their receptors." *Nat Rev Immunol* 3: 36–50.

Andersen, P. S., A. Stryhn, B. E. Hansen, L. Fugger, J. Engberg, et al. 1996. "A recombinant antibody with the antigen-specific, major histocompatibility complex-restricted specificity of T cells." *Proc Natl Acad Sci U S A* 93: 1820–4.

Anderson, N. G., J. L. Gerin and N. L. Anderson. 2003. "Global screening for human viral pathogens." *Emerg Infect Dis* 9: 768–74.

Andrews, B., H. Adari, G. Hannig, E. Lahue, M. Gosselin, et al. 1996. "A tightly regulated high level expression vector that utilizes a thermosensitive lac repressor: Production of the human T cell receptor V beta 5.3 in *Escherichia coli*." *Gene* 182: 101–9.

Arnold, F. H. and G. Georgiou, (eds.). (2003). *Directed Enzyme Evolution: Screening and Selection Methods*. Totowa, New Jersey, Humana Press.

Atkins, M. B. 2002. "Interleukin-2: Clinical applications." *Semin Oncol* 29: 12–7.

Bahr, M., T. Kolter, G. Seipke and J. Eckel. 1997. "Growth promoting and metabolic activity of the human insulin analogue [GlyA21,ArgB31,ArgB32]insulin (HOE 901) in muscle cells." *Eur J Pharmacol* 320: 259–65.

Bajpai, A. and P. S. Menon. 2005. "Growth hormone therapy." *Indian J Pediatr* 72: 139–44.

Barbosa, M. D. and E. Celis. 2007. "Immunogenicity of protein therapeutics and the interplay between tolerance and antibody responses." *Drug Discov Today* 12: 674–81.

Beals, J. M. and A. B. Shanafelt. 2006. "Enhancing exposure of protein therapeutics." *Drug Discov Today* 3: 87–94.

Beck, M. 2007. "New therapeutic options for lysosomal storage disorders: Enzyme replacement, small molecules and gene therapy." *Hum Genet* 121: 1–22.

Belmont, H. J., S. Price-Schiavi, B. Liu, K. F. Card, H. I. Lee, et al. 2006. "Potent antitumor activity of a tumor-specific soluble TCR/IL-2 fusion protein." *Clin Immunol* 121: 29–39.

Bernal, F., A. F. Tyler, S. J. Korsmeyer, L. D. Walensky and G. L. Verdine. 2007. "Reactivation of the p53 tumor suppressor pathway by a stapled p53 peptide." *J Am Chem Soc* 129: 2456–7.

Bhatia, M., V. Davenport and M. S. Cairo. 2007. "The role of interleukin-11 to prevent chemotherapy-induced thrombocytopenia in patients with solid tumors, lymphoma, acute myeloid leukemia and bone marrow failure syndromes." *Leuk Lymphoma* 48: 9–15.

Boder, E. T., J. R. Bill, A. W. Nields, P. C. Marrack and J. W. Kappler. 2005. "Yeast surface display of a noncovalent MHC class II heterodimer complexed with antigenic peptide." *Biotechnol Bioeng* 92: 485–91.

Boder, E. T., K. S. Midelfort and K. D. Wittrup. 2000. "Directed evolution of antibody fragments with monovalent femtomolar antigen-binding affinity." *Proc Natl Acad Sci U S A* 97: 10701–5.

Boder, E. T. and K. D. Wittrup. 1997. "Yeast surface display for screening combinatorial polypeptide libraries." *Nat Biotechnol* 15: 553–557.

Boder, E. T. and K. D. Wittrup. 1998. "Optimal screening of surface-displayed polypeptide libraries." *Biotechnol Prog* 14: 55–62.

Borden, E. C., G. C. Sen, G. Uze, R. H. Silverman, R. M. Ransohoff, et al. 2007. "Interferons at age 50: Past, current and future impact on biomedicine." *Nat Rev Drug Discov* 6: 975–90.

Boulter, J. M., M. Glick, P. T. Todorov, E. Baston, M. Sami, et al. 2003. "Stable, soluble T-cell receptor molecules for crystallization and therapeutics." *Protein Eng* 16: 707–11.

Bowley, D. R., A. F. Labrijn, M. B. Zwick and D. R. Burton. 2007. "Antigen selection from an HIV-1 immune antibody library displayed on yeast yields many novel antibodies compared to selection from the same library displayed on phage." *Protein Eng Des Sel* 20: 81–90.

Brideau-Andersen, A. D., X. Huang, S. C. Sun, T. T. Chen, D. Stark, et al. 2007. "Directed evolution of gene-shuffled IFN-alpha molecules with activity profiles tailored for treatment of chronic viral diseases." *Proc Natl Acad Sci U S A* 104: 8269–74.

Brooks, S. A. 2006. "Protein glycosylation in diverse cell systems: Implications for modification and analysis of recombinant proteins." *Expert Rev Proteomics* 3: 345–59.

Brophy, S. E., P. D. Holler and D. M. Kranz. 2003. "A yeast display system for engineering functional peptide-MHC complexes." *J Immunol Methods* 272: 235–46.

Caliceti, P. and F. M. Veronese. 2003. "Pharmacokinetic and biodistribution properties of poly(ethylene glycol)-protein conjugates." *Adv Drug Deliv Rev* 55: 1261–77.

Carter, P. 2001. "Improving the efficacy of antibody-based cancer therapies." *Nat Rev Cancer* 1: 118–29.

Ceaglio, N., M. Etcheverrigaray, R. Kratje and M. Oggero. 2008. "Novel long-lasting interferon alpha derivatives designed by glycoengineering." *Biochimie* 90: 437–49.

Cheng, P. N., T. L. Lam, W. M. Lam, S. M. Tsui, A. W. Cheng, et al. 2007. "Pegylated recombinant human arginase (rhArg-peg5,000mw) inhibits the in vitro and in vivo proliferation of human hepatocellular carcinoma through arginine depletion." *Cancer Res* 67: 309–17.

Chirino, A. J., M. L. Ary and S. A. Marshall. 2004. "Minimizing the immunogenicity of protein therapeutics." *Drug Discov Today* 9: 82–90.

Chlewicki, L. K., P. D. Holler, B. C. Monti, M. R. Clutter and D. M. Kranz. 2005. "High-affinity, peptide-specific T cell receptors can be generated by mutations in CDR1, CDR2 or CDR3." *J Mol Biol* 346: 223–39.

Cox, G. N., M. S. Rosendahl, E. A. Chlipala, D. J. Smith, S. J. Carlson, et al. 2007. "A long-acting, mono-PEGylated human growth hormone analog is a potent stimulator of weight gain and bone growth in hypophysectomized rats." *Endocrinology* 148: 1590–7.

Crawford, F., E. Huseby, J. White, P. Marrack and J. W. Kappler. 2004. "Mimotopes for alloreactive and conventional T cells in a peptide-MHC display library." *PLoS Biol* 2: E90.

Crawford, F., K. R. Jordan, B. Stadinski, Y. Wang, E. Huseby, et al. 2006. "Use of baculovirus MHC/peptide display libraries to characterize T-cell receptor ligands." *Immunol Rev* 210: 156–70.

De Groot, A. S. and L. Moise. 2007. "Prediction of immunogenicity for therapeutic proteins: State of the art." *Curr Opin Drug Discov Devel* 10: 332–40.

De Groot, A. S. and D. W. Scott. 2007. "Immunogenicity of protein therapeutics." *Trends Immunol* 28: 482–90.

DeFilippis, V., C. Raggo, A. Moses and K. Fruh. 2003. "Functional genomics in virology and antiviral drug discovery." *Trends Biotechnol* 21: 452–7.

Denkberg, G. and Y. Reiter. 2006. "Recombinant antibodies with T-cell receptor-like specificity: Novel tools to study MHC class I presentation." *Autoimmun Rev* 5: 252–7.

Dillman, R. O. 1999. "What to do with IL-2?" *Cancer Biother Radiopharm* 14: 423–34.

Dorner, A. J., S. J. Goldman and J. C. Keith, Jr. 1997. "Interleukin-11." *BioDrugs* 8: 418–29.

Eckardt, K. and J. Eckel. 2008. "Insulin analogues: Action profiles beyond glycaemic control." *Arch Physiol Biochem* 114: 45–53.

Egrie, J. C. and J. K. Browne. 2001. "Development and characterization of novel erythropoiesis stimulating protein (NESP)." *Br J Cancer* 84 Suppl 1: 3–10.

Eng, C. 2007. "Combining targeted therapies to enhance the effectiveness of chemotherapy in patients with treatment-refractory colorectal cancer." *Clin Colorectal Cancer* 6 Suppl 2: S53–9.

Esteban, O. and H. Zhao. 2004. "Directed evolution of soluble single-chain human class II MHC molecules." *J Mol Biol* 340: 81–95.

Filikov, A. V., R. J. Hayes, P. Luo, D. M. Stark, C. Chan, et al. 2002. "Computational stabilization of human growth hormone." *Protein Sci* 11: 1452-61.

FitzGerald, K. 2000. "In vitro display technologies—new tools for drug discovery." *Drug Discov Today* 5: 253–258.

Flamant, M. and A. Bourreille. 2007. "Biologic therapies in inflammatory bowel disease: Anti-TNF and new therapeutic targets." *Rev Med Interne* 28: 852–61.

Georgiou, G., C. Stathopoulos, P. S. Daugherty, A. R. Nayak, B. L. Iverson, et al. 1997. "Display of heterologous proteins on the surface of microorganisms: From the screening of combinatorial libraries to live recombinant vaccines." *Nat Biotechnol* 15: 29–34.

Gergely, P. and B. Fekete. 2007. "New possibilities of treating patients with autoimmune disorders." *Orv Hetil* 148 Suppl 1: 58–62.

Gillies, S. D., E. B. Reilly, K. M. Lo and R. A. Reisfeld. 1992. "Antibody-targeted interleukin 2 stimulates T-cell killing of autologous tumor cells." *Proc Natl Acad Sci U S A* 89: 1428–32.

Grabowski, G. A. and R. J. Hopkin. 2003. "Enzyme therapy for lysosomal storage disease: Principles, practice, and prospects." *Annu Rev Genomics Hum Genet* 4: 403–36.

Groves, M., S. Lane, J. Douthwaite, D. Lowne, D. G. Rees, et al. 2006. "Affinity maturation of phage display antibody populations using ribosome display." *J Immunol Methods* 313: 129–39.

Hamilton, S. R., P. Bobrowicz, B. Bobrowicz, R. C. Davidson, H. Li, et al. 2003. "Production of complex human glycoproteins in yeast." *Science* 301: 1244–6.

Holleman, F. and J. B. Hoekstra. 1997. "Insulin lispro." *N Engl J Med* 337: 176–83.

Holler, P. D., P. O. Holman, E. V. Shusta, S. O'Herrin, K. D. Wittrup, et al. 2000. "In vitro evolution of a T cell receptor with high affinity for peptide/MHC." *Proc Natl Acad Sci U S A* 97: 5387–92.

Holler, P. D., A. R. Lim, B. K. Cho, L. A. Rund and D. M. Kranz. 2001. "CD8(-) T cell transfectants that express a high affinity T cell receptor exhibit enhanced peptide-dependent activation." *J Exp Med* 194: 1043–52.

Home, P. D. and S. G. Ashwell. 2002. "An overview of insulin glargine." *Diabetes Metab Res Rev* 18 Suppl 3: S57–63.

Hopkins, A. L. and C. R. Groom. 2002. "The druggable genome." *Nat Rev Drug Discov* 1: 727–30.

Hopkins, A. L. and C. R. Groom. 2003. "Target analysis: A priori assessment of druggability." *Ernst Schering Res Found Workshop* 11–7.

Johnson-Leger, C., C. A. Power, G. Shomade, J. P. Shaw and A. E. Proudfoot. 2006. "Protein therapeutics—lessons learned and a view of the future." *Expert Opin Biol Ther* 6: 1–7.

Jones, L. L., S. E. Brophy, A. J. Bankovich, L. A. Colf, N. A. Hanick, et al. 2006. "Engineering and characterization of a stabilized alpha1/alpha2 module of the class I major histocompatibility complex product Ld." *J Biol Chem* 281: 25734–44.

KaloramaInformation (2006). The Protein Therapeutics Market: The Science and Business of a Growing Sector, MarketResearch.com.

Keeffe, E. B. and F. B. Hollinger. 1997. "Therapy of hepatitis C: Consensus interferon trials. Consensus Interferon Study Group." *Hepatology* 26: 101S–7S.

Kefalides, P. T. 1998. "New methods for drug delivery." *Ann Intern Med* 128: 1053–5.

Key, N. S. and C. Negrier. 2007. "Coagulation factor concentrates: past, present, and future." *Lancet* 370: 439–48.

Kieke, M. C., E. V. Shusta, E. T. Boder, L. Teyton, K. D. Wittrup, et al. 1999. "Selection of functional T cell receptor mutants from a yeast surface-display library." *Proc Natl Acad Sci U S A* 96: 5651–6.

Kondo, A. and M. Ueda. 2004. "Yeast cell-surface display—applications of molecular display." *Appl Microbiol Biotechnol* 64: 28–40.

Kubetzko, S., C. A. Sarkar and A. Pluckthun. 2005. "Protein PEGylation decreases observed target association rates via a dual blocking mechanism." *Mol Pharmacol* 68: 1439–54.

Kurokawa, M. S., S. Ohoka, T. Matsui, T. Sekine, K. Yamamoto, et al. 2002. "Expression of MHC class I molecules together with antigenic peptides on filamentous phages." *Immunol Lett* 80: 163–8.

Le Doussal, J., B. Piqueras, I. Dogan, P. Debre and G. Gorochov. 2000. "Phage display of peptide/major histocompatibility complex." *J Immunol Methods* 241: 147–58.

Lee, D. L., I. Sharif, S. Kodihalli, D. I. Stewart and V. Tsvetnitsky. 2008. "Preparation and characterization of monopegylated human granulocyte-macrophage colony-stimulating factor." *J Interferon Cytokine Res* 28: 101–12.

Lee, K. D., H. W. Chen, C. C. Chen, Y. C. Shih, H. K. Liu, et al. 2006. "Construction and characterization of a novel fusion protein consisting of anti-CD3 antibody fused to recombinant interleukin-2." *Oncol Rep* 15: 1211–6.

Li, Y., R. Moysey, P. E. Molloy, A. L. Vuidepot, T. Mahon, et al. 2005. "Directed evolution of human T-cell receptors with picomolar affinities by phage display." *Nat Biotechnol* 23: 349–54.

Lobo, E. D., R. J. Hansen and J. P. Balthasar. 2004. "Antibody pharmacokinetics and pharmacodynamics." *J Pharm Sci* 93: 2645–68.

Logtenberg, T. 2007. "Antibody cocktails: next-generation biopharmaceuticals with improved potency." *Trends Biotechnol* 25: 390–4.

Lucas, A. and G. McFadden. 2004. "Secreted immunomodulatory viral proteins as novel biotherapeutics." *J Immunol* 173: 4765–74.

Marshall, S. A., G. A. Lazar, A. J. Chirino and J. R. Desjarlais. 2003. "Rational design and engineering of therapeutic proteins." *Drug Discov Today* 8: 212–21.

Mecham, J. O., D. Rowitch, C. D. Wallace, P. H. Stern and R. M. Hoffman. 1983. "The metabolic defect of methionine dependence occurs frequently in human tumor cell lines." *Biochem Biophys Res Commun* 117: 429–34.

Miles, J. J., S. L. Silins and S. R. Burrows. 2006. "Engineered T cell receptors and their potential in molecular medicine." *Curr Med Chem* 13: 2725–36.

Molloy, P. E., A. K. Sewell and B. K. Jakobsen. 2005. "Soluble T cell receptors: novel immunotherapies." *Curr Opin Pharmacol* 5: 438–43.

Moore, J. A., C. G. Rudman, N. J. MacLachlan, G. B. Fuller, B. Burnett, et al. 1988. "Equivalent potency and pharmacokinetics of recombinant human growth hormones with or without an N-terminal methionine." *Endocrinology* 122: 2920–6.

Mosnier, L. O., A. J. Gale, S. Yegneswaran and J. H. Griffin. 2004. "Activated protein C variants with normal cytoprotective but reduced anticoagulant activity." *Blood* 104: 1740–4.

Mosnier, L. O., X. V. Yang and J. H. Griffin. 2007. "Activated protein C mutant with minimal anticoagulant activity, normal cytoprotective activity, and preservation of thrombin activable fibrinolysis inhibitor-dependent cytoprotective functions." *J Biol Chem* 282: 33022–33.

Ni, Y., U. Schwaneberg and Z. H. Sun. 2008. "Arginine deiminase, a potential anti-tumor drug." *Cancer Lett* 261: 1–11.

Osborn, B. L., L. Sekut, M. Corcoran, C. Poortman, B. Sturm, et al. 2002. "Albutropin: a growth hormone-albumin fusion with improved pharmacokinetics and pharmacodynamics in rats and monkeys." *Eur J Pharmacol* 456: 149–58.

Osusky, M., L. Teschke, X. Wang, K. Wong and J. T. Buckley. 2008. "A chimera of interleukin 2 and a binding variant of aerolysin is selectively toxic to cells displaying the interleukin 2 receptor." *J Biol Chem* 283: 1572–9.

Pasut, G., M. Sergi and F. M. Veronese. 2008. "Anti-cancer PEG-enzymes: 30 years old, but still a current approach." *Adv Drug Deliv Rev* 60: 69–78.

Pipe, S. W. 2005. "The promise and challenges of bioengineered recombinant clotting factors." *J Thromb Haemost* 3: 1692–701.

Purbhoo, M. A., D. H. Sutton, J. E. Brewer, R. E. Mullings, M. E. Hill, et al. 2006. "Quantifying and imaging NY-ESO-1/LAGE-1-derived epitopes on tumor cells using high affinity T cell receptors." *J Immunol* 176: 7308–16.

Rao, B. M., I. Driver, D. A. Lauffenburger and K. D. Wittrup. 2004. "Interleukin 2 (IL-2) variants engineered for increased IL-2 receptor alpha-subunit affinity exhibit increased potency arising from a cell surface ligand reservoir effect." *Mol Pharmacol* 66: 864–9.

Rao, B. M., I. Driver, D. A. Lauffenburger and K. D. Wittrup. 2005. "High-affinity CD25-binding IL-2 mutants potently stimulate persistent T cell growth." *Biochemistry* 44: 10696–701.

Rao, B. M., A. T. Girvin, T. Ciardelli, D. A. Lauffenburger and K. D. Wittrup. 2003. "Interleukin-2 mutants with enhanced alpha-receptor subunit binding affinity." *Protein Eng* 16: 1081–7.

RNCOS (2007). Global Protein Therapeutics Market Analysis.

Rosenberg, M. and A. Goldblum. 2006. "Computational protein design: A novel path to future protein drugs." *Curr Pharm Des* 12: 3973–97.

Rothe, A., R. J. Hosse and B. E. Power. 2006. "In vitro display technologies reveal novel biopharmaceutics." *Faseb J* 20: 1599–610.

Rutgeerts, P., G. Van Assche and S. Vermeire. 2006. "Review article: Infliximab therapy for inflammatory bowel disease—seven years on." *Aliment Pharmacol Ther* 23: 451–63.

Sadrzadeh, N., M. J. Glembourtt and C. L. Stevenson. 2007. "Peptide drug delivery strategies for the treatment of diabetes." *J Pharm Sci* 96: 1925–54.

Sarkar, C. A., K. Lowenhaupt, T. Horan, T. C. Boone, B. Tidor, et al. 2002. "Rational cytokine design for increased lifetime and enhanced potency using pH-activated 'histidine switching'." *Nat Biotechnol* 20: 908–13.

Segura, J., J. A. Pascual and R. Gutierrez-Gallego. 2007. "Procedures for monitoring recombinant erythropoietin and analogues in doping control." *Anal Bioanal Chem* 388: 1521–9.

Sergeeva, A., M. G. Kolonin, J. J. Molldrem, R. Pasqualini and W. Arap. 2006. "Display technologies: application for the discovery of drug and gene delivery agents." *Adv Drug Deliv Rev* 58: 1622–54.

Sethuraman, N. and T. A. Stadheim. 2006. "Challenges in therapeutic glycoprotein production." *Curr Opin Biotechnol* 17: 341–6.

Shechter, Y., M. Mironchik, S. Rubinraut, H. Tsubery, K. Sasson, et al. 2008. "Reversible pegylation of insulin facilitates its prolonged action in vivo." *Eur J Pharm Biopharm* 70: 19–28.

Shusta, E. V., P. D. Holler, M. C. Kieke, D. M. Kranz and K. D. Wittrup. 2000. "Directed evolution of a stable scaffold for T-cell receptor engineering." *Nat Biotechnol* 18: 754–9.

Shusta, E. V., M. C. Kieke, E. Parke, D. M. Kranz and K. D. Wittrup. 1999. "Yeast polypeptide fusion surface display levels predict thermal stability and soluble secretion efficiency." *J Mol Biol* 292: 949–56.

Sinclair, A. M. and S. Elliott. 2005. "Glycoengineering: The effect of glycosylation on the properties of therapeutic proteins." *J Pharm Sci* 94: 1626–35.

Starwalt, S. E., E. L. Masteller, J. A. Bluestone and D. M. Kranz. 2003. "Directed evolution of a single-chain class II MHC product by yeast display." *Protein Eng* 16: 147–56.

Stocks, M. R. 2004. "Intrabodies: Production and promise." *Drug Discov Today* 9: 960–6.

Subramanian, G. M., M. Fiscella, A. Lamouse-Smith, S. Zeuzem and J. G. McHutchison. 2007. "Albinterferon alpha-2b: A genetic fusion protein for the treatment of chronic hepatitis C." *Nat Biotechnol* 25: 1411–9.

Takagi, A., N. Yamashita, T. Yoshioka, Y. Takaishi, K. Sano, et al. 2007. "Enhanced pharmacological activity of recombinant human interleukin-11 (rhIL11) by chemical modification with polyethylene glycol." *J Control Release* 119: 271–8.

Teichman, S. L., A. Neale, B. Lawrence, C. Gagnon, J. P. Castaigne, et al. 2006. "Prolonged stimulation of growth hormone (GH) and insulin-like growth factor I secretion by CJC-1295, a long-acting analog of GH-releasing hormone, in healthy adults." *J Clin Endocrinol Metab* 91: 799–805.

Thomson, A. H. 1995. "Human recombinant DNase in cystic fibrosis." *J R Soc Med* 88 Suppl 25: 24–9.

Tortorella, D., B. E. Gewurz, M. H. Furman, D. J. Schust and H. L. Ploegh. 2000. "Viral subversion of the immune system." *Annu Rev Immunol* 18: 861–926.

Van Gelder, T., M. Warle and R. G. Ter Meulen. 2004. "Anti-interleukin-2 receptor antibodies in transplantation: What is the basis for choice?" *Drugs* 64: 1737–41.

Veronese, F. M. and G. Pasut. 2005. "PEGylation, successful approach to drug delivery." *Drug Discov Today* 10: 1451–8.

Vest Hansen, N., L. Ostergaard Pedersen, A. Stryhn and S. Buus. 2001. "Phage display of peptide / major histocompatibility class I complexes." *Eur J Immunol* 31: 32–8.

Vilcek, J. and M. Feldmann. 2004. "Historical review: Cytokines as therapeutics and targets of therapeutics." *Trends Pharmacol Sci* 25: 201–9.

Wang, Y., A. Rubtsov, R. Heiser, J. White, F. Crawford, et al. 2005. "Using a baculovirus display library to identify MHC class I mimotopes." *Proc Natl Acad Sci U S A* 102: 2476–81.

Weber, K. S., D. L. Donermeyer, P. M. Allen and D. M. Kranz. 2005. "Class II-restricted T cell receptor engineered in vitro for higher affinity retains peptide specificity and function." *Proc Natl Acad Sci U S A* 102: 19033–8.

Wen, F., O. Esteban and H. Zhao. 2008. "Rapid identification of CD4+ T-cell epitopes using yeast displaying pathogen-derived peptide library." *J Immunol Methods* 336: 37–44.

Willcox, B. E., G. F. Gao, J. R. Wyer, C. A. O'Callaghan, J. M. Boulter, et al. 1999. "Production of soluble alphabeta T-cell receptor heterodimers suitable for biophysical analysis of ligand binding." *Protein Sci* 8: 2418–23.

Williams, S. A., R. F. Merchant, E. Garrett-Mayer, J. T. Isaacs, J. T. Buckley, et al. 2007. "A prostate-specific antigen-activated channel-forming toxin as therapy for prostatic disease." *J Natl Cancer Inst* 99: 376–85.

Wilde, M. I. and D. Faulds. 1998. "Oprelvekin: a review of its pharmacology and therapeutic potential in chemotherapy-induced thrombocytopenia." *BioDrugs* 10: 159–71.

Wishart, D. S., C. Knox, A. C. Guo, D. Cheng, S. Shrivastava, et al. 2008. "DrugBank: A knowledgebase for drugs, drug actions and drug targets." *Nucleic Acids Res* 36: D901–6.

Wong, B. Y., S. A. Gregory and N. H. Dang. 2007. "Denileukin diftitox as novel targeted therapy for lymphoid malignancies." *Cancer Invest* 25: 495–501.

Yoon, C. Y., Y. J. Shim, E. H. Kim, J. H. Lee, N. H. Won, et al. 2007. "Renal cell carcinoma does not express argininosuccinate synthetase and is highly sensitive to arginine deprivation via arginine deiminase." *Int J Cancer* 120: 897–905.

Zafir-Lavie, I., Y. Michaeli and Y. Reiter. 2007. "Novel antibodies as anticancer agents." *Oncogene* 26: 3714–33.

Zhao, Y., A. D. Bennett, Z. Zheng, Q. J. Wang, P. F. Robbins, et al. 2007. "High-affinity TCRs generated by phage display provide CD4+ T cells with the ability to recognize and kill tumor cell lines." *J Immunol* 179: 5845–54.

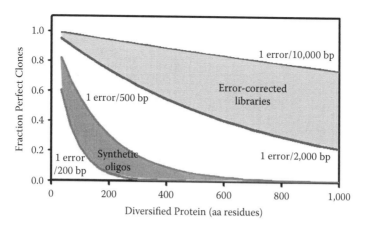

**FIGURE 4.5** Comparison in library quality with and without error correction. The fraction of clones in a library that conform to the intended library design drop exponentially with the length of the fragment being diversified. Orange: error rates in standard commercially available synthetic oligonucleotides, and fraction of perfect clones when a library is assembled from standard oligonucleotides. Green: error rates in synthetic oligonucleotides that have been error-corrected by consensus filtering (Figure 4.4, the section titled "Library Quality"), and fraction of perfect clones when a library is assembled from error-corrected oligonucleotides.

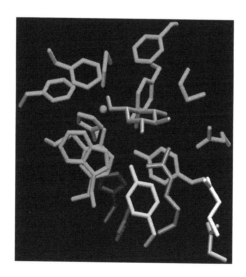

**FIGURE 4.6** Active site of galactose oxidase. Galactose (in gray) was modeled into the experimentally determined crystal structure of the enzyme [PDB ID: 1GOF (Ito et al. 1994)] using the program Schrödinger (Glide, version 4.5, and Prime, version 1.6, Schrödinger, LLC, New York, NY, 2007) (Friesner et al. 2004) and displayed using VMD (Humphrey et al. 1996); for clarity, none of the hydrogens are shown. Active site residues 194, 227, 228, 272, 290, 405, 406, 463, 464, 495, 496, 581, and the copper ion are shown in green. Side-chains of six residues that were mutated in a library designed to change the specificity of galactose oxidase to glucose (Table 4.1) are shown in color: Asp 324 (cyan), Gln 326 (white), Tyr 329 (yellow), Arg 330 (orange), Asn 333 (red), and His 334 (blue).

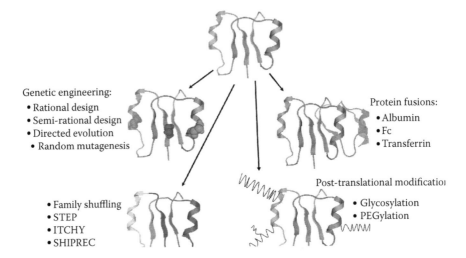

**FIGURE 7.1** Strategies for designing effective protein therapeutics. StEP = staggered extension process, ITCHY = incremental truncation for the creation of hybrid enzymes, SHIPREC = sequence homology-independent protein recombination (see Chapter 4 for a detailed description of each method).

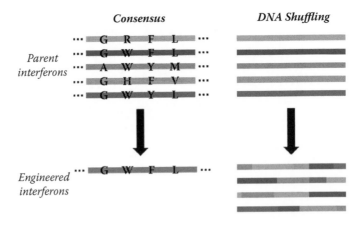

**FIGURE 7.2** Comparison of genetic engineering strategies used to create improved interferon-α mutants. The consensus method (left) was used to create highly active consensus interferon (CIFN), while DNA shuffling (right) was used to create interferon-α mutants tailored for treatment of chronic viral illness.

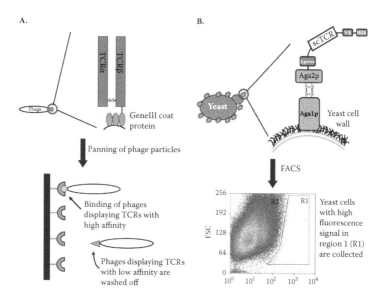

**FIGURE 7.3** TCR engineering using (A) phage display and (B) yeast display, adapted from (Boder and Wittrup 1997). A phage display library of TCRs is usually screened by panning phage particles on a surface/matrix coated with purified pMHC complexes, while a yeast display library can be rapidly screened by FACS using fluorophore-labeled (such as phycoerythrin, PE) pMHC complexes.

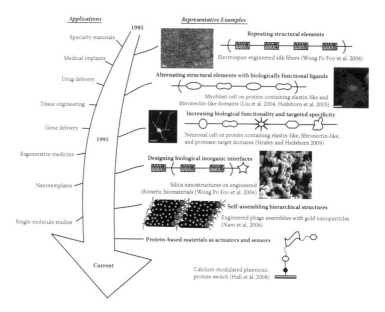

**FIGURE 8.1** Representative examples of engineered protein-based materials that illustrate development of the field. (Scale bars = 10 μm.)

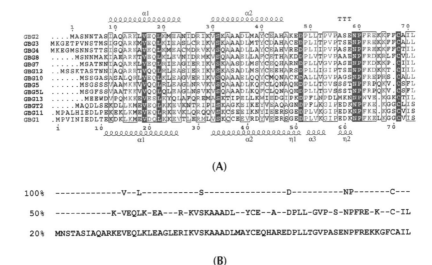

(A)

100%    -------------V--L-----------S----------------D----------NP-------C---

50%     -----------K-VEQLK-EA---R-KVSKAAADL--YCE--A--DPLL-GVP-S-NPFRE-K--C-IL

20%     MNSTASIAQARKEVEQLKLEAGLERIKVSKAAADLMAYCEQHAREDPLLTGVPASENPFREKKGFCAIL

(B)

**FIGURE 10.1** Multiple sequence alignment and consensus sequence determination. (A) A multiple sequence alignment highlights key similarities between the sequences. Here, a red background indicates 100% identity across all sequences, blue boxes indicate largely conserved residues, with similar residue types in red, and black indicates unconserved residues. The known regions of helical secondary structure for the top and bottom sequences are also shown. Figure generated with ESPRIPT (Gouet et al. 1999). (B) From the multiple sequence alignment, consensus sequences can be derived. Only seven residues are strictly conserved across all sequences (100%), while for 39 positions, there is one amino acid that occurs in the majority of sequences (50%). A complete sequence of 69 residues can be obtained by choosing the most common amino at each position (20%, the chosen amino acids all occur in at least three of the sequences). The sequences displayed are the set of γ-subunits from human heterotrimeric G-proteins.

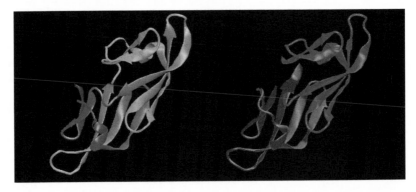

**FIGURE 10.2** Homology building can allow a model structure to be built from the structure of a related sequence. Shown here is a homology model of a *de novo* designed dimeric complex (right), based on the structure of a monomeric protein (cyanovirin-N, left). The sequence of the dimer is 62% identical to that of the reference structure. Notice that all the key structural features are duplicated in the model, despite significant differences in sequence. Model was generated with MODELLER (Sali and Blundell 1993; Fiser and Sali 2003); figures were generated with VMD (Humphrey et al. 1996).

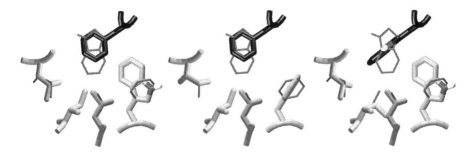

**FIGURE 10.3** Affinities of protein complexes can be modulated by introducing perturbations in steric complementarity guided by visual analysis. Here we show a portion of the binding interface between calmodulin (gray) and the M13 CaM-binding peptide (black). **Left:** Reversing the positions of the Trp and Phe from the wild-type structure (shown in magenta) is expected to preserve near-optimal packing. **Center:** Changing only the Trp on M13 (to Phe) is expected to leave an unsatisfied void in the interface. **Right:** Changing only the Phe on CaM (to Trp) is expected to produced steric clashes and unfavorable structural rearrangements. Figures generated with VMD (Humphrey et al. 1996).

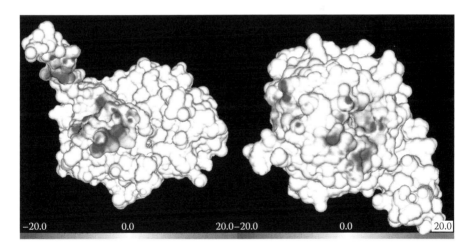

**FIGURE 10.4** The residual potential graphically displays deviations from perfect electrostatic complementarity. Displayed is the interface between the α (left) and βγ (right) subunits of a heterotrimeric G-protein. Regions of white indicate complementary surfaces (including surfaces not involved in the interface), while colored regions indicate the deviation from optimal complementarity. Blue indicates that the chemical groups underlying that region are either too positive or not negative enough, while red indicates groups that are either too negative or not positive enough. In the left-hand figure, it is clear that there are both regions that are excessively positive and excessively negative on this surface.

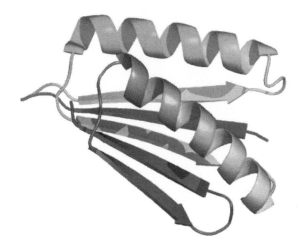

**FIGURE 11.2** Ribbon diagram of the crystal structure of TOP7, a protein designed to fold into a novel topology not seen previously in nature (Kuhlman et al. 2003). The observed structure is remarkably similar (RMS deviation = 1.2 A) to the designed model. Figure provided by Brian Kuhlman.

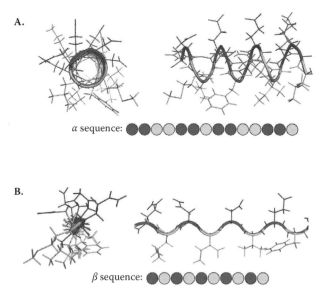

**FIGURE 11.3** Binary pattern strategy applied to (A) an α-helix and (B) a β-strand. Nonpolar residues are represented by yellow circles and polar residues are represented by red circles. In the α-helix design, nonpolar residues are placed every three to four positions, while in the β-strand design a nonpolar residue is placed at every other position. In each figure, a head-on view is shown on the left and a side view on the right.

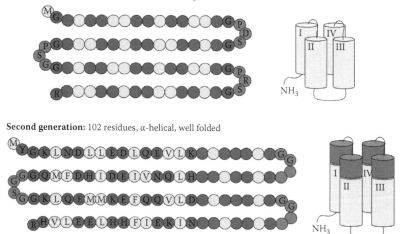

**First generation:** 74 residues, α-helical, molten globular

**Second generation:** 102 residues, α-helical, well folded

**FIGURE 11.5** Design of the first- and second-generation libraries of binary patterned four-helix bundles. The first generation library was designed to have 74 residues. In the second generation, each helix was extended by one helical turn leading to a 102-residue library. The design templates of the first- and second-generation four-helix bundle libraries differ in terms of number of total residues and the number and placement of combinatorially diverse residues. Polar residues are represented by red circles, nonpolar residues by yellow circles, and turn residues by blue circles. Residues that were held constant for the purposes of gene construction are coded with the single letter code, while combinatorial positions are colored but contain no letter.

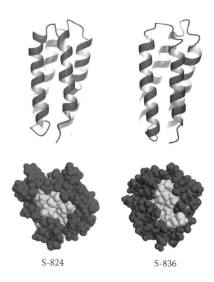

S-824          S-836

**FIGURE 11.6** NMR solution structures of protein S-824 and S-836. Top: Ribbon diagram of the four-helix bundle proteins. Bottom: Space-filling head-on view of the proteins with polar residues on the exterior (red or blue) and nonpolar residues in the core (yellow).

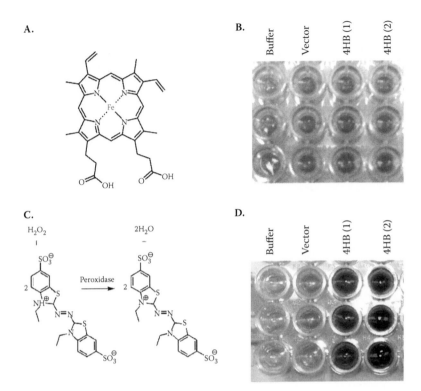

**A.**

**B.**

Buffer  Vector  4HB (1)  4HB (2)

**C.**

H₂O₂

2H₂O

Peroxidase

**D.**

Buffer  Vector  4HB (1)  4HB (2)

**FIGURE 11.7** Summary of heme binding and peroxidase activity exhibited by *de novo* four-helix bundle proteins. (A) Heme. (B) Heme binding assay. Each column shows a different sample performed in triplicate. Column 1 is heme alone in buffer. Column 2 is heme incubated with lysates from cells containing empty vector, which represents the activity of background *E. coli* proteins. Both columns 1 and 2 are negative controls and show a brown color. Columns 3 and 4 are cell lysates from cells expressing two different four-helix bundle (4HB) proteins incubated with heme showing various intensities of red. (C) The peroxidase reaction catalyzed by *de novo* four-helix bundle proteins using heme as cofactor and 2,2'-azino-di(3-ethyl-benzthiazoline-6-sulfonic acid) (ABTS) as the reducing agent. (D) Peroxidase assay. Each column is a different sample performed in triplicate. Column 1 shows the background activity of buffer alone and Column 2 shows the background activity of endogenous *E. coli* proteins. Both controls show a light teal color. Columns 3 and 4 are two different four-helix bundle (4HB) proteins exhibiting peroxidase activity giving rise to a dark teal color.

A.

B.

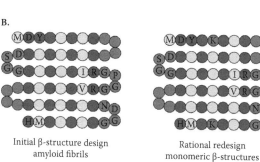

Initial β-structure design          Rational redesign
amyloid fibrils                monomeric β-structures

**FIGURE 11.8**   Design of binary coded libraries of β-sheet proteins. (A) Schematic representation of the six-stranded β-sheet target structure. Yellow coloring designates the nonpolar
face of the β-sheets, facing the interior of the protein. Red coloring designates the polar faces
of β-sheets, which are exposed to aqueous solvent. Turns are indicated in blue. (B) The design
templates of the initial library and of the redesigned proteins differ only in the placement of
lysine residues into the middle of the first and last β-strands. Residues that were held constant
for the purposes of gene construction are coded with the single letter code.

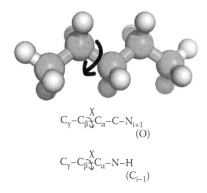

$$C_\gamma - C_\beta \overset{X}{\rightleftarrows} C_\alpha - C - N_{i+1}$$
$$(O)$$

$$C_\gamma - C_\beta \overset{X}{\rightleftarrows} C_\alpha - N - H$$
$$(C_{i-1})$$

**FIGURE 13.1**   Conformation of pentane. Shown are the positions of the carbons (colored
green) within pentane, illustrating which atoms in proteins can affect $\chi_1$ angle.

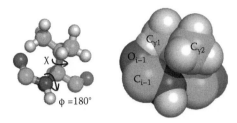

**FIGURE 13.2** Val conformation represented in ball and stick (left) and space filling (right) models. Carbon is shown in green, oxygen in red, nitrogen in blue, and hydrogen in white. When φ angle is around 180°, the χ angle of Val rarely takes the gauche- (χ = −60°) conformation because Cγ₁ has a clash with the carbonyl O atom of the preceding residue.

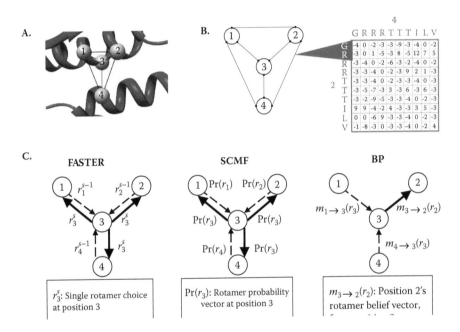

**FIGURE 14.2** Comparison of iterative update techniques (FASTER, SCMF, BP) for computational protein design. (A) Toy protein design problem, consisting of four fully interconnected design positions. (B) The corresponding graph structure of this design problem, where each node corresponds to a position and edges to interactions between them (pairwise rotamer-rotamer energy matrices). (C) The calculations performed for position 3 at a given iteration, where dashed arrows indicate relevant incoming data and solid arrows the results of the update calculations performed for position 3. Note that for both FASTER and SCMF, a single calculation is performed for each position, the results of which are then forwarded to all other positions. In BP, however, a distinct calculation is performed for each neighbor of position 3.

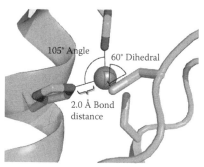

**FIGURE 17.1** Parameters for geometric definition of a metal-binding site. The primary coordination sphere is defined by the distances between the coordinating atoms and the metal, the angles around the metal (e.g., the angle formed by $N\varepsilon_{His}$-$Zn^{2+}$-$N\varepsilon_{His}$), and the dihedrals that define the relationship of the coordinating side chains to each other (in this example, the dihedral formed by the $C\beta_{Cys}$-$S\gamma_{Cys}$ bond relative to the $Zn^{2+}$-$N\varepsilon_{His}$ bond, about the $S\gamma_{Cys}$-$Zn^{2+}$ axis). Typically, the parameters are represented as a range of values. If an ideal $N$-$Zn^{2+}$-$N$ angle is 109.5°, then the designer could define an acceptable angle as being from 100° to 120°.

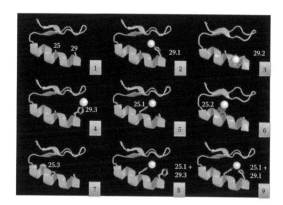

**FIGURE 17.2** Snapshots of a $His_4$-$Zn^{2+}$ binding site search algorithm in progress. For simplicity, analysis of only two positions is illustrated. Frame 1: Individual amino-acid positions are evaluated for their ability to participate in a $His_4$-$Zn^{2+}$ binding site. At this point, residues 25 and 29 are being considered. Frames 2, 3, 4: Position 29 is being considered as a potential zinc-coordinating histidine residue. Frame 2: Rotamer #1 is accepted because it does not clash with the backbone. Frame 3: Rotamer #2 is rejected because it clashes with the backbone. Frame 4: Rotamer #3 of position 29 is, like rotamer #1, accepted. Frames 5, 6, 7: Position 25 is being considered as a potential zinc-coordinating histidine residue. Frame 5: Rotamer #1 of position 25 is accepted. Frame 6: Histidine rotamer #2 is rejected because the zinc clashes with the backbone. Frame 7: Rotamer #3 is accepted. Frames 8, 9: Combinations of sterically acceptable rotamers are queried for their potential to form the prescribed geometry (defined in Figure 17.1). Frame 8: The combination of 25.1 + 29.3 is rejected because it fails the geometric requirements. Frame 9: The combination of 25.1 + 29.1 meets the geometric criteria and is accepted. While only a few snapshots of the process are shown here, the search algorithm processes all positions to find candidate sites, and then checks all permutations of candidate sites for geometric compatibility. The output is a table of mutations predicted to form the desired geometry, which can be rank ordered by nearness to ideal geometry.

| Position n | n Self energy | Position p | n–p Interaction energy | Position q | n–q Interaction energy | Position r | n–r Interaction energy |
|---|---|---|---|---|---|---|---|
| Rotamer j | −2 | Rotamer $p_1$ | −2 | Rotamer $q_1$ | 5 | Rotamer $r_1$ | −2 |
| | | Rotamer $p_2$ | −1 | Rotamer $q_2$ | 1 | Rotamer $r_2$ | −2 |
| | | Rotamer $p_3$ | 4 | Rotamer $q_3$ | 2 | Rotamer $r_3$ | 0 |
| | | | | Rotamer $q_4$ | 3 | Rotamer $r_4$ | −3 |
| | | | | Rotamer $q_5$ | 2 | | |
| Rotamer k | −1 | Rotamer $p_1$ | −3 | Rotamer $q_1$ | 0 | Rotamer $r_1$ | −6 |
| | | Rotamer $p_2$ | −1 | Rotamer $q_2$ | −5 | Rotamer $r_2$ | −5 |
| | | Rotamer $p_3$ | −4 | Rotamer $q_3$ | −3 | Rotamer $r_3$ | −7 |
| | | | | Rotamer $q_4$ | −2 | Rotamer $r_4$ | −6 |
| | | | | Rotamer $q_5$ | −2 | | |

Sum(Best Energies for $n_j$) = (−2 + −2 + 1 + −3) = −6
Sum(Worst Energies for $n_k$) = (−1 + −1 + 0 + −5) = −7

**FIGURE 17.3** Example of dead-end elimination (DEE). The basic principle of the DEE algorithm is that some rotamers can be proven never to be in the GMEC, by identification of another rotamer to which it would always be favorable to switch. In the simplest form, rotamer $j$ at position $n$ ($n_j$) can be eliminated by rotamer $k$ at position $n$ ($n_k$) if the best energy conformation of $n_j$ is worse than the worst energy conformation of $n_k$. This calculation is easy to perform. First, a pair-wise matrix of energies is tabulated. In this table, the individual inter-action energies of $n_j$ and $n_k$ with each possible rotamer at every other position is calculated, along with the self energy (the energy of $n_j$ or $n_k$ with the backbone template). In this example, there are three other positions: $p$, $q$, and $r$. Then the best possible total energy for $n_j$ is calcu-lated, regardless of the interaction energies among $p$, $q$, and $r$. The worst possible total energy for $n_k$ is calculated, also regardless of the interactions among $p$, $q$, and $r$. In this example, $n_k$ eliminates $n_j$ from the GMEC, because the energetically worst combination of $p$, $q$, $r$ for $n_k$ ($p_2$, $q_1$, $r_2$) is still better than the best combination of $p$, $q$, $r$ for $n_j$ ($p_1$, $q_2$, $r_4$). The computational power of DEE can be seen by the fact that this table required only pairwise 24 energies to be calculated, whereas enumeration would require 120 (2*3*5*4) energies to be calculated.

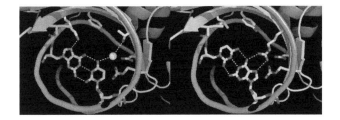

**FIGURE 17.4** Designed DNA-binding protein (Ashworth et al. 2006). The left panel shows the structure of the wild-type (MsoI) enzyme in complex with wild-type DNA (from 2FLD. pdb). The G-C base pair is shown in the center, and important side chains from the protein are shown as sticks. The right panel shows the designed enzyme in complex with the mutant DNA (from 1M5X.pdb). The G-C base pair has been mutated to C-G. Notice that the Lys and Thr in the wild-type (yellow) that form hydrogen bonds with the DNA bases are mutated to Val and Arg in the designed enzyme to provide new van der Waals and hydrogen bonds.

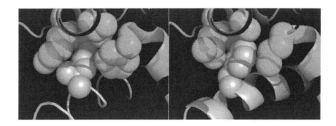

**FIGURE 17.5** Design of increased specificity for calmodulin (green) and a peptide (pink) (Palmer et al. 2006). The left panel shows the wild-type structure (from 1CDL.pdb). The right panel shows a model of the redesigned interface. On the peptide, Ile 14 has been mutated to Phe, which would form large steric clashes with the wild-type calmodulin. Three mutations were introduced in CaM (F19L, V35A, M36L) to create a compensatory "hole."

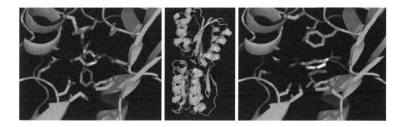

**FIGURE 17.6** Design of ribose-binding protein to TNT-binding protein (Looger et al. 2003). The center panel shows the overall structure of closed, ligand-bound ribose-BP (from 2DRI.pdb). Like all periplasmic binding proteins, the ligand-binding site is located in a cleft between two globular domains (top and bottom) separated by a hinge. In the unbound state, the two domains open up like a Venus flytrap. The left panel shows a close-up of the ribose (white) in the binding pocket with pertinent side chains shown in sticks. The right panel shows a model of TNT and the appropriate compensatory mutations required to turn ribose-BP into TNT-BP. Notice the π-sandwiching of the TNT ring by the two phenylalanine rings and the hydroxyl side chains properly positioned to form hydrogen bonds with the nitro-groups.

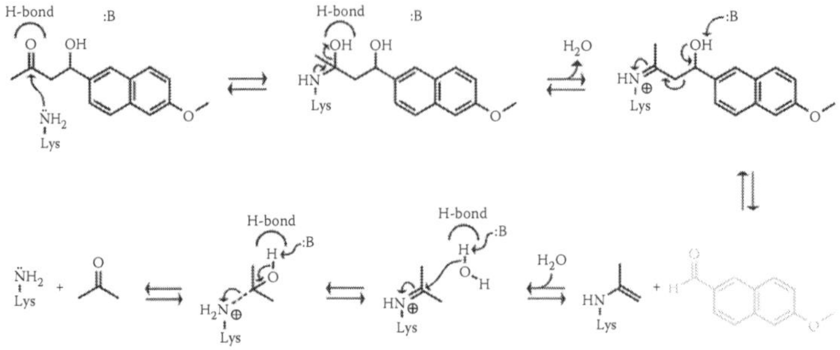

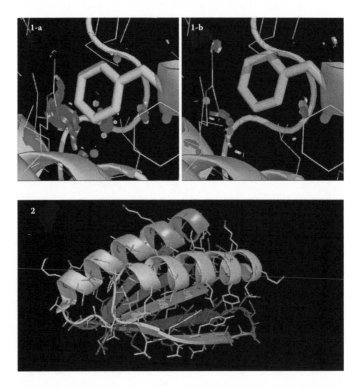

**FIGURE 17.7** Definition of the Retro-Aldol reaction. First, nucleophilic attack of lysine on the ketone of the substrate forms an imine intermediate, and a water molecule is eliminated. Then carbon-carbon bond cleavage is triggered by deprotonation of the β-alcohol and release of the fluorescent ketone (blue). Finally, the enzyme is returned to its prereactive state by nucleophilic attack by water on the imine. Adapted from Jiang et al. 2008.

**FIGURE 18.1** The fallacy of fixed-backbone protein design. (1) Mutation of residue 59 of 434 cro from Leu to Phe is disruptive and is not allowed by fixed-backbone design (Desjarlais and Handel 1999) (a). When the backbone is allowed to move the clashes seen with fixed backbone are alleviated (b). (2) Crystal structure of Top7, the first *de novo* design of a novel backbone fold. The design of Top7 showed that to select a designable backbone, backbone flexibility is necessary. Clashes are shown as red disks. All figures were generated using PyMOL (Delano 2002).

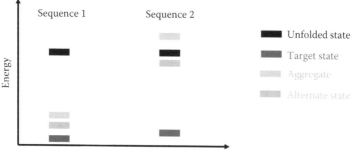

**FIGURE 18.2** The energy diagram of different states for two hypothetical designed sequences. Sequence 1 is designed using positive design only, whereas Sequence 2 utilizes both positive and negative design. Although the energy of the target structure is lower for Sequence 1, Sequence 2 is preferable as the energy difference between the target and alternative states is larger.

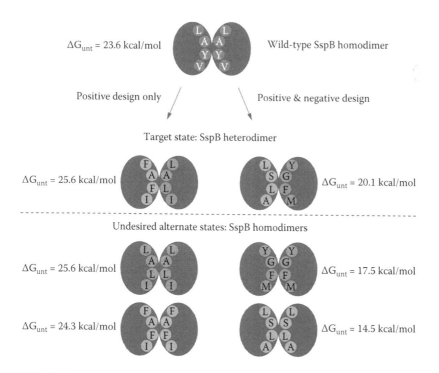

**FIGURE 18.3** Effect of explicit multistate design on specificity and stability (Bolon et al. 2005). Positive design produces stable but not specific heterodimers, whereas inclusion of negative design results in less stable but specific heterodimers.

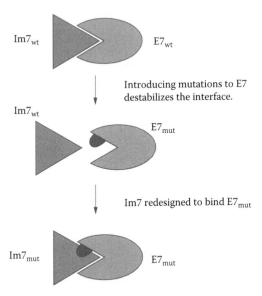

**FIGURE 18.4** Computational second-site suppressor strategy (Kortemme et al. 2004). A specific Im7-E7 interaction was designed by introducing interface-destabilizing mutations to E7 followed by the redesign of Im7 interface residues to compensate for the loss of binding. The interaction between $Im7_{mut}$ and $E7_{mut}$ was 40-fold stronger than $Im7_{mut}$-$E7_{wt}$ interaction, but several orders of magnitude weaker than $Im7_{wt}$-$E7_{wt}$ pair.

# 8 Protein Engineered Biomaterials

*Cheryl Wong Po Foo and Sarah C. Heilshorn*

## CONTENTS

## MOTIVATION

Proteins exhibit precisely defined sequences, exact biochemical compositions, and a dazzling array of complex conformational structures. Compared to their synthetic polymer analogs, which generally result in a statistical distribution of macromolecules with varied sequence, composition, and conformation, proteins are designed and produced as exactly specified macromolecules. This ability to design macromolecules with molecular-level precision can be exploited to engineer protein-based materials with unique properties. While the physical rules that govern protein folding into specific conformations and further self-assembly into hierarchical structures are still being elucidated, much has been learned already by mimicking protein sequences that are commonly found in nature to create new protein-based materials. Although this strategy was first widely described about two decades ago, the diversity of potential amino-acid sequences and range of promising materials applications are only just beginning to be explored more broadly. This chapter will give a brief historical overview of the field, present several case studies of interesting

179

and important engineered protein-based biomaterials, and provide an introduction to some of the current challenges and exciting directions the field is beginning to investigate. The chapter will be limited to materials synthesized from engineered recombinant proteins, as several excellent reviews exist on synthetic peptide-based materials (Holmes 2002, Woolfson and Ryadnov 2006).

## HISTORICAL PERSPECTIVE

Inspired by the exquisite morphological structures and outstanding properties of many naturally occurring biological materials, the field of protein-based materials evolved in the late 1980s (Figure 8.1). Early investigations generally focused on materials that were abundant in nature, had interesting mechanical properties, and were based on short repeating peptide sequences, such as β-sheets (Krejchi et al. 1994; McGrath et al. 1990), silkworm silk (Cappello et al. 1990), elastin (McPherson et al. 1992), and collagen (Goldberg et al. 1989) (see selected Case Studies in this chapter). As necessitated by the subject, these pioneering studies were quite interdisciplinary and combined experimental methods and characterization techniques from biochemistry, molecular biology, protein science, polymer science, and materials science.

Although these new protein-based materials were not specifically designed for medical applications, their potential use as engineered substrates to interface with biology was quickly recognized. One of the first nonstructural peptide motifs to

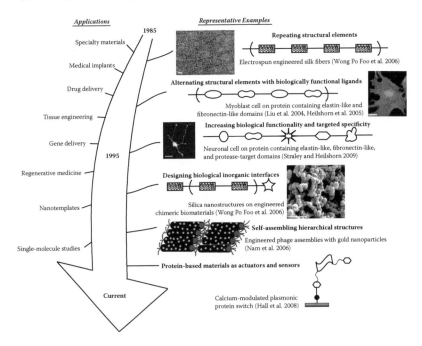

**FIGURE 8.1 (see color insert following page 178)** Representative examples of engineered protein-based materials that illustrate development of the field. (Scale bars = 10 μm.)

be included in protein-based materials was the tripeptide arginine-glycine-aspartic acid (RGD) cell surface integrin-binding sequence (Nicol et al. 1992; Pierschbacher and Ruoslahti 1984). Interspersing this sequence (or variations thereof) into the repeats of peptide structural domains resulted in materials that promoted the adhesion and spreading of many cell types. Consequently, these materials began to be heavily investigated for potential medical and clinical applications as implantable materials, drug delivery vehicles, tissue engineering scaffolds, and gene delivery vehicles. Protein-engineered materials offer significant advantages over traditional biomaterials for these types of medical applications. Traditional biomaterials include both synthetic polymers such as poly(lactic-co-glycolic acid) (PLGA) and harvested natural polymers including collagen and hyaluronic acid. While synthetic polymers allow for systematic control of material structure and properties through well-developed chemistries, they are often biologically inert and/or may degrade into toxic fragments. On the other hand, natural polymers are often intrinsically biologically active; however, their properties cannot be easily controlled, their sources are limited, and purification protocols are often expensive and variable from batch to batch. With protein engineering technology, it is possible to draw from the advantages of both synthetic and natural polymers to design and genetically engineer new classes of artificial proteins with precisely controlled physical, chemical, and biological properties that mimic a wide variety of natural biomaterials.

Building upon this concept, a greater diversity of potential biofunctional domains were incorporated into protein-engineered materials. Enzymatic crosslinking sites (McHale et al. 2005; Murphy et al. 2004), proteolytic degradation sites (Sakiyama-Elbert et al. 2001; Straley and Heilshorn 2009), domains to initiate cell signaling events (Liu et al. 2003), and recombinant growth factors (Sakiyama-Elbert et al. 2001) all began to be explored. With the inclusion of these peptide domains in protein-engineered materials, the concept of *modular protein design* became a key principle of the field. The underlying assumption of the modular protein design strategy is that individual peptide building blocks will maintain some degree of their original functionalities when fused together as multiple repeating sequences. As the field of synthetic polymeric biomaterials progressed toward more highly functionalized scaffolds using grafted peptides, protein-engineered materials also became more complex and specialized for specific target applications. This exquisite level of control and specialization may be key to developing scaffolds for regenerative medicine therapies that can communicate with stem cells and progenitor cells.

While much of the diversity of peptide building blocks included in protein-engineered materials to date have been chosen for cellular functions, more recent work has begun to include an even wider array of peptide building blocks with noncellular activity. These include peptide sequences to nucleate inorganic materials (Nam et al. 2006, Wong Po Foo et al. 2006), to induce assembly into hierarchical nanostructures (Nam et al. 2006; Padilla et al. 2001), to serve as nanomechanical actuators (Diehl et al. 2006), and to serve as sensors of local environmental conditions (Hall et al. 2008; Murphy et al. 2007; Topp et al. 2006). Potential applications of these new protein-engineered materials are quite diverse and include templates for nanostructured materials for energy storage (Nam et al. 2006), medical applications (Topp et al. 2006; Wong Po Foo et al. 2006), and fundamental single molecule studies (Diehl et al. 2006;

Hall et al. 2008). Due to their exact molecular-level specification, modular designed proteins are ideal materials for many applications in nanoscience and may be particularly useful in serving as bridges between the inorganic and biological interface.

## DESIGN STRATEGY

The modular protein design strategy is highly versatile and can employ peptide sequences derived from wild-type proteins, computationally guided amino-acid sequences, and/or peptide motifs discovered using high-throughput screening (Figure 8.2). While the majority of protein-based materials described to date have used wild-type amino-acid sequences, there are notable examples of engineered sequences as well. For example, high-throughput screening of peptides was used to determine the relative proteolytic susceptibility of several amino-acid sequences in response to the enzyme urokinase plasminogen activator (uPA) (Harris et al. 2000). Interspersing these engineered proteolytic-sensitive modules with wild-type modules resulted in a family of protein-based materials that display identical cell-binding characteristics, identical initial mechanical strengths, and tunable degradation rates (Straley and Heilshorn 2009). Recently, engineered peptide modules that were originally designed through computational modeling have also been included in protein-based materials. These materials utilize the molecular recognition between WW-domains and polyproline-rich peptide modules to induce gelation of protein-based physical hydrogels (Wong Po Foo and Heilshorn 2007). Using computational design, Russ et al. predicted an amino-acid sequence that folds into a WW-domain and binds polyproline-rich peptide modules with increased affinity compared to wild-type WW-domains (Russ et al. 2005). Replacement of the wild-type WW-domain

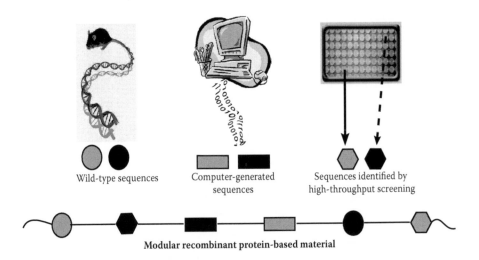

**FIGURE 8.2** Peptide sequences for engineered protein-based materials can be wild-type, computer-generated, or identified by high-throughput screening for specific functionalities. Protein engineering allows any of these peptide domains to be chosen as modules and fused together using recombinant DNA technologies.

with the engineered WW-domain in the protein-based material resulted in stiffer hydrogels with tunable mechanical properties (Wong Po Foo and Heilshorn 2007).

Interspersing peptide modules from a variety of sources can yield protein-based materials with highly specialized and tunable functionalities (Figure 8.3). When fusing together multiple modules into a single engineered recombinant protein, several important design parameters must be considered. First, the modular protein design strategy often requires the use of multiple DNA cloning steps resulting in highly repetitive DNA sequences. Although one may be tempted to simply use the most prevalent genetic codon for each encoded amino acid, this strategy greatly increases the possibility of genetic recombination events where the DNA becomes truncated during replication. Furthermore, recent studies have demonstrated that this strategy does not typically yield the highest translational efficiency (Villalobos et al. 2006). Therefore, a strategy that includes a diversity of codon sequences is preferred, and new algorithms exist to help predict useful genetic coding sequences for specific recombinant hosts (Villalobos et al. 2006). Second, when designing engineered amino-acid sequences, it is important to screen the encoding DNA and RNA sequences for potential sites that may induce unwanted secondary structure, may bind transcriptional or translational regulators, or may result in sequences with

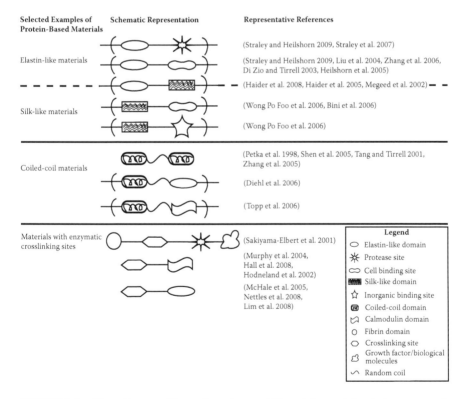

**FIGURE 8.3**  The wide versatility and precise tunability of the modular design strategy is illustrated through schematic representation of various engineered protein-based materials.

unusually high melting temperatures. These types of sequences may be difficult to work with during molecular cloning and/or result in diminished yields of expressed protein. Similarly, computer packages are now being developed to help screen for these types of DNA and RNA sequences for specific recombinant hosts (Villalobos et al. 2006).

Finally, one must specify what (if any) linking amino-acid residues will be included in the engineered protein to fuse the modules together. Through the careful design of clever molecular cloning strategies, it is often possible to engineer modular proteins without linking amino-acid residues (Prince et al. 1995). Indeed, the sequence of linking amino acids is often chosen to enable efficient molecular cloning of the encoding DNA. However, often the peptide modules require a certain degree of conformational flexibility to retain their biological and structural functionalities. For example, the cell-binding domain tripeptide RGD is well-known to exhibit variable integrin-binding functionality depending on the identity of the flanking amino acids and the ability of the peptide to form a small, exposed loop. Strategies to enhance conformational flexibility include utilizing longer portions of the primary sequence for wild-type derived modules (Heilshorn et al. 2003) and sandwiching functional modules between random-coil sequences (Topp et al. 2006). As an interesting counter-point, sometimes conformational rigidity may be desired in modular engineered proteins. For example, the relative orientation of multiple structural domains was fixed by linking them together with an alpha-helical coiled fusion sequence (Padilla et al. 2001). Molecular modeling of the fusion region allowed prediction of the orientation angles of the semirigid modular proteins, which dictated hierarchical self-assembly into specific nanoarchitectures. Once the structural requirements of the linking amino acids are determined, the primary amino-acid sequence of the entire modular protein should also be considered. For example, the pKa and hydrophobicity of the resulting engineered protein can be estimated using various models to help predict solubility and secondary structure under specific solvent conditions (Gasteiger et al. 2003; Kyte and Doolittle 1982).

## CASE STUDIES

### ELASTIN MOTIFS

Elastin-like polypeptides have been widely studied as potential injectable biomaterials and implantable scaffolds for regenerative medicine and drug delivery (Chilkoti et al. 2002; Rodriguez-Cabello et al. 2007). Potential applications include cartilage and invertebral disk repair (Betre et al. 2006, Betre et al. 2002), small diameter vascular grafts (Liu et al. 2004), and spinal cord repair (Straley and Heilshorn 2009). Elastin motifs are attractive components of materials for tissue engineering because of their biocompatibility in *in vivo* studies (Nettles et al. 2008; Rincon et al. 2006), high expression levels in recombinant systems (Chow et al. 2006), and inverse phase transition, which allows simple purification (Chow et al. 2006). At low temperatures, proteins containing elastin-like motifs are soluble in aqueous solvents; however, at higher temperatures the proteins form a polymer-rich coacervate phase. This temperature sensitivity can be modified by altering the primary amino-acid sequence of the engineered protein (Urry et al. 1991). Temperature sensitivity has been used to trigger *in situ* gelation and drug release from elastin-like gels. These biopolymers are being explored as drug delivery vehicles for the sustained release of pharmacological agents to articular joints (Betre et al. 2006) and dorsal root ganglion (Shamji et al. 2008), thereby reducing the side effects usually associated with systemic drug delivery.

The temperature sensitivity of elastin-like proteins has also been used to prepare scaffold-free cell sheets for implantation studies (Zhang et al. 2006). By nonspecifically adsorbing engineered elastin-like proteins with cell-binding domains onto standard cell culture substrates, human amniotic epithelial and mesenchymal cells were able to form cohesive cell monolayers. These cellular sheets were then incubated for a short time at 4°C; this causes the elastin-like protein to swell and releases the cells from the surface as a cohesive cell sheet that can be transferred to another culture surface or used for implantation.

To form mechanically robust monolithic structures, elastin-like proteins have been electrospun into fibers (Huang 2000) and covalently crosslinked using bifunctional chemical crosslinkers (Di Zio and Tirrell 2003) or enzymatic crosslinkers (McHale et al. 2005). Two recent crosslinking strategies for elastin-like proteins are of particular interest. First, by incorporating a photoactive amino acid, *p*-azido-phenylalanine, into the elastin-like sequence, Carrico et al. created a material that could be photocrosslinked into various patterns using standard lithography techniques (Carrico et al. 2007). Second, Lim et al. have included lysine amino acids with elastin-like proteins to enable rapid crosslinking with hydroxymethylphosphines (HMP) at physiological conditions (Lim et al. 2008). This novel crosslinking chemistry allows for *in situ* gelation of elastin-like proteins for cell encapsulation. The mechanical properties, and hence the cellular microenvironment, can be tuned by altering the location and number of lysine reactive sites contained within the engineered proteins.

Engineered elastin-like proteins have also been suggested as useful materials for potential mechanical applications. Atomic force microscopy studies of elastin-like proteins have been used to characterize the force required to stretch a single macromolecule (Flamia et al. 2004). This knowledge was exploited by Diehl et al. to create a mechanical bridge that tethered together multiple molecular motors (Diehl et al. 2006). In another example, through modification with additional chemical moieties, elastin-like proteins may be useful as mechanically active light sensors (Carriedo et al. 2000; Strzegowski et al. 1994). Taken together, these results suggest that elastin-like polymers make an easily tunable system that tolerates substantial modifications of the primary amino-acid sequence to control the resulting material properties.

## SILK MOTIFS

Natural silk has outstanding mechanical properties; however, silks can be very difficult to harvest from natural sources. For example, the strongest silks tend to be dragline silk secreted by canabalistic spiders that are unable to breed in captivity. Therefore, great efforts have been made to engineer recombinant silk materials. Synthetic genes encoding spider dragline silk from *Nephila clavipes* were successfully constructed, cloned, and expressed by Prince et al. (Prince et al. 1995). Synthetic gene technology has been widely used to control silk protein size, allowing the study of sequence length and structure-function relationships (Wong Po Foo et al. 2006), and to create engineered proteins with novel compositions such as silk-elastin block copolymers (Cappello et al. 1990) for a variety of applications including scaffolds for tissue engineering (Bini et al. 2006; Haider et al. 2008), gene delivery (Haider et al. 2005), and drug delivery (Cappello et al. 1998; Megeed et al. 2002). A multitude of silk-types from various species (for example, spiders *N. clavipes*, *Nephila edulis*, *Argiope aurantia*, *Nephilengys cruentata*, *Euprosthenops australis* and silkworms *Bombyx mori* and *Samia cynthica ricini*) have now been recombinantly expressed in a variety of host systems, the most common being the bacteria *Escherichia coli* (Fahnestock and Irwin 1997; Lewis et al. 1996) and yeast *Pichia pastoris* (Fahnestock and Bedzyk 1997).

Much of the current work with silk motifs is focused on creating materials with tunable mechanical and biophysical characteristics based on the wide diversity of silk motifs found in nature. For example, a chimeric silkworm silk that included sequences from the domestic silkworm *B. mori* and the wild silkworm *S. c. ricini* was found to have improved solubility and an α-helical conformation unlike the β-sheet secondary structures present in native silk proteins (Asakura et al. 2003). Generally, engineered silks require the use of strong solvents and careful processing to overcome the tendency of the hydrophobic amino-acid sequences to aggregate into large coacervates. However, in a surprising discovery, Johansson et al. have recently expressed repetitive silk motifs from *E. australis* together with a nonrepetitive silk sequence and a hydrophilic tag to enhance solubility. By simply cleaving the hydrophilic tag from the engineered silk in aqueous solvent, the spontaneous formation of meter-long silk fibers was achieved (Stark et al. 2007).

A second goal in the study of silk-mimetic materials is to create novel chimeras that combine superior silk-like mechanical properties with specific biofunctionalities. As an example, Wong Po Foo et al. used the complex mineralized composite systems found in diatoms as inspiration for the design of materials with remarkable morphological and nanostructural details (Wong Po Foo et al. 2006). These materials featured multimeric RGD-functionalized domains of the major ampullate spidroin 1 (MaSp1) of *N. clavipes* spider dragline silk fused to the silica precipitating R5 peptide derived from the silaffin protein of the diatom *Cylindrotheca fusiformis*. Purified fusion proteins were either cast into films or electrospun into fibers, and silicification reactions were conducted under mild conditions (pH 5.5 and room temperature) to yield substrates decorated with silica nanoparticles. Human bone marrow–derived mesenchymal stem cells (hMSCs) were able to grow and differentiate on these engineered silk materials due to the added biofunctionality of the RGD cell-binding motif. This study was the first to combine a structural domain with a functional cell-binding domain and an inorganic binding domain.

## COILED-COIL MOTIFS

Based on the common characteristics of physical hydrogels where rigid and soft segments coexist, Petka et al. created a multidomain artificial self-assembling protein (Petka et al. 1998), which consists of leucine zipper motifs as the rigid association segments and flanking flexible, water-soluble polyelectrolyte domains as the soft segments necessary to promote swelling. Leucine zipper motifs form part of a sub-category of coiled-coil domains found widely in nature and play key roles in the dimerization and DNA-binding of transcriptional regulatory proteins (O'Shea et al. 1989). These coiled-coils comprise six heptad repeats and fold into amphiphilic α-helices, which multimerize through electrostatic interactions. Hydrophobic inter-actions between interspersed nonpolar side chains further drive their association into oligomeric clusters (O'Shea et al. 1991). When two leucine zipper domains are linked together by a random coil polyelectrolyte domain, these triblock copolymers self-assemble into a three-dimensional coiled-coil polymer network at near-neutral pH (Petka et al. 1998).

Gelation is reversible in response to environmental fluctuations in pH, tem-perature, and ionic strength that disrupt the self-assembly of the leucine zip-per domains. While the reversibility of these coiled-coil interactions is a useful property to design materials with controllable sol-gel transitions, the sensitivity of these physical interactions to environmental changes often results in slightly unstable and weak hydrogels. Several strategies were subsequently developed to stabilize the assembly of these coiled-coil domains into engineered protein physi-cal hydrogels including the addition of cysteine residues (Shen et al. 2005), and the incorporation of noncanonical fluorinated amino-acid residues (Tang and Tirrell 2001) within the leucine zipper domains. Shen et al. successfully stabi-lized these protein-engineered hydrogels through the addition of cysteine residues, which bind to each other through covalent disulfide bonds (Shen et al. 2005). An alternative strategy used by Tang et al. was the successful incorporation of non-canonical hexafluoroleucine into the recombinant coiled-coil proteins through synthesis within an engineered bacterial host with a modified leucyl-tRNA syn-thetase (Tang and Tirrell 2001). Molecular dynamics simulation and experimental spectroscopic studies have shown that the hexafluorinated side chains resulted in more stable coiled-coil domain interactions, thereby increasing the thermal stabil-ity of the coiled-coil structures (Tang and Tirrell 2001; Yoder and Kumar 2002). Furthermore, the addition of noncanonical amino acids into engineered proteins adds novel chemical functionalities and physical properties to the materials. As an example, Zhang et al. introduced photoactive moieties within the elastin-like protein domains fused to coiled-coil domains via a hydrophilic spacer sequence (Zhang et al. 2005). The photoactive characteristics of the elastin-like domain allowed for its immobilization to surfaces while the coiled-coil domain remained available as a tethering site for target proteins with leucine zipper motifs. By fus-ing leucine zipper motifs onto various target proteins such as growth factors or peptide pharmaceuticals, hierarchical self-assembly of protein scaffolds can be created for various applications.

## CALMODULIN MOTIFS

Calmodulin motifs represent one of the more recent additions to the list of pep-tide domains included in a variety of protein-engineered materials. Calmodulin

(CaM) is a 16.5 kDa protein that regulates many $Ca^{2+}$-sensitive pathways including neuronal communication and muscle contraction (Zhang and Yuan 1998). CaM undergoes a conformational change upon binding of four $Ca^{2+}$ ions that allows it to reversibly associate with CaM-binding domains present in over 100 different proteins. Topp et al. used CaM motifs together with other peptide modules to create stimuli-responsive biomaterials that reversibly self-assemble (Topp et al. 2006). These biomaterials were designed to be sensitive to specific environmental chemical cues including pH and ionic concentration. Several tri-block modular proteins were designed using CaM as the sensory motif, leucine zippers as the self-assembling motif, random-coil hydrophilic spacers to confer flexibility and solubility, and various CaM-binding domains as the actuator motif. Depending on their molecular architecture, the hydrogels respond with predictable changes in their physical characteristics and mechanical properties to environmental stimuli. This study highlights a key advantage of modular protein engineering techniques, where a "genetic toolbox" of the various motifs are synthesized and used to create numerous materials of exact composition and known $Ca^{2+}$ sensitivity using a systematic combinatorial approach. The ability to quickly synthesize a family of materials with exact molecular-level alterations should be quite useful both in fundamental studies of structure-function relationships in macromolecular systems and in optimization studies of material properties for specific applications.

Hall et al. have also taken advantage of the distinct functionality of CaM as a biosensor to make a calcium-modulated plasmonic switch (Hall et al. 2008). In this novel construct, a CaM domain was flanked by an inactive N-terminal cutinase domain and an active C-terminal cutinase domain. The latter covalently binds to a phosphonate functional group that decorates nanoparticles on a coated monolayer surface. This particular design allows CaM to be unidirectionally oriented and evenly spaced on the surface. The added inactive N-terminal cutinase domain provides an additional mass, which results in a more significant change in overall protein packing density. Using a high-resolution localized surface plasmon resonance spectrometer, the real-time changes in dynamics, orientation, and structure of CaM with changes in $Ca^{2+}$ ion concentration can be studied over long periods of time without photobleaching due to the absence of a fluorescence label. Additionally, due to the precise tunability of protein engineering, the CaM domain and the inactive N-terminal cutinase domain can be substituted by any peptide drug or protein of interest, such as a growth factor, to immobilize the latter on a surface and to present active site-directed ligands (Hodneland et al. 2002; Murphy et al. 2004). This immobilization approach of any engineered protein provides a powerful technique to study dynamics and conformation changes of an unlabeled protein in real-time and to immobilize specific biological molecules at known densities on surfaces for fundamental cell biology, tissue engineering, and protein science studies.

# CURRENT AND FUTURE CHALLENGES

## EXPANDING THE LIBRARY OF PEPTIDE MODULES

One of the leading challenges in the field of protein-based materials is to expand the scope of potential functional amino-acid sequences that can be successfully incorporated into engineered recombinant proteins. As described in the case studies, the majority of peptide sequences explored to date primarily exhibit mechanical functions such as elasticity, association sites, crosslinking sites, and assembly sites. To these sequences, a more limited number of biofunctional peptides have been added, including cell-binding domains and proteolytic domains. In choosing these domains, scientists have attempted to select amino-acid sequences that are relatively short (hence, they do not require complicated folding protocols) and easy to express in bacterial systems (therefore, they do not require post-translational modifications). However, the wide diversity of the proteome represents an enormous potential for expanding the types of functional materials that can be created. Tables 8.1 and 8.2 represent lists of sequences and functions that either have been incorporated into designed materials already or are exciting candidates for future exploration. Three hurdles to incorporating a wider diversity of peptide modules into protein-based materials include predictive methods to link sequence choice to activity, theoretical and experimental methods to improve folding of recombinant multidomain proteins, and experimental systems to control post-translational modifications within complex peptide domains.

As demonstrated by the protein engineering and protein design fields, small amino-acid sequence changes quite distal from the putative active site can often cause dramatic effects on the protein function (Carbone and Arnold 2007). Similar observations have been made with protein-based materials, where even relatively short biofunctional sequences without complicated secondary folds can experience large fluctuations in biological efficacy based on the choice of distal amino-acid sequences. For example, identical 22-amino-acid sequences incorporating the CS5 cell-binding domain of fibronectin demonstrated very different cell adhesion properties when interspersed with two different elastin-like sequences differing in amino-acid content by only 3.4% (Heilshorn et al. 2005). Indeed, because small changes in amino-acid sequence can alter the electrostatic charge and relative hydrophobicity, and hence solubility and conformation, of the larger macromolecule, predictive methods to link amino-acid choice to resulting bioactivity are difficult to develop. The current requirement of complete amino-acid specification in engineered recombinant proteins can at times seem a burden, because so little predictive capability currently exists. However, as a range of experimental observations are gathered and as theoretical models of proteins continue to develop, it will become possible to fine-tune designed material properties through use of simple amino-acid mutations.

Similarly, as peptide modules with more complicated protein folds or required post-translational modifications are investigated, new synthetic techniques will need to be developed. Perhaps the most widely studied peptide module that exemplifies this need is the relatively simple collagen structure, which requires multiple

**TABLE 8.1**

**Representative Domains with Mechanical Functions for Use in Protein Engineering**

| Mechanical Function | Representative Domains | References |
|---|---|---|
| Crosslinking sites | Serine esterase cutinase | Murphy et al. 2004 |
| | Transglutaminase sites | Hu and Messersmith 2003, McHale et al. 2005 |
| | Lysl oxidase sites | Elbjeirami et al. 2003 |
| Elasticity | Elastin | Cappello et al. 1990, Cappello et al. 1998, Di Zio and Tirrell 2003, Diehl et al. 2006, Haider et al. 2008, Haider et al. 2005, Liu et al. 2003, Megeed et al. 2002, Rodriguez-Cabello et al. 2006, Rodriguez-Cabello et al. 2007, Urry et al. 1991 |
| | Titin | |
| | Resilin | |
| | Glutenin | |
| | | Labeit and Kolmerer 1995 |
| | | Elvin et al. 2005 |
| | | Wellner et al. 2006 |
| Adhesion | Mussel adhesive protein | Hwang et al. 2007 |
| Fibrous | Silk: Spider dragline | Bini et al. 2006, Fahnestock and Irwin 1997, Prince et al. 1995, Scheller et al. 2001, Wong Po Foo et al. 2006 |
| | Silk: Silkworm | |
| | Keratin | Asakura et al. 2003 |
| | | Vasconcelos et al. 2008 |
| Association | Leucine zippers | Carrico et al. 2007, Petka et al. 1998, Shen et al. 2005 |
| | WW domains/polyproline repeats | Macias et al. 2002, Sudol et al. 2005 |
| | Calmodulin | Murphy et al. 2007, Topp et al. 2006 |
| Protease sites | tPA and uPA sites | Harris et al. 2000, Straley and Heilshorn 2009 |
| | MMP sites | Lutolf et al. 2003 |
| | Plasmin sites | Schense and Hubbell 1999 |
| Flexibility: Sequences with random coil 2° structure | PEGAEG | Petka et al. 1998 |
| | GGGS | Porumb et al. 1994 |
| | GAGQGEA | Patch and Barron 2003 |
| | GAGQGSA | Patch and Barron 2003 |
| Assembly | Virion coat proteins | Nam et al. 2006 |
| | β-sheets | Krejchi et al. 1994, Smeenk et al. 2005 |
| | Collagen | Martin et al. 2003 |
| | Heat shock protein cage | Suci et al. 2006 |
| | Chaperonin | McMillan et al. 2005 |
| | α-helical coils | Padilla et al. 2001, Yeates and Padilla 2002 |
| Inorganic binders | Silica | Belton et al. 2008, Cha et al. 2000, Naik et al. 2003, Sumper and Kroger 2004, Wong Po Foo et al. 2006 |
| | Gold, silver, copper titanium, platinum, cobalt | Nam et al. 2006, Peelle et al. 2005, Stevens et al. 2004 |
| | | Whaley et al. 2000 |

## TABLE 8.1 (continued)
## Representative Domains with Mechanical Functions for Use in Protein Engineering

| Mechanical Function | Representative Domains | References |
| --- | --- | --- |
| Triggers/Actuators | Elastin | Diehl et al. 2006 |
| | Influenza hemagglutinin HA2 | Mavroidis et al. 2005 |
| pH responsive | Self-assembling peptide TZ1H | Zimenkov et al. 2006 |
| | | Stevens et al. 2004 |
| | Coiled coils | Murphy et al. 2007, Topp et al. 2006 |
| | Calmodulin-based protein | |
| Temperature responsive | Elastin-like polymers | Betre et al. 2006, Mackay and Chilkoti 2008, Rodriguez-Cabello et al. 2006, Rodriguez-Cabello et al. 2007 |
| Light responsive | Glycolipid-based supramolecular hydrogel | Matsumoto et al. 2008 |

## TABLE 8.2
## Representative Domains with Biochemical Functions for Use Protein Engineering

| Biochemical Function | Representative Domains | References |
| --- | --- | --- |
| Cell-cell mimics | Cadherins | Leckband 2008 |
| | Cell adhesion molecules | Hubbell 1999, Lutolf et al. 2003 |
| DNA/RNA adhesion | Coiled coils: GCN4 | Tang and Tirrell 2001 |
| | Positively charged sequences | Segura et al. 2003 |
| Growth factors | Vascular endothelial growth factor | Sakiyama-Elbert et al. 2001 |
| Antimicrobial | AMP antimicrobial peptide: pexiganan | Statz et al. 2008 |
| Cell penetrating | Tat and penetratin | Stewart et al. 2008 |
| Cell signaling | Notch ligands | Liu et al. 2003 |
| | Inhibitory sequences | Unkeless and Jin 1997 |
| ECM mimics | Cell-adhesive peptides | Meiners and Mercado 2003 |
| Fluorescence | Cyan and yellow fluorescent proteins | Evers et al. 2007 |

hydroxyproline post-translational modifications at specific amino-acid sites. Multiple expression systems have been explored to efficiently incorporate this domain into recombinant materials, including multivector bacterial systems (Lee and Gelvin 2008), eukaryotic *in vitro* expression systems (Kito et al. 2002; Ruggiero et al. 2000; Scheller et al. 2001), and whole organism mammalian expression systems such as excretion in goat's milk (Lazaris et al. 2002). As new peptide modules with longer amino-acid sequences and more complicated folds begin to be investigated, new folding techniques to allow proper folding of multidomain proteins will also need to be developed. For example, growth factors, fluorescent and luminescent proteins, and cell signaling proteins often require a series of proper domain folds to enable functional activity. In some instances, protein engineering and design techniques, such as those described in other chapters in this volume, can be employed to distill out shorter peptide sequences that recapitulate the functional activity of these larger protein domains. For example, short peptide sequences have been identified to recapitulate action of nerve growth factor protein (LeSauteur et al. 1995) and to specifically bind $\alpha_v\beta_3$ integrin cell-surface receptors with high affinity (Silverman et al. 2008). However, in other instances, inclusion of the full protein domain will undoubtedly be required to elicit the desired activity. These types of modules represent an enormous challenge and opportunity to the protein-based materials community.

Finally, much work has been done to develop methods to incorporate amino acids with chemical functionality beyond the 20 canonical amino acids into recombinant proteins (Chapter 9). For example, proteins including amino acids with unsaturated bonds, photoactive groups, and halides have all been reported (Link et al. 2003; Xie and Schultz 2006). However, only few examples of protein-based materials with noncanonical amino acids are currently published (Carrico et al. 2007; Tang and Tirrell 2001; Zhang et al. 2005). These cases involve slight modifications of previously described materials, where the novel chemical moieties serve to alter the properties of the resulting material. Depending on the technology employed to synthesize the recombinant protein, the noncanonical amino acid often competes and replaces one of the canonical amino acids in the protein. This technique limits the inclusion of multiple noncanonical amino-acid moieties in a single recombinant protein and also limits the resolution of the primary amino-acid sequence. However, new molecular biology strategies are being developed that overcome both of these limitations. The ability to incorporate novel chemical functionality into protein-based materials affords a great deal of versatility and creativity into the design process that is only just beginning to be explored.

## OPTIMIZATION OF COST-EFFICIENT SYNTHESIS

A further challenge for the adaptation of protein engineered materials remains the development of cost-efficient ways of synthesizing and purifying them. Indeed, although recombinant proteins have the advantages of "green" synthesis utilizing renewable resources and relatively low temperatures and pressures compared to traditional polymerization synthetic schemes, optimization of scaled-up biosynthetic processes to make them economical is still a relatively new field. In part, this is due to the lab-scale, batch-style processes commonly used to develop new recombinant proteins. However, as the use of recombinant proteins (both wild-type and engineered) continues to increase in the pharmaceutical field, industrial methods to increase productivity will be systematically developed (Baneyx 1999; Sarikaya et al. 2003). In the short-term, protein-based materials are most likely to be applied in biomedical applications, where knowledge from the pharmaceutical field about good manufacturing processes can be easily translated and higher cost-margins are acceptable for new products that are truly transformative in terms of their medical benefit. For these types of applications, it may be suitable to simply optimize standard variables using a systematic, although tedious, experimental system. For example, simple changes in medium recipes have been shown to result in a dramatic 30-fold increase in the production of elastin-like materials (Chow et al. 2006). Similarly, optimization of the vector through promoter engineering (Swartz 2001), use of self-cleaving tags (Gillies et al. 2008), codon optimization (Burgess-Brown et al. 2008), and co-expression with other proteins (Makrides 1996) have all been capable of increasing expression levels.

In the slightly longer term, applications utilizing protein-based materials for their nanoscale features will require production schemes financially competitive with those of the materials they are replacing. Toward this goal, the field of metabolic engineering will offer useful insights. For example, in the quest to identify enzymes that can be harnessed for alternative energy applications, the research community is currently investing heavily in microorganism synthetic strategies (Ajikumar et al. 2008; Kourist et al. 2008). Combining a systems-level approach with molecular-level microorganism engineering has been very successful in the optimization of recombinant protein expression (Hunt 2005). As high-throughput methods to map out metabolic pathways become more standardized, these can be used to improve the synthesis of protein-based materials.

Currently, most recombinant protein purification techniques rely on batch-level manipulations such as chromatography. In contrast, synthetic polymers are commonly purified through continuous separation processes that are more cost-efficient. This is primarily because synthetic polymers can generally be purified without concern for their conformation, while proteins must be purified in a manner that yields a properly folded conformation in order to retain structure-function relationships. A few examples of protein-based materials that can be purified through thermodynamic separations exist (Chilkoti et al. 2002; Urry et al. 1991), although to date these have not been exploited to allow continuous purification processes. Most importantly, relying on thermodynamic properties such as altered solubility in various solvents or environmental conditions greatly restricts the types of secondary structures and protein folds that can be included in the engineered proteins. Nevertheless, the number of wild-type recombinant proteins commercially available continues to increase at a rapid pace, and the diversity of technologies to synthesize and purify recombinant proteins utilizing a variety of expression systems (including prokaryotic microorganisms, eukaryotic cultures, transgenic animals, and cell-free culture systems) holds great promise.

## IMMUNOGENICITY

Specifically for protein-based biomaterials for medical applications, much further work in potential immunogenicity of the materials is required. Recent studies have highlighted new connections between the innate and adaptive immune responses that occur upon implantation of various biomaterials (Babensee 2008; Meinel et al. 2005). It appears that many biomaterials can serve as an adjuvant for the immune system, complicating potential immunosuppressive techniques. Therefore, it is highly likely that both the innate and adaptive responses will need to be managed interdependently for protein-based materials. To minimize potential immune effects, most recombinant proteins for biomedical implants are based on amino-acid sequences derived from human proteins (e.g., fibronectin, elastin, and laminin) or from nonhuman proteins with a history of use in human patients (e.g., silk). While initial studies have been quite promising (Nettles et al. 2008; Rincon et al. 2006; Wang et al. 2008), still relatively little is known about the immune system interactions with these materials. A current clinical trial utilizing a silk-elastin–like hybrid protein as a spinal disc material should provide some interesting details.

Potentially much can be learned and applied from the vaccine development community, where a large body of work on multiple fusion proteins has been conducted. For these vaccines, it is important that the individual domains retain their antibody-generating function, while the amino-acid sequences present at the fusion sites do not represent a potential trigger for immune reactivity. Currently, the engineering of these fusion sites is often performed through random mutagenesis and systematic screening; however, new algorithms are beginning to be developed that help to predict sequences that may be recognized by antigen presenting cells (Hon et al. 2008). As the characteristics of amino-acid sequences that serve as antigen-presenting substrates becomes more widely known, these sequence characteristics can be avoided in engineered sequences.

In addition, because these materials are synthesized in recombinant hosts, care must also be taken to adequately purify the engineered proteins from contaminants present in the expression system. For example, in Gram negative bacterial cultures, such as the commonly used *E. coli*, a key component of the membrane wall, lipopolysaccharide (also known as endotoxin), can stimulate a severe innate immune response (Trent 2004). In eukaryotic systems, there is a limited but finite potential to transfer diseases between hosts. As in cost-efficient synthesis, knowledge from the pharmaceutical industry, where recombinant proteins such as insulin are commonly manufactured and marketed, will be critical.

## Dynamic and Adaptive Materials

As the field of biomaterials evolves from implantation of inert materials to implantation of scaffolds that promote tissue regeneration, a fundamental understanding of how biomaterials adapt to the dynamic tissue microenvironment is critical. Materials scientists must develop methods to precisely control biomaterial properties during the entire service-life of a biomaterial: from implantation to adaptation to degradation and resorption. Furthermore, the ideal biomaterial properties immediately following implantation may not be identical to the ideal biomaterial properties at a later stage of healing. Due to their exact molecular-level precision and the ability to incorporate biologically functional modules directly into the design, engineered proteins are excellent candidates for dynamic and adaptive biomaterials.

Dynamic biomaterials (including both engineered-protein materials and synthetic polymeric materials) generally rely on enzyme-responsive peptide modules to alter their properties over time. These enzyme-responsive modules have been used to modify growth factor delivery rates, biomaterial mechanical properties and structure, and cell adhesivity in response to cell-secreted enzymes. For example, a composite biomaterial made of harvested naturally occurring fibrin and an engineered protein was developed to release human β-nerve growth factor (β-NGF) in response to local neuronal growth (Sakiyama-Elbert et al. 2001). The β-NGF module was linked to a plasmin-degradable module and a module that enabled enzymatic crosslinking to the fibrin matrix. By dynamically responding to the local cellular microenvironment, these biomaterials enhanced neurite extension 50% compared to scaffolds with wild-type β-NGF in solution. It is highly likely that many regenerative medicine applications will require the delivery of multiple growth factors with distinct temporal delivery profiles and concentration ranges from a single biomaterial (Chan and Mooney 2008). For example, guided angiogenesis and directed stem cell differentiation would both necessitate use of multiple growth factors. Utilizing the modular design strategy of protein-based materials, these types of scaffolds with multiple payloads being delivered at distinct temporal rates will be possible. Furthermore, this cell-mediated release strategy may be extended to other deliverables such as RNAi, neuroprotective peptides, antibacterial peptides, anti-inflammatory peptides, and so on.

As an example of altered mechanical properties, enzyme-responsive modules of varying proteolytic sensitivity were interspersed with elastin-like modules and cell-binding domain modules (Straley 2009). By tuning the relative concentrations of the enzyme-responsive modules in a homogeneous biomaterial, scaffolds were created with specific dynamic mechanical property profiles that adapt over time to the cellular microenvironment. For example, the biomaterial may be designed to initially have high mechanical strength optimized for surgical manipulation and implantation, followed by a rapid loss of mechanical strength immediately after implantation, a prolonged period of intermediate mechanical strength optimized for enhanced cell migration, and eventual complete degradation of the scaffold. These same engineered proteins can also be used to create composite scaffolds with microstructured regions that undergo either fast or slow biodegradation. These enzyme-responsive materials result in scaffolds with predetermined voids and cavities that form over time to direct the migration of cells (Straley 2009).

Finally, dynamic enzyme-responsive modules have also been used to adapt the cell-adhesivity of biomaterials (Salinas and Anseth 2008). High densities of the cell-binding tripeptide RGD are known to enhance the viability of human mesenchymal stem cells and the early stages of chondrogenic differentiation. However, these same high RGD densities are also thought to be detrimental to later stages of chondrogenesis and chondrocyte maturation. Therefore, a scaffold was designed with an initial high density of RGD sequences; upon cell secretion of matrix metalloprotease-13, which occurs during early stages of chondrogenesis, the scaffold responds by shedding RGD sequences, resulting in enhanced chondrogenic differentiation. To date, temporal regulation of cell adhesivity has been demonstrated only in synthetic polymeric materials functionalized with grafted peptides; however, a similar strategy can be employed in protein-based biomaterials. The engineering of adaptive protein-engineered materials is an exciting development that will enable the design of scaffolds that engage in two-way cell communication.

## Hierarchical Assembly

While many engineered protein-based materials (including those built from silk, elastin, and coiled-coil modules) rely on molecular self-assembly processes to tune the resulting material physical properties, examples of engineered protein systems with higher-orders of assembly are much less common. Understanding the hierarchical assembly of engineered proteins will not only yield fundamental insights into protein biochemistry and biophysics, but will also enable design and synthesis of novel materials with predictable nanoarchitectures (Sarikaya et al. 2003). As evidenced by the diversity of hierarchically assembled protein materials designed to date, the potential applications for these types of nanoarchitectures are wide ranging and include both medical and nonmedical uses. The discovery of new peptide sequences that can bind specific inorganic materials opens the door to using protein-based materials as templates for conducting nanostructures. For example, simultaneous genetic engineering of multiple viral-coat proteins on the M13 virus was used to guide the one-dimensional assembly of the rod-like virus particles and their subsequent decoration with inorganic nanoparticles of gold and cobalt oxide (Whaley et al. 2000). Flexible thin films of these protein-engineered materials have been used as electrodes in lithium ion batteries (Nam et al. 2006).

Moving beyond one-dimensional assemblies, protein-engineered materials have also been designed to form two- and three-dimensional hierarchical assemblies. In an early example, Yu et al. demonstrated that the monodispersity of semi-rigid polymers synthesized from engineered protein precursors allowed assembly into rare smectic liquid crystalline phases with nanoscale precision (Yu et al. 1997). Based on geometric design constraints, Padilla et al. identified several peptide modules that could be rearranged and modified to yield structural proteins with designed assembly properties (Padilla et al. 2001). These structural modules were fused together at various angles using a semirigid alpha-helical linker to yield proteins that predictably self-assemble into long filaments and cage-like structures (Padilla et al. 2001). As computational modeling becomes capable of handling larger macromolecules, the design of even more complex three-dimensional hierarchical assemblies will be enabled. These macromolecular materials can be further functionalized using inorganic-binding peptides (as described previously) or other peptide modules (Table 8.1) to result in a spectacular array of multifunctional, nanostructured materials. Given the increasing diversity of functional domains that can now be successfully designed into engineered protein materials, these new macromolecules are certain to find utility in a wide range of creative new applications.

## REFERENCES

Ajikumar, P. K., K. Tyo, S. Carlsen, O. Mucha, T. H. Phon, et al. 2008. "Terpenoids: Opportunities for biosynthesis of natural product drugs using engineered microorganisms." *Mol Pharm* 5: 167–90.

Asakura, T., K. Nitta, M. Yang, J. Yao, Y. Nakazawa, et al. 2003. "Synthesis and characterization of chimeric silkworm silk." *Biomacromolecules* 4: 815–20.

Babensee, J. E. 2008. "Interaction of dendritic cells with biomaterials." *Semin Immunol* 20: 101–8.

Baneyx, F. 1999. "Recombinant protein expression in *Escherichia coli*." *Curr Opin Biotechnol* 10: 411–21.

Belton, D. J., S. V. Patwardhan, V. V. Annenkov, E. N. Danilovtseva and C. C. Perry. 2008. "From biosilicification to tailored materials: Optimizing hydrophobic domains and resistance to protonation of polyamines." *Proc Natl Acad Sci U S A* 105: 5963–8.

Betre, H., W. Liu, M. R. Zalutsky, A. Chilkoti, V. B. Kraus, et al. 2006. "A thermally responsive biopolymer for intra-articular drug delivery." *J Controlled Release* 115: 175–82.

Betre, H., L. A. Setton, D. E. Meyer and A. Chilkoti. 2002. "Characterization of a genetically engineered elastin-like polypeptide for cartilaginous tissue repair." *Biomacromolecules* 3: 910–6.

Bini, E., C. Wong Po Foo, J. Huang, V. Karageorgiou, B. Kitchel, et al. 2006. "RGD-functionalized bioengineered spider dragline silk biomaterial." *Biomacromolecules* 7: 3139–45.

Burgess-Brown, N. A., S. Sharma, F. Sobott, C. Loenarz, U. Oppermann, et al. 2008. "Codon optimization can improve expression of human genes in *Escherichia coli*: A multi-gene study." *Protein Expr Purif* 59: 94–102.

Cappello, J., J. Crissman, M. Dorman, M. Mikolajczak, G. Textor, et al. 1990. "Genetic engineering of structural protein polymers." *Biotechnol Prog* 6: 198–202.

Cappello, J., J. W. Crissman, M. Crissman, F. A. Ferrari, G. Textor, et al. 1998. "In-situ self-assembling protein polymer gel systems for administration, delivery, and release of drugs." *J Controlled Release* 53: 105–17.

Carbone, M. N. and F. H. Arnold. 2007. "Engineering by homologous recombination: Exploring sequence and function within a conserved fold." *Curr Opin Struct Biol* 17: 454–9.

Carrico, I. S., S. A. Maskarinec, S. C. Heilshorn, M. L. Mock, J. C. Liu, et al. 2007. "Lithographic patterning of photoreactive cell-adhesive proteins." *J Am Chem Soc* 129: 4874–5.

Carriedo, G. A., F. J. G. Alonso, P. G. Elipe, J. L. Garcia-Alvarez, M. P. Tarazona, et al. 2000. "Spectroscopic and solution properties of phenoxyphosphazene random copolymers containing optically active binaphthoxy groups." *Macromolecules* 33: 3671–9.

Cha, J. N., G. D. Stucky, D. E. Morse and T. J. Deming. 2000. "Biomimetic synthesis of ordered silica structures mediated by block copolypeptides." *Nature* 403: 289–92.

Chan, G. and D. J. Mooney. 2008. "New materials for tissue engineering: Towards greater control over the biological response." *Trends Biotechnol* 26:382–92.

Chilkoti, A., M. R. Dreher and D. E. Meyer. 2002. "Design of thermally responsive, recombinant polypeptide carriers for targeted drug delivery." *Adv Drug Deliv Rev* 54: 1093–111.

Chow, D. C., M. R. Dreher, K. Trabbic-Carlson and A. Chilkoti. 2006. "Ultra-high expression of a thermally responsive recombinant fusion protein in *E. coli*." *Biotechnol Prog* 22: 638–46.

Di Zio, K. and D. A. Tirrell. 2003. "Mechanical properties of artificial protein matrices engineered for control of cell and tissue behavior." *Macromolecules* 36: 1553–8.

Diehl, M. R., K. Zhang, H. J. Lee and D. A. Tirrell. 2006. "Engineering cooperativity in biomotor-protein assemblies." *Science* 311: 1468–71.

Elbjeirami, W. M., E. O. Yonter, B. C. Starcher and J. L. West. 2003. "Enhancing mechanical properties of tissue-engineered constructs via lysyl oxidase crosslinking activity." *J Biomed Mater Res A* 66: 513–21.

Elvin, C. M., A. G. Carr, M. G. Huson, J. M. Maxwell, R. D. Pearson, et al. 2005. "Synthesis and properties of crosslinked recombinant pro-resilin." *Nature* 437: 999–1002.

Evers, T. H., M. A. Appelhof, P. T. de Graaf-Heuvelmans, E. W. Meijer and M. Merkx. 2007. "Ratiometric detection of Zn(II) using chelating fluorescent protein chimeras." *J Mol Biol* 374: 411–25.

Fahnestock, S. R. and L. A. Bedzyk. 1997. "Production of synthetic spider dragline silk protein in *Pichia pastoris*." *Appl Microbiol Biotechnol* 47: 33–9.

Fahnestock, S. R. and S. L. Irwin. 1997. "Synthetic spider dragline silk proteins and their production in *Escherichia coli*." *Appl Microbiol Biotechnol* 47: 23–32.

Flamia, R., P. A. Zhdan, M. Martino, J. E. Castle and A. M. Tamburro. 2004. "AFM study of the elastin-like biopolymer poly(ValGlyGlyValGly)." *Biomacromolecules* 5: 1511–8.

Gasteiger, E., A. Gattiker, C. Hoogland, I. Ivanyi, R. D. Appel, et al. 2003. "ExPASy: The proteomics server for in-depth protein knowledge and analysis." *Nucleic Acids Res* 31: 3784–8.

Gillies, A. R., J. F. Hsii, S. Oak and D. W. Wood. 2008. "Rapid cloning and purification of proteins: Gateway vectors for protein purification by self-cleaving tags." *Biotechnol Bioeng* 101: 229–40.

Goldberg, I., A. J. Salerno, T. Patterson and J. I. Williams. 1989. "Cloning and expression of a collagen-analog-encoding synthetic gene in *Escherichia coli*." *Gene* 80: 305–14.

Haider, M., J. Cappello, H. Ghandehari and K. W. Leong. 2008. "In vitro chondrogenesis of mesenchymal stem cells in recombinant silk-elastinlike hydrogels." *Pharm Res* 25: 692–9.

Haider, M., V. Leung, F. Ferrari, J. Crissman, J. Powell, et al. 2005. "Molecular engineering of silk-elastinlike polymers for matrix-mediated gene delivery: Biosynthesis and characterization." *Mol Pharm* 2: 139–50.

Hall, W. P., J. N. Anker, Y. Lin, J. Modica, M. Mrksich, et al. 2008. "A calcium-modulated plasmonic switch." *J Am Chem Soc* 130: 5836–7.

Harris, J. L., B. J. Backes, F. Leonetti, S. Mahrus, J. A. Ellman, et al. 2000. "Rapid and general profiling of protease specificity by using combinatorial fluorogenic substrate libraries." *Proc Natl Acad Sci U S A* 97: 7754–9.

Heilshorn, S. C., K. A. DiZio, E. R. Welsh and D. A. Tirrell. 2003. "Endothelial cell adhesion to the fibronectin CS5 domain in artificial extracellular matrix proteins." *Biomaterials* 24: 4245–52.

Heilshorn, S. C., J. C. Liu and D. A. Tirrell. 2005. "Cell-binding domain context affects cell behavior on engineered proteins." *Biomacromolecules* 6: 318–23.

Hodneland, C. D., Y. S. Lee, D. H. Min and M. Mrksich. 2002. "Selective immobilization of proteins to self-assembled monolayers presenting active site-directed capture ligands." *Proc Natl Acad Sci U S A* 99: 5048–52.

Holmes, T. C. 2002. "Novel peptide-based biomaterial scaffolds for tissue engineering." *Trends Biotechnol* 20: 16–21.

Hon, L. S., Y. Zhang, J. S. Kaminker and Z. Zhang. 2008. "Computational prediction of the functional effects of amino acid substitutions in signal peptides using a model-based approach." *Hum Mutat* In print DOI: 10.1002/humu.20798.

Hu, B. H. and P. B. Messersmith. 2003. "Rational design of transglutaminase substrate peptides for rapid enzymatic formation of hydrogels." *J Am Chem Soc* 125: 14298–9.

Huang, L., R. A. McMillan, R. P. Apkarian, B. Pourdeyhimi, V. P. Conticello, and E. L. Chaikof. 2000. "Generation of synthetic elastin-mimetic small diameter fibers and fiber networks." *Macromolecules* 33: 2989–97.

Hubbell, J. A. 1999. "Bioactive biomaterials." *Curr Opin Biotechnol* 10: 123–9.

Hunt, I. 2005. "From gene to protein: A review of new and enabling technologies for multi-parallel protein expression." *Protein Expr Purif* 40: 1–22.

Hwang, D. S., Y. Gim, D. G. Kang, Y. K. Kim and H. J. Cha. 2007. "Recombinant mussel adhesive protein Mgfp-5 as cell adhesion biomaterial." *J Biotechnol* 127: 727–35.

Kito, M., S. Itami, Y. Fukano, K. Yamana and T. Shibui. 2002. "Construction of engineered CHO strains for high-level production of recombinant proteins." *Appl Microbiol Biotechnol* 60: 442–8.

Kourist, R., P. Dominguez de Maria and U. T. Bornscheuer. 2008. "Enzymatic synthesis of optically active tertiary alcohols: Expanding the biocatalysis toolbox." *Chembiochem* 9: 491–8.

Krejchi, M. T., E. D. Atkins, A. J. Waddon, M. J. Fournier, T. L. Mason, et al. 1994. "Chemical sequence control of beta-sheet assembly in macromolecular crystals of periodic polypeptides." *Science* 265: 1427–32.

Kyte, J. and R. F. Doolittle. 1982. "A simple method for displaying the hydropathic character of a protein." *J Mol Biol* 157: 105–32.

Labeit, S. and B. Kolmerer. 1995. "Titins: Giant proteins in charge of muscle ultrastructure and elasticity." *Science* 270: 293–6.

Lazaris, A., S. Arcidiacono, Y. Huang, J. F. Zhou, F. Duguay, et al. 2002. "Spider silk fibers spun from soluble recombinant silk produced in mammalian cells." *Science* 295: 472–6.

Leckband, D. 2008. "Beyond structure: Mechanism and dynamics of intercellular adhesion." *Biochem Soc Trans* 36: 213–20.

Lee, L. Y. and S. B. Gelvin. 2008. "T-DNA binary vectors and systems." *Plant Physiol* 146: 325–32.

LeSauteur, L., L. Wei, B. F. Gibbs and H. U. Saragovi. 1995. "Small peptide mimics of nerve growth factor bind TrkA receptors and affect biological responses." *J Biol Chem* 270: 6564–9.

Lewis, R. V., M. Hinman, S. Kothakota and M. J. Fournier. 1996. "Expression and purification of a spider silk protein: A new strategy for producing repetitive proteins." *Protein Expr Purif* 7: 400–6.

Lim, D. W., D. L. Nettles, L. A. Setton and A. Chilkoti. 2008. "In situ cross-linking of elastin-like polypeptide block copolymers for tissue repair." *Biomacromolecules* 9: 222–30.

Link, A. J., M. L. Mock and D. A. Tirrell. 2003. "Non-canonical amino acids in protein engineering." *Curr Opin Biotechnol* 14: 603–9.

Liu, C. Y., U. Westerlund, M. Svensson, M. C. Moe, M. Varghese, et al. 2003. "Artificial niches for human adult neural stem cells: Possibility for autologous transplantation therapy." *J Hematother Stem Cell Res* 12: 689–99.

Liu, J. C., S. C. Heilshorn and D. A. Tirrell. 2004. "Comparative cell response to artificial extracellular matrix proteins containing the RGD and CS5 cell-binding domains." *Biomacromolecules* 5: 497–504.

Lutolf, M. P., J. L. Lauer-Fields, H. G. Schmoekel, A. T. Metters, F. E. Weber, et al. 2003. "Synthetic matrix metalloproteinase-sensitive hydrogels for the conduction of tissue regeneration: Engineering cell-invasion characteristics." *Proc Natl Acad Sci U S A* 100: 5413–8.

Lutolf, M. P., G. P. Raeber, A. H. Zisch, N. Tirelli and J. A. Hubbell. 2003. "Cell-responsive synthetic hydrogels." *Advanced Materials* 15: 888–+.

Macias, M. J., S. Wiesner and M. Sudol. 2002. "WW and SH3 domains, two different scaffolds to recognize proline-rich ligands." *FEBS Lett* 513: 30–7.

Mackay, J. A. and A. Chilkoti. 2008. "Temperature sensitive peptides: Engineering hyperthermia-directed therapeutics." *Int J Hyperthermia*: 1–13.

Makrides, S. C. 1996. "Strategies for achieving high-level expression of genes in *Escherichia coli*." *Microbiol Rev* 60: 512–38.

Martin, R., L. Waldmann and D. L. Kaplan. 2003. "Supramolecular assembly of collagen triblock peptides." *Biopolymers* 70: 435–44.

Matsumoto, S., S. Yamaguchi, S. Ueno, H. Komatsu, M. Ikeda, et al. 2008. "Photo gel-sol/sol-gel transition and its patterning of a supramolecular hydrogel as stimuli-responsive biomaterials." *Chemistry* 14: 3977–86.

Mavroidis, C., J. Nikitczuk, B. Weinberg, G. Danaher, K. Jensen, et al. 2005. "Smart portable rehabilitation devices." *J Neuroeng Rehabil* 2: 18.

McGrath, K. P., D. A. Tirrell, M. Kawai, T. L. Mason and M. J. Fournier. 1990. "Chemical and biosynthetic approaches to the production of novel polypeptide materials." *Biotechnol Prog* 6: 188–92.

McHale, M. K., L. A. Setton and A. Chilkoti. 2005. "Synthesis and in vitro evaluation of enzymatically cross-linked elastin-like polypeptide gels for cartilaginous tissue repair." *Tissue Eng* 11: 1768–79.

McMillan, R. A., J. Howard, N. J. Zaluzec, H. K. Kagawa, R. Mogul, et al. 2005. "A self-assembling protein template for constrained synthesis and patterning of nanoparticle arrays." *J Am Chem Soc* 127: 2800–1.

McPherson, D. T., C. Morrow, D. S. Minehan, J. Wu, E. Hunter, et al. 1992. "Production and purification of a recombinant elastomeric polypeptide, G-(VPGVG)19-VPGV, from *Escherichia coli*." *Biotechnol Prog* 8: 347–52.

Megeed, Z., J. Cappello and H. Ghandehari. 2002. "Genetically engineered silk-elastinlike protein polymers for controlled drug delivery." *Adv Drug Deliv Rev* 54: 1075–91.

Meinel, L., S. Hofmann, V. Karageorgiou, C. Kirker-Head, J. McCool, et al. 2005. "The inflammatory responses to silk films in vitro and in vivo." *Biomaterials* 26: 147–55.

Meiners, S. and M. L. Mercado. 2003. "Functional peptide sequences derived from extracellular matrix glycoproteins and their receptors: Strategies to improve neuronal regeneration." *Mol Neurobiol* 27: 177–96.

Murphy, W. L., W. S. Dillmore, J. Modica and M. Mrksich. 2007. "Dynamic hydrogels: Translating a protein conformational change into macroscopic motion." *Angew Chem Int Ed Engl* 46: 3066–9.

Murphy, W. L., K. O. Mercurius, S. Koide and M. Mrksich. 2004. "Substrates for cell adhesion prepared via active site-directed immobilization of a protein domain." *Langmuir* 20: 1026–30.

Naik, R. R., P. W. Whitlock, F. Rodriguez, L. L. Brott, D. D. Glawe, et al. 2003. "Controlled formation of biosilica structures in vitro." *Chem Commun (Camb)*: 238–9.

Nam, K. T., D. W. Kim, P. J. Yoo, C. Y. Chiang, N. Meethong, et al. 2006. "Virus-enabled synthesis and assembly of nanowires for lithium ion battery electrodes." *Science* 312: 885–8.

Nettles, D. L., K. Kitaoka, N. A. Hanson, C. M. Flahiff, B. A. Mata, et al. 2008. "In situ cross-linking elastin-like polypeptide gels for application to articular cartilage repair in a goat osteochondral defect model." *Tissue Eng Part A* 14: 1133–40.

Nicol, A., D. C. Gowda and D. W. Urry. 1992. "Cell adhesion and growth on synthetic elastomeric matrices containing Arg-Gly-Asp-Ser-3." *J Biomed Mater Res* 26: 393–413.

O'Shea, E. K., J. D. Klemm, P. S. Kim and T. Alber. 1991. "X-ray structure of the GCN4 leucine zipper, a two-stranded, parallel coiled coil." *Science* 254: 539–44.

O'Shea, E. K., R. Rutkowski and P. S. Kim. 1989. "Evidence that the leucine zipper is a coiled coil." *Science* 243: 538–42.

Padilla, J. E., C. Colovos and T. O. Yeates. 2001. "Nanohedra: Using symmetry to design self assembling protein cages, layers, crystals, and filaments." *Proc Natl Acad Sci U S A* 98: 2217–21.

Patch, J. A. and A. E. Barron. 2003. "Helical peptoid mimics of magainin-2 amide." *J Am Chem Soc* 125: 12092–3.

Peelle, B. R., E. M. Krauland, K. D. Wittrup and A. M. Belcher. 2005. "Design criteria for engineering inorganic material-specific peptides." *Langmuir* 21: 6929–33.

Petka, W. A., J. L. Harden, K. P. McGrath, D. Wirtz and D. A. Tirrell. 1998. "Reversible hydrogels from self-assembling artificial proteins." *Science* 281: 389–92.

Pierschbacher, M. D. and E. Ruoslahti. 1984. "Cell attachment activity of fibronectin can be duplicated by small synthetic fragments of the molecule." *Nature* 309: 30–3.

Porumb, T., P. Yau, T. S. Harvey and M. Ikura. 1994. "A calmodulin-target peptide hybrid molecule with unique calcium-binding properties." *Protein Eng* 7: 109–15.

Prince, J. T., K. P. McGrath, C. M. DiGirolamo and D. L. Kaplan. 1995. "Construction, cloning, and expression of synthetic genes encoding spider dragline silk." *Biochemistry* 34: 10879–85.

Rincon, A. C., I. T. Molina-Martinez, B. de Las Heras, M. Alonso, C. Bailez, et al. 2006. "Biocompatibility of elastin-like polymer poly(VPAVG) microparticles: In vitro and in vivo studies." *J Biomed Mater Res A* 78: 343–51.

Rodriguez-Cabello, J. C., S. Prieto, F. J. Arias, J. Reguera and A. Ribeiro. 2006. "Nanobiotechnological approach to engineered biomaterial design: The example of elastin-like polymers." *Nanomed* 1: 267–80.

Rodriguez-Cabello, J. C., S. Prieto, J. Reguera, F. J. Arias and A. Ribeiro. 2007. "Biofunctional design of elastin-like polymers for advanced applications in nanobiotechnology." *J Biomater Sci Polym Ed* 18: 269–86.

Ruggiero, F., J. Y. Exposito, P. Bournat, V. Gruber, S. Perret, et al. 2000. "Triple helix assembly and processing of human collagen produced in transgenic tobacco plants." *FEBS Lett* 469: 132–6.

Russ, W. P., D. M. Lowery, P. Mishra, M. B. Yaffe and R. Ranganathan. 2005. "Natural-like function in artificial WW domains." *Nature* 437: 579–83.

Sakiyama-Elbert, S. E., A. Panitch and J. A. Hubbell. 2001. "Development of growth factor fusion proteins for cell-triggered drug delivery." *Faseb J* 15: 1300–2.

Salinas, C. N. and K. S. Anseth. 2008. "The enhancement of chondrogenic differentiation of human mesenchymal stem cells by enzymatically regulated RGD functionalities." *Biomaterials* 29: 2370–7.

Sarikaya, M., C. Tamerler, A. K. Jen, K. Schulten and F. Baneyx. 2003. "Molecular biomimetics: Nanotechnology through biology." *Nat Mater* 2: 577–85.

Scheller, J., K. H. Guhrs, F. Grosse and U. Conrad. 2001. "Production of spider silk proteins in tobacco and potato." *Nat Biotechnol* 19: 573–7.

Schense, J. C. and J. A. Hubbell. 1999. "Cross-linking exogenous bifunctional peptides into fibrin gels with factor XIIIa." *Bioconjug Chem* 10: 75–81.

Segura, T., M. J. Volk and L. D. Shea. 2003. "Substrate-mediated DNA delivery: Role of the cationic polymer structure and extent of modification." *J Controlled Release* 93: 69–84.

Shamji, M. F., L. Whitlatch, A. H. Friedman, W. J. Richardson, A. Chilkoti, et al. 2008. "An injectable and in situ-gelling biopolymer for sustained drug release following perineural administration." *Spine* 33: 748–54.

Shen, W., R. G. H. Lammertink, J. K. Sakata, J. A. Kornfield and D. A. Tirrell. 2005. "Assembly of an artificial protein hydrogel through leucine zipper aggregation and disulfide bond formation." *Macromolecules* 38: 3909–16.

Silverman, A. P., A. M. Levin, J. L. Lahti and J. R. Cochran. 2008. "Engineered cystine-knot peptides that bind alphav beta3 integrin with antibody-like affinities." *J Mol Biol* In print.

Smeenk, J. M., M. B. Otten, J. Thies, D. A. Tirrell, H. G. Stunnenberg, et al. 2005. "Controlled assembly of macromolecular beta-sheet fibrils." *Angew Chem Int Ed Engl* 44: 1968–71.

Stark, M., S. Grip, A. Rising, M. Hedhammar, W. Engstrom, et al. 2007. "Macroscopic fibers self-assembled from recombinant miniature spider silk proteins." *Biomacromolecules* 8: 1695–701.

Statz, A. R., J. P. Park, N. P. Chongsiriwatana, A. E. Barron and P. B. Messersmith. 2008. "Surface-immobilised antimicrobial peptoids." *Biofouling* 24: 439–48.

Stevens, M. M., S. Allen, J. K. Sakata, M. C. Davies, C. J. Roberts, et al. 2004. "pH-dependent behavior of surface-immobilized artificial leucine zipper proteins." *Langmuir* 20: 7747–52.

Stevens, M. M., N. T. Flynn, C. Wang, D. A. Tirrell and R. Langer. 2004. "Coiled-coil peptide-based assembly of gold nanoparticles." *Adv Mater* 16: 915–918.

Stewart, K. M., K. L. Horton and S. O. Kelley. 2008. "Cell-penetrating peptides as delivery vehicles for biology and medicine." *Org Biomol Chem* 6: 2242–55.

Straley, K., S. C. Heilshorn. 2009. "Independent tuning of multiple biomaterials properties using tissue engineering." *Soft Matter* 5:114–124.

Straley, K., C. Wong Po Foo and S. C. Heilshorn. 2007. "Cell-adaptable protein scaffolds for spinal cord nerve regeneration." *Mater Res Soc Proc* 1060: NN03–05.

Strzegowski, L. A., M. B. Martinez, D. C. Gowda, D. W. Urry and D. A. Tirrell. 1994. "Photomodulation of the inverse temperature transition of a modified elastin poly(pentapeptide)." *J Am Chem Soc* 116: 813–814.

Suci, P. A., M. T. Klem, F. T. Arce, T. Douglas and M. Young. 2006. "Assembly of multilayer films incorporating a viral protein cage architecture." *Langmuir* 22: 8891–6.

Sudol, M., C. C. Recinos, J. Abraczinskas, J. Humbert and A. Farooq. 2005. "WW or WoW: The WW domains in a union of bliss." *IUBMB Life* 57: 773–8.

Sumper, M. and N. Kroger. 2004. "Silica formation in diatoms: The function of long-chain polyamines and silaffins." *J Mater Chem* 14: 2059–2065.

Swartz, J. R. 2001. "Advances in *Escherichia coli* production of therapeutic proteins." *Curr Opin Biotechnol* 12: 195–201.

Tang, Y. and D. A. Tirrell. 2001. "Biosynthesis of a highly stable coiled-coil protein containing hexafluoroleucine in an engineered bacterial host." *J Am Chem Soc* 123: 11089–90.

Topp, S., V. Prasad, G. C. Cianci, E. R. Weeks and J. P. Gallivan. 2006. "A genetic toolbox for creating reversible Ca2+-sensitive materials." *J Am Chem Soc* 128: 13994–5.

Trent, M. S. 2004. "Biosynthesis, transport, and modification of lipid A." *Biochem Cell Biol* 82: 71–86.

Unkeless, J. C. and J. Jin. 1997. "Inhibitory receptors, ITIM sequences and phosphatases." *Curr Opin Immunol* 9: 338–43.

Urry, D. W., C. H. Luan, T. M. Parker, D. C. Gowda, K. U. Prasad, et al. 1991. "Temperature of polypeptide inverse temperature transition depends on mean residue hydrophobicity." *J Am Chem Soc* 113: 4346–8.

Vasconcelos, A., G. Freddi and A. Cavaco-Paulo. 2008. "Biodegradable materials based on silk fibroin and keratin." *Biomacromolecules* 9: 1299–305.

Villalobos, A., J. E. Ness, C. Gustafsson, J. Minshull and S. Govindarajan. 2006. "Gene Designer: A synthetic biology tool for constructing artificial DNA segments." *BMC Bioinformatics* 7: 285.

Wang, Y., D. D. Rudym, A. Walsh, L. Abrahamsen, H. J. Kim, et al. 2008. "In vivo degradation of three-dimensional silk fibroin scaffolds." *Biomaterials* 29: 3415–28.

Wellner, N., J. T. Marsh, A. W. Savage, N. G. Halford, P. R. Shewry, et al. 2006. "Comparison of repetitive sequences derived from high molecular weight subunits of wheat glutenin, an elastomeric plant protein." *Biomacromolecules* 7: 1096–103.

Whaley, S. R., D. S. English, E. L. Hu, P. F. Barbara and A. M. Belcher. 2000. "Selection of peptides with semiconductor binding specificity for directed nanocrystal assembly." *Nature* 405: 665–8.

Wong Po Foo, C., E. Bini, J. Huang, S. Y. Lee and D. L. Kaplan. 2006. "Solution behavior of synthetic silk peptides and modified recombinant silk proteins." *Appl Phys A-Mater Sci & Process* 82: 193–203.

Wong Po Foo, C. and S. C. Heilshorn. 2007. "Protein-based hydrogels for cell transplantation under constant physiological conditions." *Mater Res Soc Proc* 1060: LL09–04.

Wong Po Foo, C., S. V. Patwardhan, D. J. Belton, B. Kitchel, D. Anastasiades, et al. 2006. "Novel nanocomposites from spider silk-silica fusion (chimeric) proteins." *Proc Natl Acad Sci U S A* 103: 9428–33.

Woolfson, D. N. and M. G. Ryadnov. 2006. "Peptide-based fibrous biomaterials: Some things old, new and borrowed." *Curr Opin Chem Biol* 10: 559–67.

Xie, J. and P. G. Schultz. 2006. "A chemical toolkit for proteins—an expanded genetic code." *Nat Rev Mol Cell Biol* 7: 775–82.

Yeates, T. O. and J. E. Padilla. 2002. "Designing supramolecular protein assemblies." *Curr Opin Struct Biol* 12: 464–70.

Yoder, N. C. and K. Kumar. 2002. "Fluorinated amino acids in protein design and engineering." *Chem Soc Rev* 31: 335–41.

Yu, S. M., V. P. Conticello, G. Zhang, C. Kayser, M. J. Fournier, et al. 1997. "Smectic ordering in solutions and films of a rod-like polymer owing to monodispersity of chain length." *Nature* 389: 167–70.

Zhang, H., M. Iwama, T. Akaike, D. W. Urry, A. Pattanaik, et al. 2006. "Human amniotic cell sheet harvest using a novel temperature-responsive culture surface coated with protein-based polymer." *Tissue Eng* 12: 391–401.

Zhang, K., M. R. Diehl and D. A. Tirrell. 2005. "Artificial polypeptide scaffold for protein immobilization." *J Am Chem Soc* 127: 10136–7.

Zhang, M. and T. Yuan. 1998. "Molecular mechanisms of calmodulin's functional versatility." *Biochem Cell Biol* 76: 313–23.

Zimenkov, Y., S. N. Dublin, R. Ni, R. S. Tu, V. Breedveld, et al. 2006. "Rational design of a reversible pH-responsive switch for peptide self-assembly." *J Am Chem Soc* 128: 6770–1.

# 9 Protein Engineering Using Noncanonical Amino Acids

*Deniz Yüksel, Diren Pamuk, Yulia Ivanova, and Krishna Kumar*

## CONTENTS

The limited set of structural and chemical functionalities represented in the side chains of the canonical amino acids necessitate post-translational processing of expressed protein products. Such processing combinatorially expands both the diversity and functional attributes of proteins. This suggests that while the extant set of twenty encoded amino acids can support the rudimentary functions of living systems, optimized function requires a larger set. Site-directed mutagenesis has proven to be an invaluable tool in tailoring structural features of enzymes (Zoller and Smith 1987) and has allowed detailed scrutiny of function and creation of enzymatic activities not found in nature. Nevertheless, the limited set of functional groups available constrains the range of possible applications. One could envision entirely new functions for proteins, or even the evolution of entire organisms endowed with extrabiological properties, if the set of functional groups present on the side chains can be expanded.

Schultz and Chamberlin pioneered the incorporation of unnatural amino acids using nonsense codon suppression (Bain et al. 1989; Noren et al. 1989). More than 150 unnatural amino acids have been incorporated by this technique both *in vitro* and *in vivo*, allowing interrogation of mechanism and structure in unsurpassed detail. This chapter focuses on applications made possible by introducing unnatural amino acids into proteins with a brief overview of methodologies of incorporation.

## METHODOLOGIES

Both chemical and biosynthetic approaches have been utilized to incorporate unnatural amino acids into proteins. Purely chemical methods of modifying intact proteins lack control over regio- or chemospecificity, and are plagued by inaccessibility of buried sites (DeSantis and Jones 1999). Solid-phase peptide synthesis (SPPS) enables incorporation of virtually any unnatural amino acid but is restricted to synthesis of peptides of size 60–75 residues. Nevertheless, short peptides containing unnatural building blocks or backbones synthesized by SPPS have yielded valuable information on how to control folding and oligomerization states of peptide ensembles. Native chemical ligation has allowed construction of proteins containing up to ~200 amino acids (Dawson et al. 1994). In this method, two peptide fragments, one with an N-terminal cysteine and another with a C-terminal thioester, are ligated yielding a native peptide bond. This basic principle has been extended in expressed protein ligation by taking advantage of intein mediated splicing of proteins (Evans Jr. et al. 1998; Muir et al. 1998), where the fragment carrying a thioester functionality is recombinantly produced. The method allows synthesis of proteins too large for native chemical ligation (Hofmann and Muir 2002). Regardless, both techniques require an N-terminal cysteine and a C-terminal thioester, which may not be suitable for every application.

The earliest method of biosynthetic incorporation of unnatural amino acids was achieved with auxotrophic bacterial strains that are unable to synthesize one or more amino acids (Hortin and Boime 1983; Van Hest et al. 2000). Here, the growth medium is bereft of a natural amino acid but supplanted with an analogue, resulting in misacylation of the corresponding tRNA. This strategy is only beneficial if global replacement of a certain amino acid within the protein is desired, because there is no control over the sites that are substituted. Moreover, no more than one unnatural analogue can be inserted into the same protein due to formation of product mixtures. Another limiting factor is that the unnatural amino acid to be incorporated should structurally resemble its natural analogue to enable recognition by the corresponding aminoacyl tRNA synthetase (aaRS). Ibba and coworkers have shown that it is possible to mutate the active site of the aaRS to accept a wider range of amino acids (Ibba et al. 1994). The Tirrell group has employed this approach to incorporate bromo, iodo, azido, cyano, ethynyl, and other side chains into proteins (Link et al. 2003).

Site-specific incorporation of unnatural amino acids using biosynthetic methods requires manipulation of the existing machinery. First, a stop codon is inserted at the desired position in the gene of interest. The amber stop codon, UAG, has been widely used for this purpose since it is the least used stop codon both in *E. coli* and

*S. cerevisiae.* Next, a suppressor tRNA, misacylated with an unnatural amino acid, is prepared to recognize the stop codon. Finally, these components are used in an *in vitro* transcription-translation system to produce the target protein (Heckler et al. 1984; Bain et al. 1989; Noren et al. 1989) (Figure 9.1). However, the initial efforts suffered from low yields of the mutant proteins because of the synthetic challenge in the production of large quantities of aminoacyl-tRNA (aa-tRNA), as it is stoichiometrically consumed.

The first example using nonsense codon suppression technology in intact cells was reported by Lester and Dougherty (Nowak et al. 1995). A chemically acylated suppressor tRNA and the corresponding modified mRNA were microinjected into *Xenopus* oocytes, leading to site-selective incorporation of unnatural amino acids into an ion channel protein. Although chemically aminoacylated suppressor tRNAs still cannot be regenerated, sensitive electrophysiological assays used to explore the mutated ion channels, where only attomol amounts of protein were needed, obviated the need for large quantities of the chemically acylated tRNA (Beene et al. 2003). More recently, the use of chemically aminoacylated suppressor tRNAs has been extended to mammalian cells (Monahan et al. 2003).

The main challenge of nonsense suppression methodology is to create an orthogonal tRNA:aaRS pair so that the tRNA does not cross react with endogenous

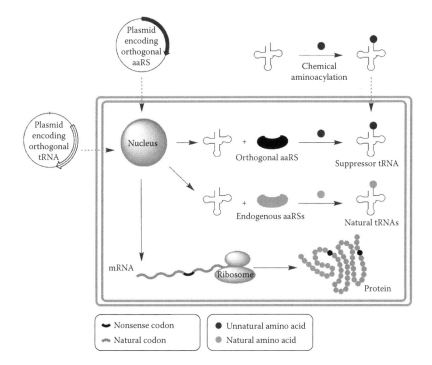

**FIGURE 9.1**  Nonsense suppression. Site-specific incorporation of unnatural amino acids into proteins requires nonsense codons to be recognized by suppressor tRNAs. Such tRNAs are obtained either by chemical acylation *in vitro* followed by delivery into cells or by direct charging by orthogonal aaRSs.

aaRSs (Figure 9.1). Such a system would eliminate the need for stoichiometric quantities of acylated tRNA. Initially, the efforts were focused on evolving orthogonal tRNA:aaRS pairs from existing bacterial pairs for use in *E. coli*. Although an orthogonal glutaminyl tRNA (tRNA$^{Gln}$) was successfully produced, the endogenous synthetase was still able to recognize the amber suppressor tRNA (Liu et al. 1997). Alternatively, tRNA:aaRS pairs from other organisms may be imported into *E. coli*. Even though an orthogonal tRNA$^{Gln}$:GlnRS pair derived from the yeast was expressed efficiently in *E. coli*, the synthetase could not be evolved to charge any unnatural amino acids (Wang et al. 2006b). The first orthogonal tRNA:aaRS pair in *E. coli* was derived from a tRNA$^{Tyr}$:TyrRS pair from the archaeon *Methanococcus jannaschii* (*Mj*) (Wang et al. 2001). *Mj*TyrRS was initially chosen because it could be efficiently expressed in *E. coli* and does not have an editing mechanism. Despite the efficient production of this pair in *E. coli*, aminoacylation of the amber suppresser tRNA by the endogenous aaRSs was observed to some extent. More recently, several orthogonal pairs in *E. coli* have been generated from other species of archaea (Wang et al. 2006b).

Directed evolution has been used to build libraries of both amber suppressor tRNAs and synthetases. The specificities of the orthogonal tRNA and synthetase candidates were altered via a series of positive and negative selections. The first evolved construct using these selection strategies was a *Mj*TyrRS mutant which introduced *O*-methyl-L-tyrosine into *E. coli* proteins in response to the amber codon. A similar double selection strategy to evolve tRNA/synthetases was also developed for *S. cerevisiae* (Wang et al. 2006b; Xie and Schultz 2006). More than 30 unnatural amino acids have been incorporated into *E. coli* proteins using this approach. The drawback of this strategy lies in the rather difficult proposition of creating unique orthogonal tRNA:synthetase pairs for *every* new unnatural amino acid to be incorporated.

Incorporation of unnatural amino acids in mammalian systems remains a challenge. Yokoyama and colleagues demonstrated that an *E. coli* TyrRS mutant and *Bacillus stearothermophilus* amber suppressor tRNA could be used together as an amber suppressor pair in mammalian cells (Sakamoto et al. 2002). However, the mutant TyrRS was not very specific and still charged the amber suppressor tRNA with tyrosine. Recently, a previously evolved tRNA:aaRS pair in *S. cerevisiae* was used in mammalian cells with higher fidelity and reduced toxicity to the cells (Liu et al. 2007). These technical advances have resulted in the introduction of many unnatural amino acids with various reactive groups including ketone, azide, acetylene, and thioester functionalities (Xie and Schultz 2006). These unique chemical handles could be further derivatized by bioorthogonal reactions, leading to installation of biophysical probes to explore protein structure, localization, and interactions with other factors within cells.

Sisido and Schultz have independently developed a frameshift methodology for incorporating unnatural amino acids in response to four- and five-base codons (Hohsaka and Sisido 2002; Wang et al. 2006b). This method allows incorporation of two different unnatural amino acids into the same protein because most four-base codon pairs are orthogonal to each other. It was also demonstrated that both UAG and a four-base codon-anticodon pair could be used simultaneously to install two different unnatural amino acids into the same protein (Anderson III et al. 2002). The frameshift suppression approach was also tested in *Xenopus* oocytes (Rodriguez et al. 2006). Even though frameshift suppressor tRNAs were reported to be less efficient than amber suppressor tRNAs, they showed reduced cross-reactivity to endogenous *Xenopus* components, allowing simultaneous incorporation of two and three different unnatural amino acids into neuroreceptors expressed in *Xenopus* oocytes.

The nonsense suppression technology is advantageous in tackling many biological questions but suffers from low efficiency. The highest suppression efficiencies reported so far are in the range of 30–40%, and then only so under optimal conditions (Köhrer et al. 2004). Chin and colleagues substantially increased incorporation efficiencies by evolving an orthogonal ribosome *in vivo* that specifically reads through amber codons (Wang et al. 2007). This new ribosome, termed *ribo-X*, translates only the UAG containing mRNA encoding the target protein and has reduced reactivity toward release factors that recognize UAG and UAA stop codons in *E. coli*. The suppression efficiency of a single UAG codon by a suppressor tRNA aminoacylated with *p*-benzoyl-L-phenylalanine increased from 24% in case of the wild-type ribosome to 62% in that of ribo-X.

## APPLICATIONS

### SIDE CHAIN PACKING

### Fluorinated Amino Acids

Our laboratory, that of David Tirrell and others, has independently shown that appropriate patterning of protein surfaces with fluorinated amino acids results in stable folds with enhanced thermal and chemical stability (Bilgiçer et al. 2001; Tang et al. 2001). Coiled coils are helical bundles formed by the supercoiling of individual strands. Their primary sequences contain a heptad repeat (*abcdefg*) with hydrophobic side chains at the *a* and *d* positions that align on one face of the helix. Exclusion of nonpolar side chains away from water provides the primary driving force for assembly. In order to make coiled-coil assemblies with fluorinated interfaces, we replaced the core residues of the dimerization domain of the yeast transcriptional activator GCN4 with trifluorovaline (**1**) and trifluoroleucine (**2**) at the *a* and *d* positions, respectively. (For structures of these and other unnatural amino acids mentioned in this chapter, see Figure 9.2). The resulting peptide ensemble was thermally and chemically more stable than the native one. The increased stability could be directly attributed to the higher hydrophobicities of the trifluoromethyl over methyl groups shielded from solvent water in the hydrophobic core. By substituting amino acids containing $CF_3$ groups, the hydrophobicity of the monomers is increased beyond what is available in the canon of the naturally occurring ones.

The phase separation proclivity of perfluorocarbons in both water and hydrocarbons suggests that fluorinated amino acids should be able to direct specific protein–protein interactions. A parallel coiled coil (**HH**) with seven leucines and a single asparagine on each strand in the core and a fluorinated version (**FF**) where all seven leucines were replaced with hexafluoroleucine (**3**) served as the test system (Bilgiçer and Kumar 2002). A disulfide exchange assay was used to evaluate the self-sorting

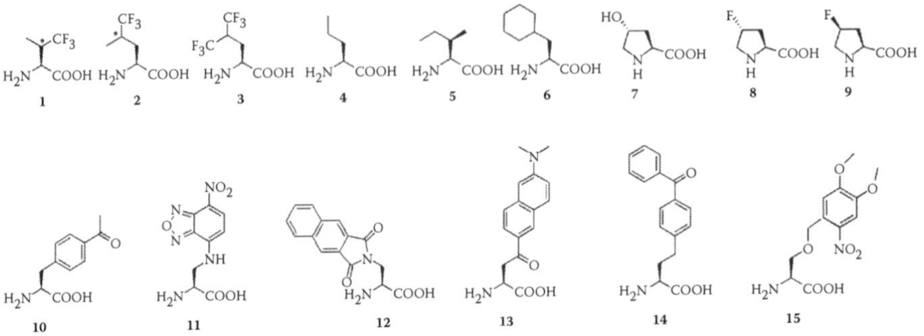

**FIGURE 9.2** Structures of unnatural amino acids used in studies discussed in the text. 1 Trifluorovaline, 2 Trifluoroleucine, 3 Hexafluoroleucine, 4 Norvaline, 5 Alloisoleucine, 6 Chx, 7 Hyp, 8 Flp, 9 flp, 10 *p*-acetyl-L-phenylalanine, 11 NBD-Dap, 12 6-DMN, 13 Aladan, 14 pBpa, 15 DMNB-Ser. * indicates unresolved stereochemistry.

behavior. Due to the limited affinity of fluorinated surfaces toward hydrocarbon surfaces, we anticipated that the fluorinated interfaces would pack against each other. Indeed, the disulfide linked heterodimer (**HF**) disproportionated nearly completely into homooligomers (**HH** and **FF**) under equilibrium conditions. The free energy of specificity ($\Delta G_{spec}$) was determined to be at least –2.1 kcal/mol.

This idea was then extended to control oligomerization of protein components in the context of the nonpolar environment of membranes (Bilgiçer and Kumar 2004; Naarmann et al. 2006). *De novo* design of membrane protein architectures presents a formidable challenge. Soluble proteins are folded with hydrophobic residues segregated inside and polar/charged residues outside. However, for the majority of integral membrane proteins, this type of asymmetry in distribution of side chains is not observed, making it difficult to design specific interaction interfaces amidst a sea of lipid molecules. Our design efforts focused on helical bundles based on the self-sorting behavior of fluorinated coiled coils. Further, by virtue of the simultaneously hydrophobic and lipophobic nature of fluorocarbons, we expected to construct higher order protein assemblies within lipid environments.

We envisioned a two-step process for the insertion and association of fluorinated transmembrane (TM) helices. First, the hydrophobic TM peptides would partition into micelles or vesicles and form α-helices by main chain hydrogen bonding. Upon secondary structure formation, one face of helices would present a highly fluorinated surface. Second, phase separation of fluorinated surfaces within hydrophobic environments would mediate helix-helix interactions, driving bundle formation (Figure 9.3). Indeed, fluorinated helices formed helical bundles where the all-hydrocarbon peptides could not. The ensembles were characterized by a battery of biophysical techniques including analytical equilibrium ultracentrifugation, Förster resonance energy transfer (FRET), SDS-PAGE, and circular dichroism. Thus, self-association of transmembrane helices with or without the aid of hydrogen bonding in membrane-like environments can be accomplished by use of unnatural amino acids. Fluorinated interfaces are orthogonal to ones found in nature and thus can be used to modulate the structure or binding of integral membrane proteins.

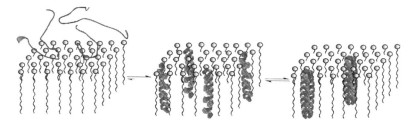

**FIGURE 9.3** Schematic depiction of the self-assembly of transmembrane peptides. Hydrophobic peptides readily partition into membranes forming α-helices. One face of the helix is lined with a string of hexafluoroleucine residues, shown in space-filling representation promoting formation of helical bundles. Only the backbone (depicted as a helix) and the core packing residues are shown for clarity. (Reproduced with permission from Bilgiçer and Kumar 2004. Copyright Proc. Natl. Acad. Sci. U.S.A.)

## Novel Folds and Control of Orientation

A long-standing goal of protein design has been the ability to engineer interaction motifs that do not have representative examples. Coiled coils form left-handed superhelical bundles because each amino acid in a normal α-helix rotates radially by 100° (3.6 residues/turn = 360°). A heptad repeat therefore lags two full turns (720°) by 20°. In the superhelical conformation, the α-helices are more tightly wound reducing the number of residues per turn to 3.5, thus creating a stripe of hydrophobic residues that align with the axis of superhelical rotation. Harbury and coworkers created a right-handed coiled coil by using an 11-amino-acid repeat (Harbury et al. 1998). Such an undecatad repeat (*abcdefghijk*) puts the *a*, *d*, and *h* positions along the superhelical axis. However, instead of lagging behind three full turns (1080°), the new repeat leads by 20° inverting the handedness of the supercoil. Dimer, trimer, and tetramer right-handed coiled coils were modeled for all possible core sequences, where alanine, valine, norvaline, leucine, isoleucine, and alloisoleucine were allowed at the *a*, *d*, and *h* positions. A family of parametric superhelix backbones were sampled, and the most frequently occurring rotamers were tested at each core position. These computations demanded an amino acid at some core positions that could project side-chain volume with a trans $\chi_1$ dihedral angle in its most prevalent rotameric form. This requirement was satisfied by alloisoleucine (**4**). Additionally, uncharged alternatives to lysine and a nonoxidizable surrogate of methionine were also needed **M** these criteria were satisfied by norvaline (**5**). Unnatural amino acids were, therefore, critical in the design of the right-handed superhelical bundle.

Steric matching of hydrophobic core side chains has enabled exquisite control in the orientation and identity of strands in coiled-coil motifs. For example, a cyclohexylalanine (Chx) (**6**) and two alanine residues make good interacting partners within the same core layer in a trimeric coiled coil (Schnarr and Kennan 2001). Specific heterotrimeric bundles containing three independent peptide strands (1:1:1 complex) could be programmed by including one Chx and two Ala residues at the same core layer position, but in different, complementary order. This steric matching strategy has allowed the design and construction of antiparallel coiled-coil ensembles (Schnarr and Kennan 2004).

## BACKBONE MUTATIONS

The backbone amides in the main secondary structure elements in proteins are hydrogen bonded. In order to estimate the contribution of main chain hydrogen bonding to protein folding, backbone modifications have been made by incorporation of unnatural amino acids that lack H-bonding capability. Strategies to achieve this have included replacement of primary with secondary amides by proline substitution, substitution of the amide with a dimethylene or a thiomethylene group, or with an ester or transalkene (Deechongkit et al. 2004a). Such changes allow examination of backbone hydrogen bonding in the stability of the fold as well as function. Substitution of an α-amino acid with an α-hydroxy acid leads to the replacement of a backbone amide to an ester, disrupting the backbone H-bonding network (Figure 9.4c). Initial studies of backbone ester containing peptides were on manually synthesized 35–40 residue fragments of the PIN WW protein, a robust β-sheet containing structure. The effects of hydrogen bonding on stability were studied by replacement of either a selected residue or a series of consecutive residues. The midpoint of thermal denaturation curves as well as free energy changes based on guanidine hydrochloride induced

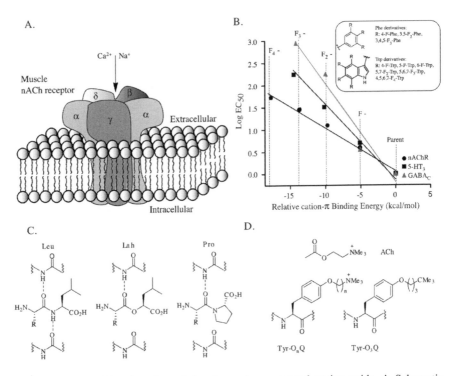

**FIGURE 9.4** Probing ion channel function using unnatural amino acids. **A** Schematic depiction of nAChR. **B** Graph illustrating the linear relationship between relative cation-π binding energies of fluorinated analogues and log $EC_{50}$ for nAChR, 5-HT$_3$, and GABA$_c$ receptors (Data from Lummis, Beene, Harrison et al. 2005 and references therein). **C** Introduction of amide-to-ester backbone mutations. **D** Structures of ACh and two tethered agonists.

unfolding at 20°C were used to extract the thermodynamic contribution of specific hydrogen bonds. Data from such experiments show that hydrogen bonding effects are more pronounced in hydrophobic rather than solvent exposed regions (Scheike et al. 2007). The highest estimated hydrogen bonding strength in the context of β-sheets was found to be 25 kJ/mol at buried sites (Deechongkit et al. 2004a; Deechongkit et al. 2004b). The amide-to-ester replacements, however, do not decouple the effects of H-bonding from the additional repulsive interactions between the amide carbonyl and the ester oxygens (Deechongkit et al. 2006). Estimation of such repulsive interactions has been achieved by use of the trans alkenes. For instance, replacing Val22-Tyr23 in the PIN WW domain with a Phe-Phe $E$-olefin dipeptide isostere reveals that the repulsion is worth 0.3 kcal/mol and the hydrogen bond between F23 and R14 is 1.3 kcal/mol (Fu et al. 2006).

## DISSECTING COLLAGEN

Collagen, the most abundant protein in animals, has a unique tertiary structure consisting of three parallel left-handed polyproline II–type helices wound together into a right-handed superhelix. Each polypeptide chain of collagen is composed of Xaa-Yaa-Gly trimer repeats where Xaa is usually Pro and Yaa is (2$S$,4$R$)-4-hydroxyproline (Hyp) (**7**) (Jenkins and Raines 2002). Earlier work showed that the Hyp content of collagen triple helix correlates with its structural stability. Hence, the hydroxyl group of Hyp was postulated to stabilize the protein through interstrand hydrogen bonding networks. In contrast, stabilities of model peptides mimicking collagen, such as (ProHypGly)$_{10}$, were higher in organic solvents compared to water, suggesting that the inductive effects of Hyp hydroxyls may contribute to stability (Engel et al. 1977).

In order to distinguish between contributions from hydrogen bonding and inductive effects, Hyp was replaced by (2$S$,4$R$)-4-fluoroproline (Flp) (**8**) (Holmgren et al. 1998; Holmgren et al. 1999). Unlike the hydroxyl group, carbon-bound fluorine does not usually form hydrogen bonds and fluorine elicits a large inductive effect. Flp residues were incorporated at Xaa position in (ProXaaGly)$_{10}$ sequence and the peptide was compared to its natural analogues: (ProHypGly)$_{10}$ and (ProProGly)$_{10}$. Sedimentation equilibrium and CD experiments showed that (ProHypGly)$_{10}$ was structurally indistinguishable from natural collagen mimics. However, the Flp-containing peptide has higher thermal stability in the order (ProFlpGly)$_{10}$ > (ProHypGly)$_{10}$ > (ProProGly)$_{10}$ (Holmgren et al. 1999). This enhanced stability was attributed to the high electronegativity of fluorine. The stability of collagen, therefore, arises from electron withdrawing effect of the hydroxyl group of Hyp rather than its involvement in hydrogen bonding.

Further studies illustrated that stereoelectronics plays a major role in the conformational stability of collagen. Even though Flp at the Yaa position forms a stable collagen mimic, replacing Flp with its diastereomer, (2$S$,4$S$)-4-fluoroproline (flp) (**9**) greatly destabilizes the triple helix (Bretscher et al. 2001). Electronegative substituents at the 4-position prefer a gauche orientation to the amide nitrogen, favoring the $C_\gamma$-exo puckered conformation of the pyrrolidine ring in the case of Flp. Since this is the pucker found in collagen, the structure is stabilized by preorganization of the individual strands before self-assembly (Raines 2007). Incorporation of unnatural amino acids in collagen has helped unravel the origins of its stability and may lead to development of metabolically stable synthetic collagen mimics of high clinical value (Goodman et al. 1998).

## FLUORESCENT AMINO ACIDS

### Seeing Is Believing

Visualization of proteins with high temporal and spatial resolution is crucial for studying their function in living organisms by determining their cellular localization, interaction partners, folding/unfolding patterns, and local structures. To that end, green fluorescent protein (GFP) and its variants have been tagged to numerous proteins to visualize biological processes (Zhang et al. 2002). However, the large size of naturally occurring fluorescent proteins (>20 kDa) may perturb the structure and function of the tagged protein. Hence, current efforts are focused on developing small fluorophores allowing *in vivo* examination of the structure and/or location of a target protein with minimal perturbation. Tsien and coworkers used a tetracysteine motif to specifically bind a membrane-permeable biarsenical dye (Griffin et al. 1998). Presumably, the tetracysteine motif is rare in the proteome of most cells. Nevertheless, arsenic based compounds are toxic and limit their use in living systems.

An even smaller perturbation would be to introduce fluorescent amino acids site specifically. For instance, a ketone-bearing unnatural amino acid, *p*-acetyl-L-phenylalanine (**10**), was incorporated site specifically into T4 lysozyme *in vitro* and was subsequently labeled with hydrazide-derivatized fluorescent dyes (Cornish et al. 1996). Recently, the method was extended to *in vivo* labeling of proteins in *E. coli* (Zhang et al. 2003). Unfortunately, *in vivo* labeling of cytosolic proteins is not very successful due to the presence of endogenous aldehydes resulting in a massive background. While further optimization of this method is required to track migration of cytosolic proteins, a well characterized membrane protein, LamB, was selectively labeled *in vivo* with success (Zhang et al. 2003). Such techniques should aid in the study of conformational changes, as well as protein–protein and protein–ligand interactions involving other membrane proteins.

### Illuminating Structure–Function Relationships

Diffraction methods are useful in the study of structural and conformational changes of proteins but are difficult to apply to analyze structural dynamics in real time. FRET and NMR can be used to evaluate static and dynamic structural properties of proteins containing unnatural amino acids.

For example, NBD-Dap (**11**) was introduced into the G-protein coupled neuro-kinin-2 receptor (NK2) by heterologous expression in *Xenopus* oocytes (Turcatti et al. 1996; Turcatti et al. 1997). NK2 has seven transmembrane segments and is believed to undergo conformational changes to selectively bind agonist or antagonist ligands. Insights into the three-dimensional structural model of the NK2-ligand complex were gained from FRET-based distance measurements between NBD-labeled receptor and tetramethylrhodamine-labeled peptide ligands. Experimental results suggested that an antagonist peptide ligand is inserted between the fifth and sixth transmembrane domains and disrupts the packing of the helices needed for receptor activation (Turcatti et al. 1996). The authors were able to investigate the NK2 ligand-receptor complex in its native environment and avoid preparation of large quantities of protein for crystallographic study. On the other hand, only few of the mutants were functionally active, implying that the NBD group perturbs the structure and folding of the receptor (Turcatti et al. 1997). Nevertheless, the NBD moiety has excellent fluorescence characteristics, making NBD-Dap a potential candidate for use in direct *in vivo* protein visualization. Introducing other amino acids bearing fluorophores via nonsense codon suppression would be a breakthrough for *in vivo* analysis of protein trafficking but so far only a few are available due to ribosomal constraints on side-chain size.

Fluorescent amino acids that respond to changes in their surrounding environment are useful to interrogate protein structure and protein–protein interactions (Cohen et al. 2002; Vázquez et al. 2005; Summerer et al. 2006). One such fluorophore, 6DMN (**12**), has a low quantum yield in aqueous solution, but becomes fluorescent in nonpolar solvents or upon interacting with hydrophobic surfaces on proteins or membranes. The fluorophore was used to monitor phosphorylation-dependent peptide–protein interactions in Src homology 2 (SH2) domains. Using SPPS, Fmoc-Dap (6DMN)-OH was incorporated into several short phosphotyrosine-containing peptides that are recognized by SH2 binding domains. When the designed peptides are incubated with SH2 binding domains, an increase in fluorescence emission is observed, indicating binding. In particular, an 11-fold change in fluorescence is observed for the Crk-bp2 and Crk-SH2 pair, illustrating that this fluorophore may be viable as a general peptidic probe (Vázquez et al. 2005).

Electrostatic interactions play a major role in protein function. The ability to measure static and dynamic electric fields within proteins would provide a deeper understanding of a protein's interactions with its environment. Aladan (**13**) undergoes a large shift in emission maximum when the polarity of a solvent changes (e.g., 409 nm in heptane to 542 nm in water). Aladan was inserted in the potassium ion channel by nonsense suppression methodology in *Xenopus* oocytes. It was also introduced at exposed, partially exposed, and two buried sites within the B1 domain of streptococcal protein G by SPPS. Both steady-state as well as time-resolved fluorescence measurements were used to estimate the dielectric constant of the protein interior ($\varepsilon = 6$; close to EtOAc) and exterior [$\varepsilon = 21$, close to $(CH_3)_2CO$] (Cohen et al. 2002). Such experiments make possible estimation of the electrostatic contribution during substrate/ligand binding and catalysis.

## PHOTOREACTIVE AMINO ACIDS

Numerous techniques exist to characterize protein–protein interactions, including affinity chromatography, immunoprecipitation of protein complexes from cell lysates, yeast two hybrid screens, and protein arrays. However, these methods often analyze protein complexes outside their native environment. Introduction of cross-linkable side chains into proteins provides a complementary approach that allows elucidation of specific interactions *in vivo*. Unnatural amino acids bearing photoreactive groups such as benzophenones, aryl azides, and diazarines become reactive radicals when irradiated with ultraviolet light and are particularly useful in photo-crosslinking proteins. For instance, a benzophenone modified amino acid, pBpa (**14**), has been incorporated into the *E. coli* proteome using nonsense suppression to map protein–protein interactions associated with glutathione S-transferase. Irradiation of cells with near-UV light and cell lysis resulted in covalent complexes of glutathione S-transferase formed within the cell (Chin and Schultz 2002).

Photocaged cysteine, serine, and tyrosine residues have also been introduced site-specifically to study proteins involved in different cellular processes, such as the dimerization of HIV-1 protease *in vitro* (Short III et al. 1999) and to study the mechanism of intein splicing (Cook et al. 1995). Alternatively, short caged peptides that selectively interfere with a protein's activity were also employed to probe protein function. In one example, two peptides designed to inhibit calmodulin and myosin light chain kinase upon photolysis of the caging moiety were injected into intact cells to determine the role of these proteins in cell motility (Walker et al. 1998). Photocaging strategies are promising in protein visualization where fluorescent amino acids are initially caged but become fluorescent upon removal of the caging moiety to track protein migration within the cell (Politz 1999; Beatty et al. 2006; Kwon and Tirrell 2007).

Incorporation of photocaged amino acids into *E. coli* and *S. cerevisiae* using nonsense suppressor tRNA:aaRS pairs has made it possible to photochemically control the activity and function of target proteins, allowing detailed analysis directly within the cell. This approach has been useful in the study of phosphorylation pathways. Earlier studies used amino acids containing an *o*-nitrobenzyl group, but removing the moiety requires high-intensity short-wavelength UV light that causes photoreactions in nucleic acids, cleavage of disulphide bonds, and other cellular damage. These limitations are overcome by using DMNB-Ser (**15**), which is activated by relatively low energy blue light. This unnatural amino acid was genetically encoded in yeast to study the role of phosphorylation in the transcription factor Pho4. The study illuminated the effects of specific serine phosphorylation on the dynamics of protein trafficking between the nucleus and cytoplasm (Lemke et al. 2007).

## PROBING ION CHANNEL FUNCTION

Being able to tune a desired side chain without large structural perturbation is crucial for systematic, in-depth investigation of structure-function relationships in proteins. Dougherty and coworkers have used unnatural amino acid mutagenesis for functional analysis of ion channels with atomistic precision where conventional site-directed mutagenesis methods had failed (Beene et al. 2003). They successfully adapted the nonsense suppression method for use with a heterologous expression system in *Xenopus* oocytes (Nowak et al. 1995) to study a ligand-gated ion channel, the nicotinic acetylcholine receptor (nAChR) (Figure 9.4A). Incorporation of modified Trp-analogues at several positions demonstrated the importance of aromatic amino acids at the agonist binding site. Cation-π interactions are known to be involved in biological recognition events in a variety of proteins (Dougherty 1996), and prevalence of several aromatic amino acids at the agonist binding site of nAChR suggested that the quaternary ammonium group of acetylcholine could bind via the agency of cation-π interactions. To explore this possibility, several fluorinated Trp-analogues were incorporated in the binding domain of nAChR, allowing substantial *electronic* but minimal *steric* changes. A systematic decrease in agonist affinity with increasing number of fluorine substitutions was observed at position Trp149 of the α subunit. Furthermore, a linear correlation was observed between experimental $EC_{50}$ values and the cation-π binding ability predicted by *ab initio* quantum mechanical calculations (Zhong et al. 1998). This approach was extended to two other ligand-gated ion channel receptors—serotonin (5-HT$_3$) and GABA$_C$ receptors—to probe cation-π interactions in a similar manner (Figure 9.4B) (Beene et al. 2002; Lummis et al. 2005). The magnitude of the electrostatic component of cation-π interaction was determined to be ~2 kcal/mol for acetylcholine at the nAChR and ~4 kcal/mol for serotonin at the 5-HT$_3$ receptor.

To establish the location and the nature of agonist binding in nAChR, a tethered-agonist approach was employed (Li et al. 2001). Previously, others have used site-directed mutagenesis to introduce cysteine residues that were then modified with probes. However, cysteine mutants did not always yield functional receptors, and some cysteines were inaccessible to labeling reagents. These problems were circumvented by incorporating a series of tyrosine derivatives with tethered quaternary ammonium groups (Figure 9.4D). Introduction of these derivatives at three different sites (α93, α149, and γ55/δ57) resulted in active receptors and channel activity correlated with tether length. For example, at position α149 only Tyr-O$_3$Q was well tolerated, whereas Tyr-O$_n$Q n = 2-5 at position α93 and Tyr-O$_n$Q n ≥3 at position γ55/δ57 all generated active receptors. These results led to a model describing the location of the agonist binding site.

Amide-to-ester backbone mutations were also used to explore ion channel function (Beene et al. 2003). Proline is the only amino acid that can act as a H-bond acceptor but not as a donor. An amide-to-ester backbone replacement results in a similar loss of the H-bond donating capability while accepting H-bonds. The role of conserved Pro residues in the transmembrane regions of two ion channels was interrogated by inserting α-hydroxy acids at positions Pro221 of nAChR and Pro256 of 5-HT$_{3A}$ (Figure 9.4C). Functional channels were obtained with an amide-to-ester mutation but not with a Gly, Ala, or Leu substitution. These results established that the unique H-bonding properties of conserved proline residues are important for gating (England et al. 1999). Another example of an amide-to-ester backbone replacement involves the selectivity filter of a K$^+$ ion channel (Lu et al. 2001). Two conserved glycine residues were mutated to their α-hydroxy analogues resulting in substitution of a potential hydrogen bond donor (–NH) by a hydrogen bond acceptor (–O–) and reduction of electronegativity of the carbonyl oxygen. Mutant channels displayed the same selectivity as the wild type despite the reduced electronegativity of the filter. Further analyses of mutant channels suggested that the gating process is governed by localized conformational changes in the ion filter through ion-ion and ion-carbonyl oxygen interactions. Detailed mapping of ion channel function illuminates drug-channel interactions, and may prove useful in development of therapeutics for related disorders.

## CONCLUSIONS AND FUTURE OUTLOOK

Incorporation of unnatural amino acids has enabled the study of mechanism, structure, location, and function of proteins in ways not possible with the genetically encoded set of amino acids. In addition, with our increasing ability to manipulate the biosynthetic machinery and an ever increasing palette of side chains that are tolerated, novel structure and function not represented in natural proteins seem within reach. Methods to expand the genetic code and the ability to biosynthetically incorporate multiple unnatural amino acids in a site-specific manner promise evolution of proteins, and even entire organisms, with new functions and capabilities.

## ACKNOWLEDGMENTS

This work was supported in part by the NIH (GM65500) and the NSF. K.K. is a DuPont Young Professor.

## REFERENCES

Anderson III, R. D., J. Zhou and S. M. Hecht. 2002. "Fluorescence resonance energy transfer between unnatural amino acids in a structurally modified dihydrofolate reductase." *J Am Chem Soc* 124: 9674–75.

Bain, J. D., C. G. Glabe, T. A. Dix, A. R. Chamberlin and E. S. Diala. 1989. "Biosynthetic site-specific incorporation of a non-natural amino acid into a polypeptide." *J Am Chem Soc* 111: 8013–14.

Beatty, K. E., J. C. Liu, F. Xie, D. C. Dieterich, E. M. Schuman, et al. 2006. "Fluorescence visualization of newly synthesized proteins in mammalian cells." *Angew Chem Int Ed Engl* 45: 7364–7.

Beene, D. L., G. S. Brandt, W. Zhong, N. M. Zacharias, H. A. Lester, et al. 2002. "Cation-Π interactions in ligand recognition by serotonergic (5-HT3A) and nicotinic acetylcholine receptors: The anomalous binding properties of nicotine." *Biochemistry* 41: 10262–9.

Beene, D. L., D. A. Dougherty and H. A. Lester. 2003. "Unnatural amino acid mutagenesis in mapping ion channel function." *Curr Opin Neurobiol* 13: 264–70.

Bilgiçer, B., A. Fichera and K. Kumar. 2001. "A coiled coil with a fluorous core." *J Am Chem Soc* 123: 4393–9.

Bilgiçer, B. and K. Kumar. 2002. "Synthesis and thermodynamic characterization of self-sorting coiled coils." *Tetrahedron* 58: 4105–12.

Bilgiçer, B. and K. Kumar. 2004. "De novo design of defined helical bundles in membrane environments." *Proc Natl Acad Sci U S A* 101: 15324–9.

Bretscher, L. E., C. L. Jenkins, K. M. Taylor, M. L. DeRider and R. T. Raines. 2001. "Conformational stability of collagen relies on a stereoelectronic effect [23]." *J Am Chem Soc* 123: 777–778.

Chin, J. W. and P. G. Schultz. 2002. "In vivo photocrosslinking with unnatural amino acid mutagenesis." *ChemBioChem* 3: 1135–7.

Cohen, B. E., T. B. McAnaney, E. S. Park, Y. N. Jan, S. G. Boxer, et al. 2002. "Probing protein electrostatics with a synthetic fluorescent amino acid." *Science* 296: 1700–03.

Cook, S. N., W. E. Jack, X. Xiong, L. E. Danley, J. A. Ellman, et al. 1995. "Photochemically initiated protein splicing." *Angew Chem Int Ed Engl* 34: 1629–30.

Cornish, V. W., K. M. Hahn and P. G. Schultz. 1996. "Site-specific protein modification using a ketone handle." *J Am Chem Soc* 118: 8150–1.

Dawson, P. E., T. W. Muir, I. Clark-Lewis and S. B. H. Kent. 1994. "Synthesis of proteins by native chemical ligation." *Science* 266: 776–9.

Deechongkit, S., P. E. Dawson and J. W. Kelly. 2004a. "Toward assessing the position-dependent contributions of backbone hydrogen bonding to β-sheet folding thermodynamics employing amide-to-ester perturbations." *J Am Chem Soc* 126: 16762–71.

Deechongkit, S., H. Nguyen, M. Jager, E. T. Powers, M. Gruebele, et al. 2006. "β-sheet folding mechanisms from perturbation energetics." *Curr Opin Struct Biol* 16: 94–101.

Deechongkit, S., H. Nguyen, E. T. Powers, P. E. Dawson, M. Gruebele, et al. 2004b. "Context-dependent contributions of backbone hydrogen bonding to β-sheet folding energetics." *Nature* 430: 101–05.

DeSantis, G. and J. B. Jones. 1999. "Chemical modification of enzymes for enhanced functionality." *Current Opinion in Biotechnology* 10: 324–30.

Dougherty, D. A. 1996. "Cation-Π interactions in chemistry and biology: A new view of benzene, Phe, Tyr, and Trp." *Science* 271: 163–8.

Engel, J., H. T. Chen, D. J. Prockop and H. Klump. 1977. "The triple helix in equilibrium with coil conversion of collagen-like polytripeptides in aqueous and nonaqueous solvents. Comparison of the thermodynamic parameters and the binding of water to (L-Pro-L-Pro-Gly)n and (L-Pro-L-Hyp-Gly)n." *Biopolymers* 16: 601–22.

England, P. M., Y. Zhang, D. A. Dougherty and H. A. Lester. 1999. "Backbone mutations in transmembrane domains of a ligand-gated ion channel: Implications for the mechanism of gating." *Cell* 96: 89–98.

Evans Jr., T. C., J. Benner and M. Q. Xu. 1998. "Semisynthesis of cytotoxic proteins using a modified protein splicing element." *Prot Sci* 7: 2256–64.

Fu, Y., J. Gao, J. Bieschke, M. A. Dendle and J. W. Kelly. 2006. "Amide-to-E-olefin versus amide-to-ester backbone H-bond perturbations: Evaluating the O-O repulsion for extracting H-bond energies." *J Am Chem Soc* 128: 15948–9.

Goodman, M., M. Bhumralkar, E. A. Jefferson, J. Kwak and E. Locardi. 1998. "Collagen mimetics." *Biopolymers* 47: 127–42.

Griffin, B. A., S. R. Adams and R. Y. Tsien. 1998. "Specific covalent labeling of recombinant protein molecules inside live cells." *Science* 281: 269–72.

Harbury, P. B., J. J. Plecs, B. Tidor, T. Alber and P. S. Kim. 1998. "High-resolution protein design with backbone freedom." *Science* 282: 1462–7.

Heckler, T. G., L. H. Chang, Y. Zama, T. Naka, M. S. Chorghade, et al. 1984. "T4 RNA ligase mediated preparation of novel "chemically misacylated" tRNAPheS." *Biochemistry* 23: 1468–73.

Hofmann, R. M. and T. W. Muir. 2002. "Recent advances in the application of expressed protein ligation to protein engineering." *Curr Opin Biotechnol* 13: 297–303.

Hohsaka, T. and M. Sisido. 2002. "Incorporation of non-natural amino acids into proteins." *Curr Opin Chem Biol* 6: 809–15.

Holmgren, S. K., L. E. Bretscher, K. M. Taylor and R. T. Raines. 1999. "A hyperstable collagen mimic." *Chem Biol* 6: 63–70.

Holmgren, S. K., K. M. Taylor, L. E. Bretscher and R. T. Raines. 1998. "Code for collagen's stability deciphered [9]." *Nature* 392: 666–7.

Hortin, G. and I. Boime. 1983. "Applications of amino acid analogs for studying co- and post-translational modifications of proteins." *Methods Enzymol* 96: 777–84.

Ibba, M., P. Kast and H. Hennecke. 1994. "Substrate specificity is determined by amino acid binding pocket size in *Escherichia coli* phenylalanyl-tRNA synthetase." *Biochemistry* 33: 7107–12.

Jenkins, C. L. and R. T. Raines. 2002. "Insights on the conformational stability of collagen." *Nat Prod Rep* 19: 49–59.

Köhrer, C., E. L. Sullivan and U. L. RajBhandary. 2004. "Complete set of orthogonal 21st aminoacyl-tRNA synthetase-amber, ochre and opal suppressor tRNA pairs: Concomitant suppression of three different termination codons in an mRNA in mammalian cells." *Nucleic Acids Res* 32: 6200–11.

Kwon, I. and D. A. Tirrell. 2007. "Site-specific incorporation of tryptophan analogues into recombinant proteins in bacterial cells." *J Am Chem Soc* 129: 10431–7.

Lemke, E. A., D. Summerer, B. H. Geierstanger, S. M. Brittain and P. G. Schultz. 2007. "Control of protein phosphorylation with a genetically encoded photocaged amino acid." *Nat Chem Biol* 3: 769–72.

Li, L., W. Zhong, N. Zacharias, C. Gibbs, H. A. Lester, et al. 2001. "The tethered agonist approach to mapping ion channel proteins—toward a structural model for the agonist binding site of the nicotinic acetylcholine receptor." *Chem Biol* 8: 47–58.

Link, A. J., M. L. Mock and D. A. Tirrell. 2003. "Non-canonical amino acids in protein engineering." *Curr Opin Biotechnol* 14: 603–09.

Liu, D. R., T. J. Magliery and P. G. Schultz. 1997. "Characterization of an 'orthogonal' suppressor tRNA derived from *E. coli* tRNA2(Gln)." *Chem Biol* 4: 685–91.

Liu, W., A. Brock, S. Chen, S. Chen and P. G. Schultz. 2007. "Genetic incorporation of unnatural amino acids into proteins in mammalian cells." *Nat Methods* 4: 239–44.

Lu, T., A. Y. Ting, J. Mainland, L. Y. Jan, P. G. Schultz, et al. 2001. "Probing ion permeation and gating in a K+ channel with backbone mutations in the selectivity filter." *Nat Neurosci* 4: 239–46.

Lummis, S. C. R., D. L. Beene, N. J. Harrison, H. A. Lester and D. A. Dougherty. 2005. "A cation-Π binding interaction with a tyrosine in the binding site of the GABAC receptor." *Chem Biol* 12: 993–7.

Monahan, S. L., H. A. Lester and D. A. Dougherty. 2003. "Site-specific incorporation of unnatural amino acids into receptors expressed in mammalian cells." *Chem Biol* 10: 573–80.

Muir, T. W., D. Sondhi and P. A. Cole. 1998. "Expressed protein ligation: A general method for protein engineering." *Proc Natl Acad Sci U S A* 95: 6705–10.

Naarmann, N., B. Bilgiçer, H. Meng, K. Kumar and C. Steinem. 2006. "Fluorinated interfaces drive self-association of transmembrane α helices in lipid bilayers." *Angew Chem Int Ed Engl* 45: 2588–91.

Noren, C. J., S. J. Anthony-Cahill, M. C. Griffith and P. G. Schultz. 1989. "A general method for site-specific incorporation of unnatural amino acids into proteins." *Science* 244: 182–8.

Nowak, M. W., P. C. Kearney, J. R. Sampson, M. E. Saks, C. G. Labarca, et al. 1995. "Nicotinic receptor binding site probed with unnatural amino acid incorporation in intact cells." *Science* 268: 439–42.

Politz, J. C. 1999. "Use of caged fluorochromes to track macromolecular movement in living cells." *Trends Cell Biol* 9: 284–7.

Raines, R. T. 2007. Hyperstable collagen based on 4-fluoroproline residues. ACS Symposium Series 949: 477–86.

Rodriguez, E. A., H. A. Lester and D. A. Dougherty. 2006. "In vivo incorporation of multiple unnatural amino acids through nonsense and frameshift suppression." *Proc Natl Acad Sci U S A* 103: 8650–5.

Sakamoto, K., A. Hayashi, A. Sakamoto, D. Kiga, H. Nakayama, et al. 2002. "Site-specific incorporation of an unnatural amino acid into proteins in mammalian cells." *Nucleic Acids Res* 30: 4692-4699.

Scheike, J. A., C. Baldauf, J. Spengler, F. Albericio, M. T. Pisabarro, et al. 2007. "Amide-to-ester substitution in coiled coils: The effect of removing hydrogen bonds on protein structure." *Angew Chem Int Ed Engl* 46: 7766–9.

Schnarr, N. A. and A. J. Kennan. 2001. "Coiled-coil formation governed by unnatural hydrophobic core side chains [9]." *J Am Chem Soc* 123: 11081–2.

Schnarr, N. A. and A. J. Kennan. 2004. "Coiled-coil surface presentation: an efficient HIV gp41 binding interface mimic." *J Am Chem Soc* 126: 10260–1.

Short III, G. F., M. Lodder, A. L. Laikhter, T. Arslan and S. M. Hecht. 1999. "Caged HIV-1 protease: Dimerization is independent of the ionization state of the active site aspartates." *J Am Chem Soc* 121: 478–9.

Summerer, D., S. Chen, N. Wu, A. Deiters, J. W. Chin, et al. 2006. "A genetically encoded fluorescent amino acid." *Proc Natl Acad Sci U S A* 103: 9785–9.

Tang, Y., G. Ghirlanda, N. Vaidehi, J. Kua, D. T. Mainz, et al. 2001. "Stabilization of coiled-coil peptide domains by introduction of trifluoroleucine." *Biochemistry* 40: 2790–6.

Turcatti, G., K. Nemeth, M. D. Edgerton, J. Knowles, H. Vogel, et al. 1997. "Fluorescent labeling of NK2 receptor at specific sites in vivo and fluorescence energy transfer analysis of NK2 ligand-receptor complexes." *Receptors and Channels* 5: 201–07.

Turcatti, G., K. Nemeth, M. D. Edgerton, U. Meseth, F. Talabot, et al. 1996. "Probing the structure and function of the tachykinin neurokinin-2 receptor through biosynthetic incorporation of fluorescent amino acids at specific sites." *J Biolog Chem* 271: 19991–8.

Van Hest, J. C. M., K. L. Kiick and D. A. Tirrell. 2000. "Efficient incorporation of unsaturated methionine analogues into proteins in vivo." *J Am Chem Soc* 122: 1282–8.

Vázquez, M. E., J. B. Blanco and B. Imperiali. 2005. "Photophysics and biological applications of the environment-sensitive fluorophore 6-N,N-dimethylamino-2,3-naphthalimide." *J Am Chem Soc* 127: 1300–06.

Walker, J. W., S. H. Gilbert, R. M. Drummond, M. Yamada, R. Sreekumar, et al. 1998. "Signaling pathways underlying eosinophil cell motility revealed by using caged peptides." *Proc Natl Acad Sci U S A* 95: 1568–73.

Wang, J., J. Xie and P. G. Schultz. 2006a. "A genetically encoded fluorescent amino acid." *J Am Chem Soc* 128: 8738–9.

Wang, L., J. Xie and P. G. Schultz. 2006b. "Expanding the genetic code." *Annu Rev Biophys Biomol Struct* 35: 225–49.

Wang, K., H. Neumann, S. Y. Peak-Chew and J. W. Chin. 2007. "Evolved orthogonal ribosomes enhance the efficiency of synthetic genetic code expansion." *Nat Biotechnol* 25: 770–7.

Wang, L., A. Brock, B. Herberich and P. G. Schultz. 2001. "Expanding the genetic code of *Escherichia coli*." *Science* 292: 498–500.

Xie, J. and P. G. Schultz. 2006. "A chemical toolkit for proteins—an expanded genetic code." *Nat Rev Mol Cell Biol* 7: 775–82.

Zhang, J., R. E. Campbell, A. Y. Ting and R. Y. Tsien. 2002. "Creating new fluorescent probes for cell biology." *Nat Rev Mol Cell Biol* 3: 906–18.

Zhang, Z., B. A. C. Smith, L. Wang, A. Brock, C. Cho, et al. 2003. "A new strategy for the site-specific modification of proteins in vivo." *Biochemistry* 42: 6735–46.

Zhong, W., J. P. Gallivan, Y. Zhang, L. Li, H. A. Lester, et al. 1998. "From ab initio quantum mechanics to molecular neurobiology: A cation-Π binding site in the nicotinic receptor." *Proc Natl Acad Sci U S A* 95: 12088–93.

Zoller, M. J. and M. Smith. 1987. "Oligonucleotide-directed mutagenesis: a simple method using two oligonucleotide primers and a single-stranded DNA template." *Methods Enzymol* 154: 329–50.

# 10 Computer Graphics, Homology Modeling, and Bioinformatics

*David F. Green*

## CONTENTS

Computational methods play a range of roles in protein engineering from the simple use of visualization to guide rational design to fully automated *de novo* design algorithms. In the next chapters, many of these approaches will be discussed. Here we will focus on the former, that is, computational methods that complement human insight in rational protein engineering. The approaches can loosely be grouped into three classes: (1) methods based on analysis of primary sequence; (2) the visual analysis of protein structure; and (3) fast estimation of mutational effects. The mechanistic details of performing sequence and structural analysis have been extensively discussed in other texts, and thus the focus here is on the application of these approaches. The approaches discussed here all involve use of software that is either available as a Web service or as a freely available, downloadable program. The Web locations of key tools are summarized in the tables.

## PRIMARY SEQUENCE ANALYSIS

Evolution provides a tremendously useful model for protein design. As seen in previous chapters, several approaches to mimicking evolution in the laboratory have been demonstrated to be powerful methods for the engineering of improved or novel function, but we may also take advantage of the results of natural evolution. Many families of proteins contain hundreds or thousands of members, spread across diverse species. By considering the common features of the sequences of these proteins, it is possible to deduce the key elements that determine protein structure and function—even in absence of any explicit structural information. In order to take this approach, several tools are needed. First, given one (or a few) sequences of a target structure, it is necessary to be able to search through the vast array of known sequences for related proteins. Second, this large family of related proteins must be aligned such that conserved positions are in register with one another. From this point, analysis of the degree of conservation at each position can give important insight applicable to protein engineering. Computational methods for this analysis of primary sequence are well established, and many tools are available through Web-based servers (Table 10.1).

### ENGINEERING THROUGH CONSENSUS MOTIFS

One of the most straightforward applications of primary sequence data in protein engineering is the use of multiple-sequence alignments to define consensus motifs for a particular structure or function. These sequence signatures focus on the common features of a class, while not corresponding to any natural sequence. As a result, the resulting sequence may be expected to share the features that all members of the family have in common (such as a particular structure) without the specific features of particular family members (such as affinity for a specific binding partner). One of the first applications of this approach was in the design of a consensus-based zinc finger protein (Krizek et al. 1991). Subsequently, it was demonstrated that frequency of occurrence in a multiple-sequence alignment was a

## TABLE 10.1
## Web Services and Databases for Primary Sequence Analysis

| Service | Web Location (URL) | Description |
|---|---|---|
| GenBank | http://www.ncbi.nlm.nih.gov/Genbank/ | Repository of all publicly available nucleotide sequences. |
| Swiss-Prot | http://ca.expasy.org/sprot/ | Annotated database of protein sequences. |
| BLAST | http:/www.ncbi.nlm.nih.gov/BLAST/ | Online service to search for related sequences. |
| ClustalW2 | http://www.ebi.ac.uk/tools/clustalw2/index.html | Online service for multiple sequence alignment. |
| ClustalW | http://www.clustal.org/ | Downloadable software for multiple sequence alignment. |

good predictor of the effects of point mutations on the stability of an immunoglobulin domain (Steipe et al. 1994). These early implementations took a fairly simple approach; relatively small numbers of sequences were used, and global differences in amino-acid frequencies were not considered. An example of the approach is shown in Figure 10.1. More recently, these approaches have been refined and applied to a number of diverse systems.

The fundamental approach is as follows.

1. Select a set of known sequences of the target protein family.
2. Use the profile of these sequences to search global sequence databases for additional family members.
3. Perform multiple sequence alignment on this large set of sequences.
4. Compute statistical enrichment measures for the occurrence of each amino acid at each position.
5. Use this information to bias the selection of sequences in an engineering context.

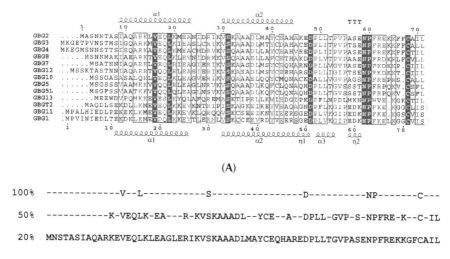

(A)

100%   -------------V--L-----------S----------------D----------NP-------C---

50%    -----------K-VEQLK-EA---R-KVSKAAADL--YCE--A--DPLL-GVP-S-NPFRE-K--C-IL

20%    MNSTASIAQARKEVEQLKLEAGLERIKVSKAAADLMAYCEQHAREDPLLTGVPASENPFREKKGFCAIL

(B)

**FIGURE 10.1 (see color insert following page 178)** Multiple sequence alignment and consensus sequence determination. (A) A multiple sequence alignment highlights key similarities between the sequences. Here, a red background indicates 100% identity across all sequences, blue boxes indicate largely conserved residues, with similar residue types in red, and black indicates unconserved residues. The known regions of helical secondary structure for the top and bottom sequences are also shown. Figure generated with ESPRIT (Gouet et al. 1999). (B) From the multiple sequence alignment, consensus sequences can be derived. Only seven residues are strictly conserved across all sequences (100%), while for 39 positions, there is one amino acid that occurs in the majority of sequences (50%). A complete sequence of 69 residues can be obtained by choosing the most common amino at each position (20% of the chosen amino acids all occur in at least three of the sequences). The sequences displayed are the set of γ-subunits from human heterotrimeric G-proteins.

Regan and coworkers have applied this approach to the design of several repeat proteins, including the tetratricopeptide repeat proteins (Main et al. 2003). The tetratricopeptide-repeat (TPR) family of proteins is a class of proteins consisting of repeated units of a small domain. In natural proteins, a TPR domain consists of 34 amino acids, and a typical TPR protein contains 3 to 16 repeats. Main et al. took the sequences of 1837 domains from 107 naturally occurring proteins and derived site-specific global amino-acid propensities based on this alignment. The most common amino acid at every position was significantly enriched over the average amino-acid frequency across all proteins, with enrichment factors ranging from 2.5 to 11.9.

Often, consensus sequences are defined by selecting positions that are conserved above a certain threshold. For example, for the TPR domain, Trp at position 4, Tyr at position 11, Gly at position 15, Tyr at position 17, Ala at position 20, Tyr at position 24, Ala at position 27, and Pro at position 32 all show enrichment of at least sixfold above the average amino-acid frequency. In terms of understanding the key sequence determinants of the protein fold, this is all the information that is needed. When designing a protein, however, one clearly needs a strategy for constructing a fully defined sequence.

Main et al. defined the sequence corresponding to the amino acid with the highest propensity at each site. Such a sequence would capture the most common features of this domain, but does not correspond to any natural protein. Repeats of one to three units of this engineered domain were constructed, with a few slight modifications: The single position where Cys was the most enriched residue was replaced with Ala (the second choice) in all repeats; a three residue helix-capping motif was added at the N-terminus of the protein; a solvating helix, corresponding to the first helix of the consensus motif with large aromatic residues, was replaced with Lys and Gln.

When they synthesized the model domain, even single domains of this protein were found to be well-structured, although they had relatively low stability. Repeats of this consensus sequence are very stable, with a repeat of three having a melting temperature of 83°C and a repeat of two melting at 74°C. In comparison, a naturally occurring three-repeat domain from PP5 has a melting temperature of 47°C, comparable to that of the single repeat consensus protein (49°C). These results suggest that the observed minimal repeat length of three (in natural systems) is not a result of stability requirements, but rather of an alternate reason, such as the requirement to bind protein targets.

Similar results have been obtained in numerous systems, including the ankyrin repeat proteins (Kohl et al. 2003), the leucine-rich repeat proteins (Binz et al. 2003), phytases (Lehmann et al. 2000), and antibodies (Knappik et al. 2000). The rationale for why this approach works is quite simple. Proteins have evolved both for stability and for function, and in a family of conserved fold, features defining protein stability will be conserved, while those that define diverse functions will not. In any given natural sequence, some of the residues that provide stability will likely be varied in order to accommodate function, but the particular residues that are varied will differ from protein to protein. Thus, the consensus sequence will define the underlying common feature that all members of the family share—an ability to stably fold into the target structure. As the consensus sequence does not contain those variations that create specific function at the expense of stability, it may form a much more stable structure.

A highly stable protein engineered through this approach may form a good starting point for engineering novel function, either through directed evolution or rational, computed-aided design. When the initial sequence is particularly stable, sequence variations that contribute strongly to a particular function at the expense of stability are more easily accommodated.

## PAIRWISE INTERACTIONS FROM SEQUENCE ANALYSIS

Consensus-based engineering assumes an independence of each position in the primary sequence. That is, it is presumed that combining the most common residues at each position will produce a stable protein. However, the contributions of residues in folded proteins (both to stability and to function) are known to be strongly coupled in many cases. More detailed analysis of sequence conservation from multiple sequence alignments can provide insight into this coupling, which can subsequently be applied in an engineering context.

One such approach is statistical coupling analysis (Lockless and Ranganathan 1999). This method quantifies the difference in amino-acid frequency at one position when sequence subsets containing only a single type of amino acid are considered at a second position. For example, if two positions contain Lys or Glu with roughly equal probabilities, the sequences with Lys at the first position are unlikely to contain Lys at the second position, and vice versa. Such statistical coupling may be due to an easily interpretable structural feature, for example, the presence of a salt-bridge, or to more subtle functional interactions.

In certain cases, these coupled interactions may be essential characteristics that must be considered when engineering a protein through sequence analysis. Ranganathan and coworkers have considered this problem in the context of the WW-domain family of proteins (Socolich et al. 2005). A set of 120 sequences was aligned, and statistical enrichments for each amino acid at each position were computed, as were the coupling parameters between each position and a number of moderately conserved sites. Two sets of novel protein sequences were generated from this data. The first were sequences randomly generated based on the site-independent enrichments; these are not consensus sequences per se, but rather sequences with similar amino-acid distributions to the natural set. The second set of sequences was generated through a computational procedure designed to match both the site-independent distributions and the pairwise coupling values. Forty-three proteins of each type were expressed in *E. coli* and tested whether they fold to a well-defined, native-like structure. Control sets of 42 natural sequences and 19 random sequences with the same overall mean amino-acid frequency as the other sets were also considered. None of the site-independent derived set were natively folded, although 70% were expressed and soluble. In comparison, 28% of the coupled-conservation set were natively folded, with a similar total number of soluble sequences. Of the random set, less than 50% were expressed and soluble (and none were folded), while 84% of the native set were soluble and 67% folded. These results clearly suggest that statistical correlations between sites are essential in sequence-based design.

At first glance, the results from these two studies seem contradictory. In the first case, a site-independent consensus sequence resulted in a highly thermostable protein, while in the second, designed proteins based on a site-independent analysis did not fold to the native state. One possible explanation is simply that each protein family may act differently. Site-independent information may completely determine the structure for some proteins, for example, zinc-fingers, phytases, antibodies, and the repeat proteins (TPR, ankyrin, and leucine-rich). For others, including the WW domains, the coupling between particular residues may be essential.

However, an alternate explanation is also possible. In the work of Ranganathan and coworkers, the site-independent sequences were not consensus sequences; that is, they did not use the most enriched amino acid at each site. Instead, they designed sequences with an overall amino-acid conservation profile similar to wild-type proteins. Thus, in the context of a given wild-type protein, the coupling between positions may be essential. That pairwise interactions can be important is, of course, well known—a deleterious mutation at one position can be rescued by a compensating mutation at a second site—and it is useful to note that a purely sequence-based analysis can aid in determining which interactions are functionally important. The consensus-based protein, however, is a single sequence with the most enriched amino acids at each position. Thus, some of the coupling information may be implicitly taken into account. Consider, for example, two positions that are strongly coupled—every sequence in the set contains KE or EK, but never KK or EE. If the KE motif is seen in just slightly more sequences, then Lys will be the preferred choice at site one while Glu is the preferred choice at site two. Thus, the consensus sequence will contain the KE motif, which is an acceptable choice. If, however, the sequences are derived randomly but biased by the independent conservation profiles, they will contain KK and EE in a significant number of cases.

Statistical coupling analysis can also be used together with consensus-based design, as was done by Magliery and Regan for the TPR-repeat designs, discussed previously (Magliery and Regan 2004). This analysis identified three strongly coupled networks of residues. While the consensus contains one choice of residues at each of these, other alternatives are possible. Additionally, the consensus-derived TRP sequence is highly negatively charged (–6), compared to natural sequences that predominantly have net charges between –3 and +3. Consideration of the statistical coupling between charged residues revealed subtle effects by which natural sequences contain compensating pairings of positive and negative residues. As these are at relatively weakly conserved sites and the pairings are not unique, this information is lost in the consensus sequence.

## GRAPHICAL ANALYSIS OF PROTEIN STRUCTURE

Many protein-engineering applications involve the creation of a small number of mutations to a naturally occurring protein so as to enhance its function in a well-defined manner. In these cases, a structural biologist's intuition is often an important tool in the design of the desired variants, an approach that may be termed *structure-based protein design* to borrow a term from the drug design field. Visualization of the known reference structure is a key component of this. For example, visualization can identify unsatisfied hydrogen bond donors or acceptors that may be mutated to increase stability or affinity. Similarly, visualizing steric interactions can help engineer interactions to discriminate among several potential binding targets.

## HOMOLOGY MODELING AND STRUCTURE VISUALIZATION

In many cases, protein engineering targets a protein whose structure has not been solved. This does not preclude the use of structure-based methods, as the known structures of related proteins can be used to create model structures though the process of homology modeling. Briefly, homology modeling consists of a number of conceptual steps, which may or may not be performed independently in a given program. The first consists of mapping the backbone of homologous residues from a protein of unknown structure onto a known structure. Next, the side chains of these residues must be packed into the structure defined by the threaded backbone. For highly homologous proteins this may use the reference structure as an aid, while for less-similar proteins an alternate search procedure must be used. In many cases, however, there are regions of nonhomologous sequence, even in highly homologous proteins. Therefore, a new backbone, with appropriately placed side chains, must be constructed for these regions. As these sequences most often occur in loop regions, this procedure is referred to as *loop building*. Finally, the initial model structure is refined using a minimization protocol based on an empirical force field. Numerous homology modeling programs are available for the purpose, including several Web-based servers (Table 10.2). Each method produces slightly different models, but generally the closer the sequence, the more accurate the homology model will be. When a model of relatively low sequence similarity is desired, the use of multiple reference structures can improve the accuracy. Figure 10.2 displays a homology model of a designed dimeric protein complex, based on the known structure of a monomeric protein from which the complex was engineered.

Numerous tools are available for visualizing protein structure, many of which are freely available (Table 10.3). While every atom in a protein may play a role in structure and function, visualization typically focuses on the use of reduced models. For example, a structure is often rendered as a cartoon with the elements of secondary structure represented abstractly and connected by loops that follow

### TABLE 10.2
### Automated Web Services for Homology Modeling

| Service | Web Location (URL) |
| --- | --- |
| Swiss-Model | http://swissmodel.expasy.org/SWISS-MODEL.html |
| Geno3D | http://geno3d-pbil.ibcp.fr/ |
| ESyPred3D | http://www.fundp.ac.be/sciences/biologie/urbm/bioinfo/esypred/ |
| What-If | http://swift.cmbi.kun.nl/WIWWWI/ |
| CPHModels | http://www.cbs.dtu.dk/services/CPHmodels/ |
| MODELLER* | http://salilab.org/modeller/modeller.html |

*Downloadable software, not a Web service*

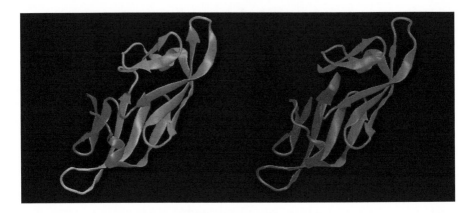

**FIGURE 10.2 (see color insert following page 178)** Homology building can allow a model structure to be built from the structure of a related sequence. Shown here is a homology model of a *de novo* designed dimeric complex (right), based on the structure of a monomeric protein (cyanovirin-N, left). The sequence of the dimer is 62% identical to that of the reference structure. Notice that all the key structural features are duplicated in the model, despite significant differences in sequence. Model was generated with MODELLER (Sali and Blundell 1993; Fiser and Sali 2003); figures were generated with VMD (Humphrey et al. 1996).

**TABLE 10.3**
**Freely Available Software for Structural Analysis**

| Program | Web Location (URL) | Description |
|---|---|---|
| VMD | http://www.ks.uiuc.edu/Research/vmd/ | Structural visualization program |
| PyMol | http://pymol.sourceforge.net/ | Structural visualization program |
| GRASP | http://wiki.c2b2.columbia.edu/honiglab_public/index.php/Software:GRASP | Visualization of protein surfaces and electrostatic potential maps |
| Residual Potential | http://web.mit.edu/tidor/www/residual/index.html | GRASP scripts for computing the residual potential |
| Probe | http://kinemage.biochem.duke.edu/software/probe.php | Software for computation of steric complementarity |

the backbone only. These cartoon models are useful for understanding the overall architecture of a protein or protein complex, and thus in identifying particular regions to include as variables in a combinatorial engineering application (either through directed-evolution or computational search). However, for the engineering of a small number of rational modifications, consideration of the chemical properties of each group can be essential. These properties are explicitly considered in the detailed force field-based calculations that will be discussed in later chapters, but can also be examined more qualitatively through visualization. The most straightforward approach involves mapping a property of interest, such as electrostatic potential or hydrophobicity, onto a surface representation of the protein.

## MODULATION OF STABILITY AND AFFINITY THROUGH STERIC COMPLEMENTARITY

While chemical complementarity may be assessed simply by visualization of surface maps, steric complementarity is more difficult for the eye to capture. Both the cores of proteins and the interface of protein–ligand and protein–protein complexes are generally well-packed with atoms forming intricate complementarity. While not the only consideration, the overall size and shape of particular amino acids play an important role. As a result, protein stability and binding affinity can be perturbed by relatively small changes targeted at modulating this steric complementarity. The addition of even a single methyl unit to an underpacked region can stabilize a protein or complex, as can removal of steric bulk from an overpacked region. The Probe program (Word et al. 1999), accessible through the MolProbity Web service (Lovell et al. 2003), is designed specifically to address this issue. Comparison of the molecular surface generated with a small probe sphere with that generated with a water-sized sphere reveals regions of suboptimal packing. Steric clashes can be detected by the overlap of two or more atoms, and regions of near-optimal packing (close van der Waals contact between two atoms) may also be defined. This allows for a direct visualization of regions of suboptimal packing in the given structure.

The rational modulation of steric complementarity has been used by Jasanoff and coworkers to create variants of calmodulin (CaM) and the M13 CaM-binding peptide (from rabbit skeletal muscle myosin) with altered specificity relative to wild type (Green et al. 2006). As seen in Figure 10.3, the interface contains a key hydrophobic interaction involving a Trp on M13, which packs into a pocket formed by CaM Phe 92 and several aliphatic hydrophobic groups. Replacement of the M13 Trp by a smaller aromatic group (either Phe or Tyr) resulted in an underpacked interface, with subsequent loss of affinity; the W→F mutant binds eightfold worse than wild type, and the W→Y mutant binds 14-fold worse. This loss of affinity was then

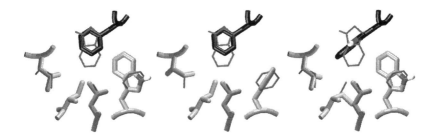

**FIGURE 10.3 (see color insert following page 178)** Affinities of protein complexes can be modulated by introducing perturbations in steric complementarity guided by visual analysis. Here we show a portion of the binding interface between calmodulin (gray) and the M13 CaM-binding peptide (black). **Left:** Reversing the positions of the Trp and Phe from the wild-type structure (shown in magenta) is expected to preserve near-optimal packing. **Center:** Changing only the Trp on M13 (to Phe) is expected to leave an unsatisfied void in the interface. **Right:** Changing only the Phe on CaM (to Trp) is expected to produced steric clashes and unfavorable structural rearrangements. Figures generated with VMD (Humphrey et al. 1996).

complemented by a corresponding mutation of Phe 92 on CaM to the larger Trp, as well as a conservative variation of Ile 125 to Leu. The mutant-mutant complex containing a W→F variation on M13 and a F→W (and I→L) substitution on CaM thus has the same total atom count as the wild type. The loss of affinity between the variant M13 and wild-type CaM is largely recovered in the complex with variant CaM. The affinity of the W→F mutant M13 for the variant CaM is only threefold different from WT and that of M13 W→Y differs by fourfold.

This work also reveals some of the challenges of taking advantage of steric complementarity in an engineering context. The site was chosen through visual analysis, with the motivation of engineering an orthogonal binding pair. That is, the goal was to create variants of both CaM and M13 that would bind to one another with affinity similar to that of the wild-type complex, but that would both have reduced affinity for the corresponding wild-type binding partner. As discussed previously, one direction of this specificity was achieved: The M13 W→F/Y variants bind preferentially to CaM FI→WL. However, specificity in the reverse direction was not achieved; in fact, the CaM FI→WL variant bound with twofold higher affinity to wild-type M13 than to the M13 mutants. This behavior is contrary to visual analysis—the pocket observed in the NMR solution structure seems well packed, and thus the replacement of Phe by the larger Trp would be expected to create steric clashes and reduce affinity. However, structural rearrangements in the pocket are able to accommodate this change, and the increase in buried hydrophobic surface in the F→W variant thus leads to increased affinity. While detailed packing calculations of the type that will be discussed in later chapters may be able to capture this behavior to some degree, consideration of the wild-type complex alone is inadequate.

## MODULATION OF AFFINITY THROUGH ELECTROSTATIC COMPLEMENTARITY

Electrostatic interactions play essential roles in protein stability, binding affinity, and catalytic activity. As a result, modification of the electrostatic properties of a protein provides a particularly useful tool for structure-based protein design.

The energetic contributions of electrostatic interactions are complicated by solvent effects; not only can solvent reduce the strength of an interaction in a single state through screening effects, but differential interactions with solvent between two states (folded and unfolded, bound and unbound) directly contribute to their relative energies. In molecular association, favorable interactions between water and polar groups at the binding interface are lost upon binding. These are replaced with direct interactions between the two binding partners, but the degree of compensation varies from complex to complex. In many cases, the net electrostatic contribution may be unfavorable. Theoretical work by Tidor and colleagues has resulted in a framework to describe the degree of electrostatic complementarity at an interface using the Poisson-Boltzmann continuum model of solvent (Lee and Tidor 1997; Kangas and Tidor 1998). While the theory provides methods for the detailed consideration of atom-by-atom contributions through intensive calculations, a simple graphical assessment of complementarity can also be defined. Briefly, the electrostatic potential due to desolvation costs is mapped onto the surface of a target protein (desolvation potential). Additionally, the electrostatic interaction potential of the protein's binding partner is similarly mapped onto the surface (interaction potential). It has been shown that in an electrostatically optimal complex the interaction potential must equal the negative of the desolvation potential across the entire protein surface. The residual potential is thus defined as the sum of the interaction potential and the desolvation potential. Deviations of the residual potential from zero highlight regions of the surface that deviate from optimal complementarity (see Figure 10.4)—a positive residual potential indicates that a reduction in positive charge (or increase in negative charge) would likely enhance affinity, and vice versa. Scripts for computing the residual potential with the GRASP software (Nicholls et al. 1993) are available online (Table 10.3).

An approach based on electrostatic analysis can also be applied to solve problems other than affinity optimization. For example, Lauffenburger and colleagues used this approach to design a modified variant of granulocyte-colony stimulating factor (GCSF) with enhanced lifetime (Sarkar et al. 2002). Cellular trafficking models had predicted that the rate of degradation depends on whether GCSF remains bound to its target receptor following internalization. In the late endosome, unbound ligands are recycled to the cell surface, while bound ligand-receptor complexes are retained for lysosomal degradation. Thus, recycling could be enhanced if a GCSF variant dissociates from the receptor in the low-pH environment of the late endosome. As the GCSFR binding surface has a positive charge, placement of histidines at the interface could produce the desired effect—at low pH, the protonation of histidine side chains on GCSF would create an electrostatic repulsion with the receptor, leading to reduced affinity. However, it is equally important to maintain affinity in the neutral

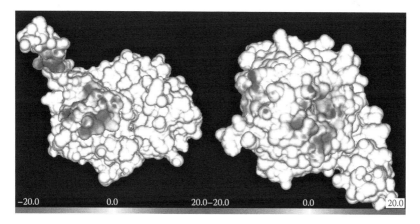

**FIGURE 10.4 (see color insert following page 178)**   The residual potential graphically displays deviations from perfect electrostatic complementarity. Displayed is the interface between the α (left) and βγ (right) subunits of a heterotrimeric G-protein. Regions of white indicate complementary surfaces (including surfaces not involved in the interface), while colored regions indicate the deviation from optimal complementarity. Blue indicates that the chemical groups underlying that region are either too positive or not negative enough, while red indicates groups that are either too negative or not positive enough. In the left-hand figure, it is clear that there are both regions that are excessively positive or excessively negative on this surface.

environment of the cell-surface. Thus, the residual potential of GCSF for binding its target receptor was plotted on the GCSF surface and regions of excess negative potential were identified. These correspond to positions where GCSF is overly negative for optimal binding to the receptor, and thus may tolerate substitution with a neutral histidine without destabilizing the complex. Three acidic and three neutral residues were identified in this manner. Additional analysis suggested two aspartates as ideal candidates for mutation; the third acidic residue (a glutamate) was deemed essential to binding at neutral pH, and the substitution at the neutral residues did not provide adequate discrimination at low pH. Experimental characterization of the two D→H mutants demonstrated that (1) the mutants had affinities at pH 7.4 within threefold of wild type; (2) the mutants had over fourfold difference in affinity at pH 7.4 and at pH 5.5, compared with a difference of less than twofold for wild type; and (3) the mutants had significantly increased lifetime in cellular proliferation assay, with a 50% increase in recycling at each cycle, leading to between a 50% and 100% increase in effectiveness after 6–8 days.

## FAST METHODS FOR MUTATIONAL EVALUATION

Visual analysis of protein structure, either with or without an energetic guide such as the residual potential, can suggest sites of modification; chemical intuition can then motivate particular amino-acid substitution. However, additional analysis is often needed due to the intricate networks of interactions typically found in proteins. In the cores of proteins, for example, a substituted residue must fit within the three-dimensional packing of residues; the substitution may, however, lead to rearrangements in this packing while maintaining stability. At binding interfaces, the same concerns with geometric packing exist, but as there are more polar groups, one must also consider balancing the desolvation costs with intermolecular interactions. The following chapters include a discussion of calculations that aim to address some of these problems. However, in some cases, it may be possible to evaluate the likely effect of a variation in a much simpler, and faster, manner.

### Affinity Enhancement through Peripheral Electrostatic Interactions

Recently, it has been suggested that protein–protein binding affinities may be perturbed in a predictable manner without resorting to detailed calculations by targeting a particular class of modifications—electrostatic interactions made by surface residues at the periphery of a binding interface (Selzer and Schreiber 2001; Joughin et al. 2005; Shaul and Schreiber 2005). Charged residues on the surface of a protein can make significant interactions with a binding partner even when located at a moderate (5 to 10 Å) distance from the interface. These have been referred to as *action-at-a-distance interactions*. As surface residues in general do not form the same intricately packed networks as core and interface groups, mutations can be introduced at surface positions with less detailed modeling. Web-based interfaces for the identification of such mutations have been developed (see Table 10.4)

These interactions seem to work through two nonexclusive mechanisms. Schreiber and coworkers have suggested a kinetic mechanism, by which these

**TABLE 10.4**

**Web Services for Structural and Energetic Analysis**

| Service | Web Location (URL) | Description |
|---------|-------------------|-------------|
| MolProbity | http://molprobity.biochem.duke.edu/ | Web service for visualization and analysis of protein structure, including H-bonding and steric complementarity. |
| HyPARE | http://bip.weizmann.ac.il/hypareb/main | Web service for "Predicting Association Rate Enhancement" mutations. |
| AAAD | http://groups.csail.mit.edu/tidor/aaad/ | Web service for prediction of thermodynamic "Action-at-a-Distance" interactions. |

peripheral residues enhance the on-rate $k_{on}$ of binding (Selzer and Schreiber 2001; Shaul and Schreiber 2005). Peripheral residues may make energetically significant interactions in the binding transition state to contribute to the rate of association, while such interactions are clearly absent in the unbound state. However, if the same interactions exist in the bound state as in the transition state, then they would have little effect on the rate of dissociation. Tidor and colleagues (including the author) have suggested an alternate mechanism, based purely on thermodynamic considerations (Joughin et al. 2005). Peripheral residues remain largely solvent exposed in the bound state, and thus pay only a very low (if any) desolvation penalty. However, the screening effects of solvent are such that small, but significant, nonspecific intermolecular interactions can persist at up to 10 Å separation. Fast methods of predicting long-distance interactions have been developed, using different methods designed to address different mechanisms of interaction. The kinetic and thermodynamic models lead to some overlap in predictions (residues that are expected to improve affinity also accelerate the kinetics of association), but many differences are seen as well. That is, some mutations are expected to increase affinity with minimal effect on the association rate, while others may increase the kinetics of association and disassociation without perturbing the overall affinity. These differences suggest that two distinct mechanisms are at play—peripheral action-at-a-distance interactions may independently be involved in modulating the affinity of an interaction as well as the kinetics of complex formation. Clearly both have applications in the engineering of protein complexes.

Mutations that add favorable interactions or remove an existing unfavorable interaction can be used to enhance binding affinity, as has been demonstrated. However, the same class of interactions could also be harnessed for additional modulations of affinity. For example, while introducing a mutation to severely reduce affinity is generally simple, destabilizing the complex by a desired degree is more challenging. Such an ability would be useful when engineering protein complexes for use as *in vivo* sensors or designing reagents with carefully tuned sensitivities. The introduction of an unfavorable action-at-a-distance interaction (or the elimination of an existing favorable interaction) could achieve this goal. The effect of individual residues in this type of interaction is relatively small, and thus engineering these interactions would be more effective in subtly modulating the stability of a complex. Similarly, the introduction of titratable groups (such as histidine) could be used to modify the pH dependence of binding in a relatively straightforward manner.

## SUMMARY

Information that is directly applicable to protein engineering can be found in the sequences and structures of known proteins. This knowledge can then be applied to new design objectives through the use of relatively simple computational approaches. Analysis of sequence conservation across a family of related proteins can be used to create hyperstable proteins through consensus-based design. Visualization of protein structure is an efficient way to identify key regions of interest, either by proximity to a site of known importance or by specifically targeting regions of suboptimal complementarity in packing or electrostatic interactions. Finally, fast, approximate methods are available for estimating the effects of mutations and have been shown to be highly successful in some cases, such as residues at the periphery of a protein–protein binding interface.

These approaches together provide a toolbox that complements the rational insight of a protein engineer in the process of design. Yet, they are not a solution to the design process in and of themselves. Rather, they act as a guide, suggesting a small number of mutations to consider or highlighting essential, conserved residues that should not be changed. Because they are simple, fast, and intuitive, they have been successfully adopted by experimental protein engineers in building a design or selection strategy.

## REFERENCES

Binz, H. K., M. T. Stumpp, P. Forrer, P. Amstutz and A. Pluckthun. 2003. "Designing repeat proteins: Well-expressed, soluble and stable proteins from combinatorial libraries of consensus ankyrin repeat proteins." *J Mol Biol* 332: 489–503.

Fiser, A. and A. Sali. 2003. "Modeller: Generation and refinement of homology-based protein structure models." *Methods Enzymol* 374: 461–91.

Gouet, P., E. Courcelle, D. I. Stuart and F. Metoz. 1999. "ESPript: Analysis of multiple sequence alignments in PostScript." *Bioinformatics* 15: 305–308.

Green, D. F., A. T. Dennis, P. S. Fam, B. Tidor and A. Jasanoff. 2006. "Rational design of new binding specificity by simultaneous mutagenesis of calmodulin and a target peptide." *Biochemistry* 45: 12547–59.

Humphrey, W., A. Dalke and K. Schulten. 1996. "VMD: Visual molecular dynamics." *J Mol Graph* 14: 33–38.

Joughin, B. A., D. F. Green and B. Tidor. 2005. "Action-at-a-distance interactions enhance protein binding affinity." *Protein Science* 14: 1363–9.

Kangas, E. and B. Tidor. 1998. "Optimizing electrostatic affinity in ligand-receptor binding: Theory, computation, and ligand properties." *J Chem Phys* 109: 7522–45.

Knappik, A., L. M. Ge, A. Honegger, P. Pack, M. Fischer, et al. 2000. "Fully synthetic human combinatorial antibody libraries (HuCAL) based on modular consensus frameworks and CDRs randomized with trinucleotides." *J Mol Biol* 296: 57–86.

Kohl, A., H. K. Binz, P. Forrer, M. T. Stumpp, A. Pluckthun, et al. 2003. "Designed to be stable: Crystal structure of a consensus ankyrin repeat protein." *Proc Natl Acad Sci U S A* 100: 1700–05.

Krizek, B. A., B. T. Amann, V. J. Kilfoil, D. L. Merkle and J. M. Berg. 1991. "A consensus zinc finger peptide—design, high-affinity metal-binding, a pH-dependent structure, and a His to Cys sequence variant." *J Am Chem Soc* 113: 4518–23.

Lee, L. P. and B. Tidor. 1997. "Optimization of electrostatic binding free energy." *J Chemical Phys* 106: 8681–90.

Lehmann, M., D. Kostrewa, M. Wyss, R. Brugger, A. D'Arcy, et al. 2000. "From DNA sequence to improved functionality: Using protein sequence comparisons to rapidly design a thermostable consensus phytase." *Protein Eng* 13: 49–57.

Lockless, S. W. and R. Ranganathan. 1999. "Evolutionarily conserved pathways of energetic connectivity in protein families." *Science* 286: 295–299.

Lovell, S. C., I. W. Davis, W. B. Adrendall, P. I. W. de Bakker, J. M. Word, et al. 2003. "Structure validation by C alpha geometry: Phi, psi and C beta deviation." *Proteins-Structure Function and Genetics* 50: 437–50.

Magliery, T. J. and L. Regan. 2004. "Beyond consensus: Statistical free energies reveal hidden interactions in the design of a TPR motif." *J Mol Biol* 343: 731–45.

Main, E. R. G., Y. Xiong, M. J. Cocco, L. D'Andrea and L. Regan. 2003. "Design of stable alpha-helical arrays from an idealized TPR motif." *Structure* 11: 497–508.

Nicholls, A., R. Bharadwaj and B. Honig. 1993. "GRASP—Graphical representation and analysis of surface-properties." *Biophys J* 64: A166.

Sali, A. and T. L. Blundell. 1993. "Comparative protein modeling by satisfaction of spatial restraints." *J Mol Biol* 234: 779–815.

Sarkar, C. A., K. Lowenhaupt, T. Horan, T. C. Boone, B. Tidor, et al. 2002. "Rational cytokine design for increased lifetime and enhanced potency using pH-activated 'histidine switching.'" *Nat Biotechnol* 20: 908–13.

Selzer, T. and G. Schreiber. 2001. "New insights into the mechanism of protein–protein association." *Proteins-Structure Function and Genetics* 45: 190–8.

Shaul, Y. and G. Schreiber. 2005. "Exploring the charge space of protein–protein association: A proteomic study." *Proteins-Structure Function and Bioinformatics* 60: 341–52.

Socolich, M., S. W. Lockless, W. P. Russ, H. Lee, K. H. Gardner, et al. 2005. "Evolutionary information for specifying a protein fold." *Nature* 437: 512–8.

Steipe, B., B. Schiller, A. Pluckthun and S. Steinbacher. 1994. "Sequence statistics reliably predict stabilizing mutations in a protein domain." *J Mol Biol* 240: 188–92.

Word, J. M., S. C. Lovell, T. H. LaBean, H. C. Taylor, M. E. Zalis, et al. 1999. "Visualizing and quantifying molecular goodness-of-fit: Small-probe contact dots with explicit hydrogen atoms." *J Mol Biol* 285: 1711–33.

# 11 Knowledge-Based Protein Design

*Michael A. Fisher,* Shona C. Patel,* Izhack Cherny,* and Michael H. Hecht*

## CONTENTS

Bottom-up approaches to protein design rely on a rational understanding of the fundamental principles that govern protein structure and stability. Such rational-based approaches contrast with methods that rely on random sequence libraries. While random libraries enable the exploration of vast amounts of sequence space, most members of randomly constructed libraries do not fold into stable protein-like structures (Mandecki 1990; Davidson et al. 1995). The goal of bottom-up rational design is to direct the formation of a desired three-dimensional structure by using knowledge-based approaches to design favorable interactions (both local secondary structures and long-range tertiary interactions) into the design of a linear amino-acid sequence.

---

* These authors contributed equally.

**239**

## MODULAR DESIGN: SELF-ASSEMBLY OF SEGMENTS OF SECONDARY STRUCTURE

One of the simplest design strategies involves the synthesis of single units of secondary structure that oligomerize to form the desired tertiary structure (Figure 11.1A). In the most basic cases, a single peptide segment of α-helix or β-strand is designed to self-assemble with several identical copies to produce a globular structure. Association of the units is typically favored by the design of amphiphilic secondary structures, which bury their hydrophobic surfaces upon formation of the desired three-dimensional structure.

Due to its simplicity, modular design was the first bottom-up design strategy to be implemented successfully. Early designs were based on simple α-helical structures. The α-helix is an appealing modular building block because it is a relatively stable structure that is maintained by local i to i+4 hydrogen bonding (Marqusee and Baldwin 1987; Shoemaker et al. 1987; Marqusee et al. 1989). β-structures, in contrast, are less modular because individual β-strands are not stable in isolation, and must form nonlocal hydrogen bonds with other strands (Hecht 1994). For these reasons, coiled coils and four-helix bundles were the first structures to be successfully designed by the modular approach.

The coiled-coil structure is composed of two α-helices wrapped around one another with a left-handed superhelical twist. Most coiled coils occur in a parallel (versus

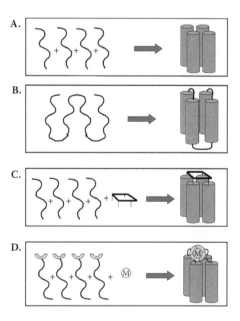

**FIGURE 11.1**  Bottom-up strategies for protein design: (A) Modular design by self-assembly of segments of secondary structure. (B) Design of single-chain proteins that fold into unique tertiary structures. (C) Template-directed assembly. (D) Ligand-induced folding by binding to a metal (M). In these figures, unfolded chains are shown as wavy lines, and α-helices are represented by cylinders.

antiparallel) orientation (O'Shea et al. 1991). In the 1980s, Hodges and coworkers designed a series of coiled-coil structures based on simple heptad repeats. They synthesized novel homodimer coiled-coil proteins with various lengths and enhanced stabilities (Lau et al. 1984; Hodges et al. 1990).

In the 1990s, Kim and coworkers showed that heterodimeric coiled coils could also be designed *de novo*. They used "negative design" to destabilize the electrostatic interactions that would occur in the alternative homodimeric structures (O'Shea et al. 1993). Further work from the groups of Kim, Alber, and Tidor extended this approach to the design of coiled coils with three or four α-helices (Harbury et al. 1993; Harbury et al. 1994).

Similar to coiled coils, the four-helix bundle structure can be assembled from individual α-helical modules (Figure 11.1A). In the late 1980s and early 1990s, DeGrado and coworkers designed a simple 16-residue amphiphilic helical peptide (denoted $\alpha_1$) that assembled into four-helix bundles in solution (Ho and DeGrado 1987; Osterhout et al. 1992).

While the modular design strategy offers a range of advantages, mainly due to its simplicity in design and synthesis, it is rather limited for more complex protein structures. The main drawback is the relatively low stability of modularly assembled structures compared with single-chain proteins (Figure 11.1B). This is due to the entropic cost associated with the intermolecular assembly of a unique three-dimensional structure from several independent subunits. Moreover, the intermolecular assembly of fragments is inherently concentration-dependent, and produces stable structures only at relatively high concentrations. This contrasts with the folding of single-chain protein sequences, which is intramolecular and—assuming there are no competing aggregation pathways—independent of concentration.

## TOWARD THE DESIGN OF SINGLE-CHAIN PROTEINS

To facilitate the formation of stable native-like *de novo* proteins, the lessons learned from the modular approach were applied to the design of single-chain proteins. This was accomplished by covalently linking together the individual elements of secondary structure (Figure 11.1B). In pioneering work, DeGrado and coworkers linked two $\alpha_1$ peptides with a short turn sequence, forming a helix-turn-helix (denoted $\alpha_2$), and showed that two $\alpha_2$ sequences could dimerize to form a four-helix bundle (Ho and DeGrado 1987). The DeGrado group proceeded to the final step of producing a single chain four-helix bundle by linking the two $\alpha_2$ peptides to form a single $\alpha_4$ chain. The resulting 74-amino-acid sequence folded into a stable α-helical bundle (Regan and DeGrado 1988).

The four helices in $\alpha_4$ were identical to one another and composed of only three amino acids—Glu, Lys, and Leu. The next step in the bottom-up approach required the design of a single-chain protein with a nonrepeating sequence. Hecht and coworkers (1990) achieved this goal by designing Felix, a novel four-helix bundle protein containing 19 of the 20 natural amino acids arranged in four different α-helices. As was done for the preceding designed helical bundles, the α-helices of Felix were designed to be amphiphilic. Special attention was given to the role of position-dependent α-helical propensities (Richardson and Richardson 1988) and further stabilization of the structure was achieved by designing a single intramolecular disulfide bond between the first and fourth helices. Moreover, to favor the desired structure, "negative design" was used to destabilize competing structures: Thus, alternating polar and nonpolar residues, which would favor β-strands (West and Hecht 1995), were avoided, turns were designed to disfavor the continuation of one long helix, and the hydrophobic faces of the helices were designed to favor a "left-turning" bundle topology (Hecht et al. 1990). The resulting protein was shown to fold into a globular helical structure. However, it was not very stable, and like $\alpha_4$, it displayed molten globule-like characteristics. For both Felix and $\alpha_4$, the dynamic nature of the structures presumably resulted from nonunique, poorly packed hydrophobic cores.

## DESIGN OF PREORGANIZED STRUCTURES

Although the entropic cost of folding a single chain is less than that required for the intermolecular association of modules, the cost of forming a unique three-dimensional structure from an unconstrained linear chain is still substantial. The unfavorable conformational entropy change associated with folding into a unique structure can be decreased by preorganizing the desired structure. Several groups have employed this strategy by using covalent cross-links, synthetic templates, or ligand binding.

Linking peptides to a synthetic template is a non-natural but effective way to covalently constrain peptides to adopt a particular structure. Mutter and coworkers pioneered this approach by developing an artificial template suitable for the formation of template-assembled synthetic proteins (TASPs; see Figure 11.1C) (Mutter and Vuilleumier 1989; Altmann and Mutter 1990). Their template was a short oligopeptide designed to form two antiparallel β-strands, stabilized by a disulfide bond at the open end. The ε-amino groups of four lysine side chains located on the template served as attachment sites for four units of secondary structure. Thus, four α-helices could be attached to form a parallel four-helix bundle with considerable stability. Ultimately, Mutter and coworkers were able to construct a variety of TASPs with different stoichiometries and secondary structures, including βαβ and 4α/4β structures (Mutter et al. 1989). Using similar strategies, Haehnel and coworkers have designed a series of template-based structures, including several with biologically relevant activities (Li et al. 2006).

In such designs, the template does not have to be a peptide. Thus, Sasaki and Kaiser used a porphyrin ring to attach four copies of an amphiphilic peptide (Sasaki and Kaiser 1989; Sasaki and Kaiser 1990). The resulting four-helix bundle heme protein (called *Helichrome*) was as stable as many natural proteins and also possessed a low level of enzymatic activity.

The introduction of ligand binding sites into secondary structural units can also be used to drive assembly into globular protein structures (Figure 11.1D). Both Lieberman and Sasaki (1991) and Ghadiri et al. (1992) designed a non-natural divalent metal-binding moiety (bipyridine) onto the N-terminus of a short amphiphilic α-helix, thereby inducing self-assembly into a three-helix bundle upon addition of metals such as Fe(II), Ni(II), Co(II), or Ru(II). Ghadiri's group elaborated on this strategy to assemble four-helix bundles by utilizing the high affinity of Ru(II) for nitrogen-containing aromatic heterocycles attached to the termini of four α-helices (Ghadiri et al. 1992b).

# FROM BOTTOM-UP STRATEGIES TO THE DESIGN
# OF UNIQUELY FOLDED STRUCTURES

The field of protein design advanced considerably in the second half of the 1990s as researchers (1) tackled the design of proteins with both alpha and beta secondary structures; and (2) began to apply the rapidly developing tools for computational searching and optimization.

In 1996, Imperiali and coworkers reported the design of a 23-residue protein with a ββα architecture similar to that found in DNA-binding zinc finger modules (Struthers et al. 1996). Their goal was to design a novel sequence that would fold into the desired structure without the use of disulfide bridges or metal-binding sites. In contrast, natural zinc fingers require metal to fold. This was accomplished by starting with a native zinc finger backbone as a template and incorporating a type II' β turn (Sibanda and Thornton 1985) and a D-proline. The resulting peptide, BBA1, was soluble and monomeric. The NMR structure showed that the protein had the desired secondary structure and type II' β turn.

Progress in protein design was greatly enhanced in the mid-1990s as researchers took advantage of newly emerging computational approaches. A year after the rational design of BBA1 by Imperiali and coworkers, Dahiyat and Mayo (1997) devised a fully automated computational algorithm to redesign the ββα zinc finger motif. A natural zinc finger module was chosen as the design template and its coordinates were used as the fixed backbone for the sequence selection algorithm. After computationally screening $10^{27}$ amino-acid sequences, the designed sequence FSD-1 was selected as the optimal sequence for the desired ββα fold. FSD-1 had low sequence identity to any known protein sequence. The design was experimentally validated by solving the NMR structure of FSD-1, which showed that this novel protein was well-folded and looked very similar to the designed structure. This landmark study demonstrated that a purely computational approach could successfully design a small, well-folded protein containing alpha, beta, and turn structural elements (Dahiyat and Mayo 1997).

Although initial attempts at computational protein design were limited to the design of new side chains onto fixed backbone structures that were "borrowed" from the coordinates of natural proteins, further advances in computational methods enabled new algorithms that allowed flexible backbones (Desjarlais and Handel 1999). While the fixed backbone algorithms were sufficient to redesign side chains onto pre-existing natural structures, the ability to incorporate backbone flexibility is essential for the computational design of truly novel folds for which no backbone template pre-exists.

## DESIGN OF NOVEL FOLDS

A major breakthrough in the design of novel protein folds was the computational design of a family of α-helical bundles with a right-handed superhelical topology (Harbury et al. 1998). Prior to this work, no structures (natural or designed) of trimeric or tetrameric helical bundles with right-handed superhelical topology were known. The right-handed superhelical twist exhibited throughout the family was prescribed by an 11-residue repeat of nonpolar (N) and polar (P) residues: NPPNPPPNPPP.

Trimeric and tetrameric bundles were designed using backbone parameterization and detailed computational modeling of core packing. Experimentally derived crystal structures of trimer and tetramer sequences that had been designed for optimal stability and specificity *in silico* met the design goals and agreed with predicted models in atomic detail (Harbury et al. 1998; Plecs et al. 2004).

The design of novel folds has progressed to a point where it is now possible to create entirely novel proteins with arbitrarily chosen three-dimensional structures. Kuhlman, Baker, and coworkers sketched out a protein topology that had not been observed in nature, and set out to design a 93-residue α/β protein with this novel fold (Kuhlman et al. 2003). Using a computational design strategy that cycled between sequence optimization and structure prediction, a refined sequence was generated that fit the design goal. Experimental characterization of this sequence revealed that the crystal structure of the protein, shown in Figure 11.2, was practically indistinguishable from the design model at atomic resolution.

## DESIGN OF MEMBRANE PROTEINS

The computational design of novel proteins is not limited to globular proteins in aqueous solution. Recently, Yin, Slusky, DeGrado, and coworkers described a computational method to design peptides that specifically target the intramembrane region of transmembrane proteins (Yin et al. 2007). Their method comprises three steps: The first step is to use examples of helix pairs in known structures to select an appropriate backbone geometry for the desired complex of the novel peptide bound to the target transmembrane helix. Next, the known sequence of the targeted transmembrane helix is threaded onto one of the two helices of the selected pair. Finally, a side-chain repacking algorithm is used to design the amino-acid sequence of the

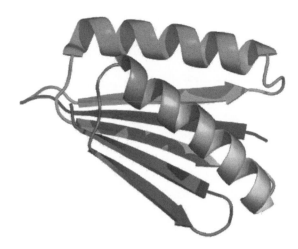

**FIGURE 11.2 (see color insert following page 178)**   Ribbon diagram of the crystal structure of TOP7, a protein designed to fold into a novel topology not seen previously in nature (Kuhlman et al. 2003). The observed structure is remarkably similar (RMS deviation = 1.2 A) to the designed model. Figure provided by Brian Kuhlman.

novel helix that best recognizes the transmembrane helix. Experimental characterization of the novel peptides showed that they bind to their target transmembrane regions with high affinity and specificity.

The development of computational methods that (1) work with either fixed backbones or flexible backbones and (2) can be applied to either soluble proteins or membrane bound proteins has revolutionized the field of protein design. Further examples of how these methods have been implemented are described throughout this book.

## DESIGNED LIBRARIES: THE INTERFACE BETWEEN RATIONAL AND COMBINATORIAL APPROACHES

The bottom-up rational design approaches summarized previously are based on the assumption that a detailed understanding of the fundamental principles of protein structure can enable the design of specific sequences that fold into predetermined structures. Yet, the relationship between sequence and structure is not "one-to-one" in both directions. While it has been known since the pioneering experiments of Anfinsen that a given sequence will fold into a unique three-dimensional structure (Haber and Anfinsen 1962), the converse is not true: A given structure can be specified by many different sequences (Axe et al. 1996; Gassner et al. 1996). Therefore, knowledge-based approaches that rely on the fundamental principles of protein structure need not be limited to the design of individual sequences one-at-a-time. Indeed, rational approaches can be used to design large libraries of sequences compatible with a chosen target structure. Such libraries provide an ideal starting point for high throughput screens and selections that enable the evolution *in vitro* of novel proteins with desired structural and functional properties.

By using a confluence of rational and combinatorial approaches, Case and McLendon (2000) designed a library of amphiphilic α-helices and then used metal-assisted protein assembly, as described in a previous section, to select the sequences in their library that had the greatest propensity to assemble into the desired three-helix bundle structure. They synthesized a collection of 20-residue peptides with a bipyridine group on the N-terminus. As demonstrated previously by Ghadiri's and Sasaki's groups (Lieberman and Sasaki 1991; Ghadiri et al. 1992a), binding of a single metal ion by the three bipyridine groups can direct the formation of a three-helix bundle. Case and McLendon allowed the sequences in their library of α-helices to compete for the metal under thermodynamically reversible conditions, and then isolated the "winner" sequences that successfully formed three-helix bundles. Through this approach they were able to select those amphiphilic helical sequences that had the greatest propensity to pack together into stable three-helix bundles.

Our own work has focused on a large-scale method for designing vast libraries of novel proteins. By constraining our combinatorial methods with features of rational design, we have been able to construct large collections of novel proteins wherein the majority of sequences fold into stable three-dimensional structures. Our approach, called the *binary code for protein design*, is summarized in the remainder of this chapter.

## THE BINARY CODE STRATEGY FOR PROTEIN DESIGN

The binary code strategy is based on the design of specific patterns of polar and nonpolar amino acids in the primary sequence. The strategy is described as *binary* because the amino acids are treated as belonging to one of two groups: polar or nonpolar. This polar/nonpolar patterning is designed to match the inherent structural periodicity of the desired secondary structure (α-helix or β-strand), and thereby specify amphiphilic units of secondary structure (Figure 11.3). When these amphiphilic secondary structures fold into the target tertiary structure, their nonpolar faces become buried against one another. Once folded, the entire sequence partitions its side chains such that hydrophobic amino acids are sequestered into the core and hydrophilic amino acids are exposed on the exterior of the protein.

The binary code strategy incorporates both rational and combinatorial features. Thus, the *positions* of polar and nonpolar residues are designed explicitly; however, the exact side-chain *identities* at each position are allowed to vary combinatorially. The combinatorial mixtures of polar or nonpolar amino acids are specified by degenerate codons (Figure 11.4) in combinatorial libraries of synthetic genes.

The binary code strategy focuses the exploration of sequence space into those regions that favor folded structures. By carefully designing template structures, it is possible to design libraries with a range of tertiary folds. In principle, all-alpha, all-beta, and combined alpha/beta structures are suitable design targets, although

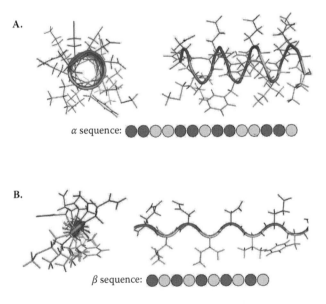

**FIGURE 11.3 (see color insert following page 178)**   Binary pattern strategy applied to (A) an α-helix and (B) a β-strand. Nonpolar residues are represented by yellow circles and polar residues are represented by red circles. In the α-helix design, nonpolar residues are placed every three to four positions, while in the β-strand design a nonpolar residue is placed at every other position. In each figure, a head-on view is shown on the left and a side view on the right.

|  | T | C | A | G |
|---|---|---|---|---|
| **T** | TTT Phe<br>TTC Phe<br>TTA Leu<br>TTG Leu | TCT Ser<br>TCC Ser<br>TCA Ser<br>TCG Ser | TAT Tyr<br>TAC Tyr<br>TAA Stop<br>TAG Stop | TGT Cys<br>TGC Cys<br>TGA Stop<br>TGG Trp |
| **C** | CTT Leu<br>CTC Leu<br>CTA Leu<br>CTG Leu | CCT Pro<br>CCC Pro<br>CCA Pro<br>CCG Pro | CAT His<br>CAC His<br>CAA Gln<br>CAG Gln | CGT Arg<br>CGC Arg<br>CGA Arg<br>CGG Arg |
| **A** | ATT Ile<br>ATC Ile<br>ATA Ile<br>ATG Met | ACT Thr<br>ACC Thr<br>ACA Thr<br>ACG Thr | AAT Asn<br>AAC Asn<br>AAA Lys<br>AAG Lys | AGT Ser<br>AGC Ser<br>AGA Arg<br>AGG Arg |
| **G** | GTT Val<br>GTC Val<br>GTA Val<br>GTG Val | GCT Ala<br>GCC Ala<br>GCA Ala<br>GCG Ala | GAT Asp<br>GAC Asp<br>GAA Glu<br>GAG Glu | GGT Gly<br>GGC Gly<br>GGA Gly<br>GGG Gly |

**FIGURE 11.4** The degeneracy of the genetic code permits subsets of polar (boxed in dashed line) and nonpolar (boxed in solid line) residues to be encoded at specific positions in combinatorial libraries of synthetic genes. Only the second base of the codons is defined: Nonpolar amino acids are encoded by the degenerate codon NTN, while polar amino acids are encoded by the degenerate codon VAN. N represents any of the nucleobases, A, G, T, or C; V represents A, G, or C. Variations can be used to avoid codon bias.

beta and combined structures present a much greater challenge. Since the proteins from a binary-coded library are designed to fold in aqueous solution, there is a high probability that these proteins will express and be soluble. This, in turn, leads to greater potential for cofactor binding or functional activity compared to randomly generated sequences, which usually lead to insoluble aggregates (Mandecki 1990; Davidson et al. 1995).

## Libraries of Binary Patterned Four-Helix Bundles

The four-helix bundle is the simplest and most common α-helical domain found in nature (Brandon and Tooze 1998). Natural four-helix bundles exhibit a variety of functions, including cofactor binding, catalysis, and protein or RNA binding (Presnell and Cohen 1989). Moreover, as described previously, the four-helix bundle has a rich history as a target for protein design.

Binary patterned amphiphilic α-helices are designed by placing a nonpolar residue (excluding proline) every three or four positions along a polypeptide chain. As shown in Figure 11.3, this is consistent with a structural repeat of 3.6 residues per turn, which is characteristic of a canonical α-helix. Connecting four such polypeptide segments with turn or loop regions allows the formation of a single-chain sequence capable of folding into a bundle that sequesters nonpolar residues in the core of the protein while exposing polar residues to aqueous solvent.

Libraries of four-helix bundle proteins are expressed from synthetic genes cloned into *E. coli*. The synthetic genes are assembled from DNA oligonucleotides consisting mostly of combinatorially degenerate codons that encode specific sets of polar or nonpolar residues. Not all possible codons are used for the combinatorial portions of the synthetic genes because their inclusion would lead to unwanted stop codons or interruptions of the specified polar/nonpolar periodicity (see Figure 11.4). The DNA oligonucleotides are assembled into full-length genes via standard techniques of molecular cloning.

Libraries of four-helix bundles have been iterated in our lab through several generations. With each generation, the design and assembly has improved (Figure 11.5). In the first generation, each protein in the binary-patterned library was designed to be 74 residues long (Kamtekar et al. 1993). To increase protein stability, a second generation library was designed in which each helix was extended by one helical turn, resulting in binary-patterned sequences with 102 residues (Wei et al. 2003a).

**First generation:** 74 residues, α-helical, molten globular

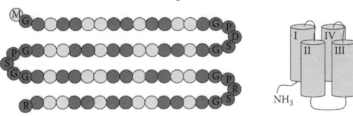

**Second generation:** 102 residues, α-helical, well folded

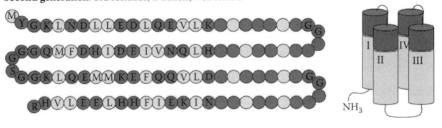

**FIGURE 11.5 (see color insert following page 178)** Design of the first- and second-generation libraries of binary patterned four-helix bundles. The first generation library was designed to have 74 residues. In the second generation, each helix was extended by one helical turn leading to a 102-residue library. The design templates of the first- and second-generation four-helix bundle libraries differ in terms of number of total residues and the number and placement of combinatorially diverse residues. Polar residues are represented by red circles, nonpolar residues by yellow circles, and turn residues by blue circles. Residues that were held constant for the purposes of gene construction are coded with the single letter code, while combinatorial positions are colored but contain no letter.

Because it is technically difficult to produce large libraries of full-length sequences devoid of frame shifts and stop codons, a preselection system was devised to weed out segments with incorrect sequences prior to the assembly of full-length genes (Bradley et al. 2005). DNA segments that were in frame and devoid of stop codons (as determined by the preselection) were assembled into full-length genes, producing a high-quality and diverse library for use in high-throughput screens and selections for functional activity.

The observed expression of most of the binary code proteins at high levels in soluble form, combined with the relatively small size (8–12 kDa) of the proteins from the two libraries, served to expedite protein purification. A straightforward freeze/thaw procedure followed by ion exchange chromatography was in many cases sufficient to yield proteins that were ≥95% pure (Johnson and Hecht 1994; Bradley et al. 2005). The relative ease with which proteins from all generations could be purified facilitated their biophysical characterization.

Circular dichroism (CD) analysis of proteins from the four-helix bundle libraries confirmed that nearly all the proteins were α-helical (Kamtekar et al. 1993; Wei et al. 2003a). CD was also used to monitor thermal or chemical denaturation and to determine the free energy of unfolding. Most of the proteins from the first-generation library have slightly lower stabilities than natural proteins, while proteins from the second library are quite stable and often have melting temperatures near 100°C (Kamtekar et al. 1993; Roy et al. 1997a; Roy and Hecht, 2000; Wei et al. 2003a).

Nuclear magnetic resonance (NMR) was used to study the structure and dynamics of the binary patterned four-helix bundles. While proteins from the initial collection (shorter helices) typically formed fluctuating structures reminiscent of molten globules (Roy et al. 1997b), the majority of proteins studied from the second-generation library were well-folded and substantially more native-like (Wei et al. 2003a). Lengthening each helix by one turn apparently was the key feature that enabled the production of a large library containing an abundance of well-folded proteins.

The NMR solution structures of two *de novo* proteins from the second-generation library were solved, and are shown in Figure 11.6 (Wei et al. 2003b; Go et al. 2008). These proteins are both well-folded four-helix bundles with a hydrophobic core and hydrophilic exterior, as per the binary code design strategy. While both proteins share the same overall scaffold, they have distinct sequences and are characterized by variations in the packing of hydrophobic side chains in the core.

Based on our results thus far, we expect that our libraries of 102-residue sequences contain proteins with an assortment of stabilities, ranging from dynamic to very well-ordered structures. It is also possible that the quaternary structure will vary from protein to protein, with some sequences taking on higher order oligomeric states rather than being monomeric. Taking into consideration the wide range of sequences, structural features, and oligomeric states, these libraries provide enormous diversity for searching for functional activity among *de novo* proteins.

## FUNCTIONAL ACTIVITIES OF BINARY PATTERNED FOUR-HELIX BUNDLES

The *de novo* four-helix bundle proteins from our binary patterned collections display various types of functional activity. Proteins from the first- and second-generation

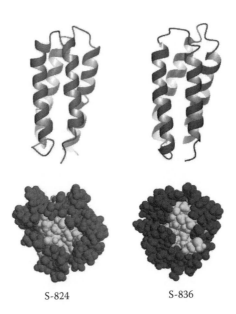

S-824                                          S-836

**FIGURE 11.6 (see color insert following page 178)**    NMR solution structures of proteins
S-824 and S-836. Top: Ribbon diagram of the four-helix bundle proteins. Bottom: Space-
filling head-on view of the proteins with polar residues on the exterior (red or blue) and nonpolar residues in the core (yellow).

libraries bound heme, a common cofactor among natural four-helix bundle proteins
(Rojas et al. 1997). As a follow-up to heme binding, proteins were probed for heme-
dependent activities; and proteins exhibiting peroxidase activity were discovered within
the library (Moffet et al. 2000). Assays for heme binding and peroxidase activity are
illustrated in Figure 11.7. The proteins were also assayed for function without the use
of a cofactor, and several were shown to possess hydrolytic activity (Wei and Hecht
2004). These findings are intriguing, as these proteins were designed only to form a
particular structure; they were not explicitly designed to have enzymatic activity.

## BINARY-PATTERNED BETA SHEET PROTEINS

Libraries of antiparallel β-sheet structures can be constructed in a manner analogous
to the construction of the four-helix bundle libraries described previously. A library of
synthetic genes is constructed by assembling synthetic DNA oligonucleotides in which
most codons are combinatorially degenerate, encoding either polar residues, nonpolar
residues, or a group of residues appropriate for turns or loops. In order to generate a
*de novo* β-sheet structure using the binary code strategy, the combinatorially degenerate codons are used to specify an alternating pattern of polar and nonpolar residues
with a periodicity of two, recapitulating the structural repeat of amphiphilic β-strands
(Figure 11.3). Negative design plays an essential role in the design of β-structures: It is
of paramount importance to disfavor undesired structures, such as long uninterrupted
amphiphilic β-strands running the length of the entire sequence.

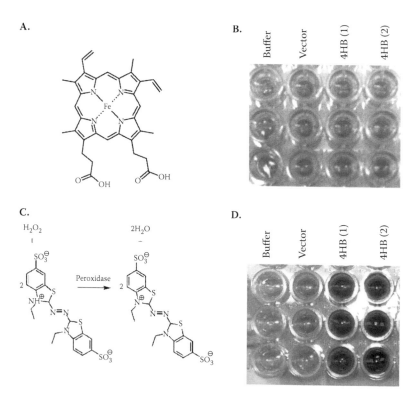

**FIGURE 11.7 (see color insert following page 178)**   Summary of heme binding and peroxidase activity exhibited by *de novo* four-helix bundle proteins. (A) Heme. (B) Heme binding assay. Each column shows a different sample performed in triplicate. Column 1 is heme alone in buffer. Column 2 is heme incubated with lysates from cells containing empty vector, which represents the activity of background *E. coli* proteins. Both columns 1 and 2 are negative controls and show a brown color. Columns 3 and 4 are cell lysates from cells expressing two different four-helix bundle (4HB) proteins incubated with heme showing various intensities of red. (C) The peroxidase reaction catalyzed by *de novo* four-helix bundle proteins using heme as cofactor and 2,2′-azino-di(3-ethyl-benzthiazoline-6-sulfonic acid) (ABTS) as the reducing agent. (D) Peroxidase assay. Each column is a different sample performed in triplicate. Column 1 shows the background activity of buffer alone and Column 2 shows the background activity of endogenous *E. coli* proteins. Both controls show a light teal color. Columns 3 and 4 are two different four-helix bundle (4HB) proteins exhibiting peroxidase activity giving rise to a dark teal color.

The binary code strategy has been applied to design β-sheet proteins and a specific design target is shown in Figure 11.8. Members of the library were designed to contain six β strands of seven residues each, connected by five turns of four residues each. As noted previously, β-structures are inherently more difficult to design than α-helical structures. Because the backbone hydrogen bonding of β-structure is *interstrand*—rather than *intra*segmental as in the i to i+4 H-bonding of α-helices—amphiphilic β-strands have a high propensity to aggregate (Hecht 1994). Therefore, it is not surprising that proteins from the original binary patterned β-sheet library

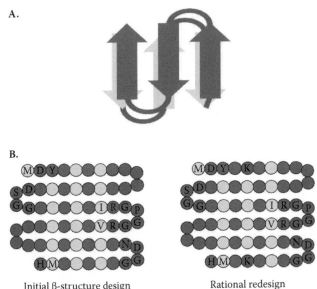

Initial β-structure design
amyloid fibrils

Rational redesign
monomeric β-structures

**FIGURE 11.8 (see color insert following page 178)** Design of binary coded libraries of β-sheet proteins. (A) Schematic representation of the six-stranded β-sheet target structure. Yellow coloring designates the nonpolar face of the β-sheets, facing the interior of the protein. Red coloring designates the polar faces of β-sheets, which are exposed to aqueous solvent. Turns are indicated in blue. (B) The design templates of the initial library and of the redesigned proteins differ only in the placement of lysine residues into the middle of the first and last β-strands. Residues that were held constant for the purposes of gene construction are coded with the single letter code.

assembled into multimeric structures that mimicked amyloid fibrils (West et al. 1999). Later work showed that incorporation of negative design features that disfavor aggregation enabled the production of monomeric *de novo* β-sheet proteins (Wang and Hecht 2002).

## CONCLUSIONS

Scientists and engineers have been trying to design proteins "from scratch" for more than two decades. Since the early attempts by Richardson and Richardson (1987) to design Betabellin from first principles, it has been clear that successful and reproducible protein design would depend upon a detailed understanding of the fundamental physical chemistry that drives a linear polypeptide sequence to fold into a unique three-dimensional protein structure.

In the intervening years, spectacular progress has occurred: It is now possible to use fully automated computational methods to design novel folds that have never been seen in nature. It is also possible to use approaches drawn from both rational and combinatorial methods to construct vast libraries of novel proteins that fold into predetermined structures and possess biologically relevant functions.

If the ability to design something "from scratch" is the ultimate test of our understanding, then recent advances in knowledge-based protein design demonstrate that the "knowledge" and understanding that form the basis of this endeavor have reached a high level of sophistication.

# REFERENCES

Altmann, K. H., and Mutter, M. 1990. A general strategy for the de novo design of proteins template assembled synthetic proteins. *Int J Biochem* 22: 947–56.

Axe, D. D. , Foster, N. W., and Fersht, A. R. 1996. Active barnase mutants with random hydrophobic cores. *Proc Natl Acad Sci* 93: 5590–4.

Bradley, L. H., Kleiner, R. E., Wang, A. F., Hecht, M. H., and Wood, D. W. 2005. An intein-based genetic selection allows the construction of a high-quality library of binary patterned de novo protein sequences. *Protein Engineering, Design & Selection* 18: 201–7.

Brandon, C., and Tooze J. 1998. *Introduction to protein structure*, 2nd ed. New York: Garland.

Case, M. A., and McLendon, G. L. 2000. A virtual library approach to investigate protein folding and internal packing. *J Am Chem Soc* 122: 8089–90.

Dahiyat, B. I., and Mayo, S. L. 1997. De novo protein design: Fully automated sequence selection. *Science* 278: 82–7.

Davidson A. R., Lumb, K. J., and Sauer R. T. 1995. Cooperatively folded proteins in random sequence libraries. *Nat Struct Biol* 2: 856–64.

Desjarlais, J. R., and Handel, T. M. 1999. Side-chain and backbone flexibility in protein core design. *J Mol Bio* 289: 305–18.

Gassner, N. C., Baase, W. A., and Matthews, B. W. 1996. A test of the "jigsaw puzzle" model for protein folding by multiple methionine substitutions within the core of T4 lysozyme. *Proc Natl Acad Sci* 93: 12155–8.

Ghadiri, M. R., Soares, C., and Choi, C. 1992a. A convergent approach to protein design—metal ion-assisted spontaneous self-assembly of a polypeptide into a triple-helix bundle protein. *J Amer Chem Soc* 114: 825–31.

Ghadiri, M. R., Soares, C. and Choi, C. 1992b. Design of an artificial 4-helix bundle metalloprotein via a novel ruthenium(II)-assisted self-assembly process. *J Amer Chem Soc* 114: 4000–02.

Go, A., Kim, S., Baum, J., and Hecht, M. H. 2008. Structure and dynamics of de novo proteins from a designed superfamily of 4-helix bundles. *Prot Sci* 17: 821–32.

Haber, E., and Anfinsen, C. B. 1962. Side-chain interactions governing the pairing of half-cystine residues in ribonuclease. *J Biol Chem* 237: 1839–44.

Harbury, P. B., Zhang, T., Kim, P. S., and Alber, T. 1993. A switch between two-, three-, and four-stranded coiled coils in GCN4 leucine zipper mutants. *Science* 262: 1401–07.

Harbury, P. B., Kim, P. S., and Alber, T. 1994. Crystal structure of an isoleucine-zipper trimer. *Nature* 371: 80–3.

Harbury, P.B., Plecs, J.J., Tidor, B., Alber, T., and Kim, P.S. 1998. High-resolution protein design with backbone freedom. *Science* 282: 1462–7.

Hecht, M. H., Richardson, J. S., Richardson, D. C., and Ogden, R. C. 1990. De novo design, expression, and characterization of Felix: A four-helix bundle protein of native-like sequence. *Science* 249: 884–91.

Hecht, M. H. 1994. De novo design of β-sheet proteins. *Proc Natl Acad Sci* 91: 8729–30.

Ho, S. P., and DeGrado, W. F. 1987. Design of a 4-helix bundle protein—synthesis of peptides which self-associate into a helical protein. *J Amer Chem Soc* 109: 6751–8.

Hodges, R. S., Zhou, N. E., Kay, C. M., and Semchuk, P. D. 1990. Synthetic model proteins: Contribution of hydrophobic residues and disulfide bonds to protein stability. *Peptide Res* 3: 123–37.

Johnson, B. H., and Hecht, M. H. 1994. Recombinant proteins can be isolated from *E. coli* cells by repeated cycles of freezing and thawing. *Bio/Technology* 12: 1357–60.

Kamtekar, S., Schiffer, J. M., Xiong, H., Babik, J. M., and Hecht, M. H. 1993. Protein design by binary patterning of polar and nonpolar amino acids. *Science* 262: 1680–85.

Kuhlman, B., Dantas, G., Ireton, G.C., Varani, G., Stoddard, B.L., and Baker, D. 2003. Design of a novel globular protein fold with atomic-level accuracy. *Science* 302: 1364–68.

Lau, S. Y., Taneja, A. K., and Hodges, R. S. 1984. Synthesis of a model protein of defined secondary and quaternary structure. Effect of chain length on the stabilization and formation of two-stranded alpha-helical coiled-coils. *J Biol Chem* 259: 13253–61.

Li, W-W, Hellwig, P., Ritter, M., and Haehnel, W. 2006. De novo design, synthesis, and characterization of quinoproteins. *Chem Eur J* 12: 7236–45.

Lieberman, M., and Sasaki, T. 1991. Iron(II) organizes a synthetic peptide into 3-helix bundles. *J Amer Chem Soc* 113: 1470–1.

Mandecki, W. 1990. A method for construction of long randomized open reading frames and polypeptides. *Protein Eng* 3: 221–6.

Marqusee, S., and Baldwin, R. L. 1987. Helix stabilization by Glu-...Lys+ salt bridges in short peptides of de novo design. *Proc Natl Acad Sci* 84: 8898–902.

Marqusee, S., Robbins, V. H., and Baldwin, R. L. 1989. Unusually stable helix formation in short alanine-based peptides. *Proc Natl Acad Sci* 86: 5286–90.

Moffet, D. A., Certain, L. K., Smith, A. J., Kessel, A. J., Beckwith, K. A., and Hecht, M. H. 2000. Peroxidase activity in heme proteins derived from a designed combinatorial library. *J Am Chem Soc* 122: 7612–13.

Mutter, M., and Vuilleumier, S. 1989. A chemical approach to protein design—template-assembled synthetic proteins (TASP). *Angew Chem Int Ed Engl* 28: 535–54.

Mutter, M., Hersperger, R., Gubernator, K., and Muller, K. 1989. The construction of new proteins. 5. A template-assembled synthetic protein (TASP) containing both a 4-helix bundle and beta-barrel-like structure. *Proteins-Structure Function and Genetics* 5: 13–21.

O'Shea, E. K., Klemm, J. D., Kim, P. S., and Alber, T. 1991. X-ray structure of the GCN4 leucine zipper, a two-stranded, parallel coiled coil. *Science* 254: 539–44.

O'Shea, E. K., Lumb, K. J., and Kim, P. S. 1993. Peptide 'Velcro': Design of a heterodimeric coiled coil. *Curr Biol* 3: 658–67.

Osterhout, J. J., Handel, T., Na, G., Toumadje, A., Long, R. C., Connolly, P. J., Hoch, J. C., Johnson, W. C., Live, D., and DeGrado, W. F. 1992. Characterization of the structural properties of $\alpha_1\beta$, a peptide designed to form a 4-helix bundle. *J Amer Chem Soc* 114: 331–7.

Plecs, J. J., Harbury, P. B., Kim, P. S., and Alber, T. 2004. Structural test of the parameterized-backbone method for protein design. *J Mol Biol* 342: 289–97.

Presnell, S. R., and Cohen, F. E. 1989. Topological distribution of four-alpha-helix bundles. *Proc Natl Acad Sci* 86: 6592–6.

Regan, L., and DeGrado, W. F. 1988. Characterization of a helical protein designed from first principles. *Science* 241: 976–8.

Richardson, J. S., and Richardson, D. C. 1987. Some design principles: Betabellin. In *Protein Engineering*; Oxender, D. L. and Fox, C. F. Eds.; Alan R. Liss: New York, pp 149–63.

Richardson, J. S., and Richardson, D. C. 1988. Amino acid preferences for specific locations at the ends of alpha helices. *Science* 240: 1648–52.

Rojas, N. R. L., Kamtekar, S., Simons, C. T., McLean, J. E., Vogel, K. M., Spiro, T. G., Farid, R. S., and Hecht, M. H. 1997. De novo heme proteins from designed combinatorial libraries. *Protein Sci* 6: 2512–24.

Roy, S., Ratnaswamy, G., Boice, J. A., Fairman, R., McLendon, G. and Hecht, M. H. 1997a. A protein designed by binary patterning of polar and nonpolar amino acids displays native-like properties. *J Am Chem Soc* 119: 5302–06.

Roy, S., Helmer, K. J., and Hecht, M. H. 1997b. Detecting native-like properties in combinatorial libraries of de novo proteins. *Folding & Design* 2: 89–92.

Roy, S., and Hecht, M. H. 2000. Cooperative thermal denaturation of proteins designed by binary patterning of polar and nonpolar amino acids. *Biochemistry* 39: 4603–07.

Sasaki, T., and Kaiser, E. T. 1989. Helichrome—synthesis and enzymatic activity of a designed hemeprotein. *J Amer Chem Soc* 111: 380–1.

Sasaki, T., and Kaiser, E. T. 1990. Synthesis and structural stability of helichrome as an artificial hemeproteins. *Biopolymers* 29: 79–88.

Shoemaker, K. R., Kim, P. S., York, E. J., Stewart, J. M., and Baldwin, R. L. 1987. Tests of the helix dipole midel for stabilization of α-helices. *Nature* 326: 563–567. Sibanda, B. L., and Thornton, J. M. 1985. β-hairpin families in globular proteins. *Nature* 316: 170–74.

Struthers, M. D., Cheng, R. P., and Imperiali, B. 1996. Design of a monomeric 23-residue polypeptide with defined tertiary structure. *Science* 271: 342–5.

Wang, W., and Hecht, M. H. 2002. Rationally designed mutations convert de novo amyloid-like fibrils into monomeric β-sheet proteins. *Proc Natl Acad Sci* 99: 2760–65.

Wei, Y., Liu, T., Sazinsky, S. L., Moffet, D. A., Pelczer, I., and Hecht, M. H. 2003a. Stably folded de novo proteins from a designed combinatorial library. *Protein Science* 12: 92–112.

Wei, Y., Kim, S., Fela, D., Baum, J., and Hecht, M. H. 2003b. Solution structure of a de novo protein from a designed combinatorial library. *Proc Natl Acad Sci* 100: 13270–73.

Wei, Y., and Hecht, M. H. 2004. Enzyme-like proteins form an unselected library of designed amino acid sequences. *Protein Engineering, Design & Selection* 17: 67–75.

West, M. W., and Hecht, M. H. 1995. Binary patterning of polar and nonpolar amino acids in the sequences and structures of native proteins. *Protein Sci* 4: 2032–9.

West, M. W., Wang, W., Patterson, J., Mancias, J. D., Beasley, J. R., and Hecht, M. H. 1999. De novo amyloid proteins from designed combinatorial libraries. *Proc Natl Acad Sci* 96: 11211–6.

Yin, H., Slusky, J.S., Berger, B.W., Walters, R.S., Vilaire, G., Litvinov, R.I., Lear, J.D., Caputo, G.A., Bennett, J.S., and DeGrado, W.F. 2007. Computational design of peptides that target transmembrane helices. *Science* 315: 1817–22.

# 12 Molecular Force Fields

*Patrice Koehl*

## CONTENTS

*C'est à ce niveau d'organisation chimique que
gît, s'il y en a un, le secret de la vie*

—Jacques Monod, 1973

Proteins, the end products of the information contained in the genome of any organism, are the active elements of life whose chemical activities regulate all cellular activities. According to Monod, it is in the protein that lies the secret of life. As a consequence, studies of protein sequences, structures, and functions occupy a central role in molecular biology. Genomics projects aim at mapping the genetic information of as many organisms as possible, which contributed to an exponential growth in our knowledge of the protein sequence space (more than 6.6 million protein sequences were available as of July 2008). In parallel, structural genomics projects (Sali 1998; Baker and Sali 2001) aim at covering the corresponding protein structure space, yielding another exponential growth in the number of protein structures derived from experimental data; currently, there are more than 48,000 protein structures in the Protein Data Bank (PDB; Berman et al. 2000). Finally, a major effort is put in assigning a function to all these proteins (the functional genomics projects). The key to the success of all these initiatives lies in unraveling the connections between the sequence, structure, and function of a protein. Experimental data at a molecular level are scarce, however; this has led to the development of many modeling initiatives to shed light on these connections. Probably the most famous is the study of the protein-folding problem—the "holy grail" for the structural biology community. Its elusive goal is to predict the detailed three-dimensional structure of a protein from its sequence. Though substantial experimental and theoretical progresses have been made, the "grail" is still out of reach (Koehl and Levitt 1999). An alternative route to this goal is to reformulate the quest as searching for protein sequences that fold into a given stable conformation. This is the inverse folding problem (Drexler 1981; Pabo 1983), which itself attracts considerable interest since it is fundamental for protein engineering and design.

All modeling initiatives require a description of the thermodynamic equilibrium properties of the protein under study, which are usually derived from a sampling of its free energy surface. The "state" of a protein structure usually corresponds to a point or patch on this surface. The native state corresponds to a large basin in this landscape; protein folding studies are mostly interested in the structure of this basin, while computational protein design studies focus on how this basin changes as the sequence of the protein is changed. In theory, the laws of quantum mechanics fully define the energetics of a molecule; in practice, however, only the simplest system such as the hydrogen atom can be solved exactly, and modelers of large molecular systems such as proteins must rely on approximations. While some simulations remain anchored in quantum mechanics (Gogonea et al. 2001; Raha et al. 2007), most protein modeling studies rely on a space-filling representation of the molecule, in which atoms are represented as hard spheres that interact through empirical or semiempirical "molecular force fields." A wide range of strategies for parameterizing these force fields exists, ranging from physically based methods to methods based on the statistical analysis of known protein structures. In this chapter, we present an overview of these strategies, focusing on the specifics of computational protein sequence design.

The chapter is organized as follows. The next section covers the physics-based empirical force fields for protein sequence design. It starts with a definition of the stability of a protein sequence, which is followed by descriptions of the internal energy of a protein and its interactions with its environment. We distinguish force fields that account for the solvent explicitly from those that use an implicit representation. Next, we discuss the force fields based on statistics and their applicability to the problem of protein sequence design. We conclude the chapter with a discussion of future research directions.

## PHYSICS-BASED FORCE FIELDS

One of the most important elements of any computational protein sequence design strategy is an ability to measure the "compatibility" of a sequence for a given target protein structure conformation, which combines stability and specificity.

Compatibility is usually measured as the free energy difference between the target structure and an unfolded conformation. We start this section by defining the free energy of a protein, and then show how it applies to sequence design. The rest of the section discusses various methods of computing the free energy.

## STABILITY OF PROTEIN SEQUENCES: APPLICATION TO SEQUENCE DESIGN

The stability of a protein sequence $P$ is a thermodynamics quantity. It is measured as the difference $\Delta G(P)$ in free energy between its native state, $N$, and an unfolded state, $U$:

$$\Delta G_{U \to N}(P) = G_N(P) - G_U(P) \tag{12.1}$$

Note that $P$ here refers to the "solvated" protein, that is, accounts for the protein and its surrounding solvent and ionic atmosphere. $G$ refers to the Gibbs free energy of the system, defined as

$$G = U - TS + pV \tag{12.2}$$

The three terms are the internal energy $U$ of the protein, its entropy $S$, which is a measure of disorder, times the temperature of the system $T$, and the product $pV$, where $p$ and $V$ are the pressure and volume of the system, respectively. Since the pressure and the volume can be considered constant for proteins in solution, Equation 12.1 can be rewritten as

$$\Delta G_{U \to N}(P) = \Delta U_{U \to N}(P) - T \Delta S_{U \to N}(P) \tag{12.3}$$

A native-like, "stable" sequence has a minimum (negative) $\Delta G(P)$: ideally this is reached when $\Delta U(P)$ is minimum and $\Delta S(P)$ is maximum. However, $\Delta U(P)$ is minimal when the native state has many stabilizing, nonlocal contacts, which requires an organized structure, whereas the entropy is maximal when the structure has a low level of organization. Stability therefore is reached through a compromise between these two effects. To estimate the free energy of a protein, we need to compute its internal energy $U$, and sample its conformational space to measure the entropy $S$. This sampling is usually performed through simulations.

A typical computational protein sequence design experiment starts from a known protein structure template $N$ and tests the "compatibility" of many sequences for this template, searching for sequences that are both stable (positive design) and specific (negative design) to the structure $N$. Two putative sequences $P0$ and $P1$ for $N$ are compared based on their stability, as defined by Equation 12.1 (Figure 12.1):

$$\Delta \Delta G_{U \to N}(P0 \to P1) = \Delta G_{U \to N}(P1) - \Delta G_{U \to N}(P0) \tag{12.4}$$

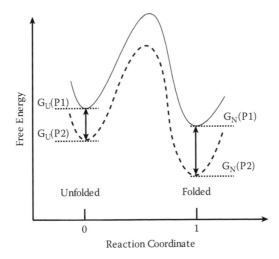

**FIGURE 12.1** Schematic representation of the thermodynamics involved in computational protein design. We assume a first-order folding for a protein sequence $P1$. The stability $\Delta G(P1)$ of $P1$ in the native conformation $N$ is the difference in free energy between $N$ and the unfolded state: $\Delta G(P1) = G_N(P1) - G_U(P1)$. When a new sequence $P2$ is tested against the target structure $N$, we need to take into account the changes in sequence at both ends of the folding process, that is, for the folded and the unfolded state.

which can be rewritten as

$$\Delta\Delta G_{U \to N}(P0 \to P1) = \left(G_N(P1) - G_N(P0)\right) - \left(G_U(P1) - G_U(P0)\right) \quad (12.5)$$

Although the quantity defined by Equation 12.5 cannot be measured experimentally, it can be computed during computational protein design.

## THE DENATURED STATE

Equation 12.5 requires computing the free energy of a sequence in the "unfolded" state, which is usually taken to be the fully extended structure. The denatured "state" of a protein is known to be a distribution of different molecular conformations [for review, see (Dill and Shortle 1991)]. Though distant residue contacts do exist in the denatured states, their free energies are dominated by local interactions within the protein and by interactions with the solvent. Therefore, the free energy of a denatured protein depends mostly on the amino-acid composition of the sequence rather than the sequence itself. In the case of a canonical sequence optimization with fixed amino-acid composition (Shakhnovich and Gutin 1993; Koehl and Levitt 1999), all sequences thus have the same free energy in the unfolded state, and Equation 12.5 can be rewritten as

$$\Delta\Delta G_{U \to N}(P0 \to P1) \approx G_N(P1) - G_N(P0) \quad (12.6)$$

for all test sequences $P0$ and $P1$. Therefore, the denatured states have little influence in the design process as long as the sequence composition is kept constant. Macrocanonical optimization (i.e., optimization with unconstrained amino-acid composition) requires more care [see (Sun et al. 1995; Vendruscolo et al. 1997)].

### COMPUTING THE FREE ENERGY WITH EXPLICIT WATER REPRESENTATION

A system with explicit solvent representation includes a protein of interest, water molecules that surround it, and the ions in the solvent (or any ligands of the protein). To compute its free energy, we need to estimate its internal energy, $U$, and its entropy, $S$.

### Internal Energy

A typical, semiempirical energy function used in classical molecular simulation has the form:

$$U = \frac{1}{2}\sum_b K_b (r_b - r_{b0})^2 + \frac{1}{2}\sum_\theta K_\theta (\theta - \theta_0)^2 + \sum_o \sum_n K_n \left(1 + \cos\left(n\phi - \delta_n\right)\right)$$

$$+ \sum_{i<j} \left(\frac{A_{ij}}{r_{ij}^{12}} - \frac{B_{ij}}{r_{ij}^6}\right) + \sum_{i<j} \frac{q_i q_j}{\varepsilon_0 r_{ij}}$$

$$(12.7)$$

This energy form implies all interactions are additive, since it includes only pairwise interactions involving only two atoms, and neglects all high-order interactions. The first three sums on the right side of Equation 12.7 represent bonded interactions (covalent bonds, valence angles, and torsions around bonds—see Figure 12.2), while the last two terms represent nonbonded interactions.

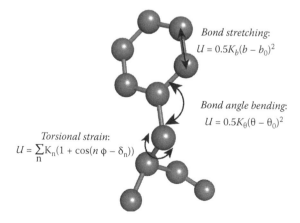

**FIGURE 12.2** The three types of bonded interactions in empirical force fields.

*Bonded Interactions*

*Bond stretching.* When a bond is stretched or compressed, its energy goes up. For small deviations from equilibrium, the functional form for this energy is approximated by Hooke's law (i.e., the energy is a quadratic function of the change in bond length). The reference bond length $b_0$ is often called the *equilibrium* bond length. Although the *equilibrium* bond length is the bond length for the minimum energy configuration of the molecule, the bond length with all other force field terms set to zero is often used as reference. The forces between bonded atoms are very high in comparison to other forces, which justifies the use of a harmonic form of the potential. For sequence design, this approximation is realistic, as bond lengths are not expected to change significantly.

*Bond angle bending.* In the ground state, the angle between two bonds attached to the same central atom assumes an equilibrium value. While this energy increases when the angle deviates from the ground state, there is no consensus on its functional form: Some force fields use a direct harmonic term as in Equation 12.7, while others use a cosine harmonic potential of the form

$$E_\theta = K_\theta \left( \cos(\theta) - \cos(\theta_0) \right)^2 \tag{12.8}$$

The difference between the two functional forms may be significant if the bond angles vary significantly; it is expected to have negligible effects on sequence design.

*Torsional stress.* Intramolecular rotations, that is, rotations about torsion angles, require energy. For example, it takes energy to flip a proline ring from an up-puckered conformation to a down-puckered conformation. This energy is different from the energies observed for bond stretching and bond angle bending: First, the energy barriers are low, and changes in torsion angles can be large; second, the energy is periodic with a period of 360°. Torsional energies are usually expressed as a Fourier series:

$$E(\phi) = \sum_n K_n \left( 1 + \cos\left( n\phi - \delta_n \right) \right) \tag{12.9}$$

$K_n$ is the torsional rotation constant; the phase angles $\delta_n$ are usually chosen so that terms with positive $K_n$ have minima at $\phi = 180°$ (i.e., $\delta_n = 0°$ for $n$ odd and 180° for $n$ even). $n$ is the multiplicity, which reflects the type of symmetry in the torsion angle. A CH3-CH3 bond, for example, should repeat its energy every 120°.

It is worth noting that the bonded interactions are local, and, to a first approximation, depend only on the amino-acid composition of the sequence of the protein of interest. In the framework of the canonical optimization with fixed amino-acid composition, bonded interactions do not change when the sequence is modified, and therefore can be ignored in the expression of U. This is not true for macrocanonical sequence design.

## Nonbonded Interactions

By definition the nonbonded interactions act between atoms that are not linked by chemical bonds. In practice, all pairs of atoms connected by one bond (1–2 interactions) and two bonds (1–3 interactions) are also discarded and accounted for only by bonded interactions (see previous discussion). In addition, interactions between atoms connected by three bonds (1–4 interactions) are often scaled down. There are two types of nonbonded interactions that make significant contributions: electrostatics and van der Waals (vdW).

*Electrostatics interactions.* As a first "classical" approximation (in opposition to quantum mechanical calculations), the interaction between two charges $q_i$ and $q_j$ is governed by Coulomb's law:

$$U_{elec} = \frac{q_i q_j}{4\pi\varepsilon_0\varepsilon_r r_{ij}} \tag{12.10}$$

where $\varepsilon_0$ is the dielectric permittivity of free space, $\varepsilon_r$ is the relative dielectric permittivity of the medium in which the two charges are placed, and $r_{ij}$ is the separation between the two charges.

When more than two charges interact, the total electrostatic energy of the system is derived as a sum of its pairwise Coulomb interactions (the superposition principle).

*vdW interactions*: The van der Waals interactions cover all interactions not accounted for by bonded and electrostatic interactions, including induced dipole-dipole interactions, London dispersion forces, and repulsion forces that prevent electrons from nonbonded atoms to occupy the same space. The attractive part of vdW interactions has a $1/r^6$ dependence and is dominated by dipole-dipole interactions. The repulsive part of vdW interactions is usually modeled with a $1/r^{12}$ dependence, although there are no physical reasons that the exponent should be 12. Together, they lead to the classical Lennard-Jones potential for vdW interactions:

$$U_{vdW} = \frac{A_{ij}}{r_{ij}^6} - \frac{B_{ij}}{r_{ij}^{12}} \tag{12.11}$$

where $A_{ij}$ and $B_{ij}$ are parameters that depend on the nature of atoms $i$ and $j$.

## Most Common Force Fields

The different force fields currently available differ in how they set the parameters of the different terms included in Equation 12.7; some of them also have slightly different functionals than the ones presented above (such as the cosine harmonic term for angle bending), or even extra terms (the CHARMM force field, for example, includes a Urey-Bradley potential to constrain bond angles, in the form of a virtual bond between the first and last atom defining the angle that is maintained with an harmonic potential). Table 12.1 lists the five most common of these force fields, providing Web addresses for accessing them.

## Entropy

The total entropy of a molecule in solution can be separated into two parts: a solvent entropy associated with solvent motions, and a solute, or configurational, entropy associated with protein motions. Computing these entropy terms requires in principle complete sampling of the conformational space accessible by the protein and the solvent molecules. This is usually achieved through molecular dynamics simulation (Meirovitch 2007).

### COMPUTING THE FREE ENERGY WITH IMPLICIT SOLVENT REPRESENTATION

Implicit solvent models reduce the protein-solvent interactions to their mean-field characteristics, which are expressed as a function of the protein degrees of freedom alone. They represent the solvent as a dielectric continuum that mimics the solvent-solute interactions, including their apolar components (vdW contacts and the entropic effects of creating a cavity in the solvent) and their polar components (mostly the screening of the electrostatics interactions). We review all three concepts, that is, potential of mean forces, and apolar and polar interactions for implicit models.

### Potential of Mean Forces

A protein in solution adopts a conformation $C$ with the probability $P$:

$$P(C,S) = \frac{e^{-\frac{U(C,S)}{kT}}}{\int \int e^{-\frac{U(X,Y)}{kT}} dXdY} \tag{12.12}$$

In this equation, $S$ represents the conformation of the water molecules. The double integral in the denominator defines the partition function, summed over all possible conformations $X$ and $Y$ of the protein and the solvent, respectively. $U$ is the internal energy of the system, which depends on both $C$ and $S$. We are interested in the

## TABLE 12.1
### The Five Most Common Force Fields and How to Access Them

| Force Field | Reference | Web Site |
|---|---|---|
| Amber | (Ponder and Case 2003) | http://amber.scripps.edu |
| CHARMM | (Brooks et al. 1983) | http://www.charmm.org |
| GROMOS | (Christen et al. 2005) | http://www.igc.ethz.ch/GROMOS/ |
| OPLS | (Jorgensen et al. 1996) | http://zarbi.chem.yale.edu/software.html |
| ECEPP/2 | (Momeny et al. 1975) | http://cbsu.tc.cornell.edu/software/eceppak/ |

protein molecule, not the water. Using a mean field approach, therefore, we derive the probability $P_P(C)$ of the conformation $C$ of the protein alone by averaging over all possible conformations of the solvent:

$$P_P(C) = \int P(C,Y)dY \qquad (12.13)$$

$P_P(C)$ can be expressed mathematically as

$$P_P(C) = \frac{e^{-\frac{W_T(C)}{kT}}}{\int e^{-\frac{W_T(X)}{kT}} dX} \qquad (12.14)$$

$W_T(C)$ is called the potential of mean forces for the conformation $C$ of the protein. It can be expressed as

$$W_T(C) = U(C) + W_{sol}(C) \qquad (12.15)$$

$U(C)$ is the internal energy of the protein with conformation $C$, which can be computed using Equation 12.7 and any of the force fields described previously. $W_{sol}(C)$ is the solvation free energy to account for the effects of the solvent on $C$. Implicit solvent models are designed to provide an accurate estimate of $W_{sol}(C)$.

For a protein with conformation $C$, $W_{sol}(C)$ accounts for three major effects: formation of a cavity in the solvent to accommodate the protein, vdW interactions between the protein and the water molecules, and changes in the electrostatics interactions between the charged atoms of the protein due to the presence of water. The first two correspond to the nonpolar or hydrophobic effect, $W_{np}$, while the last is the polar contribution of water, $W_{pol}$. These contributions are computed separately and then combined (Figure 12.3).

### $W_{np}$: The Nonpolar Contribution to $W_{sol}$

*The surface area model.* Eisenberg and McLachlan (Eisenberg and McLachlan 1986) computed the nonpolar part of the free energy of solvation as the sum of the contributions from all atoms of a protein. The individual contributions are computed as the accessible surface area (*ASA*; Richards 1977) of the atom multiplied by a surface tensor factor referred to as *atomic solvation parameter*, or *ASP*:

$$W_{np}(C) = \sum_i ASP_i * ASA_i \qquad (12.16)$$

*ASP* is positive for nonpolar atoms and negative for polar atoms.

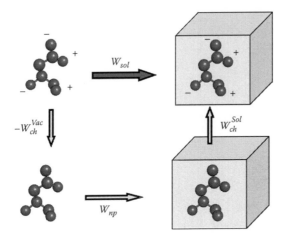

**FIGURE 12.3** The solvation free energy. In implicit solvent models, the solvation free energy $W_{sol}$ is a mean force potential that measures the energy required to solvate a molecule. There are two distinct contributions to $W_{sol}$: a polar contribution from the screening of the charges of the solute by the water molecule plus the reaction of the solute to the polarization of water; and a nonpolar contribution that accounts for the formation of a cavity in water to accommodate the molecule as well as the vdW interactions between the solute and the solvent. These two contributions can be clearly identified if we follow a thermodynamics cycle. First, we remove the charges on the protein *in vacuo*, then we solvate the neutral molecule, and finally we add the charges back to the protein in solution. This cycle leads to the following expression for $W_{sol} = W_{np} + W_{ch}^{sol} - W_{ch}^{Vac}$, where $W_{ch}^{sol}$ and $W_{ch}^{Vac}$ are the free energies required to charge the molecule in the solvent and *in vacuo*, respectively. $W_{np}$ is computed using a surface area model, $W_{ch}^{Vac}$ is computed using a Coulomb potential, and $W_{ch}^{sol}$ is computed using a modified Coulomb potential, the Poisson-Boltzmann equation, or the generalized Born potential (see text).

This model, referred to as SA (for surface area), is supported indirectly by the observed linearity between the Gibbs free energy and the surface area for transferring small compounds from nonaqueous liquids to water. Similarly, the free energy of transfer correlates with the sum of the transfer free energies of the constituent atomic groups. SA has become the method of choice for computing the hydrophobic effects on proteins.

It is interesting to recall that $W_{np}$ accounts for cavity formation in water as well as the vdW interactions between the protein and the solvent molecules. The latter occurs within the first hydration shell around the protein, and therefore is expected to be proportional to the accessible surface area of the protein. Cavity formation, however, is proportional to the volume of the protein. This apparent contradiction between a surface area model and a volume model is part of the debate on the geometric nature of the nonpolar solvation energy. Lum, Chandler, and Weeks have unified these two models by showing that $W_{np}$ scales with the volume of the solute for small solutes, and is proportional to the surface area for large solutes (Lum et al. 1999). Their theory of hydrophobicity adds to the validation of the surface area model for proteins.

*Variant to the SA model.* Equation 12.16 does not take into account interactions within the protein. In the SA formalism, the surface area of polar atoms buried upon folding yields an unfavorable contribution to $W_{np}$, *irrespective* of the environment of the polar atoms in the core of the folded protein. However, polar atoms may interact with other polar atoms in the core of the proteins, for example, by forming hydrogen bonds and secondary structures, which is different from burying isolated polar atoms. Similarly, atomic surface areas of nonpolar atoms buried upon folding always contribute favorably to $W_{np}$, even when they contact polar atoms, which seems counterintuitive. Koehl and Delarue introduced an extension to the SA model by defining a free energy of environment, $W_{env}$, which takes into account the full environment of each atom of the protein, including the solvent and other protein atoms (Koehl and Delarue 1994):

$$W_{env} = \sum_i A_i * (ASA_i + PCA_i) + B_i * NPCA_i \qquad (12.17)$$

The summation extends over all atoms $i$ of the protein; $ASA_i$ is the accessible surface area of $i$; $PCA_i$ and $NPCA_i$ are the surface areas of $i$ occluded by polar and nonpolar atoms, respectively (also defined as the polar and nonpolar contact areas of $i$). $A_i$ and $B_i$ are surface tension factors for polar and nonpolar interactions, similar to the $ASP$ of Eisenberg and McLachlan (Eisenberg and McLachlan 1986). The environment free energy has been used with success for computational protein sequence design (Koehl and Levitt 1999; Koehl and Levitt 1999).

*Computing the nonpolar solvation energy.* All internal energy terms discussed in the previous sections involve an atomic pairwise potential that can be readily computed from the Cartesian coordinates of the atoms. The nonpolar solvation free energy is the first exception to this rule, since it requires the calculation of surface area for each atom of the molecule. Note, however, that there are several potential definitions of surface area. Figure 12.4 illustrates three common definitions of surface areas, of which the accessible surface area remains the most widely used.

The original approach of Lee and Richards computed the accessible surface area by first cutting the molecule with a set of parallel planes (Lee and Richards 1971). The intersection of a plane with an atom is a circle that can be partitioned into accessible arcs on the boundary and occluded arcs in the interior. The accessible surface area of an atom is then the sum of the contribution of all its accessible arcs. This method was originally implemented in the program ACCESS, and later in NACCESS (http://wolf.bms.umist.ac.uk/naccess/). Shrake and Rupley proposed an alternative approach based on numerical integration of the surface area using a Monte Carlo method (Shrake and Rupley 1973). Implementations of their method include applications of lookup tables (Legrand and Merz 1993), vectorized algorithms (Wang and Levinthal 1991), and parallel algorithms (Futamura et al. 2004).

The surface area computed by numerical integration is not accurate. To improve the accuracy of numerical methods, analytical approximations to the accessible surface area were developed by treating multiple overlaps probabilistically (Wodak and Janin 1980; Cavallo et al. 2003) or ignoring them altogether (Street and Mayo 1998).

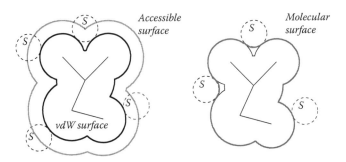

**FIGURE 12.4** Different definitions of protein surface. *The vdW surface* of a molecule is the surface of the union of balls representing the atoms, with radii set at the vdW radii. *The accessible surface* is the surface generated by the center of a probe sphere (marked *S*) rolling on the vdW surface. The radius of the probe sphere is usually set to 1.4 Å, the approximate radius of a water molecule. *The molecular surface* is the envelope generated by the rolling sphere. It differs from the vdW surface as it covers a portion of the volume around the vdW surface that is inaccessible to the rolling sphere.

Better analytical methods describe the molecule as a geometric union of spheres and analytically compute the surface area (Connolly 1983; Richmond 1984; Dodd and Theodorou 1991; Irisa 1996). Yet another approach (Gibson and Scheraga 1987) uses the inclusion-exclusion formula (Gibson and Scheraga 1987) and applies a theorem, which states that overlaps of order 5 and above can always be reduced to overlaps of order 4 or below (Kratky 1978), but doing the reduction correctly and efficiently is a difficult task. An exact solution was later obtained by using the alpha shape theory of Edelsbrunner (Edelsbrunner 1995).

## $W_{pol}$: The Polar Contribution to $W_{sol}$

Being a polar medium, water responds to the presence of the charges of a solutes, and in return this polarization of water modifies the interactions between charges. The polar solvation energy, $W_{pol}$, is designed to mimic this effect for implicit solvent models. I will briefly review three main approaches used to compute $W_{pol}$: a modified Coulomb approach, the Poisson-Boltzmann model, and the generalized Born model.

### A Modified Coulomb Formula for Implicit Solvent Models

The simplest way to evaluate the electrostatics energy in a continuum solvent is to introduce a dielectric constant for the medium, $\varepsilon_r$ ($\approx 80$ for water), in Coulomb's law, with the net effect of shielding the interacting charges. This expression is valid for point charges in the solvent but is not applicable for large solutes with a complex interface with the solvent. One improvement to the approach is to use a "distance dependent" dielectric, in which $\varepsilon_r$ is set proportional to the intercharge distances. This model is computationally efficient but is purely empirical and physically unsatisfying. Also, the model is insensitive to the actual environment of the charges. The strategy of modifying the dielectric constant was recently applied to an electrostatic model of hydrogen bond, in which hydrogen bonds are treated using a position-dependent dielectric, while all other electrostatic interactions are treated using a region- and distance-dependent dielectric (Wisz and Hellinga 2003). In this model, proteins are divided into three regions (core, boundary, and surface), and the dielectric parameters in each region are refined against a set of experimental pKa values of the amino acids in proteins. Schwarzl and coworkers (Schwarzl et al. 2005) further generalized the concept by introducing different dielectric constants for every charge pair in the molecule. Yet, its simplicity and efficiency makes the distance-dependent dielectric model the most commonly used approximation in computational protein sequence design.

### The Poisson Boltzmann Approach

Assuming that the solvent can be accurately represented as a dielectric continuum, the electrostatics potential $\varphi$ for a charge density $\rho$ is given by Poisson's law:

$$\nabla \bullet \varepsilon(x) \nabla \varphi(x) = -4\pi\rho(x)$$

(12.18)

where $\varepsilon$ is the position-dependent dielectric constant. The charge density is the sum of the charges in a protein (fixed) and of the ions in the solvent (mobile). According to the Debye-Huckel theory, the charge density is given by

$$\rho(x) = \sum_{i=1}^{Ncharges} q_i \delta(x - r_i) + F \sum z_i c_i^{\infty} \exp\left(-\frac{z_i F \varphi(x)}{RT}\right)$$

(12.19)

where $q_i$ and $r_i$ are the charges and their positions in the solute, $F$ is the Faraday constant, $c_i^{\infty}$ and $z_i$ are the bulk concentration and valence of ion species $i$, respectively, and $\delta(x - y) = 1$ if $x = y$, 0 otherwise. Substituting Equation 12.19 in Equation 12.18 yields the Poisson-Boltzmann (PB) equation. To compute the polar solvation energy for the protein, we solve the Poisson-Boltzmann twice, once in the presence of the solvent, once *in vacuo*, to derive $\varphi(x)$ and $\varphi_{vac}(x)$, respectively. The value of $W_{pol}$ is obtained by

$$W_{pol} = \frac{1}{2} \sum_{i=1}^{Ncharges} q_i \left( \varphi(r_i) - \varphi^{vac}(r_i) \right)$$

(12.20)

Methods for solving the Poisson-Boltzmann equation have been recently reviewed in a comparison paper by Feig and coauthors (Feig et al. 2004) and by Baker (Baker 2005). The Poisson-Boltzmann equation can be solved numerically using the finite difference (FD) methods, finite element (FE) methods, and boundary element methods (BEM), of which the FD methods are most commonly used due to computational efficiency. The FD method has also enabled incorporation of PB during molecular dynamics simulations (David et al. 2000; Prabhu et al. 2004). The success of FD methods has spurred interest for continuum model for electrostatics calculations in computational biology in general, and recently in computational protein sequence design. Marshall and coworkers (Marshall et al. 2005) have developed approximate formulations of FD Poisson-Boltzmann that are pairwise decomposable by side chains in order to apply it to computational protein sequence design. While their approach has not yet been tested in an actual protein design experiment, it is very promising and is expected to play an important role in the design of proteins for which stability and function are directly related to electrostatics properties.

### The Generalized Born Model

Poisson theory provides a rigorous framework for calculating electrostatic interactions in a continuum model, but its application remains limited, as solving the PB equation is computationally demanding. This has led to the development of approximate continuum solvation models, among which the generalized Born (GB) formalism is the most promising alternatives to PB solver (Bashford and Case 2000; Feig et al. 2004). GB theories compute the solvent-induced reaction field energy as a pairwise sum over the interactive charges, $q_i$ (Still et al. 1990):

$$W_{pol} = \frac{1}{8\pi}\left(\frac{1}{\varepsilon_o} - \frac{1}{\varepsilon_w}\right)\sum_{i,j}\frac{q_i q_j}{\sqrt{r_{ij}^2 + b_i b_j \exp\left(-r_{ij}^2/\left(Cb_i b_j\right)\right)}} \tag{12.21}$$

where $\varepsilon_0$ and $\varepsilon_\omega$ are the interior and solvent dielectric constants, respectively, $r_{ij}$ is the distance between atoms $i$ and $j$, and $b_i$ is the so-called Born radius of atom $i$. The factor $C$ may range from 2 to 10, with 4 being the most common value. Equation 12.21 is exact for both extreme values of the interatomic distance: When $r_{ij} = 0$, Equation 12.21 gives the Born energy of a solvated charge, while Equation 12.21 approximates to Coulomb's law for large $r_{ij}$ (Figure 12.5). For intermediate $r_{ij}$ values, Equation 12.7 is purely empirical. Atomic Born radius reflects the distance of an atomic charge to the boundary between the solute and the continuum solvent. Theoretically, they can be calculated by solving the Poisson equation, but in practice this is never done and many approximation schemes are available to compute these radii efficiently with reasonable accuracy. I will not review these schemes here and refer the readers to excellent recent papers on this subject (Bashford and Case 2000; Feig et al. 2004).

## STATISTICAL POTENTIALS

An alternative approach to deriving potentials from physics and chemistry is to rely on the observation of known protein structures. The rules derived from existing structures can be used to infer properties of unknown proteins or to design proteins with new functions. The interest in a statistical approach to understanding protein stability is growing nearly at the same speed as the databanks of protein structures. Several methods exist to extract the statistical (or knowledge-based) potentials from databanks (for review, see Sippl 1993; Poole and Ranganathan 2006). For example, knowledge-based potentials can be optimized by fitting to known protein sequence and structure data. Maiorov and Crippen (Maiorov and Crippen 1994) defined energy parameters that solve the threading problem; that is, the native structure of a sequence is clearly recognized from among a large

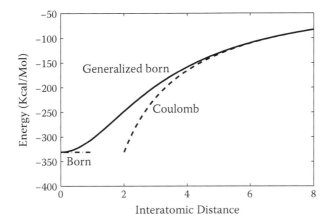

**FIGURE 12.5** The generalized Born energy between two equal charges. When the two charges coincide, the GB energy matches with the sum of the Born energies of both charges. When the distance between the two charges is large, the GB energy behaves like Coulomb. The functional for the GB energy is designed so that it goes smoothly (continuously) between these two extremes.

set of decoys. Goldstein and collaborators (Goldstein et al. 1992) developed an analytical method based on spin glass theory for determining the parameters that maximize the stability of the native structure relative to an average alternative structure, which they refer to as the *foldability* of the protein sequence. Note that the concept of foldability is related to the concept of specificity: A protein sequence will fold into a given target structure if it is stable for that structure and specific to that structure, that is, incompatible with any competing conformations. To account for both effects, we need to ensure that the energy of the sequence in the target conformation remains lower than in other competing conformations, which requires that all competing folds are known. Alternatively, Goldstein and coworkers introduced an "average" competing fold, and defined the difference in energy for a given sequence between its target or native structure and this competing fold as its foldability (Goldstein et al. 1992). This approach has been further extended to include other definitions of foldability, depending on how the competing folds are defined (Sasai 1995; Mirny and Shakhnovich 1996; Thomas and Dill 1996). Alternatively, putative energy terms can be derived from amino-acid pairing frequencies observed in known protein structures. This approach was initially proposed by Tanaka and Scheraga (Tanaka and Scheraga 1976) and subsequently extended by Miyazawa and Jernigan to account for solvent effects (Miyazawa and Jernigan 1985). Sippl and coworkers (Hendlich et al. 1990; Sippl 1990) noted that the conformation of small protein fragments is often defined by their flanking residues. To explain this observation, they introduced a residue-based energy function that depends on the separation of residues in both 3D space and along the protein sequence. Using this energy function, they modeled why certain protein fragments with identical sequences adopt different conformations in different proteins. These potentials are referred to as either log-odd potentials, if only statistical information is considered, or potentials of mean force, if a physical model based on statistical mechanics is used.

## THE POTENTIAL OF MEAN FORCE

For a state variable $X$ of a physical system in equilibrium, the probability that $X$ takes the value $x$ is given by the Boltzmann law:

$$P(X=x) = \frac{\exp\left[-\frac{E(x)}{k_B T}\right]}{Z} \tag{12.22}$$

where $E(x)$ is the energy of the system, $k_B$ is the Boltzmann's constant, $T$ the temperature, and $Z$ the partition function. If $x^*$ is the value of $X$ for which the interaction energy of the system is zero, then

$$P(X=x^*) = \frac{1}{Z} \tag{12.23}$$

and

$$P(X=x) = P(X=x^*)\exp\left[-\frac{E(x)}{k_B T}\right] \tag{12.24}$$

Conversely, we can derive the energy of the system described by $x$ by measuring the probability density functions $P$:

$$E(x) = -k_B T \ln\left[\frac{P(X=x)}{P(X=x^*)}\right] \tag{12.25}$$

$E(x)$ given in Equation 12.25 is referred to as the potential of mean force.

For proteins, potentials of mean force are mainly concerned with pairwise interaction energies between any two amino-acid types $a$ and $b$, and one is mostly interested in knowing the probability of observing two amino acids $a$ and $b$ at distance $r$, $P_{ab}(X = r)$. In the limit of a dilute idealized gas phase in which pairwise interactions dominate, Equation 12.25 can be written as

$$E_{ab}(r) = -k_B T \ln\left[\frac{P_{ab}(X=r)}{P_{ab}(X=+\infty)}\right] \tag{12.26}$$

since the interaction energy between $a$ and $b$ is zero at infinite separation only.

## THE REFERENCE STATE

As a first approximation, the marginal probabilities $Pab$ are derived from the discrete frequencies observed in known protein structures. The definition of the "reference state" or "zero-energy state" is not straightforward, since two residues are always within a finite distance of each other. This is sometimes referred to as the *partition function problem* (in reference to Equation 12.23). Miyazawa and Jernigan (Miyazawa and Jernigan 1985) developed a solution to this problem by considering only local interactions; that is, the probability to observe an amino-acid pair is integrated for $r < r_c$, where the cut-off value $rc \sim 6.5$ Å. In their reference state, pairwise contacts depend only on the concentration of the amino acids. Later potentials included corrections for amino-acid types (e.g., Ala-Gly versus Met-Trp) and for the distributions of distances observed (i.e., number of pairs observed at 10 Å versus number of pairs observed at 50 Å) (Sippl 1990). By normalizing the probability for amino-acid types, we get

$$P_{ab}\left(X=r\right)=\frac{P\left(X=r/T=\left(a,b\right)\right)}{P\left(T=\left(a,b\right)\right)} \tag{12.27}$$

$$E_{ab}\left(r\right)=-k_BT\ln\left[\frac{P\left(X=r/T=\left(a,b\right)\right)}{P_{a,b}\left(X=+\infty\right)P\left(T=\left(a,b\right)\right)}\right] \tag{12.28}$$

and similarly for a nonspecific amino-acid type, noted $E(r)$. Finally, the distance-dependent net pair potential $\Delta Eab(r)$ is

$$\Delta E_{ab}\left(r\right)=E_{ab}\left(r\right)-E\left(r\right)=-k_BT\ln\left[\frac{P\left(X=r/T=\left(a,b\right)\right)}{P\left(X=r\right)P\left(T=\left(a,b\right)\right)}\right]-k_BT\ln\left[\frac{P\left(X=+\infty\right)}{P_{ab}\left(X=+\infty\right)}\right] \tag{12.29}$$

The second term of the right-hand side of Equation 12.29 is independent of the conformation, and depends only on the nature of $a$ and $b$, which is constant for a fixed sequence $S$, and can be ignored for fold recognition. Then,

$$\Delta E'_{ab}\left(r\right)=-k_BT\ln\left[\frac{P\left(X=r/T=\left(a,b\right)\right)}{P\left(X=r\right)P\left(T=\left(a,b\right)\right)}\right] \tag{12.30}$$

This is illustrated in Figure 12.6.

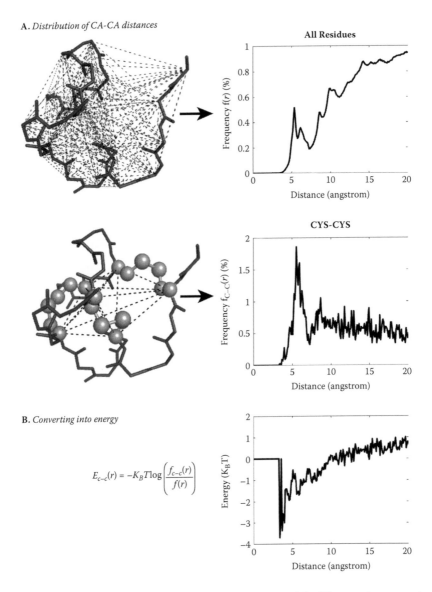

A. *Distribution of CA-CA distances*

**All Residues**

B. *Converting into energy*

$$E_{c-c}(r) = -K_B T \log\left(\frac{f_{c-c}(r)}{f(r)}\right)$$

**FIGURE 12.6** Distance-dependent knowledge-based potentials. We start by computing the distribution of all CA-CA distances in all proteins in the database of protein structures, excluding distances between neighboring residues; this distribution serves to define the reference state. The procedure is then repeated for all types of amino-acid pairs; here we show the example of the Cys-Cys pair. Most Cys-Cys pairs are observed at small distances; this is expected and corresponds to cysteines involved in disulphide bridges. The distributions are converted into energies using the Boltzmann law (see text). The energy for Cys-Cys pair is shown; it quantifies the effect just described, namely that there is a preference to observe Cys-Cys pairs at low distance separations.

The net potential energy $\Delta E'(S,C)$ of a protein of sequence $S$ that adopts a conformation $C$ is obtained by summing over the individual residue pair contributions from Equation 12.30. Rooman and Wodak have shown that $\Delta E'(S,C)$ directly approximates the difference between the free energy of $S$ in conformation $C$ and that of $S$ in a denatured-like state and can therefore be used for computational sequence design experiments even if the amino-acid composition is not fixed (Rooman and Wodak 1995).

## OTHER STATISTICAL POTENTIALS

Most knowledge-based potentials used for protein structure prediction or protein sequence design are based on Sippl's formalism and measure distributions of amino acids or atoms as a function of distance. New energy functions, however, have been introduced based on other statistical terms, such as residue triplets (Goldstein et al. 1992), residue quadruplets (Singh et al. 1991), solvent accessibility (Jones et al. 1992), atomic environment (Delarue and Koehl 1995; Eisenhaber 1996), and combinations of these terms (Park and Levitt 1996). Most of these potentials have been used for error recognition in protein models derived from x-ray data, NMR data, or modeling, for *ab initio* folding calculation, for threading or fold recognition problems, and for protein sequence design.

## CONCLUSIONS AND PERSPECTIVES

Protein structure and function are most often modeled using physical potentials. Many force-field variants exist that implement Equation 12.7 (see Table 12.1). They have been validated within the context of structure prediction studies or molecular dynamics simulations, but their accuracy in computational protein design, however, is more difficult to quantify. There is no unique solution to the inverse folding problem, as many sequences can stabilize a protein fold. It is not clear that the current force fields are accurate enough to allow us to study the whole sequence space compatible with a protein fold.

Protein sequence design is CPU intensive, which necessitates the development of a computationally tractable energy model. Balancing accuracy and speed is essential for computational sequence design. In principle, force fields such as those listed in Table 12.1 may be implemented as is for sequence design, but approximations are frequently used in practice, for example, by combining physics-based energy functions with statistical potentials. Other models involve modifications of the original force fields. For example, the vdW radii of atoms are often reduced to compensate for discrete rotamer states (Dahiyat and Mayo 1997), and empirical energy terms are introduced to account for hydrogen bonds (Dahiyat et al. 1997). Computational sequence design most often uses implicit solvent models, as explicit solvent models are intractable. Even implicit solvation models can be computationally costly, especially the polar contribution to solvation. Current implementations of continuum models for protein sequence design either use simple models, such as those involving a modified Coulomb equation or the generalized Born equation, or apply a simplified

Poisson-Boltzmann equation. Designing sequences for systems in which electrostatics plays a crucial role, such as designing ion channels, or highly charged active sites, will need to address the approximations in the current force fields.

A careful accounting of thermodynamics is key to successful protein sequence design. A well-designed sequence stabilizes the target structure and is incompatible with competing structures, including denatured conformation. The internal energy of the protein can be computed using Equation 12.7, whereas methods to compute the entropy of a system remains a challenge for future computational sequence designs.

## ACKNOWLEDGMENTS

PK wishes to thank the SLOAN foundation as well as the NIH for financial support.

## REFERENCES

Baker, D. and A. Sali. 2001. "Protein structure prediction and structural genomics." *Science* 294: 93–6.

Baker, N. A. 2005. "Improving implicit solvent simulations: A Poisson-centric view." *Curr Opin Struct Biol* 15: 137–43.

Bashford, D. and D. A. Case. 2000. "Generalized Born models of macromolecular solvation effects." *Annu Rev Phys Chem* 51: 129–52.

Berman, H. M., J. Westbrook, Z. Feng, G. Gilliland, T. N. Bhat, et al. 2000. "The Protein Data Bank." *Nucl Acids Res* 28: 235–42.

Brooks, B., R. Bruccoleri, B. Olafson, D. States, S. Swaminathan, et al. 1983. "CHARMM: A program for macromolecular energy minimization and dynamics calculations." *J Comput Chem* 4: 187–217.

Cavallo, L., J. Kleinjung and F. Fraternali. 2003. "POPS: A fast algorithm for solvent accessible surface areas at atomic and residue level." *Nucl Acids Res* 31: 3364–6.

Christen, M., P. H. Hunenberger, D. Bakowies, R. Baron, R. Burgi, et al. 2005. "The GROMOS software for biomolecular simulation: GROMOS05." *J Comput Chem* 26: 1719–51.

Connolly, M. 1983. "Analytical molecular surface calculation." *J Appl Cryst* 16: 548–58.

Dahiyat, B. I., D. B. Gordon and S. L. Mayo. 1997. "Automated design of the surface positions of protein helices." *Protein Sci* 6: 1333–7.

Dahiyat, B. I. and S. L. Mayo. 1997. "Probing the role of packing specificity in protein design." *Proc Natl Acad Sci U S A* 94: 10172–7.

David, L., R. Luo and M. Gilson. 2000. "Comparison of generalized Born and Poisson models: Energetics and dynamics of HIV protease." *J Comput Chem* 21: 295–309.

Delarue, M. and P. Koehl. 1995. "Atomic environment energies in proteins defined from statistics of accessible and contact surface areas." *J Mol Biol* 249: 675–90.

Dill, K. A. and D. Shortle. 1991. "Denatured states of proteins." *Annu Rev Biochem* 60: 795–825.

Dodd, L. and D. Theodorou. 1991. "Analytical treatment of the volume and surface area of molecules formed by an arbitrary collection of unequal spheres intersected by planes." *Mol Physics* 72: 1313–45.

Drexler, K. E. 1981. "Molecular engineering: An approach to the development of general capabilities for molecular manipulation." *Proc Natl Acad Sci U S A* 78: 5275–8.

Edelsbrunner, H. 1995. "The union of balls and its dual shape." *Discrete Comput Geom* 13: 415–40.

Eisenberg, D. and A. McLachlan. 1986. "Solvation energy in protein folding and binding." *Nature (London)* 319: 199–203.

Eisenhaber, F. 1996. "Hydrophobic regions on protein surfaces. Derivation of the solvation energy from their area distribution in crystallographic protein structures." *Protein Sci* 5: 1676–86.

Feig, M., A. Onufriev, M. S. Lee, W. Im, D. A. Case, et al. 2004. "Performance comparison of generalized Born and Poisson methods in the calculation of electrostatic solvation energies for protein structures." *J Comput Chem* 25: 265–84.

Futamura, N., S. Alura, D. Ranjan and B. Hariharan. 2004. "Efficient parallel algorithms for solvent accessible surface area of proteins." *IEEE Trans Parallel Dist Syst* 13: 544–55.

Gibson, K. and H. Scheraga. 1987. "Exact calculation of the volume and surface area of fused hard-sphere molecules with unequal atomic radii." *Mol Phys* 62: 1247–65.

Gogonea, V., D. Suarez, A. van der Vaart and K. M. Merz, Jr. 2001. "New developments in applying quantum mechanics to proteins." *Curr Opin Struct Biol* 11: 217–23.

Goldstein, R. A., Z. A. Luthey-Schulten and P. G. Wolynes. 1992. "Protein tertiary structure recognition using optimized hamiltonians with local interactions." *Proc Natl Acad Sci U S A* 89: 9029–33.

Hendlich, M., P. Lackner, S. Weitckus, H. Floeckner, R. Froschauer, et al. 1990. "Identification of native protein folds amongst a large number of incorrect models." *J Mol Biol* 216: 167–80.

Irisa, M. 1996. "An elegant algorithm of the analytical calculation for the volume of fused spheres with different radii." *Comp Phys Comm* 98: 317–38.

Jones, D. T., W. R. Taylor and J. M. Thornton. 1992. "A new approach to protein fold recognition." *Nature (London)* 358: 86–9.

Jorgensen, W. L., D. S. Maxwell and J. Tirado-Rives. 1996. "Development and testing of the OPLS all-atom force-field on conformational energetics and properties of organic liquids." *J Am Chem Soc* 118: 11225–36.

Koehl, P. and M. Delarue. 1994. "Polar and non-polar atomic environment in the protein core: Implications for folding and binding." *Proteins: Struct Funct Genet* 20: 264–78.

Koehl, P. and M. Levitt. 1999. "A brighter future for protein structure prediction." *Nature Struct Biol* 6: 108–11.

Koehl, P. and M. Levitt. 1999. "De novo protein design. I. In search of stability and specificity." *J Mol Biol* 293: 1161–81.

Koehl, P. and M. Levitt. 1999. "De novo protein design. II. Plasticity in sequence space." *J Mol Biol* 293: 1183–93.

Kratky, K. W. 1978. "Area of intersection of n equal circular disks." *J Phys A: Math Gen* 11: 1017–24.

Lee, B. and F. M. Richards. 1971. "Interpretation of protein structures: Estimation of static accessibility." *J Mol Biol* 55: 379–400.

Legrand, S. M. and K. M. Merz. 1993. "Rapid approximation to molecular-surface area via the use of Boolean logic and look-up tables." *J Comp Chem* 14: 349–52.

Lum, K., D. Chandler and J. D. Weeks. 1999. "Hydrophobicity at small and large length scales." *J Phys Chem B* 103: 4570–7.

Maiorov, V. N. and G. M. Crippen. 1994. "Learning about protein folding via potential functions." *Proteins: Struct Funct Genet* 20: 167–73.

Marshall, S. A., C. L. Vizcarra and S. L. Mayo. 2005. "One and two-body decomposable Poisson-Boltzmann methods for protein design calculations." *Protein Sci* 14: 1293–1304.

Meirovitch, H. 2007. "Recent developments in methodologies for calculating the entropy and free energy of biological systems by computer simulation." *Curr Opin Struct Biol* 17: 181–6.

Mirny, L. A. and E. I. Shakhnovich. 1996. "How to derive a protein folding potential: A new approach to an old problem." *J Mol Biol* 264: 1164–79.

Miyazawa, S. and R. L. Jernigan. 1985. "Estimation of effective inter-residue contact energies from protein crystal structures: Quasi-chemical approximation." *Macromolecules* 18: 534–52.

Momeny, F. A., R. F. McGuire, A. W. Burgess and H. A. Scheraga. 1975. "Energy parameters in polypeptides. VII. Geometric parameters, partial atomic charges, non-bonded interactions, hydrogen bond interactions, and intrinsic torsional potentials for the naturally occurring amino acids." *J Phys Chem* 79: 2361–81.

Pabo, C. 1983. "Designing proteins and peptides." *Nature* 301: 200.

Park, B. H. and M. Levitt. 1996. "Energy functions that discriminate x-ray and near-native folds from well-constructed decoys." *J Mol Biol* 258: 367–92.

Ponder, J. W. and D. A. Case. 2003. "Force fields for protein simulations." *Adv Protein Chem* 66: 27–85.

Poole, A. M. and R. Ranganathan. 2006. "Knowledge-based potentials in protein design." *Curr Opin Struct Biol* 16: 508–13.

Prabhu, N. V., P. Zhu and K. A. Sharp. 2004. "Implementation and testing of stable, fast implicit solvation in molecular dynamics using the smooth-permittivity finite difference Poisson-Boltzmann method." *J Comput Chem* 25: 2049–64.

Raha, K., M. B. Peters, B. Wang, N. Yu, A. M. Wollacott, et al. 2007. "The role of quantum mechanics in structure-based drug design." *Drug Discov Today* 12: 725–31.

Richards, F. M. 1977. "Areas; volumes; packing; and protein-structure." *Ann Rev Biophys Bioeng* 6: 151–76.

Richmond, T. J. 1984. "Solvent accessible surface-area and excluded volume in proteins. Analytical equations for overlapping spheres and implications for the hydrophobic effect." *J Mol Biol* 178: 63–89.

Rooman, M. J. and S. J. Wodak. 1995. "Are database-derived potentials valid for scoring both forward and inverted protein folding?" *Prot Eng* 8: 849–58.

Sali, A. 1998. "100,000 protein structures for the biologist." *Nat Struct Biol* 5: 1029–32.

Sasai, M. 1995. "Conformation, energy, and folding ability of selected amino acid sequences." *Proc Natl Acad Sci U S A* 92: 8438–42.

Schwarzl, S. M., D. Huang, J. C. Smith and S. Fisher. 2005. "Nonuniform charge scaling (NUCS): A practical approximation of solvent electrostatic screening in proteins." *J Comput Chem* 26: 1359–71.

Shakhnovich, E. I. and A. M. Gutin. 1993. "A new approach to the design of stable proteins." *Protein Eng* 6: 793–800.

Shrake, A. and J. Rupley. 1973. "Environment and exposure to solvent of protein atoms in lyzozyme and insulin." *J Mol Biol* 79: 351–71.

Singh, R. K., A. Tropsha and I. I. Vaisman. 1991. "Delaunay tessalation of proteins: Four body nearest-neighbor propensities of amino acid residues." *Comput Aided Geom Des* 8: 123–42.

Sippl, M. 1990. "Calculation of conformational ensembles from potentials of mean force: An approach to the knowledge-based prediction of local structures in globular proteins." *J Mol Biol* 1990: 859–83.

Sippl, M. J. 1993. "Boltzmann's principle, knowledge-based mean fields and protein folding. An approach to the computational determination of protein structures." *J Comput Aided Mol Des* 7: 473–501.

Still, W. C., A. Tempczyk, R. C. Hawley and T. Hendrickson. 1990. "Semi-analytical treatment of solvation for molecular mechanics and dynamics." *J Am Chem Soc* 112: 6127–9.

Street, A. G. and S. L. Mayo. 1998. "Pairwise calculation of protein solvent-accessible surface areas." *Folding & Design* 3: 253–8.

Sun, S., R. Brem, H. S. Chan and K. A. Dill. 1995. "Designing amino acid sequences to fold with good hydrophobic cores." *Prot Eng* 8: 1205–13.

Tanaka, S. and H. A. Scheraga. 1976. "Medium- and long-range interaction parameters between amino acids for predicting three-dimensional structures of proteins." *Macromolecules* 9.

Thomas, P. D. and K. A. Dill. 1996. "An iterative method for extracting energy-like quantities from protein structures." *Proc Natl Acad Sci U S A* 93: 11628–33.

Vendruscolo, M., A. Maritan and J. R. Banavar. 1997. "Stability threshold as a selection principle for protein design." *Phys Rev Lett* 78: 3967–70.

Wang, H. and C. Levinthal. 1991. "A vectorized algorithm for calculating the accessible surface area of macromolecules." *J Comp Chem* 12: 868–71.

Wisz, M. S. and H. W. Hellinga. 2003. "An empirical model for the electrostatic interactions in proteins incorporating multiple geometry-dependent dielectric constants." *Proteins: Struct Funct Genet* 51: 360–77.

Wodak, S. J. and J. Janin. 1980. "Analytical approximation to the accessible surface-area of proteins." *Proc Natl Acad Sci U S A* 77: 1736–40.

# 13 Rotamer Libraries for Molecular Modeling and Design of Proteins

*Hidetoshi Kono*

## CONTENTS

Although the 3D structures of more than 50,000 proteins have been determined and recorded in the Protein Data Bank (PDB) to date (Berman et al. 2000), accurate structural modeling of unknown proteins remains one of the most challenging problems in molecular biology. Modeling the protein structure, which involves determining the optimum side-chain conformations as well as backbone geometry, is important to understand molecular function because the activity of a protein often depends critically on its structure. For example, to understand enzymatic catalysis at a molecular level, one must first understand the structural aspects of the active site, including the elaborate tertiary arrangement of side chain and backbone atoms.

A typical approach to constructing an all-atom model of a protein structure involves initially generating an ensemble of backbone conformations, to which side-chain-heavy atoms are systematically added. This approach is often used because the separate treatment of main chain and side chain can significantly reduce the total number of conformational combinations that must be examined, thus simplifying the computational task involved. However, even against a fixed-backbone geometry, the number of possible combinations of side-chain conformations often remains impossibly large. This number becomes even larger when one is designing proteins, since the residues are then allowed to vary in both identity and in conformation.

When modeling a protein tertiary structure for a given amino-acid sequence, homology modeling is most often used. This strategy is efficient if the protein has significant sequence similarity (i.e., more than 25%) to another protein whose tertiary structure is known. If the similarity is extremely high, say, more than 90%, it is reasonable to expect that the constituent side chains will assume similar conformations in both proteins. However, in case the similarity is relatively low, say, ~20–25%, the side chain may have substantially different conformations (Summers et al. 1987). Fortunately, from analyzing numerous protein structures, we know that side chains have structural preferences. This means that their positions can be grouped into a relatively small number of statistically likely orientations referred to as rotamers. The term *rotamer* comes from *rotational isomer* and highlights the usefulness of dihedral angles in defining side-chain conformation. The first complete rotamer library was constructed by Ponder and Richards based on the few structures known at the time (Ponder and Richards 1987). Since then, the reliability and completeness of rotamer libraries have increased significantly as more high-quality 3D structures became available.

Analysis of protein 3D structures showed that a protein interior is condensed as much as solid and its side chains are tightly packed together, reducing the number of possible conformations (Ponder and Richards 1987). Moreover, although they may have conformational degrees of freedom, side chains often have preferred conformations that are in accord with the principles of structural chemistry. For instance, Leu residues have two dihedral angles, $\chi_1$ and $\chi_2$, and nine possible conformers, but two of the conformers ($g^+$-t and t-g-) account for 88% of all occurrences (Dunbrack and Karplus 1993). By incorporating the observed preferences of side-chain conformations, therefore, it is possible to construct a rotamer library that can be used to model side-chain conformations during structural modeling. Such a rotamer library would result in improved performance compared to a model based on continuous changes in dihedral angles. It is also noteworthy that the conformations included in the rotamer libraries correspond well to energy minima that are calculated in the form of an isolated dipeptide (Gelin and Karplus 1975).

In this chapter, I will discuss the structural basis of the side-chain conformational bias, introduce published rotamer libraries, and discuss the strategies of using rotamer libraries to model side-chain conformations.

## STRUCTURAL PREFERENCES OF SIDE CHAINS

The principles of structural chemistry predict that the side-chain conformations of amino acids would be severely restricted to a few conformations. For example, a chain of four consecutive heavy atoms with tetrahedral carbons at the center would adopt either one of two potential gauche conformations or the *trans* conformation, in order to minimize the conformation energy. The available conformations are further limited in the interior of proteins, where both backbone and side-chain atoms are tightly packed. Early studies showed that the distributions of side-chain dihedral angles ($\chi$s) of amino acids tend to cluster around particular values (Ponder and Richards 1987) and subsequent analyses further revealed that side-chain rotamer preferences may also depend on the backbone dihedral angles $\phi$ and $\varphi$ (Dunbrack and Karplus 1994). As done in an analysis of protein backbone conformation by Ramachandran and coworkers (Ramachandran et al. 1963), Dunbrack and Karplus found that the steric factors observed in butane and pentane can take account for most of the trends observed in rotamers (Dunbrack and Karplus 1994). In particular, the distribution of the first side-chain dihedral angle $\chi_1$ of residue $i$ is mainly affected by repulsive interactions between the side-chain atoms and the local main chain atoms, including $C_{i-1}$, $N_i$, $C_i$, $O_i$, and $N_{i+1}$, as well as hydrogen binding interactions with $HN_i$, as shown in Figure 13.1. Because the positions of $C_{i-1}$, $O_i$, $N_{i+1}$, and $HN_i$ depend on the backbone dihedral angles $\phi$ and $\varphi$, the distribution of the rotamers also changes with the backbone-dependent conformation (Dunbrack and Karplus 1993; Dunbrack and Cohen 1997). As one may expect, the side-chain conformations of branched amino acids, that is, residues having two C$\gamma$ heady atoms (Thr, Val,

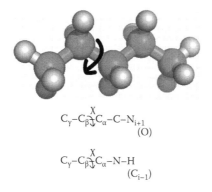

$$C_\gamma-C_\beta\overset{\chi}{\downarrow}C_\alpha-C-N_{i+1}$$
$$(O)$$

$$C_\gamma-C_\beta\overset{\chi}{\downarrow}C_\alpha-N-H$$
$$(C_{i-1})$$

**FIGURE 13.1 (see color insert following page 178)**  Conformation of pentane. Shown are the positions of the carbons (colored green) within pentane, illustrating which atoms in proteins can affect $\chi_1$ angle.

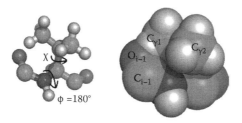

**FIGURE 13.2 (see color insert following page 178)**  Val conformation represented in ball and stick (left) and space filling (right) models. Carbon is shown in green, oxygen in red, nitrogen in blue, and hydrogen in white. When $\phi$ angle is around 180°, the $\chi$ angle of Val rarely takes the gauche- ($\chi = -60°$) conformation because C$\gamma_1$ has a clash with the carbonyl O atom of the preceding residue.

and Ile), are even more restricted. For example, Val rarely assumes the gauche- ($\chi_1$ = −60°) conformation when $\phi$ is 180° (Figure 13.2). Hydrocarbon analysis can also be applied to analyze the $\chi_1$ and $\chi_2$ rotamer combinations. For linear side chains such as those of Met, Glu, Gln, Lys, and Arg, the combinations of $\chi_1$ and $\chi_2$ of (60°, 60°), (60°, −60°), (180°, −60°), and (−60°, 60°) are structurally forbidden due to steric hindrance (*syn*-pentane effect), and thus are rarely found in the crystal structures. Nonetheless, these rotamers do exist in protein structures with deviations up to 10–30° from the canonical values. In addition, hydrogen bonds and electrostatic interactions can perturb the side-chain orientation, especially for Ser, Asp, and Glu, resulting in noncanonical values.

## ROTAMER LIBRARIES

### LIBRARIES

Rotamer libraries can be backbone-independent or depend on the backbone conformation of the secondary structure (this was discussed in Dunbrack 2002). Table 13.1 lists the published rotamer libraries. The original motivation for using a rotamer library was to make the side-chain conformational space as small as possible in order to reduce computer memory requirements. Because 3D structural data were limited, the initial rotamer libraries were created without considering backbone conformation. The first rotamer library (Ponder and Richards 1987) was useful when discriminating sequences that are compatible with the known backbone conformation in the protein core. This led to a number of independent groups to successfully design the protein core *de novo* (Desjarlais and Handel 1995; Harbury et al. 1995; Dahiyat and Mayo 1997; Kono et al. 1998; Desjarlais and Handel 1999).

However, the desire to accurately model the side chains has resulted in significantly larger rotamer libraries in recent years. In particular, as more structures were determined, it became clear that the backbone and side-chain conformations were correlated. This led to the creation of the first backbone-dependent rotamer library that reflects the principles of structural chemistry (Dunbrack and Karplus 1993). Also, with more structures to choose from, one could impose stricter criteria when selecting data to create libraries. At the present, most of the libraries are created using only high-resolution crystal structures (2.0 Å or better). In addition, Lovell and coworkers (Lovell et al. 2000) further eliminated the side chains with high B-factors (B factor of < 40Å$^2$ was recommended), steric clashes, or ambiguous amide or histidine ring orientations. They also used the peak values of fitted Gaussian functions rather than the means to avoid strained conformations due to skewed angle distributions. Among the available rotamer libraries today, the most widely used are the backbone-dependent library developed by Dunbrack and coworkers (Dunbrack and Cohen 1997) and the backbone-independent library developed by Lovell and coworkers (Lovell et al. 2000), both of which can be downloaded from the Web.* Other rotamer libraries derived from these original libraries are also available.

---

* http://dunbrack.fccc.edu/bbdep/bbdepformat.php and http://kinemage.biochem.duke.edu/databases/rotamer.php, respectively.

**TABLE 13.1**
**Published Rotamer Libraries***

| Authors | Library Type | Number of Protein Structures | Resolution of Structures | Total Number of Rotamers | Year |
|---|---|---|---|---|---|
| Ponder and Richards | Independent | 19 | 2.0 | 84 | 1987 |
| McGregor et al. | Secondary str. dependent | 61 | 2.0 | — | 1987 |
| Tuffery et al. | Independent | 53 | 2.0 | 110 or 214 | 1991 |
| Dunbrack and Karplus | Independent/dependent | 132 | 2.0 | — | 1993 |
| Schrauber et al. | Independent / Secondary str. dependent | 70 | 2.0 | 81/109 | 1993 |
| Kono and Doi | Independent | 103 | NA | 141–693 | 1996 |
| De Maeyer et al. | Dependent | 19 | 2.0 | 213–859 | 1997 |
| Dunbrack and Cohen (1997) | Independent/dependent | 850 | 1.7 | 341 (independent) | 1997–2002 |
| Lovell et al. (2000) | Independent/dependent / Secondary str. dependent | 240 | 1.7 / B-factor < 40 / Van der Waals overlap < 0.4Å | 153 | |
| Xiang and Honig | Independent | 135–2312 | 2.0 | 7562 | 2001 |
| Shetty et al. | Independent/dependent | 500 | 1.7 | 446–5988 | 2003 |
| Peterson et al.** | Independent | NA | 1.7 | 49042 | 2004 |
| Xiang et al. | Independent | 648 | 1.6 | 831–6737 | 2007 |
| Jiang et al. | Solvated rotamer | Add water positions on the rotamer library of Dunbrack & Cohen | | | 2005 |

* After 1987

** Systematically expand the backbone independent library of Dunbrack & Cohen

Note that when using the original backbone-dependent rotamer library of Dunbrack and Cohen, it was necessary for $\chi_1$ of Asn and $\chi_2$ of His and Gln to be split into two new rotamers, one with the original torsion angle and the one with the angle flipped by 180° (Dunbrack and Cohen 1997). This is because the difference in the N and O atoms were not considered to be uniquely determined in crystal structures and was ignored when generating the original library. This problem has been fixed in the latest version of the library (Canutescu et al. 2003).

## Off-Rotamer Conformations

Of course, conformations in rotamer libraries cannot completely cover side-chain conformations observed in protein crystal structures. Petrella and Karplus examined such nonrotameric conformations using CHARMM energy function (Petrella and Karplus 2001). They found that 36.2% of the nonrotameric conformations observed in 1709 nonpolar side chains from 24 proteins were not in a local energy minimum in the context of the crystal environment, while 97.6% of on-rotamer conformations corresponded to local minima. In addition 10% off-rotamer fraction was observed in the x-ray structures examined and about 6% off-rotamer fraction was observed in the trajectories of umbrella-sampled molecular dynamics simulation. They concluded that one-third or more of nonrotameric conformations are likely to be artifacts from the x-ray refinement process.

## Solvated Rotamers

When considering protein–protein and protein–nucleic acid interfaces, water-mediated hydrogen bonds are an important factor for predicting binding free energy changes. To that end, Jiang and coworkers expanded the definition of a rotamer library by introducing "solvated rotamers," which include modeled water molecules bound to the charged and polar side chains (Ser, Thr, Tyr, Trp, Lys, Asn, Asp, Gln, Glu, Arg, and His) and to polar main chain atoms (Jiang et al. 2005) (Figure 13.3). For example, Ser, which has three rotamers, would have 24 rotamers if a water molecule is allowed to make a hydrogen bond to the hydroxyl oxygen at three positions (two as donor and one as acceptor), with three additional hydrogen bonds to the carbonyl oxygen and the amide nitrogen (see Table I and II in the reference Jiang et al., 2005). In practical terms, however, the explicit inclusion of water molecules produced little improvement in the prediction of binding free energy changes upon mutation, as compared to predictions made without water molecules. It was therefore concluded that the benefit of correctly predicted water-mediated hydrogen bonds is offset by spurious water-mediated hydrogen bonds. However, the accuracy of prediction is significantly improved if crystal water molecules are included in the calculation, suggesting that there is some room for improvement using solvated rotamers.

## SIDE-CHAIN MODELING AND MODEL ACCURACY

To model side-chain conformations on a fixed backbone, one must decide on a rotamer library to use, the energy functions to quantify fitness, and the algorithm

| Classes | Illustration of water position[a] | Distance $d$ (Å) | Angle $\theta$ | Description |
|---|---|---|---|---|
| Carbonyl oxygen (backbone, Asp, Glu, Gln, and Asn) | 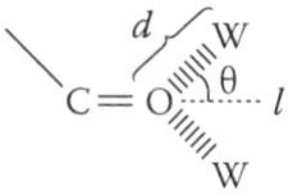 | Mainchain O: 2.70<br>O (Gly): 3.20<br>Sidechain O: 2.80 | 50°<br>50°<br>50° | |
| Amide nitrogen (backbone, Lys, Arg, Gln, and Asn) | 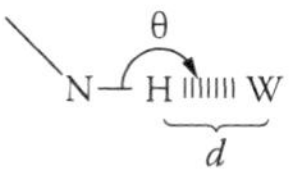 | Mainchain N: 1.95<br>(Gly): 2.30<br>Sidechain N: 1.95 | 180°<br>180°<br>180° | |
| Hydroxyl oxygen (in SER, THR) | 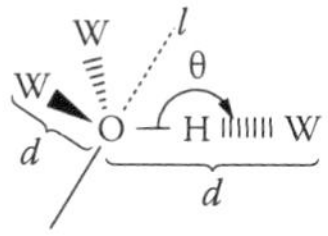 | 2.80 | 180° | Water molecules are placed every 120° along the $l$ axis |
| Hydroxyl oxygen (in TYR) | 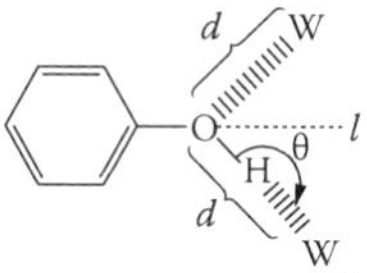 | 2.80 | 180° | |
| Aromatic nitrogen (in HIS) | 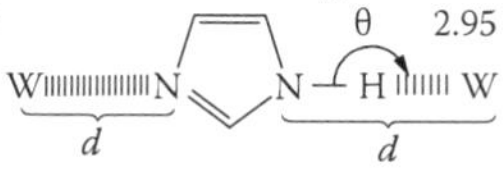 | 2.95 | 180° | |

[a] The water is in the plane of the functional group, unless otherwise noted.

**FIGURE 13.3** Solvated rotamers. Placement of water molecules around functional groups. (From Jiang, L. et al., *Proteins*, 58, 2005. With permission.)

for finding the minimum energy states. The energy functions and search algorithms are described in Chapter 14 and 16, respectively, so only a cursory discussion is presented here.

## ENERGY FUNCTIONS

The energy, or scoring, functions used typically include the van der Waals interactions, electrostatic interactions, hydrogen bond, torsion energy, and desolvation energy. Usually, the bond lengths and angles are fixed to the standard values obtained from a molecular mechanics force field. Among the various contributions, the most critical is the van der Waals term, while other terms are sometimes omitted to simplify the calculation. Because rotamer libraries have only a finite number of discrete side-chain conformations, atomic clashes occur frequently during modeling. In severe cases, no rotamer may be assigned to a residue due to severe volume overlap. Strategies that are commonly used to avoid this situation involve using van der Waals radii that are reduced by 10–20% (Dahiyat and Mayo 1996; Looger and Hellinga 2001; Shetty et al. 2003; Peterson et al. 2004) or truncating the maximum van der Waals energy (Koehl and Delarue 1994; Kono and Doi 1996; Mendes et al. 1999). This may be taken as an implicit incorporation of backbone flexibility, which is difficult to model explicitly. The van der Waals term is effective for modeling buried hydrophobic residues but not for polar residues. The terms of electrostatic interactions and hydrogen bond energy play a critical role in deciding the accurate rotamer state of both exposed and buried polar side chains. In particular, $\chi_1$ of Asn, $\chi_2$ of Gln, and $\chi_1$ of His have been shown to be difficult to model based on van der Waals interactions alone, because they all have multiple rotamer states with similar packing density. Incorporating the hydrogen bonding potential can help resolve the ambiguity (Word et al. 1999). Recently, Marshall and coworkers presented a computationally efficient formulation for a finite difference Poisson-Boltzmann model that uses one- and two-body terms to describe electrostatic interactions (Marshall et al. 2005). The torsion energy can be described using either a molecular mechanics function or a pseudoenergy function in the form of $-w \log (p_i)$, where $w$ is a constant and $p_i$ is the probability of rotamer $i$. The pseudoenergy is derived from known protein structures to favor the rotamers that are observed most often. The inclusion of the term was shown to increase the prediction accuracy (Bower et al. 1997; Liang and Grishin 2002). Finally, the desolvation energy is important to model hydrophobicity (Kono and Saven 2001; Liang and Grishin 2002), but at the moment there is no method that is consistently used to model the effect.

## Methods for Placing the Side Chains

Several algorithms have been developed to find optimal side-chain combinations for a given, fixed backbone using the energy functions mentioned in the previous section, including a systematic search for rotamers that avoid steric clashes (Dunbrack and Karplus 1993; Wilson et al. 1993; Bower et al. 1997), Monte Carlo search (Holm and Sander 1992; Vasquez 1995; Liang and Grishin 2002; Jain et al. 2006), genetic algorithms (Tuffery et al. 1991), neural networks (Hwang and Liao 1995; Kono and Doi 1996), mean field optimization (Koehl and Delarue 1994; Mendes et al. 1999; Kono and Saven 2001), dead-end elimination (Desmet et al. 1992; De Maeyer et al. 1997; Voigt et al. 2000; Looger and Hellinga 2001), and graph theory algorithm (Canutescu et al. 2003). Not all algorithms are equally effective and they all have various limitations. For example, the dead-end elimination has been shown to find the optimal combinations of side chains in a number of examples, but the method may not be compatible with a fine-grained rotamer library composed of a significantly larger number of rotamer states. Similarly, Monte Carlo methods can easily implement multibody interactions that may be important for modeling solvation, but the methods do not guarantee a global energy minimum state. Some of these algorithms are discussed in Chapter 16.

## Current Model Accuracy

In general, the choice of search algorithm does not seem to affect the accuracy of prediction as much as the energy function or the rotamer library. Model accuracy is typically evaluated in terms of the root mean square deviation (RMSD) of the side chains and the deviation of the torsion angles, where a deviation of less than 20° or 40° from the crystal structure is typically regarded as correct. The residues with a rotational symmetry axis (Asp, Glu, Phe, and Tyr) are evaluated by considering both symmetric conformations and choosing the one with a lower RMSD.

When modeling the side chains, it is useful to be mindful of the upper limits of the accuracy achievable given the choice of a rotamer library. This is because rotamer libraries are usually created so that a minimum number of rotamers (typically, 100 to 200 rotamers) can account for as wide a population of crystal structures as possible, and such rotamers cannot adequately model the tightly packed side chains in the protein cores (Shetty et al. 2003). For example, when using the backbone-dependent rotamer library (Dunbrack and Cohen 1997), the average RMSD for the side chains of 30 high-quality proteins tested could reach 0.56 Å, and the percentages of correct prediction (i.e., within 40° deviation) for $\chi_1$ and a combination of $\chi_1$ and $\chi_2$ ($\chi_{1+2}$) from the crystal structures were 98.5% and 94.3%, respectively (Liang and Grishin 2002). The gap between the upper limit values and those predicted may be mostly due to the side-chain contacts of buried residues. However, others have shown that a rotamer library with as many as 7560 distinct rotamers can be used to obtain accuracies of about 0.7 Å RMSD for buried residues, or accuracies of about 94% and 89% in terms of $\chi_1$ and $\chi_{1+2}$ (within 20° from native), respectively (Xiang and Honig 2001). And the use of a much larger rotamer library containing 49,042 rotamers can lower the RMSD for buried residues further down to 0.21 Å (Peterson et al. 2004). These studies suggest that more finely grained rotamer libraries should be used for buried residues and different accuracy criteria should be used for buried and exposed residues during model evaluation.

## VARIATION IN THE MAGNITUDE OF THE FLEXIBILITY OF DIFFERENT RESIDUE TYPES

Buried residues are more accurately modeled than exposed residues, because they are surrounded by large numbers of atoms that sterically restrict their conformations, while exposed residues experience fewer restrictions. The accuracy for modeled buried residues may have reached a maximum because there has been essentially little improvement in recent years (Eyal et al. 2004; Hartmann et al. 2007). However, additional work is required to improve the accuracy of modeled exposed residues. Comparison of several different crystal structures of the same proteins revealed that residues exposed on the protein surface are flexible and that the magnitude of their flexibility is residue type–dependent (Zhao et al. 2001). In another words, the ambiguity of a modeled side-chain conformation is residue type–dependent, particularly for residues on the protein surface. For example, Ser was found to be the most flexible amino acid, followed by Lys, Glu, Gln, Arg, and Met, in that order. Taking their distinct flexibilities into account, it may be necessary to adopt different angular thresholds for different residue types when evaluating a model rather than a common angular threshold for all residue types (Zhao et al. 2001).

## SIMULTANEOUS MODELING OF MAIN CHAIN AND SIDE CHAINS

Separately modeling the main chain and side chains frequently leads to model structures with severe van der Waals overlaps. Introducing backbone flexibility can alleviate the problem to a certain degree, but even this approach has a limitation. Especially in the protein cores, a simultaneous modeling of both main chain and side chain may be necessary to achieve a desired degree of accuracy. Additionally, a more detailed rotamer library may be necessary to model a region where atoms are tightly packed. For example, in one study (Shetty et al. 2003), a library about 40 times larger than typical libraries of around 100 to 200 rotamers was used to model a buried loop *ab initio* without atomic clashes.

## A RECOMMENDED PROCEDURE

Based on the discussion presented in this chapter, it may be useful to think of a procedure one may follow when modeling side chains. Starting with the backbone-independent library of Lovell and coworkers or the backbone-dependent library of Dunbrack and coworkers, one would first generate additional rotamers around the original ones for buried residues. Whether a site is buried or exposed can be roughly determined by counting the number of Cβ atoms within a certain radius from the center of mass of side chain (Kono and Saven 2001). Alternatively, an 8 Å probe sphere centered on the Cα atoms can be used to generate a working definition of buried residues (Marshall and Mayo 2001). This information is very useful because it can be used to limit the amino-acid types at each residue site, thus reducing the number of amino-acid sequences that need be considered. One needs to use an energy function with a minimum of van der Waals interactions but preferably with additional terms to account for the electrostatic interactions, hydrogen bonding energy, and torsion energy (or pseudoenergy-based rotamer probability). Energy minimization can be achieved using any one of a number of available algorithms, such as dead-end elimination, Monte Carlo, or mean field optimization. They are likely to optimize side-chain conformations with reasonably low energy.

## ACKNOWLEDGMENT

This work was supported in part by Grand-in-Aid for Scientific Research (B) (2000103) from Japan Society for the Promotion of Science (JSPS).

## REFERENCES

Berman, H. M., J. Westbrook, Z. Feng, G. Gilliland, T. N. Bhat, et al. 2000. "The Protein Data Bank." *Nucl Acids Res* 28: 235–42.

Bower, M. J., F. E. Cohen and R. L. Dunbrack, Jr. 1997. "Prediction of protein side-chain rotamers from a backbone-dependent rotamer library: A new homology modeling tool." *J Mol Biol* 267: 1268–82.

Canutescu, A. A., A. A. Shelenkov and R. L. Dunbrack, Jr. 2003. "A graph-theory algorithm for rapid protein side-chain prediction." *Protein Sci* 12: 2001–14.

Dahiyat, B. I. and S. L. Mayo. 1996. "Protein design automation." *Protein Sci* 5: 895–903.

Dahiyat, B. I. and S. L. Mayo. 1997. "Probing the role of packing specificity in protein design." *Proc Natl Acad Sci U S A* 94: 10172–7.

De Maeyer, M., J. Desmet and I. Lasters. 1997. "All in one: A highly detailed rotamer library improves both accuracy and speed in the modelling of sidechains by dead-end elimination." *Fold Des* 2: 53–66.

Desjarlais, J. R. and T. M. Handel. 1995. "De novo design of the hydrophobic cores of proteins." *Protein Sci* 4: 2006–18.

Desjarlais, J. R. and T. M. Handel. 1999. "Side-chain and backbone flexibility in protein core design." *J Mol Biol* 290: 305–18.

Desmet, J., M. De Maeyer, B. Hazes and I. Lasters. 1992. "The dead-end elimination theorem and its use in protein side-chain positioning." *Nature* 356: 539–42.

Dunbrack, R. L., Jr. 2002. "Rotamer libraries in the 21st century." *Curr Opin Struct Biol* 12: 431–40.

Dunbrack, R. L., Jr. and F. E. Cohen. 1997. "Bayesian statistical analysis of protein side-chain rotamer preferences." *Protein Sci* 6: 1661–81.

Dunbrack, R. L., Jr. and M. Karplus. 1993. "Backbone-dependent rotamer library for proteins. Application to side-chain prediction." *J Mol Biol* 230: 543–74.

Dunbrack, R. L., Jr. and M. Karplus. 1994. "Conformational analysis of the backbone-dependent rotamer preferences of protein sidechains." *Nat Struct Biol* 1: 334–40.

Eyal, E., R. Najmanovich, B. J. McConkey, M. Edelman and V. Sobolev. 2004. "Importance of solvent accessibility and contact surfaces in modeling side-chain conformations in proteins." *J Comput Chem* 25: 712–24.

Gelin, B. R. and M. Karplus. 1975. "Sidechain torsional potentials and motion of amino acids in proteins: Bovine pancreatic trypsin inhibitor." *Proc Natl Acad Sci U S A* 72: 2002–6.

Harbury, P. B., B. Tidor and P. S. Kim. 1995. "Repacking protein cores with backbone freedom: Structure prediction for coiled coils." *Proc Natl Acad Sci U S A* 92: 8408–12.

Hartmann, C., I. Antes and T. Lengauer. 2007. "IRECS: A new algorithm for the selection of most probable ensembles of side-chain conformations in protein models." *Protein Sci* 16: 1294–307.

Holm, L. and C. Sander. 1992. "Fast and simple Monte Carlo algorithm for side chain optimization in proteins: Application to model building by homology." *Proteins* 14: 213–23.

Hwang, J. K. and W. F. Liao. 1995. "Side-chain prediction by neural networks and simulated annealing optimization." *Protein Eng* 8: 363–70.

Jain, T., D. S. Cerutti and J. A. McCammon. 2006. "Configurational-bias sampling technique for predicting side-chain conformations in proteins." *Protein Sci* 15: 2029–39.

Jiang, L., B. Kuhlman, T. Kortemme and D. Baker. 2005. "A 'solvated rotamer' approach to modeling water-mediated hydrogen bonds at protein–protein interfaces." *Proteins* 58: 893–904.

Koehl, P. and M. Delarue. 1994. "Application of a self-consistent mean field theory to predict protein side-chains conformation and estimate their conformational entropy." *J Mol Biol* 239: 249–75.

Kono, H. and J. Doi. 1996. "A new method for side-chain conformation prediction using a Hopfield network and reproduced rotamers." *J Comp Chem* 17: 1667.

Kono, H., M. Nishiyama, M. Tanokura and J. Doi. 1998. "Designing the hydrophobic core of *Thermus flavus* malate dehydrogenase based on side-chain packing." *Protein Eng* 11: 47–52.

Kono, H. and J. G. Saven. 2001. "Statistical theory for protein combinatorial libraries. Packing interactions, backbone flexibility, and the sequence variability of a main-chain structure." *J Mol Biol* 306: 607–28.

Liang, S. and N. V. Grishin. 2002. "Side-chain modeling with an optimized scoring function." *Protein Sci* 11: 322–31.

Looger, L. L. and H. W. Hellinga. 2001. "Generalized dead-end elimination algorithms make large-scale protein side-chain structure prediction tractable: Implications for protein design and structural genomics." *J Mol Biol* 307: 429–45.

Lovell, S. C., J. M. Word, J. S. Richardson and D. C. Richardson. 2000. "The penultimate rotamer library." *Proteins* 40: 389–408.

Marshall, S. A. and S. L. Mayo. 2001. "Achieving stability and conformational specificity in designed proteins via binary patterning." *J Mol Biol* 305: 619–31.

Marshall, S. A., C. L. Vizcarra and S. L. Mayo. 2005. "One- and two-body decomposable Poisson-Boltzmann methods for protein design calculations." *Protein Sci* 14: 1293–304.

McGregor, M. J., S. A. Islam and M. J. Sternberg. 1987. "Analysis of the relationship between side-chain conformation and secondary structure in globular proteins." *J Mol Biol* 198: 295–310.

Mendes, J., C. M. Soares and M. A. Carrondo. 1999. "Improvement of side-chain modeling in proteins with the self-consistent mean field theory method based on an analysis of the factors influencing prediction." *Biopolymers* 50: 111–31.

Peterson, R. W., P. L. Dutton and A. J. Wand. 2004. "Improved side-chain prediction accuracy using an ab initio potential energy function and a very large rotamer library." *Protein Sci* 13: 735–51.

Petrella, R. J. and M. Karplus. 2001. "The energetics of off-rotamer protein side-chain conformations." *J Mol Biol* 312: 1161–75.

Ponder, J. W. and F. M. Richards. 1987. "Tertiary templates for proteins: Use of packing criteria in the enumeration of allowed sequences for different structural classes." *J Mol Biol* 193: 775–91.

Ramachandran, G. N., C. Ramakrishnan and V. Sasisekharan. 1963. "Stereochemistry of polypeptide chain configurations." *J Mol Biol* 7: 95–9.

Schrauber, H., F. Eisenhaber and P. Argos. 1993. "Rotamers: To be or not to be? An analysis of amino acid side-chain conformations in globular proteins." *J Mol Biol* 230: 592–612.

Shetty, R. P., P. I. De Bakker, M. A. DePristo and T. L. Blundell. 2003. "Advantages of fine-grained side chain conformer libraries." *Protein Eng* 16: 963–9.

Summers, N. L., W. D. Carlson and M. Karplus. 1987. "Analysis of side-chain orientations in homologous proteins." *J Mol Biol* 196: 175–98.

Tuffery, P., C. Etchebest, S. Hazout and R. Lavery. 1991. "A new approach to the rapid determination of protein side chain conformations." *J Biomol Struct Dyn* 8: 1267–89.

Vasquez, M. 1995. "An evaluation of discrete and continuum search techniques for conformational analysis of side chains in proteins." *Biopolymers* 36: 53–70.

Voigt, C. A., D. B. Gordon and S. L. Mayo. 2000. "Trading accuracy for speed: A quantitative comparison of search algorithms in protein sequence design." *J Mol Biol* 299: 789–803.

Wilson, C., L. M. Gregoret and D. A. Agard. 1993. "Modeling side-chain conformation for homologous proteins using an energy-based rotamer search." *J Mol Biol* 229: 996–1006.

Word, J. M., S. C. Lovell, J. S. Richardson and D. C. Richardson. 1999. "Asparagine and glutamine: Using hydrogen atom contacts in the choice of side-chain amide orientation." *J Mol Biol* 285: 1735–47.

Xiang, Z. and B. Honig. 2001. "Extending the accuracy limits of prediction for side-chain conformations." *J Mol Biol* 311: 421–30.

Xiang, Z., P. J. Steinbach, M. P. Jacobson, R. A. Friesner and B. Honig. 2007. "Prediction of side-chain conformations on protein surfaces." *Proteins* 66: 814–23.

Zhao, S., D. S. Goodsell and A. J. Olson. 2001. "Analysis of a data set of paired uncomplexed protein structures: New metrics for side-chain flexibility and model evaluation." *Proteins* 43: 271–9.

# 14 Search Algorithms

*Julia M. Shifman and Menachem Fromer*

## CONTENTS

## THE CHALLENGE

The complexity of the computational protein design problem is very large (Park et al. 2004, 2005; Rosenberg and Goldblum 2006; Butterfoss and Kuhlman 2006). Even for a small protein of 100 residues, there are $20^{100} = 10^{130}$ different sequence possibilities. This collection of sequences, each constructed in a single copy, would occupy a space larger than the whole universe. Additional complexity arises if one tries to model protein flexibility. It remains intractable to perform full-scale molecular dynamics simulations within the protein design calculation. Hence, most protein design studies consider only movement of protein side chains while the protein backbone remains fixed. Flexibility of amino-acid side chains is typically modeled by using a discrete set of statistically significant empirical conformations, called *rotamers* (see Chapter 13). With a larger number of rotamers used to represent each amino acid, the movement of side chains is modeled more accurately; but clearly the design problem becomes more complex. Even if only three rotamers were to be used for each amino acid, this exponentially increases the complexity of designing a 100-residue protein to $60^{100} = 10^{178}$ different rotameric assignments.

In protein design, the goal is to search over this huge sequence space and to find the best (lowest energy) sequence for the particular protein scaffold. The input to the computational protein design problem usually consists of a protein backbone structure, $N$ sequence positions to be designed, the amino acids (and their respective rotamers) permitted at each position, and an energy function. The energy function, used to evaluate candidate protein sequences, is generally pairwise and thus consists of two basic components, corresponding to rotamer-template and rotamer-rotamer interactions. The template can include the fixed backbone atoms, residues not subject to subsequent optimization, and atoms within the rotamer (for which pseudoenergies are derived from rotamer library statistics).

The energy of a rotameric assignment $r = (r_1,\ldots,r_N)$ is calculated by summing up energies of rotamer $r_i$ at position $i$ interacting with the fixed template $[E(r_i)]$ and energies of rotamer $r_i$ interacting with a rotamer $r_j$ at a neighboring position $j$ $[E(r_i,r_j)]$:

$$E(r) = \sum_i E(r_i) + \sum_{i,j} E(r_i,r_j) \tag{14.1}$$

In protein design calculations, these energy terms are precomputed to implicitly generate a discrete energy landscape, where each point represents a rotameric assignment and the energy designated for it. The problem is thus reduced to searching this landscape and finding the lowest energy sequence(s) compatible with the target structure (the global minimum energy configuration, GMEC) (Rosenberg and Goldblum 2006).

An alternative goal of computational protein design is the calculation of probabilities of the various amino acids at each design position (over all possible sequences compatible with the structure) (Park et al. 2005). Such probabilities essentially provide a position-specific amino acid score matrix (PSSM) and thus can be useful in creating biased combinatorial design libraries (see Chapter 4). Formally, amino-acid probabilities at design position $i$ are sought:

$$\Pr(S_i = a) = \sum_{S:S_i=a} \Pr(S) \tag{14.2}$$

where $S_i$ denotes the amino-acid assignment at sequence position $i$ and $a$ is a particular amino acid; the right-hand side sums the probabilities of all sequences $S$ for which amino acid $a$ is chosen at position $i$.

In computational protein design, the number of possible rotamer choices increases exponentially with the sequence length and the landscape possesses a very large number of local minima. Clearly, simple enumeration and evaluation of each rotameric assignment remains infeasible. Moreover, protein design has been demonstrated to belong to the computational class of NP-hard combinatorial optimization problems for which no guaranteed polynomial time algorithms are known (Pierce and Winfree 2002). Facing these challenges, computational protein design researchers must resort to algorithms that may compromise on exactness in more difficult cases but nevertheless have been shown to provide satisfying results in practice.

Finally, we note that the ultimate success of the algorithms developed for protein design strongly depends on the input of the problem, including the expressions for the energy function (see Chapter 12) and the size and quality of the rotamer library used (see Chapter 13).

In this chapter, we detail the major protein design algorithms previously applied. We then note some general preprocessing steps often taken in computational protein design. Lastly, we summarize these methods, so that their advantages and weaknesses are explicitly compared.

## APPROACHES FOR COMPUTATIONAL PROTEIN DESIGN

Numerous search algorithms have been developed to search the energy landscape for low energy sequences and their preferred amino acids at each position. These algorithms are divided into two classes: stochastic and deterministic. Stochastic algorithms use probabilistic trajectories, where the resulting sequence depends on initial conditions and a random number generator. Stochastic algorithms do not guarantee finding the GMEC sequence, but they can always find an approximate solution or a set of solutions. This may be sufficient, considering that simplifying assumptions in the energy function and in modeling protein flexibility inevitably result in uncertainty in defining the best protein sequence. In contrast, deterministic algorithms always produce the same solution given the same parameters. Many, but not all, of the deterministic algorithms are guaranteed to find the GMEC sequence if they converge. However, convergence is not guaranteed and the frequency of convergence is reduced with increasing problem size. In the following sections, we describe in detail several search algorithms that have been used in protein design studies, and mention some experimental studies in which they were utilized:

- Stochastic algorithms: Monte Carlo, genetic algorithms, FASTER
- Deterministic algorithms: dead-end elimination, self-consistent mean field, belief propagation, linear programming techniques

### MONTE CARLO (MC)

The Monte Carlo (MC) method (Metropolis et al. 1953; Kirckpatrick et al. 1983) was one of the first search techniques applied to side-chain prediction and protein design, and it often performs well on difficult energy landscapes. In MC, each sequence position is initially assigned a random rotamer. Randomized substitution attempts are then performed in an iterative manner. At a given step, mutation to a different rotamer is made at a random position. The energy of the new sequence is computed and compared to that of the previous sequence, so that this mutation is accepted or rejected as determined by the Metropolis criterion (Metropolis et al. 1953):

1. If the energy of the new sequence is less than the previous energy, the mutation is accepted.
2. If the energy of the new sequence is greater than the previous energy, the mutation still has a probability of being accepted. This (Boltzmann) probability is defined as $p = e^{-\beta \cdot (E_{new} - E_{previous})}$, where $\beta = 1/kT$, $k$ is Boltzmann's constant and $T$ is the system temperature.

The strategy of accepting higher energy mutations helps to overcome energy barriers in the sequence landscape. To strengthen this effect, the search system is typically annealed from high to low temperatures (for example, 4000 and 150 K, respectively), in a process known as Monte Carlo simulated annealing (MCSA). MCSA can be run for an arbitrary number of temperature annealing cycles (in which the temperature is successively lowered and raised) and an arbitrary number of substitution attempts per such cycle. In the end, the program generates a list of the lowest energy sequences observed throughout.

For rugged energy landscapes, conventional MC methods tend to become trapped in local minima, thus impeding the sampling of low energy sequences. To circumvent this problem, a few modifications to MC approaches have been suggested. One such modification is quenched MC (MCQ), wherein the algorithm quenches the energy of each solution by calculating the lowest energy rotamer for each position for a given sequence while holding all other rotamers fixed (as in Equation 14.7 in the FASTER algorithm). This quench step has been shown to produce a large improvement in performance (Voigt et al. 2000).

Another enhancement to the standard MC method was introduced in biased Monte Carlo (BMC) (Cootes et al. 2000) and mean field biased Monte Carlo (MFBMC) (Zou and Saven 2003). In these approaches, the amino-acid substitution at a particular position $i$ is not chosen uniformly at random, as in standard MC, but with certain biases meant to approximate $\Pr(S_i = a)$ (see Equation 14.2). In BMC, these biases are computed for each substitution attempt by evaluating a local energy function at design position $i$ (derived from interactions with other positions in the protein). In MFBMC, on the other hand, they are precalculated before performing the MC sampling by using the self-consistent mean field (SCMF) theory approach (see later in this chapter for a full description).

Further improvement to the standard MC approach resulted from the incorporation of a replica exchange method (REM) into MC, producing a combined algorithm termed MCREM (Yang and Saven 2005), which can also be utilized to predict site-specific amino-acid probabilities. This approach was previously applied to protein folding (Hansmann 1997; Sugita and Okamoto 1999). In MCREM, the MC procedure is simultaneously run on multiple replicas of the same protein, where the simulation for each replica is performed at a different temperature. Throughout the simulation, pairs of rotameric assignments from different replicas are exchanged with some transition probability. This probability, similar to that of the Metropolis criterion, is intended to maintain each temperature's equilibrium ensemble distribution. MCREM was shown to enable the system to surpass high energy barriers and thus reach low energy sequences more efficiently than the standard MC procedure.

MC has been repeatedly applied to both side-chain prediction and protein design problems. Among the most notable examples are the design of a protein with a novel fold (Kuhlman et al. 2003), redesign of endonuclease DNA binding and cleavage specificity (Ashworth et al. 2006), and design of a protein that can switch between two conformations (Ambroggio and Kuhlman 2006).

## GENETIC ALGORITHMS (GAs)

Genetic algorithms (GAs) were designed to mimic natural evolution, wherein genes of proteins are subjected to rounds of random mutation, recombination, and selection (Holland 1992). Initially, a population of random sequences is generated. Mutations are then randomly applied to each sequence at a rate of 1–2 mutations per sequence. The energies of the mutant sequences are evaluated and the sequences are ranked. The top several sequences are chosen for recombination. For each pair of parental sequences, the sequences are recombined by choosing a rotamer from either of the two sequences with equal probability.

The population of sequences for the next round of the GA is generated by randomly choosing $S$ sequences from the library and comparing their energies. The sequence with the lowest energy is propagated to the next round. This selection process is repeated a number of times to produce a final pool of sequences that will continue to the next round of mutation, recombination, and selection. The selection strength, represented by the number of sequences that undergo selection ($S$), is analogous to the temperature in MC simulations. By starting at low $S$ and annealing to high $S$, the population distribution in the sequence space starts out very broad and then narrows after each round to finally converge to a single sequence.

The advantages of using GAs are that the population dynamics can overcome large barriers in the sequence space. Beneficial mutations in different sequences can be combined together onto a single sequence, increasing the number of paths that circumvent local minima. However, GAs tend to perform poorly on highly coupled systems such as sequences of protein cores. This is due to the fact that the crossovers performed can result in the existence of a single deleterious mutation that had been beneficial before the crossover (when coupled with another mutation). In fact, GAs were shown to perform comparatively worse than MCQ on several side-chain placement problems (Voigt et al. 2000).

GA was first proposed as a tool for protein design calculations in a purely computational study (Jones 1994) and later applied to designs that were verified experimentally. These include the redesign of core residues of the phage 434 cro protein (Desjarlais and Handel 1995), ubiquitin (Lazar et al. 1997), and a native-like three-helix bundle (Kraemer-Pecore et al. 2003). In addition, GAs have been applied to optimize interaction specificity in coiled coils (Havranek and Harbury 2003).

## DEAD-END ELIMINATION (DEE)

Many existing protocols for side-chain prediction as well as for protein design are based on the dead-end elimination (DEE) theorem. DEE seeks to eliminate all rotamers or combinations of rotamers that are incompatible with the GMEC. The elimination proceeds until a single solution is obtained or until the algorithm cannot eliminate any more rotamers. DEE is proven to find the GMEC sequence if it converges. The original DEE theorem (Desmet et al. 1992) compares the best possible interaction (minimum energy) of the candidate rotamer $r_i$ at a given residue position $i$ with the worst interaction (maximum energy) of the candidate rotamer $s_i$ at the same position (Figure 14.1A). If it can be shown that the minimum energy of rotamer $r_i$ is greater than the maximum energy of rotamer $s_i$ (no matter which rotamer $t_j$ is chosen at all other positions $j$), then rotamer $r_i$ cannot be compatible with the GMEC and is hence eliminated from further search. Mathematically, this criterion is formulated as follows:

$$E(r_i)+\sum_{j\neq i} \min_{t_j} E(r_i,t_j)> E(s_i)+\sum_{j\neq i}\max_{t_j} E(s_i,t_j) \tag{14.3}$$

This original elimination criterion was later modified to allow more powerful elimination of rotamers. It is possible that the energy contribution of rotamer $s_i$ is always lower than $r_i$, without the maximum of $s_i$ being lower than the minimum of $r_i$. To address this problem, the Goldstein modification of the criterion (Goldstein 1994) determines if the energy profiles of the two rotamers cross. That is, it verifies that any point on the $r_i$ profile is always higher in energy than the corresponding point on the $s_i$ profile (Figure 14.1B):

$$E(r_i)- E(s_i)+\sum_{j\neq i} \min_{t_j}\left[E(r_i,t_j)- E(s_i,t_j)\right]>0 \tag{14.4}$$

If this inequality holds, the higher-energy rotamer $r_i$ can be determined to be dead-ending. Using a similar argument, Equation 14.4 can be extended to the elimination of pairs of rotamers inconsistent with the GMEC. This is done by determining that a pair of rotamers, $r_i$ at position $i$ and $t_j$ at position $j$, always contribute higher energies than the two rotamers, $s_i$ and $u_j$, with all possible rotamer combinations occupying the rest of the sequence:

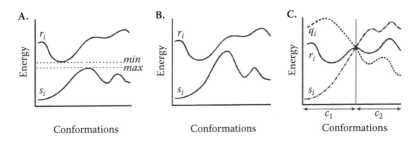

**FIGURE 14.1**   Rotamer elimination using different DEE criteria. (A) Original DEE criterion (Equation 14.3). Rotamer $r_i$ is eliminated by rotamer $s_i$ since the energy of rotamer $r_i$ is greater than that of rotamer $s_i$ for the whole conformational space. (B) Goldstein DEE criterion (Equation 14.4). The maximum of the energy of rotamer $s_i$ is greater than the minimum of the energy of rotamer $r_i$. However, at each point in the conformational space, the energy of rotamer $r_i$ is greater than that of rotamer $s_i$; hence $r_i$ can be eliminated. (C) Split DEE criterion. Neither rotamer $s_i$ nor rotamer $q_i$ display lower energy than rotamer $r_i$ if the whole conformational space is considered. However, if the conformational space is divided into two subspaces, $c_1$ and $c_2$, rotamer $r_i$ can be eliminated by rotamer $s_i$ in $c_1$ and by rotamer $q_i$ in $c_2$.

$$\left[E(r_i)+E(t_j)+E(r_i,t_j)\right]-\left[E(s_i)+E(u_j)+E(s_i,u_j)\right]$$
$$+\sum_{k\neq i,j}\min_{v_k}\left\{\left[E(r_i,v_k)+E(t_j,v_k)\right]-\left[E(s_i,v_k)+E(u_j,v_k)\right]\right\}>0 \qquad (14.5)$$

This doubles criterion leads to elimination of the pair of rotamers $\{r_i,t_j\}$ without completely eliminating the individual rotamers. The doubles elimination step reduces the number of possible pairs that need to be evaluated in Equation 14.4, allowing more rotamers to be individually eliminated. Iterative cycling between the singles (Equation 14.4) and the doubles (Equation 14.5) elimination steps progressively reduces the number of remaining rotamers.

Several modifications of DEE have been proposed to further enhance performance. Rotamers from multiple residues can be combined into so-called super-rotamers to prompt further eliminations (Merz Jr. and Le Grand 1994; Goldstein 1994). This has the advantage of eliminating multiple rotamers in a single step.

Split DEE (Pierce et al. 2000; Looger and Hellinga 2001) was introduced to enable further elimination of rotamers by partitioning the conformational space into subspaces. Rotamer $r_i$ can be eliminated if for every subspace at all other positions, there exists a (possibly distinct) rotamer $s_i$ that can eliminate it (Figure 14.1C). Thus, it may be the case that $r_i$ can be eliminated by rotamer $s_i$ in one subspace and by rotamer $q_i$ in another subspace. Usually, the number of splitting positions is kept relatively small since increasing the number of splitting positions results in a substantial rise in calculation time, negating the advantage of split DEE. The magic-bullet splitting approach introduced a heuristic strategy for picking the split position that would result in the highest probability of elimination of rotamer $r_i$, thus minimizing the computation time (Pierce et al. 2000). A procedure analogous to doubles DEE—called *splitflagging*—further enhances performance of split DEE. Consider a scenario where rotamer $r_i$ cannot be eliminated because there are some partitions in which no rotamer $s_i$ dominates $r_i$. It may be still possible to identify dead-ending pairs during the process of discovering this negative result. These pairs of rotamers are flagged as dead-ending and are not considered in further calculations. The computational cost of finding such pairs is negligible, and the overall calculation time can be reduced by about one-half (Pierce et al. 2000).

Another enhancement to the DEE algorithm comes from introducing a bounding expression (Gordon et al. 2003). Rather than eliminating rotamers by comparing them to other rotamers at the same position, bounding expressions for particular rotamers (at a subset of positions) seek to produce a lower bound on the total energy of any rotameric assignment ($E(r)$) that includes these rotamers. If this bound is higher than the energy of some known reference sequence, then these rotamers cannot be compatible with the GMEC. The reference energy can be computed using a MC search over the current conformational space. Because the MC is repeated periodically as rotamers are eliminated, the searches are performed on a shrinking conformational space. Therefore, the reference energy will decrease or stay the same, so that it should become easier to eliminate more rotamers. Bounding criteria can be used to eliminate individual rotamers and also to flag pairs of rotamers. The flagging of rotamer pairs with a bounding expression is a procedure orthogonal to the DEE criteria and hence has the potential to independently augment the DEE eliminations, further maximizing algorithm performance.

To overcome one of the major limitations in protein design, the fixed-backbone approximation, a provably accurate DEE-based algorithm, was developed to incorporate backbone flexibility (Geogiev and Donald 2007). The authors permit backbone movement by allowing for a range of backbone angles with real-space restraint volumes. The modified DEE criterion uses upper and lower bounds on rotameric interaction energies, within the specified ranges of backbone angles, to eliminate rotamers that are provably not part of the flexible-backbone GMEC. This algorithm was tested on two biological examples and in both cases found sequences with lower computed energies than those found by the standard fixed-backbone DEE procedure.

In some protein design problems, it is of interest to determine not only the lowest energy sequence but to compile a list of sequences close to the GMEC. For this purpose, the DEE algorithm was combined with the A* algorithm (Leach and Lemon 1998). In this approach, DEE is first used with a less stringent criterion that excludes the rotamers incompatible with states within a given energy threshold from the GMEC. Then, the tree-based A* search algorithm is applied to systematically enumerate the rotamer assignments in order of increasing energy. The choice of an appropriate energy threshold value, however, is not trivial, since it is necessary to prune as much of the rotamer space as possible while maintaining the desired number of lowest energy sequences.

More recently, an alternative algorithm for providing a gap-free list of the lowest energy states, termed *extended DEE* (X-DEE), was developed (Kloppmann et al. 2007). The basic idea of X-DEE is to exclude a list of states from the search space explored by DEE. If a gap-free list of $M$ rotameric assignments $\{r^l,...,r^M\}$ is known, rotamer assignment $M + 1$ is found by applying DEE to a large number of disjoint subspaces. These subspaces are designed to exclude the previously determined lowest energy assignments (or those given as input). If the input list of rotamer assignments is not gap-free, the algorithm fills in the gaps, starting from the lower energy states. Although DEE/A* and X-DEE provide a provable method of obtaining the list of the $M$ lowest energy sequences, both algorithms have the disadvantage of high computational time and hence become infeasible for medium and large protein design problems (Fromer and Yanover 2009).

Historically, DEE was developed as an algorithm for prediction of side-chain conformations (Desmet et al. 1992). Subsequently, it has been successfully applied to numerous computational and experimental protein design studies. The most notable of these include the full sequence design of a zinc finger domain (Dahiyat and Mayo 1997), a hyperthermophilic protein (Malakauskas and Mayo 1998), biologically active enzymes (Bolon and Mayo 2001; Jiang et al. 2008), a novel sensor protein (Looger et al. 2003), and stabilization of a protein in an active ligand-bound conformation (Shimaoka et al. 2000; Marvin and Hellinga 2001). DEE remains the best provably exact search algorithm available for protein design thus far.

## FASTER

The fast and accurate side-chain topology and energy refinement (FASTER) method (Desmet et al. 2002) is a heuristic iterative optimization method inspired by the previous success of the dead-end elimination (DEE) techniques. It attempts to take advantage of the elimination power of DEE without requiring its large computational run-times, while trading off a certain degree of optimality. In general, it does so by assuming a given rotamer assignment at all but one protein position and then exhaustively optimizing the remaining position [i.e., first-order "quenching" (Voigt et al. 2000)]. This method is related to DEE in the following way: If all positions except one are already assigned the respective rotamer from the GMEC, this quenching approach will correctly obtain the GMEC rotamer for the remaining position.

In detail, FASTER iteratively proceeds in attempting to approach the GMEC. Initially, pairwise interactions are ignored and it trivially optimizes each rotamer position individually, considering only the rotamer self-energies:

$$r_i^0 = \arg\min_{r_i} \left[ E(r_i) \right] \tag{14.6}$$

yielding the starting rotamer assignment at position $i$ $(r_i^0)$. At a subsequent iteration $s$, the rotamer choice at position $i$ $(r_i^s)$ is reoptimized within the context of the previous iteration's rotamer configuration ($r^{s-1}$, the intermediate minimum energy configuration, IMEC):

$$r_i^s = \arg\min_{r_i} \left[ E(r_i) + \sum_{j \neq i} E(r_i, r_j^{s-1}) \right] \tag{14.7}$$

This update rule (Figure 14.2C) is employed to simultaneously make new rotamer choices at all positions $i$.

There exist various optimizations for FASTER (both integral and optional steps). These include the perturbation-relaxation-evaluation (PRE) procedure, where a given residue is fixed to a given rotamer choice ("perturbation") and then Equation 14.7 is used to update all positions *other* than the currently fixed position ("relaxation"). This is performed for all residues and for each possible rotamer choice, where for each such per-residue rotamer fixation, the rotamer assignment yielded is kept only if it has obtained the minimal energy observed thus far ("evaluation"). Furthermore, PRE can additionally be performed for higher orders (two or more fixed positions and application of Equation 14.7 for nonfixed positions), with an accompanying increase in run-time (e.g., cubic time for two fixed positions).

Additional preprocessing and midprocessing (e.g., considering only pairs of proximate residues in second-order PRE) conditions are applied to obtain further computational speedups. Further significant run-time enhancements (up to two orders of magnitude) were provided in improvements of the initialization method by using MC-derived low energy configurations instead of Equation 14.6. Improvements were

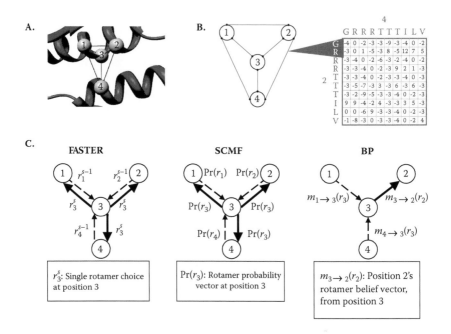

**FIGURE 14.2 (see color insert following page 178)** Comparison of iterative update techniques (FASTER, SCMF, BP) for computational protein design. (A) Toy protein design problem, consisting of four fully interconnected design positions. (B) The corresponding graph structure of this design problem, where each node corresponds to a position and edges to interactions between them (pairwise rotamer-rotamer energy matrices). (C) The calculations performed for position 3 at a given iteration, where dashed arrows indicate relevant incoming data and solid arrows the results of the update calculations performed for position 3. Note that for both FASTER and SCMF, a single calculation is performed for each position, the results of which are then forwarded to all other positions. In BP, however, a distinct calculation is performed for each neighbor of position 3.

also suggested to the relaxation step of the PRE scheme by exclusively relaxing positions that, energetically, interact strongly with the perturbed positions (Allen and Mayo 2006).

FASTER is the overall fastest search algorithm developed thus far and it was in fact shown to obtain the GMEC in many problems (Desmet et al. 2002; Allen and Mayo 2006). Moreover, it has been enhanced to provide sequences satisfying specific biological constraints, such as fixed amino-acid composition (Horn and Mayo 2006), in order to keep reference state energies relatively constant. This method also provides a collection of low energy rotamer assignments, which can also be utilized to approximate positional rotamer and amino-acid probabilities.

The FASTER algorithm has been applied in a number of experimental protein design studies. Some examples include the development of a monoclonal antibody directed against the von Willebrand factor that inhibits its interaction with fibrillar collagen (Staelens et al. 2006) (though it was used only for side-chain optimization of fixed mutants), the full-sequence design of the engrailed homeodomain (Shah et al. 2007), and the comparison of computationally designed single-site substitutions in the triple-hairpin structure of human immunodeficiency virus gp41 with patient data (Boutonnet et al. 2002).

## SELF-CONSISTENT MEAN FIELD (SCMF)

The methodologies based on self-consistent mean field (SCMF) theory (Lee 1994; Koehl and Delarue 1994; Delarue and Koehl 1997; Voigt et al. 2000) calculate positional rotamer probabilities for the protein design problem. The mean field approximation is derived under the assumption that the probability distribution for a full rotamer assignment is the product of positional rotamer probabilities (Yedidia et al. 2005). We note that SCMF was the first method to specifically focus on the calculation of rotamer probabilities (and not specific rotamer assignments), thus setting it aside from previously applied techniques.

Intuitively, for a given position $i$, its multiple interactions with all other positions is summarized into an average interaction (mean field, MF), based on the current conformational probability vectors at interacting positions. Mathematically, the mean field applied to position $i$ in rotamer state $r_i$ is

$$MF(r_i) = E(r_i) + \sum_{j \in N(i)} \sum_{r_j} Pr(r_j) \cdot E(r_i, r_j) \qquad (14.8)$$

where $N(i)$ denotes the set of positions neighboring (interacting with) position $i$ and $Pr(r_j)$ the probability of rotamer $r_j$ at position $j$. This averaging approximation reduces the computational complexity of the problem, since for a given mean field energy applied to a position, the rotamer probabilities at that position can be directly determined using Boltzmann's law (Huang 1987; Yedidia et al. 2005):

$$Pr(r_i) = \alpha_i \cdot e^{\frac{-MF(r_i)}{kT}} \qquad (14.9)$$

where $\alpha_i$ is a normalization factor, $T$ is the system temperature, and $k$ is Boltzmann's constant. Equations 14.8 and 14.9 are iteratively applied at all positions until convergence (Figure 14.2C). At termination, the amino-acid probabilities are calculated from the rotamer probabilities:

$$Pr(S_i = a) = \sum_{r_i \text{ is a rotamer of aa } a} Pr(r_i) \qquad (14.10)$$

where $Pr(S_i = a)$ is the probability of amino acid $a$ at position $i$.

Since SCMF provides amino-acid probabilities and it is always guaranteed to converge (albeit to a local optimum in the approximated probability space), it has unique advantages over alternative design methods. There do exist richer and more general mean field approximations [e.g., (Jordan et al. 1999)], which could provide better solutions in practice. However, these have yet to be actively explored or assessed in the world of computational protein design. Nonetheless, similar approaches have been further refined in the context of protein design. For example, the SCADS (statistical computationally assisted design strategy) (Kono and Saven 2001; Calhoun et al. 2003) and related methods (Zou and Saven 2003; Biswas et al. 2005) generalize mean field theory and apply a self-consistent statistical entropy approach to atomic level protein design in order to predict these probabilities, in a way that is often robust to minor backbone changes.

We note that SCMF, which is designed to directly calculate amino-acid probabilities, could also be used to obtain low energy sequences, by *independently* choosing the most probable amino acids at each position. However, it has been shown that this has only limited success in yielding the GMEC (Voigt et al. 2000). Nonetheless, the SCADS method was successfully applied in the design of 88 residues of a monomeric helical dinuclear metalloprotein, shown to be structured in both the apo and the holo forms (Calhoun et al. 2003). This method was further used to design water-soluble analogues of the potassium channel KcsA (Slovic et al. 2004), a four-helix bundle protein that selectively binds a nonbiological cofactor (Cochran et al. 2005), a redox-active minimal rubredoxin mimic (Nanda et al. 2005), an ultrafast folding Trp-cage mutant (Bunagan et al. 2006), and ferritin-like proteins with hydrophobic cavities (Swift et al. 2006).

## BELIEF PROPAGATION (BP)

Belief propagation (BP) (Pearl 1988; Yedidia et al. 2005) is an approximate inference technique applied to a formulation of protein design known as probabilistic graphical models (Lauritzen 1996). Here, we describe the form of BP utilized to calculate positional amino-acid probabilities (Moore and Maranas 2003; Fromer and Yanover 2008). For calculation of low energy sequences, a corresponding algorithm also exists (Hong and Lozano-Perez 2006; Yanover et al. 2006; Fromer and Yanover 2009).

Briefly, a graph is built where nodes represent design positions, node values correspond to rotamer choices, and edges between nodes represent pairwise rotamer interactions between respective positions (Figure 14.2A,B) (Hong and Lozano-Perez 2006; Fromer and Yanover 2009). In belief propagation, messages are passed between interacting positions, whose contents describe one position's "belief" about its neighbor—the relative likelihood of each rotamer at the neighboring position (Figure 14.2C).

A message from one node to a neighboring one is proportional to their pairwise energy (as a Boltzmann probability) and to the product of all incoming messages. Specifically, the message passed from position $i$ to position $j$ takes the form

$$m_{i \to j}(r_j) = \sum_{r_i} \left( e^{\frac{-E(r_i)-E(r_i,r_j)}{kT}} \cdot \prod_{n \in N(i) \setminus j} m_{n \to i}(r_i) \right) \qquad (14.11)$$

where the contents of $m_{i \to j}$ indicate the relative likelihood of each rotameric state at position $j$. $N(i)$ denotes the set of positions neighboring position $i$, $T$ is the system temperature, and $k$ is Boltzmann's constant.

After convergence of BP, rotamer probabilities at each design position are assigned as follows:

$$\Pr(r_i) = \alpha_i \cdot e^{\frac{-E(r_i)}{kT}} \cdot \prod_{n \in N(i)} m_{n \to i}(r_i) \tag{14.12}$$

where $\alpha_i$ is a normalization factor. Note that these probabilities attempt to account for all possible sequences at all other designed positions, and are used to compute amino-acid probabilities as in Equation 14.10.

One of the major advantages of BP is that it has been shown to outperform other methods in certain protein design settings: in quickly and accurately predicting amino acid probabilities (Fromer and Yanover 2008) and in precisely discovering an ensemble of low energy sequences (Fromer and Yanover 2009). It thus provides a flexible framework to directly compute either low energy sequences or amino-acid probabilities, or both. As an example of its usefulness, BP was applied to identify residue-residue clashes in computational hybrids of the *E. coli* and human versions of glycinamide ribonucleotide (GAR) transformylase, with a large degree of success (Moore and Maranas 2003).

However, a major hurdle for BP is the lack of convergence observed, especially for larger design problems (Fromer and Yanover 2009). We note that the tree-reweighted belief propagation (TRBP) algorithm is a variant of BP that *guarantees* to find the GMEC if it converges to a unique solution at each position (Yanover et al. 2006; Weiss et al. 2007). Intensifying research to find both convergent and accurate BP algorithms (Weiss et al. 2007; Globerson and Jaakkola 2007) indicates that BP will continue to be a promising method for accurate large-scale protein design in the future, even when using relatively sophisticated energy functions.

We now find it instructive to compare the FASTER, SCMF, and BP algorithms, all three of which perform some form of iterative rotamer updates, which are exchanged between positions. In FASTER, at a given iteration each design position is assigned a specific rotamer, which is then used to update the rotamer choices at other positions. However, in both SCMF and BP, each position maintains (or receives) a probability vector over rotamer choices. These probabilities are then used to update the probability vectors at other positions. In contrast to BP, in both FASTER and SCMF, each position decides on specific rotamer choices (a single choice in FASTER or a probability vector in SCMF) and identically presents those choices to *all* other positions (Figure 14.2C). In BP, however, each position sends different rotamer probability vectors to its various neighbors; this subtlety may partially contribute to the success of BP, since it bypasses some of the inherent problems associated with such iterative updating performed on cycles of positions (e.g., Figure 14.2A,B).

## LINEAR PROGRAMMING (LP)

Linear programming (LP) is a general mathematical technique for global optimization of a linear objective function, subject to linear equality constraints. Integer linear programming (ILP) adds the additional constraint that the solution must contain integer values, but makes the computational problem intricately harder.

Inspection of Equation 14.1 indicates that it essentially decomposes the energy of a complete rotamer configuration into a linear sum of energies for positions and pairs of positions. The LP/ILP approach takes advantage of this fact and casts the search for the minimal energy rotamer assignment as an integer linear programming (ILP) problem (Eriksson et al. 2001; Kingsford et al. 2005). Variables $x_{r_i}$ and $x_{r_i,r_j}$ are introduced for each rotamer $r_i$ at position $i$ and for each rotamer pair $r_i,r_j$ at positions $i,j$, respectively. The objective function for minimization through linear programming is then

$$\sum_i \sum_{r_i} x_{r_i} \cdot E(r_i) + \sum_{i,j} \sum_{r_i,r_j} x_{r_i,r_j} \cdot E(r_i,r_j) \qquad (14.13)$$

The ILP formulation for exact calculation of the GMEC is obtained by constraining the variables to (1) be binary integers (can assume only the values 0 or 1), and (2) consistently assign a unique rotamer at each position. A value of 1 for a variable indicates that the corresponding rotamer (or rotamer pair) is chosen in the ILP solution. The LP formulation simply ignores the requirement that the variables be integers and allows them to vary continuously in the range between 0 and 1. This linear programming (LP) relaxation is efficiently solved, possibly yielding a solution in which all variables are integers. Otherwise, a computationally more intensive ILP solver is required. Thus, the LP/ILP approach is guaranteed to obtain the lowest energy rotamer assignment, albeit with no guarantee of a nonexponential run-time. In fact, even when using a simple energy function (van der Waals interactions and a statistical rotamer self-energy), the ILP solver was required for most design problems tested, resulting in very large run-times (Kingsford et al. 2005). Nonetheless, similar approaches have been taken to perform protein library redesign, as computationally demonstrated for a 16-member library of *E. coli/B. subtilis* dihydrofolate reductase hybrids (Saraf et al. 2006).

## REDUCTION OF THE SEARCH SPACE BY PREPROCESSING

Several preprocessing algorithms have been developed to be run prior to any search algorithm and to substantially reduce the conformational search space, with only minimal or no loss of accuracy. A simple preprocessing step eliminates rotamers (or pairs of rotamers) that bear a self-energy (or rotamer-rotamer energy) above a certain cutoff value (Desmet et al. 2002; Dahiyat and Mayo 1997).

A slightly more elaborate procedure prunes rotamers that are outside of a certain cutoff from the energy required by the GMEC. For example, the Vegas algorithm described by Shah and coworkers (Shah et al. 2004) considers each position $i$ and fixes each rotamer $r_i$ at this position. For the fixation of rotamer $r_i$, an MC search is utilized to approximate the optimal sequence conformation for the rest of the molecule. The energy of this approximated GMEC (with position $i$ fixed to $r_i$) is defined as the score of $r_i$. At position $i$, all such scores are compared and only the lowest scoring rotamers (those with a score within a predefined cutoff of the best score) are retained.

Another rotamer preprocessing step, called *type-dependent DEE*, has been introduced for multistate protein design problems (Yanover et al. 2007). In multistate design, a single sequence is evaluated in the context of two or more protein backbone conformations. Type-dependent DEE eliminates rotamers that are incompatible with either structure, while nevertheless preserving the minimal energy conformational assignment for any choice of amino-acid sequence in any of the backbone conformations considered. This is performed by allowing only rotamers of the same amino-acid type to eliminate one another.

## SUMMARY

At this point, we perform a high-level comparison of the computational protein design techniques described in the previous sections. Table 14.1 compares the main characteristics of the methods, in terms of the tasks for which they are intended and of their theoretical exactness and convergence guarantees.

To select the computational protein design method to be applied to a particular problem, the final goal must be kept in mind. If one is interested in obtaining a small set of low energy protein sequences, as is typically the case, then all methods except for self-consistent mean field (SCMF) can be readily applied. On the other hand, for combinatorial library design, only the SCMF and BP algorithms were originally

## TABLE 14.1
## Comparison of Various Techniques for Computational Protein Design

|  |  | Low Energy Sequence(s) | Positional Probabilities | Provably Exact | Guaranteed Convergence |
|---|---|---|---|---|---|
| Stochastic | MC | + | $-^b$ | − | + |
|  | GA | + | $-^b$ | − | + |
|  | FASTER | + | $-^b$ | − | + |
| Elimination | DEE | + | $-^b$ | + | − |
| Probabilistic | SCMF | $-^a$ | + | − | + |
| Framework | BP | + | + | $-/+^c$ | − |
| Constrained Optimization | LP | + | $-^b$ | + | − |

[a] Low-energy sequences can be obtained by choosing high probability rotamers independently for each position.

[b] Probabilities can be estimated from the frequencies in an ensemble of low energy sequences.

[c] The special case of tree-reweighted BP (TRBP) will yield the GMEC, if it converges (Yanover et al. 2006).

intended to directly calculate amino-acid probabilities at each position. Nonetheless, an ensemble of low energy sequences calculated by other methods can be utilized to estimate these probabilities as well.

As shown in Table 14.1, there exists a computational trade-off between exactness and convergence, which derives from the computational hardness of protein design (Pierce and Winfree 2002). Therefore, no algorithm can, in all cases, provide optimal results in reasonable time.

The stochastic algorithms may require multiple runs to ensure near-optimal results, due to the random choices made during the algorithm run. Nonetheless, they possess the distinct advantage of obtaining solutions for even very large problems, albeit without a guarantee of optimality.

All other algorithms are deterministic, and some are guaranteed to provide optimal energetic solutions, if they converge. Dead-end elimination (DEE) techniques and their generalizations have been shown to be successful even in relatively large design problems. Similarly to DEE, linear programming (LP) will always yield the lowest energy sequence(s) upon convergence, but has the disadvantage of very significant run-times.

In contrast to DEE and LP, both self-consistent mean field (SCMF) and belief propagation (BP) consider the computational design problem from a probabilistic viewpoint. While SCMF is guaranteed to obtain a solution, this is not true for BP. Nonetheless, certain BP methods are guaranteed to find the exact GMEC, if they converge.

We conclude by noting that the stochastic FASTER algorithm currently seems to be the most powerful algorithm for computational protein design. This algorithm is indeed the fastest and although not provably exact, it very often obtains GMEC solutions.

## REFERENCES

Allen, B. D., and S. L. Mayo. 2006. Dramatic performance enhancements for the FASTER optimization algorithm. *J Comp Chem* 27: 1071–5.

Ambroggio, X., and B. Kuhlman. 2006. Computational design of a single amino acid sequence that can switch between two distinct protein folds. *J Am Chem Soc* 128(4): 1154–61.

Ashworth, J., J. J. Havranek, C. M. Duarte, D. Sussman, R. J. Monnat, B. L. Stoddard, and D. Baker. 2006. Computational redesign of endonuclease DNA binding and cleavage specificity. *Nature* 441: 656–9.

Biswas, P., J. Zou, and J. G. Saven. 2005. Statistical theory for protein ensembles with designed energy landscapes. *J Chem Phys* 123(15): 15490–8.

Bolon, D. N., and S. L. Mayo. 2001. Enzyme-like proteins by computational design. *Proc Natl Acad Sci U S A* 98(25): 14274–9.

Boutonnet, N., W. Janssens, C. Boutton, J.-L. Verschelde, L. Heyndrickx, E. Beirnaert, G. van der Groen, and I. Lasters. 2002. Comparison of predicted scaffold compatible sequence variation in the triple-hairpin structure of human immunodeficiency virus type f gp4f with patient data. *J Virol* 76(I5): 7595–7606. http://jvi.asm.org/cgi/reprint/76/15/7595.pdf.

Bunagan, M., X. Yang, J. Saven, and F. Gai. 2006. Ultrafast folding of a computationally designed trp-cage mutant: $Trp^2$-cage. *J Phys Chem B* 110(8): 3759–63.

Butterfoss, G. L., and B. Kuhlman. 2006. Computer-based design of novel protein structures. *Ann Rev Biophys Biomol Struct* 35(1): 49–65. http://arjournals.annualreviews.org/doi/pdf/10.1146/annurev.biophys.35.040405.102046.

Calhoun, J. R., H. Kono, S. Lahr, W. Wang, W. F. DeGrado, and J. G. Saven. 2003. Computational design and characterization of a monomeric helical dinuclear metalloprotein. *J Mol Biol* 334: 1101–15.

Cochran, F., S. Wu, W. Wang, V. Nanda, J. Saven, M. Therien, and W. DeGrado. 2005. Computational de novo design and characterization of a four-helix bundle protein that selectively binds a nonbiological cofactor. *J Am Chem Soc* 127(5): 1346–47.

Cootes, A. P., P. M. G. Curmi, and A. E. Torda. 2000. Biased Monte Carlo optimization of protein sequences. *J Chem Phys* 113(6): 2489–96.

Dahiyat, B. I., and S. L. Mayo. 1997. De novo protein design: Fully automated sequence selection. *Science* 278: 82–7.

Delarue, M., and P. Koehl. 1997. The inverse protein folding problem: Self consistent mean field optimisation of a structure specific mutation matrix. In *Pacific symposium on biocomputing,* ed. R.B. Altman, A.K. Dunker, L. Hunter, and T. Klein, 109–21. Singapore: World Scientific.

Desjarlais, J. R., and T. M. Handel. 1995. De novo design of the hydrophobic cores of proteins. *Protein Sci* 4(10): 2006–18.

Desmet, J., M. De Maeyer, B. Hazes, and I. Lasters. 1992. The dead-end elimination theorem and its use in protein side-chain positioning. *Nature* 356: 539–42.

Desmet, J., J. Spriet, and I. Lasters. 2002. Fast and accurate side-chain topology and energy refinement (FASTER) as a new method for protein structure optimization. *Proteins: Struc Funct Genet* 48: 31–43.

Eriksson, O., Y. Zhou, and A. Elofsson. 2001. Side chain-positioning as an integer programming problem. In *Wabi 01: Proceedings of the first international workshop on algorithms in bioinformatics,* 128–41. London, UK: Springer-Verlag.

Fromer, M., and C. Yanover. 2008. A computational framework to empower probabilistic protein design. *Bioinformatics* 24(13): i214–22. http://bioinformatics.oxfordjournals.org/cgi/reprint/24/13/i214.pdf.

Fromer, M., and C. Yanover. 2009. Accurate prediction for atomic-level protein design and its application in diversifying the near-optimal sequence space. *Proteins: Struct Funct Bioinform* 75(3): 682–705.

Georgiev, I., and B. R. Donald. 2007. Dead-end elimination with backbone flexibility. *Bioinformatics* 23(13): il85–94. http://bioinformatics.oxfordjournals.org/cgi/reprint/23/13/il85.pdf.

Globerson, A., and T. Jaakkola. 2007. Fixing max-product: Convergent message passing algorithms for MAP LP-relaxations. In *Advances in neural information processing systems 21.*

Goldstein, R. F. 1994. Efficient rotamer elimination applied to protein side-chains and related spin glasses. *Biophys J* 66(5): 1335–40. http://www.biophysj.org/cgi/reprint/66/5/1335.pdf.

Gordon, D. B., G. K. Horn, S. L. Mayo, and N. A. Pierce. 2003. Exact rotamer optimization for protein design. *J Comput Chew,* 24(2): 232–43.

Hansmann, U. H. E. 1997. Parallel tempering algorithm for conformational studies of biological molecules. *Chem Phys Lett* 281(1–3): 140–50.

Havranek, J. J., and P. B. Harbury. 2003. Automated design of specificity in molecular recognition. *Nat Struct Mol Biol* 10: 45–52.

Holland, J. H. 1992. *Adaptation in natural and artificial systems.* Cambridge, MA: MIT Press.

Horn, G. K., and S. L. Mayo. 2006. A search algorithm for fixed-composition protein design. *J Comp Chem* 27: 375–8.

Hong, E.-J., and T. Lozano-Perez. 2006. Protein side-chain placement through MAP estimation and problem-size reduction. In *Wabi,* ed. Philipp Bucher and Bernard M. E. Moret, vol. 4175 of *Lecture Notes in Computer Science,* 219–230. Springer.

Huang, K. 1987. *Statistical mechanics.* New York: John Wiley and Sons, Inc.

Jiang, L., E. A. Althoff, F. R. Clemente, L. Doyle, D. Rothlisberger, A. Zanghellini, J. L. Gallaher, J. L. Betker, F. Tanaka, I. Barbas, Carlos F., D. Hilvert, K. N. Houk, B. L. Stoddard, and D. Baker. 2008. De Novo Computational Design of Retro-Aldol Enzymes. *Science* 319(5868): 1387–91. http://www.sciencemag.org/cgi/reprint/319/5868/1387.pdf.

Jones, D. T. 1994. De novo protein design using pairwise potentials and a genetic algorithm. *Protein Sci* 3(4): 567–74.

Jordan, M. I., Z. Ghahramani, T. S. Jaakkola, and L. K. Saul. 1999. *Learning in graphical models,* 105–161. Cambridge, MA: MIT Press.

Kingsford, C. L., B. Chazelle, and M. Singh. 2005. Solving and analyzing side-chain positioning problems using linear and integer programming. *Bioinformatics* 21(7): 1028–39. http://bioinformatics.oxfordjournals.org/cgi/reprint/21/7/1028.pdf.

Kirckpatrick, S., C. D. Gelatt, and M. P. Vecchi. 1983. Optimization by simulating annealing. *Science* 220(4598): 671–80.

Kloppmann, E., G. M. Ullmann, and T. Becker. 2007. An extended dead-end elimination algorithm to determine gap-free lists of low energy states. *J Comp Chem* 28: 2325–35.

Koehl, P., and M. Delarue. 1994. Application of a self-consistent mean field theory to predict protein side-chains conformation and estimate their conformational entropy. *J Mol Biol* 239: 249–75.

Kono, H., and J. G. Saven. 2001. Statistical theory for protein combinatorial libraries, packing interactions, backbone flexibility, and sequence variability of main-chain structure. *J Mol Biol* 306: 607–28.

Kraemer-Pecore, C. M., J. T. J. Lecomte, and J. R. Desjarlais. 2003. A de novo redesign of the WW domain. *Prot Sci* 12(10): 2194–205.

Kuhlman, B., G. Dantas, G. C. Ireton, G. Varani, B. L. Stoddard, and D. Baker. 2003. Design of a novel globular protein fold with atomic-level accuracy. *Science* 302(5649): 1364–68. http://www.sciencemag.org/cgi/reprint/302/5649/1364.pdf.

Lauritzen, S. 1996. *Graphical models.* Oxford, UK: Oxford University Press.

Lazar, G. A., J. R. Desjarlais, and T. M. Handel. 1997. De novo design of the hydrophobic core of ubiquitin. *Protein Sci* 6(6): 1167–78.

Leach, A. R., and A. P. Lemon. 1998. Exploring the conformational space of protein side chains using dead-end elimination and the A* algorithm. *Prot: Struct Funct Genet* 33: 227–39.

Lee, C. 1994. Predicting protein mutant energetics by self-consistent ensemble optimization. *J Mol Biol* 236: 918–39.

Looger, L. L., M. A. Dwyer, J. J. Smith, and H. W. Hellinga. 2003. Computational design of receptor and sensor proteins with novel functions. *Nature* 423(6936): 185–90.

Looger, L. L., and H. W. Hellinga. 2001. Generalized dead-end elimination algorithms make large-scale protein side-chain structure prediction tractable: Implications for protein design and structural genomics. *J Mol Biol* 307: 429–45.

Malakauskas, S. M., and S. L. Mayo. 1998. Design, structure and stability of a hyperthermophilic protein variant. *Nat Struct Biol* 5(6): 470–5.

Marvin, J. S., and H. W. Hellinga. 2001. Manipulation of ligand binding affinity by exploitation of conformational coupling. *Nat Struct Biol* 8(9): 795–8.

Merz Jr., K. M., and S. M. Le Grand, eds. 1994. *The protein folding problem, and tertiary structure prediction.* Birkhauser.

Metropolis, N., A. Rosenbluth, M. Rosenbluth, A. H. Teller, and E. Teller. 1953. Equations of state calculations by fast computing machines. *J Chem Phys* 21: 1087–92.

Moore, G. L., and C. D. Maranas. 2003. Identifying residue-residue clashes in protein hybrids by using a second-order mean-field approach. *Proc Natl Acad Sci* 100(9): 5091–6. http://www.pnas.org/cgi/reprint/100/975091.pdf.

Nanda, V., M. Rosenblatt, A. Osyczka, H. Kono, Z. Getahun, P. Dutton, J. Saven, and W. DeGrado. 2005. De novo design of a redox-active minimal rubredoxin mimic. *J Amer Chem Soc* 127(16): 5804–05.

Park, S., H. Kono, W. Wang, E. T. Boder, and J. G. Saven. 2005. Progress in the development and application of computational methods for probabilistic protein design. *Comp Chem Eng* 29: 407–21.

Park, S., X. Yang, and J. G. Saven. 2004. Advances in computational protein design. *Curr Opin Struct Biol* 14: 487–94.

Pearl, J. 1988. *Probabilistic reasoning in intelligent systems: Networks of plausible inference.* San Francisco, CA: Morgan Kaufmann.

Pierce, N. A., J. A. Spriet, J. Desmet, and S. L. Mayo. 2000. Conformational splitting: A more powerful criterion for dead-end elimination. *J Comp Chem* 21: 999–1009.

Pierce, N. A., and E. Winfree. 2002. Protein design is NP-hard. *Protein Eng* 15(10): 779–82. http://peds.oxfordjournals.org/cgi/reprint/15/10/779.pdf.

Rosenberg, M., and A. Goldblum. 2006. Computational protein design: A novel path to future protein drugs. *Curr Pharm Des* 12: 3973–97.

Saraf, M. C, G. L. Moore, N. M. Goodey, V. Y. Cao, S. J. Benkovic, and C. D. Maranas. 2006. IPRO: An iterative computational protein library redesign and optimization procedure. *Biophys J* 90(ll): 4167–80. http://www.biophysj.org/cgi/reprint/90/ll/4167.pdf.

Shah, P. S., G. K. Horn, and S. L. Mayo. 2004. Preprocessing of rotamers for protein design calculations. *J Comp Chem* 25(14): 1797–1800.

Shah, P. S., G. K. Hom, S. A. Ross, J. K. Lassila, K. A. Crowhurst, and S. L. Mayo. 2007. Full-sequence computational design and solution structure of a thermostable protein variant. *J Mol Biol* 372: 1–6.

Shimaoka, M., J. M. Shifman, H. Jing, J. Takagi, S. L. Mayo, and T. A. Springer. 2000. Computational design of an integrin I domain stabilized in the open high affinity conformation. *Nature Struc Biol* 7(8): 674–8.

Slovic, A. M., H. Kono, J. D. Lear, J. G. Saven, and W. F. DeGrado. 2004. From the cover: Computational design of water-soluble analogues of the potassium channel KcsA. *Proc Natl Acad Sci* 101(7): 1828–33. http://www.pnas.org/cgi/reprint/101/771828.pdf.

Staelens, S., J. Desmet, T. H. Ngo, S. Vauterin, I. Pareyn, P. Barbeaux, I. Van Rompaey, J.-M. Stassen, H. Deckmyn, and K. Vanhoorelbeke. 2006. Humanization by variable domain resurfacing and grafting on a human IgG4, using a new approach for determination of non-human like surface accessible framework residues based on homology modelling of variable domains. *Mol Immun* 43: 1243–57.

Sugita, Y., and Y. Okamoto. 1999. Replica-exchange molecular dynamics method for protein folding. *Chem Phys Lett* 314(1–2): 141–51.

Swift, J., W. Wehbi, B. Kelly, X. Stowell, J. Saven, and I. Dmochowski. 2006. Design of functional ferritin-like proteins with hydrophobic cavities. *J Am Chem Soc* 128(20): 6611–9.

Voigt, C. A., D. B. Gordon, and S. L. Mayo. 2000. Trading accuracy for speed: A quantitative comparison of search algorithms in protein sequence design. *J Mol Biol* 299: 789–803.

Weiss, Y., C. Yanover, and T. Meltzer. 2007. MAP estimation, linear programming and belief propagation with convex free energies. In *The 23rd conference on uncertainty in artificial intelligence.*

Yang, X., and J. G. Saven. 2005. Computational methods for protein design and protein sequence variability: Biased Monte Carlo and replica exchange. *Chem Phys Lett* 401: 205–10.

Yanover, C, M. Fromer, and J. M. Shifman. 2007. Dead-end elimination for multistate protein design. *J Comp Chem* 28: 2122–9.

Yanover, C, T. Meltzer, and Y. Weiss. 2006. Linear programming relaxations and belief propagation—an empirical study. *J Mach Learn Res* 7: 1887–1907.

Yedidia, J. S., W. T. Freeman, and Y. Weiss. 2005. Constructing free-energy approximations and generalized belief propagation algorithms. *Information Theory, IEEE Transactions on* 51: 2282–312.

Zou, J., and J. G. Saven. 2003. Using self-consistent fields to bias Monte Carlo methods with applications to designing and sampling protein sequences. *J Chem Phys* 118(8): 3843–54.

# 15 Modulating Protein Structure

*M.S. Hanes, T.M. Handel, and A.B. Chowdry*

## CONTENTS

In the structural biology of proteins, the dogma is that sequence defines structure, which in turn defines function. In order to design functional proteins, therefore, the first step is to fill in the gap between sequence and structure; if a sequence can be designed for an arbitrary structure, then function should follow. To this end, many design algorithms have focused on the ability to reliably predict sequences that can fold, specifically into a desired tertiary structure. This has proven to be a robust test of our understanding of the forces that govern protein folding.

More than a test, however, protein design has been able to increase the stability of biological proteins, adjust their solubility and oligomerization behavior, and in one case, produce an entirely novel fold. At the same time, design of specific scaffolds and structures is still not trivial, and most successes have been restricted to small scaffolds, subsets of larger proteins, or simple topologies such as helical bundles. This barrier is primarily due to problems with force-field accuracy, the ability of search algorithms to find physically reasonable solutions, and the need to make approximations and compromises in both due to limited computational resources.

## THE CORE REPACKING PROBLEM

The burial of hydrophobic groups is the dominant force determining the folding of an unstructured polypeptide to a functional protein (Dill 1990). Protein cores are composed of nonsequential residues distributed throughout the polypeptide chain that are isolated from exterior solvent and interact primarily via van der Waals (vdW) and hydrophobic forces in the interior of the protein. Cores are characteristically well-packed and lack cavities, and mutational data have shown that mutations in the core often affect thermostability while maintaining the tertiary structure of the protein (Lim et al. 1992; Eriksson et al. 1993; Milla et al. 1994; Munson et al. 1996; Vlassi et al. 1999). The importance of the core to stability and its observed ability to withstand mutation made it an ideal subject for the earliest work in protein design (Hurley et al. 1992; Desjarlais and Handel 1995; Lazar et al. 1997). Restricting the design to core positions and hydrophobic residues reduces the combinatorial problem considerably, as well as simplifying or eliminating the need to model complex solvent interactions present at the protein surface. The dominance of hydrophobic interactions in the core made a simple vdW term sufficient for some problems, which could be modeled using a standard Lennard-Jones 6-12 potential.

## PREDICTING NATIVE PROTEIN CORE SEQUENCES

With few exceptions, the native sequence of a protein can be assumed to be among the most stable sequences for a given fold. Though evolution is driven by additional factors besides thermostability, and sequence variation between species strongly demonstrates that a given fold is compatible with many sequence solutions, the ability of a design algorithm to predict the native sequence in its solution set is an excellent first test of its accuracy. If the native sequence does not score favorably, then it is likely that the force field used in the design has missed an important considerations for the system and must be revised.

One of the first efforts at computationally predicting alternate protein core sequences explicitly considered the need for native-like sequences in the solution set (Ponder and Richards 1987). Their model consisted of a template with fixed main-chain and β-carbon atoms, and side-chain conformations given in a rotamer library. The rotamer library (discussed in Chapter 13) was derived from the existing protein structure database. Protein sequences compatible with the target structure were identified by passing two filter criteria: a vdW and a core volume calculation. A hard-sphere model for vdW repulsions excluded structures with clashes in the core, while a side-chain volume constraint maintained that the new side chains should occupy a similar volume as the wild type. The latter constraint reflects the experimental observation that cavities in the core were typically destabilizing (Eriksson et al. 1992). This eliminated many obviously incorrect sequences, but the resulting set was still too large to experimentally investigate.

Though modern protein design algorithms do not explicitly bias their results toward native sequences, it is still important that the native sequence and similar sequences are scored favorably, even if the algorithm rarely recapitulates the native sequence as the lowest energy configuration.

## EARLY CORE DESIGNS (CRO, UBIQUITIN, AND T4 LYSOZYME)

Early design studies focused on small, single domain proteins with contiguous cores, such as the phage 434 cro protein and ubiquitin. These proteins were useful model systems to study the ability of early design force fields to properly pack the protein core, as they were small enough to be searched deterministically, and predicted mutants could be easily expressed and characterized biophysically.

The first *de novo* core redesign where the resulting protein displayed native-like properties utilized the phage 434 cro protein (Figure 15.1). Desjarlais and Handel used an energy function consisting of only a vdW potential and allowed only hydrophobic residues (V, I, L, F, and W) at core positions (Desjarlais and Handel 1995). Designed proteins that contained as many as eight mutations in the hydrophobic core were experimentally characterized to be stable, with cooperative thermal denaturation transitions. By comparison, a randomly generated control sequence was found to be unstructured. Going one step further in an equivalent study on the core of ubiquitin, this group determined the solution structure of a ubiquitin core variant by nuclear magnetic resonance (NMR) spectroscopy (Figure 15.2) (Lazar et al. 1997). The data confirmed that the backbone structure and side-chain placement were in close agreement with predictions.

A more elaborate force field including solvation and entropic terms was used by Kono and coworkers to redesign the core of malate dehydrogenase (MDH) from *Thermus flavus* (Kono et al. 1998). This protein is significantly larger than other early scaffolds: Two multidomain MDH monomers, each consisting of 327 amino acids, associate to form the functional homodimeric enzyme. Unlike the early single-domain protein model systems, MDH contains multiple hydrophobic cores, and as a result the designed positions were restricted to a single core. Out of 11

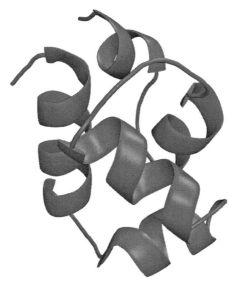

**FIGURE 15.1**   The 434 cro repressor protein is a small, globular, helical protein that was used as a model system for core design.

**FIGURE 15.2**  The fold of ubiquitin contains both helical and sheet structure surrounding one contiguous core. Designed core variant side chains are shown in black side chains while the wild-type counterparts are in gray.

possible core positions, they found that only three were mutated in the majority of their high-scoring sequences, with one being predicted to have higher stability than wild type. Chemical denaturation experiments on three variants showed that they all had increased stability compared to wild type. All of the mutations were conservative and increased side-chain volume relative to wild type, which likely enhanced vdW interactions in the protein core.

Extending this success to the bacteriophage T4 lysozyme illustrated some important limitations in protein design (Mooers et al. 2003). Using an energy function including vdW, hydrogen bonding, electrostatic, and solvation terms, 26 core positions were considered for mutation. The designed variants retained similar thermostability to the native lysozyme, but x-ray crystallography revealed backbone deviations up to 2.8 Å from wild type. Although these changes are not large enough to have a significant impact on the fold of the protein, they violate an important assumption of the protein design model. Backbone deviations of even 1 Å are known to have significant effects on the predictions of design force fields, which depend on accurate atomic positioning. Another observation from this study was that large, flexible residues such as methionine were preferentially selected, due to their ability to create extensive vdW interactions and pack into many configurations.

Though there have been many successes with computational repacking and stabilization of protein cores, these have been contingent on the simplicity of the problem. The effects of using a rotamer library and fixed backbone are minimized, while the hydrophobic nature of the core makes it easy to estimate, or even ignore, solvation effects. In order to progress from core designs and apply computational methods to the structural design of whole proteins, each of these approximations must be overcome.

## OVERCOMING LIMITATIONS: THE ROTAMER LIBRARY

The use of discrete rotamers to approximate the orientations of side chains is reasonable in the core of a protein, as crystallographic data has suggested that most buried residues adopt rotameric positions. The surrounding space for a given residue is highly constrained by its neighbors and backbone. Steric restrictions help to guide a design algorithm to an obvious energetic minimum. Still, the size and quality of the rotamer library is important to accurately orient side chains in the protein core.

The details of rotamer library use and construction have been discussed (see Chapter 13). Regardless of the library used, discrete rotamers will never be able to represent all possible orientations of side-chain residues in a protein core, and atomic clashes are unavoidable. As a result, sometimes potential design candidates are eliminated due to artificially unfavorable energy scores, when in fact, minor rotamer adjustments would result in favorable solutions. Under a fixed backbone, there are two approaches used to alleviate clashes caused by the inexact positioning of rotamers: local minimization and softening the vdW potential.

Given unlimited computational power, each choice of residue identity would be locally minimized, relative to its neighbors, to find the lowest energy conformation. The torsion angles can be optimized via conjugate gradient, but this method is computationally expensive. When designing systems with more than a few variable positions, local minimization of this sort is currently intractable. It is, however, widely used in situations where there are few sequences to explore, such as protein fold prediction and docking simulations. Its use in protein design may grow as computational power increases.

Almost universal in its prevalence is the use of softened vdW potentials during protein design calculations (Dahiyat and Mayo 1997; Lazar et al. 1997; Kuhlman and Baker 2000; Havranek and Harbury 2003; Pokala and Handel 2005). There are many approaches, but the simplest involves scaling the equilibrium radius of the atoms. This has an obvious physical interpretation of shrinking the size of each atom, allowing it to pack next to its neighbors, instead of making an unfavorable and repulsive steric clash under an unmodified potential.

The effectiveness of this approach was demonstrated in a study where various values for this scaling factor ($\alpha$) were investigated on the core of streptococcal G $\beta$1 domain (GB1) (Figure 15.3) (Dahiyat and Mayo 1997). Three regions of behavior were identified. Between $\alpha > 0.90$ and $\alpha < 1.05$, packing constraints dominated the design, and sequences were similar to wild type GB1. Values for $\alpha < 0.90$ allowed larger residues to insert into the core, and hydrophobic interaction dominating packing constraints. Conversely, using $\alpha > 1.05$ was too rigid, and many positions were chosen as alanine. Experimental characterization of all the designs showed that only those using $\alpha = 0.90$ resulted in stable, well-folded proteins. Larger $\alpha$ values produced unfolded chains, due to an inability to pack the core, while smaller $\alpha$ values produced proteins with greater mobility and loose structure.

**FIGURE 15.3** GB1 is a small, globular protein containing both helical and sheet secondary structure.

## OVERCOMING LIMITATIONS: CIRCUMVENTING THE FIXED BACKBONE

The fixed backbone simplifies the size of the design problem by reducing it to a choice of residue types and orientations along a rigid frame. It also removes the reliance on the design force field to discriminate between competing backbone structures. Since core designs start with a wild-type protein template, the choice of backbone position is obvious and likely to be reasonable, as nature has already evolved toward it. This assumption breaks down, however, when the design template is not a natural protein and the initial backbone position may be suboptimal or impossible. It is also problematic when stabilizing residue choices require backbone shifts in order to pack into the protein core, which tends to be the rule rather than the exception.

One of the earliest attempts to incorporate backbone flexibility into computational design was the work of Harbury and coworkers on coiled coils (Harbury et al. 1993). This system has a high degree of symmetry and simple topology that allowed the use of algebraic parameterization to predict an ensemble of backbone conformations. A fixed number of backbone orientations and side-chain identities were searched. Dimer, trimer, and tetramer configurations were designed and characterized via circular dichroism, sedimentation equilibrium, and, in the case of the tetramer, crystallography. Each of the designs adopted the expected oligomerization state and the tetramer differed from the predicted structure by only 0.2 Å. Though these designs were extraordinarily successful, this is a rare case in protein design, as most natural protein topologies are too complex to be parameterized in this fashion.

The first attempt at backbone flexibility in a more general protein design algorithm was the SoftROC program by Desjarlais and Handel (Desjarlais and Handel 1999). The intent of their work was to address the problem of compensating for unfavorable core mutations by relaxation of the backbone. Starting from a known crystallographically determined structure, backbone torsion angles were randomly adjusted up to 3° from their original positions to generate a population of conformations (Figure 15.4).

Though the movement of each torsion angles was small, the compounded effect over many bonds led to a wide variety of backbones that preserved the overall topology of the protein.

SoftROC was tested on the bacteriophage protein 434 cro by forcing a disruptive point mutation into the core and having the program design stabilizing mutations to compensate for it (Desjarlais and Handel 1999). When compared to the results of its fixed-backbone counterpart, SoftROC produced new sequences that folded cooperatively and shared similar stability to the wild-type 434 cro protein. This suggested that there were false negatives in the context of the fixed backbone approach and additional solutions become available when a flexible backbone is permitted.

The real goal of flexible backbone design, however, was realized when Kuhlman and coworkers designed Top7, an α/β protein with a novel topology not yet observed in nature (Kuhlman et al. 2003). The initial backbone of the desired fold was built from secondary structure elements derived from existing structures and optimized by iterative rounds of sequence and backbone structure minimization. The predicted structure was verified by crystallography and deviated by only 1.17Å from the actual coordinates while bearing no significant resemblance to any protein in the PDB (Figure 15.5. Also see color Figure 11.2).

The success of the Top7 design is promising for the potential of structural design, but it remains the only example of a fold being designed *de novo* via computational means. From the perspective of design being a validation of our knowledge of the forces involved in protein folding, there is also room for improvement. The current success requires knowledge-based potentials that use existing structural data to bias design results. Being able to achieve similar results using an *ab initio* potential

**FIGURE 15.4**  The structure of 434 cro (black) and two alternate backbone conformations that are available to SoftROC (gray). Adjusting the backbone allows SoftROC to explore additional conformations to relieve false-negative clashes between rotamers, as in this hypothetical example.

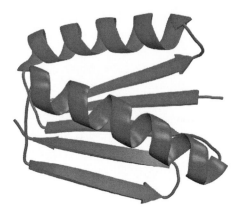

**FIGURE 15.5**  Structure of the designed protein Top7 is shown. Top7 contains both helical and sheet secondary structure elements, with a topology not observed in any other existing protein structure.

would solidify our understanding of protein structure and likely aid in our ability to predict folds from sequences.

## FULL REPACKS AND SURFACE DESIGN

Solvation is one of the hardest effects to quantify in protein design and remains an important hurdle for accurate predictions, especially for cavity and boundary positions. A hydrophobic protein core can be accurately modeled as a continuous, low dielectric medium, making electrostatic effects easy to calculate. Calculating these effects for solvent-exposed residues or buried polar residues, however, becomes complex and is one area where design force fields must make important choices between accuracy and speed.

The simplest observation to be made about the difference between a protein's core and surface is amino-acid composition. Generally, protein cores are hydrophobic in nature to aid in folding and stability, while surface and boundary positions are evenly distributed between polar and nonpolar amino acids to ensure solubility and prevent aggregation. Deviations from this trend are inspired by functional necessities such as oligomer formation or the burial of an active site.

Binary patterning is a method that exploits this observation by restricting the choice of amino acids based on the location of the position. This places little responsibility on the force field to discriminate between favorable and unfavorable solvent effects, as only favorable residues are allowed at solvent exposed positions. One illustration of the power of binary patterning was the design of a library of four-helix bundles by randomly allowing polar residues on the surface and nonpolar residues on the interior of the bundle (Hecht et al. 2004). Short (74-residue) bundles designed with this naive strategy contained a majority of stable, soluble proteins with alpha-helical secondary structure. NMR spectroscopy revealed that most of the proteins had fluctuating cores and more closely resembled molten globule intermediates than well-folded proteins.

Simply extending the bundles (102-residue) caused a majority of the tested proteins to have well-resolved NMR spectra indicative of well-ordered side chains (Hecht et al. 2004). In this case, binary patterning was sufficient to specify a simple protein fold, without high-resolution structural data to guide residue selection. The dependence on length suggests that a given tertiary structure has a minimum size in order to be stabilized. In the case of protein design, patterning can be combined with existing force fields to improve the stability and solubility properties of complex folds.

The first fully automated and rational design of an entire protein, a zinc finger domain, used this technique (Dahiyat and Mayo 1997). Only polar residues were allowed at surface positions, while only hydrophobic residues were allowed in the core. In the boundary positions between surface and core, however, both sets of residues were allowed. Though the potential included an atomic solvation model, it tended to select hydrophobic residues, given the choice between the two, necessitating the patterning approach to ensure solubility. The surface and boundary positions were additionally designed with a hydrogen-bonding potential to maximize specific contacts that would specify the fold.

The resulting protein shared only 21% identity with the wild type domain, the majority of which were buried, consistent with the steric restrictions of the core (Dahiyat and Mayo 1997). Thermal denaturation showed a cooperative unfolding transition that was completely reversible, as expected for a small hydrophobic domain. Structural analysis by solution NMR revealed that the designed protein's backbone packed in the expected topology, with a backbone deviation of 1.98Å.

Since the design of the zinc finger, design potentials have attempted to elaborate on simple patterning by developing their solvation models to accurately distinguish between surface and core positions. As a rigorous test of whether a modern design potential could accurately design full-length proteins without the use of patterning, Dantas and coworkers investigated nine globular proteins of varying secondary structure composition (Dantas et al. 2003). An implicit solvent model was used to discriminate between positions and favor hydrophobic placement in the core. The redesigned sequences were 35% similar to wild type, on average, and 50% similar when only core residues were considered.

When expressed and characterized, six of the nine designed proteins appeared monomeric and folded. Furthermore, their stabilities (as measured by chemical denaturation) were comparable to, or, in a few cases, better than the wild type proteins (Dantas et al. 2003). The improvement in stability is not surprising, as the designed proteins are free to abolish function in favor of stability, which is a luxury the wild type protein is not afforded. Still, it is encouraging that the success of Mayo's zinc finger design can be extended to multiple proteins with widely varying structures.

## HYDROGEN BONDING AND POLAR RESIDUES IN THE CORE

Natural protein cores are not always composed exclusively of hydrophobic residues. Though polar residues are often found on protein surfaces, making favorable interactions with solvent, they have also been observed within the core. For a buried polar residue to be favorable, it must contribute enough stabilizing interactions to overcome the penalty for removing its interactions with solvent. These interactions must be satisfied by intraprotein hydrogen bonds to the backbone, other polar residues nearby, or buried water molecules. These hydrogen bonding patterns are necessarily specific for the fold and likely contribute to discriminating the proper fold from decoys in the energy landscape.

When allowing polar residues into a protein core, design algorithms require a strict potential that guarantees a satisfied network of hydrogen bonds. Based on analysis of hydrogen bonding in existing protein structures, Bolon and coworkers generated rules for allowing polar amino acids in the design of protein cores: Inclusion of a filter criteria disallowed polar rotamers not forming a minimum number of hydrogen bonds, and a hydrogen bonding potential component was added to the energy function used to score results (Bolon et al. 2003). The energetic contribution of hydrogen bonds was approximated by a geometry-based function describing orientation of the hydrogen bond donor and acceptor atoms. Using this method, the core of *E. coli* thioredoxin was redesigned, achieving a variant with 2.5 kcal/mol increased stability over wild type. In comparison, similar designs permitting only hydrophobic residues in the core increased the stability by only 2.2 kcal/mol. Though the difference was small, it did suggest that a hydrogen bonding potential could contribute to the design of well-ordered and specific protein cores, including polar residues.

## ALTERING PROTEIN FOLDS

Protein design typically considers only one structural template. Although multiple sequences are usually identified for a desired structure, alternate structures closely compatible with a given sequence are not always considered. The specificity of a fold and its stability are often competing goals: The residue interactions that affect the denaturation properties are not identical to those specifying the three-dimensional fold. Explicitly modeling the alternate structures of each solution sequence has not been adequately addressed in protein design. In the interest of furthering our understanding of the determinants of protein fold specificity, Rose and Creamer issued the Paracelsus Challenge: to alter the fold of one protein into the fold of another without changing more than half the sequence (Rose and Creamer 1994).

Several groups have taken steps toward understanding this problem, but those credited with meeting the challenge used a knowledge-based approach incorporating sequence information, secondary structure prediction, and computational modeling to evaluate possible solutions (Dalal et al. 1997). The group chose to transform the mostly β-sheet structure of 56-residue GB1, into the four-helix bundle fold of 63-residue protein Rop (Figure 15.6). The resulting protein, Janus, contains 50% sequence similarity to GB1, but adopted the fold of Rop. Janus was characterized

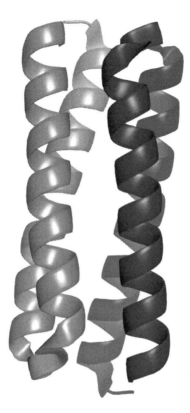

**FIGURE 15.6** While retaining 50% of the native sequence, the structure of GB1 was chosen to be converted to that of Rop1 (structure shown here) by the winners of the Paracelsus Challenge.

experimentally and shown to be native-like with cooperative melting transitions and structured by NMR.

## EXPERIMENTALLY EVALUATING SUCCESS

Experimental validation is important to evaluate the success of a design. Designed proteins are compared to their wild-type counterparts using biophysical properties such as stability or structure. Stability is measured using chemical or thermal denaturation, which can provide free energies of unfolding and allow intuitive comparison to computationally predicted values. Observing a reversible unfolding transition with independent probes such as fluorescence and circular dichroism (CD) is a good indicator of a well-behaved, folded protein. The degree of binding to the hydrophobic dye 8-anilino-1-naphthalenesulfonic acid, which cannot access the solvent-protected cores of native proteins, can also be used to test for properly folded proteins.

CD can be used to evaluate the basic secondary structure of a designed variant, but ideally high-resolution structural data should be obtained via nuclear magnetic resonance (NMR) or x-ray crystallography. This is the most robust test of a design, where each predicted rotamer choice can be validated and unexpected changes in backbone orientation or overall fold can be discovered.

## CONCLUSION

Substantial progress has been made in the design of protein structure. Observations from natural proteins and mutagenesis experiments, as well as learning from failed protein design attempts, have guided our current understanding. As these examples illustrate, the iteration between computational design and experimental validation is crucial toward improving our models. The approximations necessary for speed in protein design come at the expense of accuracy, and methods for circumventing these limitations will no doubt continue to be improved and invented. Design of protein structure remains challenging: Though some solutions have been found, few guarantees of success currently exist.

## REFERENCES

Bolon, D. N., J. S. Marcus, S. A. Ross and S. L. Mayo. 2003. "Prudent modeling of core polar residues in computational protein design." *J Mol Biol* 329: 611–22.

Dahiyat, B. I. and S. L. Mayo. 1997. "De novo protein design: Fully automated sequence selection." *Science* 278: 82–7.

Dahiyat, B. I. and S. L. Mayo. 1997. "Probing the role of packing specificity in protein design." *Proc Natl Acad Sci U S A* 94: 10172–7.

Dalal, S., S. Balasubramanian and L. Regan. 1997. "Protein alchemy: Changing beta-sheet into alpha-helix." *Nat Struct Biol* 4: 548–52.

Dantas, G., B. Kuhlman, D. Callender, M. Wong and D. Baker. 2003. "A large scale test of computational protein design: Folding and stability of nine completely redesigned globular proteins." *J Mol Biol* 332: 449–60.

Desjarlais, J. R. and T. M. Handel. 1995. "De-novo design of the hydrophobic cores of proteins." *Protein Sci* 4: 2006–18.

Desjarlais, J. R. and T. M. Handel. 1999. "Side-chain and backbone flexibility in protein core design." *J Mol Biol* 290: 305–18.

Dill, K. A. 1990. "Dominant forces in protein folding." *Biochemistry* 29: 7133–55.

Eriksson, A. E., W. A. Baase and B. W. Matthews. 1993. "Similar hydrophobic replacements of Leu99 and Phe153 within the core of T4 lysozyme have different structural and thermodynamic consequences." *J Mol Biol* 229: 747–69.

Eriksson, A. E., W. A. Baase, X. J. Zhang, D. W. Heinz, M. Blaber, et al. 1992. "Response of a protein structure to cavity-creating mutations and its relation to the hydrophobic effect." *Science* 255: 178–83.

Harbury, P. B., T. Zhang, P. S. Kim and T. Alber. 1993. "A switch between two-, three-, and four-stranded coiled coils in GCN4 leucine zipper mutants." *Science* 262: 1401–07.

Havranek, J. and P. Harbury. 2003. "Automated design of specificity in molecular recognition." *Nat Struct Mol Biol* 10: 45–52.

Hecht, M. H., A. Das, A. Go, L. H. Bradley and Y. Wei. 2004. "De novo proteins from designed combinatorial libraries." *Protein Sci* 13: 1711–1723.

Hurley, J. H., W. A. Baase and B. W. Matthews. 1992. "Design and structural analysis of alternative hydrophobic core packing arrangements in bacteriophage T4 lysozyme." *J Mol Biol* 224: 1143–59.

Kono, H., M. Nishiyama, M. Tanokura and J. Doi. 1998. "Designing the hydrophobic core of *Thermus flavus* malate dehydrogenase based on side-chain packing." *Protein Eng* 11: 47–52.

Kuhlman, B. and D. Baker. 2000. "Native protein sequences are close to optimal for their structures." *Proc Natl Acad Sci* 97: 10383–8.

Kuhlman, B., G. Dantas, G. C. Ireton, G. Varani, B. L. Stoddard, et al. 2003. "Design of a novel globular protein fold with atomic-level accuracy." *Science* 302: 1364–8.

Lazar, G. A., J. R. Desjarlais and T. M. Handel. 1997. "De novo design of the hydrophobic core of ubiquitin." *Protein Sci* 6: 1167–78.

Lim, W. A., D. C. Farruggio and R. T. Sauer. 1992. "Structural and energetic consequences of disruptive mutations in a protein core." *Biochemistry* 31: 4324–33.

Milla, M. E., B. M. Brown and R. T. Sauer. 1994. "Protein stability effects of a complete set of alanine substitutions in Arc repressor." *Nat Struct Mol Biol* 1: 518–23.

Mooers, B. H. M., D. Datta, W. A. Baase, E. S. Zollars, S. L. Mayo, et al. 2003. "Repacking the core of T4 lysozyme by automated design." *J Mol Biol* 332: 741–56.

Munson, M., S. Balasubramanian, K. G. Fleming, A. D. Nagi, R. O'Brien, et al. 1996. "What makes a protein a protein? Hydrophobic core designs that specify stability and structural properties." *Protein Sci* 5: 1584–93.

Pokala, N. and T. M. Handel. 2005. "Energy functions for protein design: Adjustment with protein–protein complex affinities, models for the unfolded state, and negative design of solubility and specificity." *J Mol Biol* 347: 203–27.

Ponder, J. W. and F. M. Richards. 1987. "Tertiary templates for proteins—use of packing criteria in the enumeration of allowed sequences for different structural classes." *J Mol Biol* 193: 775–91.

Rose, G. D. and T. P. Creamer. 1994. "Protein-folding—predicting predicting." *Protein Struct Funct Genet* 19: 1–3.

Vlassi, M., G. Cesareni and M. Kokkinidis. 1999. "A correlation between the loss of hydrophobic core packing interactions and protein stability." *J Mol Biol* 285: 817–27.

# 16 Modulation of Intrinsic Properties by Computational Design

*Vikas Nanda, Fei Xu, and Daniel Hsieh*

## CONTENTS

Proteins are malleable molecules and tolerate many mutations with marginal effects on structure and function. A limited number of three-dimensional folds are recycled repeatedly for different functions. This malleability results from surviving hundreds of millions of years of mutation and recombination. Natural proteins thus provide a gold mine of starting materials for those interested in modifying proteins to suit their needs. Unlike nature, however, protein engineers try to evolve molecules on a much shorter time scale. To that end, combinatorial gene synthesis and *in vitro* evolution methods (Stemmer 1994; Giver et al. 1998) have been developed to accelerate the mutation and selection process with impressive results. Rational protein engineering is another approach that applies our knowledge of intermolecular forces and protein structure toward the identification of mutations that confer the desired properties on a target protein (Marshall et al. 2003). Computation automates much of the design process, allowing larger, more complex targets to be rationally engineered.

This chapter reviews the recent progress in computational protein engineering, with a focus on how knowledge of protein folding and function is applied to modulate the "intrinsic" properties of proteins (e.g., stability, substrate binding, and catalysis). Each of these properties presents unique technical and conceptual challenges. Some may be overcome simply by brute force. For example, as computational costs become cheaper, larger computational tasks become accessible to average protein designers. Other challenges require the development of more sophisticated computational methods or efficient optimization algorithms. We will see several examples of both throughout this chapter. The field of computational protein design continues to advance, and the examples presented here represent only a snapshot of what is possible. Computational methods combined with *in vitro* evolution will give protein engineers increasing control over the design process.

## DESIGNING THERMOSTABLE PROTEINS

Thermostable proteins have a wide variety of applications in biotechnology and industry. A well known example is TAQ Polymerase, a DNA polymerase isolated from deep-sea hydrothermal vent bacteria *Thermus aquaticus* (Saiki et al. 1988). The protein is active at temperatures as high as 97°C and has been instrumental in making the polymerase chain reaction a practical method for DNA amplification. There are many opportunities for thermostable proteins as industrial catalysts for high-temperature reactions. Since many proteins undergo irreversible unfolding at elevated temperatures, increasing the melting temperature ($T_m$) or the free energy of folding can increase the shelf life of an industrial enzyme. Increased stability would also increase the *in vivo* half-life of a protein-based therapeutic (Marshall et al. 2003; van den Burg and Eijsink 2002).

In order to rationally design thermostability into a protein of interest, it is necessary to understand the physical basis behind increasing stability. A number of mutational strategies exist, including engineered disulfides, salt bridges between charged residues, improved side-chain packing interactions in the hydrophobic core, reduced surface loops, and added hydrogen bonds. Side-by-side comparisons of related proteins from mesophiles (room temperature organisms) and thermophiles reveal few universal trends, although many of the above mechanisms are found on a case-by-case basis (Kumar et al. 2000; Szilagyi and Zavodszky 2000; Vogt and Argos 1997). The stability of a protein determines the equilibrium between the native, folded state and an ensemble of misfolded and unfolded states. Protein stabilization strategies can be broadly divided into two categories: stabilizing the native state and destabilizing the unfolded states. We will explore the molecular basis behind these strategies and investigate how they may be incorporated in automated protein design.

### Stabilizing the Native State

One strategy for increasing the stability of a protein involves stabilizing the native state by optimizing tertiary interactions in structure. One can potentially increase the number of salt bridges, improve core packing, or add disulfides. No single mechanism explains the thermal stability of all thermophilic proteins (Kumar et al. 2000; Szilagyi and Zavodszky 2000). As such, a number of different approaches have been developed for using computation to stabilize native state interactions.

Some of the earliest attempts at computational protein design involve correlating core packing with the stability and activity of well-studied proteins such as thioredoxin and the phage λ-repressor (Hellinga and Richards 1994; Lee and Levitt 1991).

Programs such as ROC (repacking of cores) (Desjarlais and Handel 1995; Lazar et al. 1997), PROSE (protein simulated evolution) (Hellinga and Richards 1994), and PERLA (protein engineering rotamer library algorithm) (Fisinger et al. 2001; Lopez de la Paz et al. 2001) were used to redesign the hydrophobic cores of proteins by evaluating the fitness of various mutations using molecular force fields. Typically, a fixed protein backbone obtained from a high-resolution structure is redecorated with new side chains. This approach resulted in the successful redesign of many proteins, including variants of native proteins with increased thermostability.

Computational design methods make a number of approximations in order to score amino-acid substitutions during sequence design. Key approximations include fixed backbone and the use of discrete side-chain rotamers. Potentially successful designs may be overlooked because they require local relaxation of the backbone conformations. Furthermore, in order to accommodate discrete side-chain conformations, the energy function is modified to allow atomic overlap. It is assumed that these overlaps would be relieved by local relaxation of side chains into off-rotamer configurations. However, it has been suggested that these scaled potentials can result in overpacking of the core, which would induce strain and thus destabilize the folded conformation. This was demonstrated with a model SH3 domain and its stability-improved variants (Ventura et al. 2002). In this case, introducing Ile $\rightarrow$ Val substitutions in the core often improved the stability even further, suggesting that the Ile mutants had higher conformational strain.

Some designs also showed changes in the folding behavior caused by stabilization of the transition states. This was shown in the *de novo* designed $\alpha_3 D$ three-helical bundle, where adding a methyl group to the core by replacing Val with Ile resulted in a dynamic, molten globule–like protein, as measured by nuclear magnetic resonance (NMR) (Walsh et al. 2001). These studies highlight the subtle balance between the hydrophobic content of a core, which stabilizes the protein by improving the heat capacity of folding, and conformational strain, which can alter the folding behavior of the protein. Programs such as CORE (Cristian et al. 2003; Jiang et al. 2000) attempt to address this directly by concurrently optimizing the predicted heat capacity of folding and minimizing steric clashes in the protein core.

The backbone degrees of freedom also play a crucial role during core repacking. A number of methods have been employed, including rigid body optimization of the secondary structural elements within a protein (Su and Mayo 1997) and generating structural ensembles from NMR or molecular dynamics, to model backbone flexibility (Larson et al. 2002). The recent design of a thermostable yeast cytosine deaminase using ROSETTA, for example, was accomplished by concurrently optimizing backbone flexibility and sequence (Korkegian et al. 2005). ROSETTA combines molecular mechanics potentials with conformational preferences derived from Protein Data Bank (PDB) structures to model backbone flexibility. These hybrid approaches are powerful because they combine our understanding of the physical forces governing structure and stability with structural biases seen in real proteins.

Although exposed positions on the protein surface are generally not targeted for stability, studies have indicated that introducing salt bridges can increase stability. The computational design of SH3 domain variants with salt bridges showed that charge pairs could increase the free energy of folding by ~2 kcal/mol and raise the

$T_m$ by around 12°C (Schweiker et al. 2007). Using a genetic algorithm, 22 positions on the surface were optimized for charge-pair interactions based on a Tanford-Kirkwood solvent accessibility model for electrostatic interactions.

## DESTABILIZING UNFOLDED AND MISFOLDED STATES

The free energy of folding for a protein is dependent on both the difference in enthalpy and entropy between the folded and unfolded states. Amino acids can be replaced to lower conformational flexibility and thus to reduce the entropic cost of folding. Glycine has the greatest conformational flexibility and proline the least, while other amino acids fall somewhere in between (Figure 16.1). The reduction in backbone entropy when a residue is mutated from one amino acid to another is related to the change in the number of available conformations. This can be approximated by the area $(A_i)$ covered on a Ramachandran plot:

$$\Delta S = R \ln\left( A_i / A_j \right) \tag{16.1}$$

where $R$ is the gas constant. Using this formulation, the change in free energy for a typical Gly $\rightarrow$ Ala mutation is around 0.8 kcal/mol (Matthews et al. 1987; Nemethy et al. 1966; Serov et al. 2005). Likewise, an Ala $\rightarrow$ Pro mutation is expected to stabilize a protein by 1.4 kcal/mol. Predicting the effects of a mutation requires high-resolution structures to ascertain the steric compatibilities of the amino-acid side chains. Glycine is strongly favored for the left-handed $\alpha$ helix conformation ($\alpha_L$) ($\_$ $= +60°$, $\_ = +45°$). Introducing a Gly $\rightarrow$ Ala mutation at such a position would destabilize the native state. Alternatively, Xxx $\rightarrow$ Gly mutations at $\alpha_L$ positions would stabilize the protein, which was confirmed using formate dehydrogenase (Serov et al. 2005).

Contrary to what one would expect based on the entropy argument presented in the previous paragraph, an analysis of all Xxx $\rightarrow$ Pro mutations in the Protherm database (Gromiha et al. 1999) found that most proline mutations destabilized the free energy of folding and lowered the $T_m$ (Prajapati et al. 2007). Yet, in studies

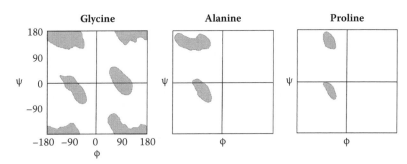

**FIGURE 16.1** The Ramachandran plots for glycine, alanine, and proline can be used to estimate the relative flexibility of each side chain based on the area of accessible conformations (shaded).

where mutations were judiciously chosen based on the analysis of high-resolution protein structures, half of attempted Xxx → Pro substitutions stabilized the protein by anywhere between 0.3 to 2.4 kcals/mol (Prajapati et al. 2007). This emphasizes the importance of considering the protein context when introducing a mutation. Automated prediction of backbone-stabilizing mutations requires an understanding of the intramolecular forces that contribute to contextual effects in the protein. Choi and Mayo used ORBIT (optimization of rotamers by iterative techniques) to predict the effects of seven proline mutations on the stability of protein G→1 (Choi and Mayo 2006). The best correlations between computed and measured stabilities were found when structures were minimized in ORBIT using the DRIEDING force field (Mayo et al. 1990) and the Lazaridis-Karplus calculation for solvation energies (Lazaridis and Karplus 1999). Interestingly, only modest improvements of correlation coefficients were observed when a backbone entropy term was included, suggesting that the primary determinants of proline contributions to stability were enthalpic and solvation related. A similar balance between entropy, solvation, and enthalpic effects was seen in molecular dynamics simulations of Gly → Ala mutations in short peptides, suggesting that peptides containing either amino acid would be effectively isoenergetic when unfolded (Scott et al. 2007). Optimizing the accuracy of free energy calculations used to evaluate the effect of amino-acid substitutions is still an important challenge in protein design.

Empirical and knowledge-based design potentials have been used to complement the molecular force fields during free energy calculations. For example, certain structural classes of turns in β-hairpins have strong preferences for glycines or prolines at specific positions. With a large number of high-quality protein structures available, it is possible to measure the propensity for an amino-acid type $i$, to be found in a particular structural context, $a$:

$$P(i,a) = \frac{N(i,a)/N(a)}{N(i)/N_{total}}$$

(16.2)

where $N(i,a)$ is the number of amino acids in a structural context across the database, $N(a)$ is the number of times the structure occurs, $N(i)$ is the total number of amino acids of type $i$ and $N_{total}$ is the size of the database. Values of $P > 1$ indicated that an amino acid is favored in a particular context, and conversely $P < 1$ indicates that the amino acid is disfavored. This type of analysis was applied to a series of β-hairpin turns in RNAse (Trevino et al. 2007). Substitutions were made at positions where $P_{WT}$ was significantly smaller than $P_{PRO}$ or $P_{GLY}$ at that location (Guruprasad and Rajkumar 2000). The approach was very successful, resulting in multiple stabilizing substitutions with $\Delta\Delta G = 0.7$–1.3 kcal/mol. Another knowledge-based potential was constructed by combining theoretical models of helix-coil transitions with a large data set of experimental studies on helical peptides (Munoz and Serrano 1995a; Munoz and Serrano 1995b). The resulting potential, AGADIR, was used to redesign solvent exposed residues on an α-helix of a human carboxypeptidase. Stability increases of 0.5–1 kcal/mol were seen for two different constructs, each containing three or four combined mutations (Villegas et al. 1995). These types of potentials are tremendously useful and often capture subtle features of the protein context that might be missed by molecular force fields.

The above examples focus mostly on backbone-centered strategies for destabilizing unfolded or misfolded states, but other options also exist. A number of studies are finding proteins where non-native tertiary contacts are crucial in stabilizing intermediates in the folding pathway (Klein-Seetharaman et al. 2002; Mok et al. 2007). Residues that are part of the transient hydrophobic core formed during protein folding may end up on the surface of the final structure. Optimizing the contacts that facilitate the "on-pathway" folding intermediates by protein engineering could be a novel method for improving their folding and stability. Although we are a long way from knowing the structures of folding intermediates, simulation methods such as molecular dynamics are beginning to fill this void (Mayor et al. 2003; White et al. 2005).

## ENGINEERING SUBSTRATE BINDING PROPERTIES

A number of experimental methods, such as phage display, are available for detecting and improving protein affinity for a substrate. The immune system is an example of optimized binding with antibodies that boast picomolar affinities for their cognate ligands. However, laboratory evolution and affinity maturation generally require detectable initial binding activity. A complementary role for computational design would involve constructing the binding sites *de novo*, which could then be enhanced by experimental methods (Chapter 17).

Designing substrate binding sites can be approached in two different ways. First, the surface amino acids of a protein can be re-engineered to facilitate the binding of a novel molecule. This requires identifying optimal sites on an existing scaffold and then choosing the appropriate amino acids to optimize binding. Alternatively, the protein fold itself can be designed from scratch to accommodate to the structural and energetic requirements for substrate binding. In such an approach, secondary structural elements such as α-helices and β-hairpins can be thought of as "Lego-blocks" that are used to build the appropriate fold. In this section, we discuss examples of protein, in which novel binding sites are constructed on existing protein scaffolds.

Some of the earliest successful designs involved introducing metal binding. Iron-sulfur (Fe-S) clusters are found in a wide variety of proteins and participate in electron transport and redox chemistry. To test whether an iron binding site can be engineered, the backbone of thioredoxin was scanned using DEZYMER (Hellinga et al. 1991; Hellinga and Richards 1991) for geometrically compatible positions for the four cysteine ligands to bind an iron (Figure 16.2) (Benson et al. 1998). The minimalist computational model optimized the bond lengths, angles, and steric compatibility with the surrounding protein matrix to construct a tetrahedral metal binding site. No backbone flexibility was included in the model. The final design used two Xxx →Cys mutations and two natural cysteines to successfully bind iron with spectroscopic properties similar to that of rubredoxin, a natural protein with similar iron coordination geometry. A number of other biologically important metal sites were similarly designed into thioredoxin including the iron center of superoxide dismutase (SOD) (Pinto et al. 1997), a cuboidal iron-sulfur center, a $Cys_2His_2Zn$ center (Wisz et

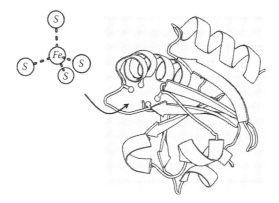

**FIGURE 16.2** An iron binding site was introduced into *E. coli* thioredoxin by searching for groups of side chains matching the tetrahedral coordination geometry of the Fe-4S cluster. Highlighted are the backbone atoms of the two wild-type and two mutant cysteines that form the metal binding site.

al. 1998), and a $FeHis_3O_2$ site (Benson et al. 2000). The effect of the protein microenvironment on binding and redox properties was tested by engineering the same metal substrate in different locations of thioredoxin. Moving the $FeHis_3O_2$ site to grooves, shallow pockets, surface-exposed positions, and buried core sites had significant effects on its metal binding and redox properties. The role of second shell interactions was studied in the SOD designs, where positively charged side chain–stabilized $O_2^-$ binding accelerated the rate limiting $Fe-O_2^-$ forming step. A similar approach was used in the program METAL_SEARCH to introduce a zinc binding site into protein Gβ1 (Clarke and Yuan 1995; Klemba et al. 1995; Regan and Clarke 1990).

Computational tools that were initially developed on metal binding sites have since been extended to include other more complex substrates. Some examples include protein–protein interactions such as enhancing the affinity of the $\alpha_2\beta_1$ integrin for collagen (Shimaoka et al. 2000), increasing the specificity of calmodulin for myosin light chain kinase (REF) (Shifman and Mayo 2002; Shifman and Mayo 2003) or programming of the target sequence specificity of PDZ domains (Reina et al. 2002). A cleft between the two domains of a bacterial periplasmic binding protein (PBP) was computationally redesigned to bind a series of small molecules including l-lactate, serotonin, and trinitrotoluene (TNT) (de Lorimier et al. 2002; Looger et al. 2003). The diversity of substrates was made possible by the ideal cleft geometry of PBP, which presented multiple positions where binding affinity and specificity could be introduced. Mutating 12 to 18 positions and considering substrate rotational and translational degrees of freedom requires sampling an immense number of conformations (on the order of $10^{70}$–$10^{80}$). In order to make this process feasible, the substrate configurations were first reduced to those compatible with a pocket consisting of alanines. These configurations were then optimized for van der Waals packing, hydrogen bonding, and other molecular forces between the substrate and protein. Sequences and side-chain rotamers were optimized using dead-end elimination (DEE). The algorithm reduces the number of permutations of $n$-interacting residues by eliminating the conformations not compatible with the global minimum based on pairwise comparison (Desmet et al. 1992; Looger and Hellinga 2001). During optimization, protein-substrate hydrogen bonding interactions were weighted strongly to address the energetic penalty of desolvating potential hydrogen bonding donors and acceptors upon binding. The PBP scaffold was functionalized with a fluorescent dye near the hinge between the two domains, allowing facile detection of binding by induced conformational change.

One of the challenges in predicting the effects of a mutation is ensuring that backbone flexibility is sufficiently accounted for. Backbone flexibility is necessary to describe the induced fit mechanism of substrate binding. As with the previous example using PBP, it is beneficial to first model flexibility at a coarse-grained level. For example in ROSETTA DOCK (Wang et al. 2007), which predicts protein–protein interactions, the sampling is carried out in three steps: (1) random perturbations of torsion angles in the backbone and rigid body motion between binding partners; (2) coarse-grained optimization of the perturbed structure with repacking of side-chain rotamers; and (3) energy minimization and local refinement of structures near the ligand binding site using off-rotamer side-chain sampling, small variations of backbone torsion angles, and rigid body sampling between proteins. Introducing sequence perturbations in step 1 could be useful in designing small molecule binding sites (Meiler and Baker 2006). As discussed later, this strategy was implemented in the design of enzyme active sites.

A number of challenges remain in the computational design of substrate binding sites in proteins. Binding sites are typically found on the surfaces of proteins, albeit in partially occluded pockets. Modeling a microenvironment where both crucial hydrophobic and polar interactions are involved is difficult. The role of water molecules in binding and structure is also a difficult problem (Reichmann et al. 2008). Continuum solvation versus explicit water molecules have their respective advantages and disadvantages when it comes to modeling. One approach is to include water as part of side chains in a rotamer library (Figure 16.3) (Jiang et al. 2005). These solvated rotamer libraries include water molecules that are permanently hydrogen bonded to donor and acceptor atoms of amino acids. While this reduces the sampling required

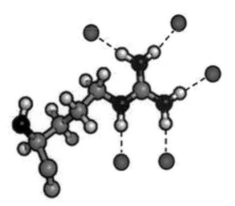

**FIGURE 16.3** A hydrated arginine residue. Water molecules (spheres) move with the side chain during conformational sampling.

to model bound waters on protein surfaces, it also significantly increases the total number of rotamers to consider during calculation (see Chapter 15).

Modest binding affinities achieved by rational or computational design can be enhanced by laboratory selection. Computationally designed antibodies were thus improved beyond natural affinity levels through judicious introduction of favorable electrostatics interactions (Lippow et al. 2007). Design and *in vitro* evolution can complement each other toward the design of high affinity, novel substrate binding sites (see Chapter 17).

## DESIGNING FUNCTION

### CATALYSIS

Designing protein function requires consideration of both stability and binding. The task is particularly challenging when designing enzymes. An enzyme not only must bind a target substrate, but also needs to stabilize the transition state and release the product (Walsh 2001). The geometry of catalytic residues must be positioned accurately to support efficiency catalysis. Designed enzymes include catalytic antibodies that are engineered by affinity maturation in the presence of a transition-state analog (Hilvert 2000). Computational design of enzymes has become possible more recently.

A series of "protozymes" were developed using ORBIT to catalyze the hydrolysis of paranitrophenol acetate (PNPA) into paranitrophenol and acetate (Bolon and Mayo 2001). In order to simulate the high-energy transition state between a histidine side-chain ligand on the protein and the PNPA substrate, the structure of a composite amino acid incorporating the bonded intermediate was modeled at various sites on the protein surface. The conformational freedom of this transition state was then explored by sampling rotamers in the histidine side chain as well as rotations between the bonded imidazole and PNPA groups. A number of candidate sites on *E. coli* thioredoxin were evaluated for their ability to accommodate the histidine-PNPA transition state. Rotamers of surrounding residues were optimized using DEE. Surrounding residues were also allowed to mutate to alanine to accommodate the substrate. Design proteins showed Kms in the micromolar range and $k_{cat}/k_{uncat}$ enhancements of up to 1000-fold.

In another study, a *de novo* catalytic protein was engineered by parameterizing four-helix bundle dinuclear metal site proteins (Kaplan and DeGrado 2004). A substrate channel was carved into one side of the bundle by replacing larger side chains with alanine. This allowed the protein to catalyze two-electron oxidation of a 4-aminophenol substrate with multiple turnovers and a 1000-fold rate enhancement over uncatalyzed reaction.

While these designs fell short of the activities of natural enzymes, they demonstrated the power of computational design methods to achieve catalysis. Two recent designs using ROSETTA have brought us even closer to realizing artificial enzymes: a retro-aldolase and a catalytic protein that carried out Kemp elimination (Jiang et al. 2008; Rothlisberger et al. 2008). One of the major obstacles in enzyme design is finding an effective way to maintain the precise network of interactions between a substrate and side-chain ligands while exploring a vast search space of potential active site positions, rotamers, and substrate configurations. The ROSETTA designs demonstrated that these obstacles can be overcome.

In order to model the transition state complex accurately, quantum mechanics calculations were performed on a truncated representation of the enzyme active site containing only those groups necessary for catalysis. This optimized structure, called the *theozyme* or *compuzyme*, was then elaborated for all torsional degrees of freedom within the substrate and those between the substrate and side-chain ligands (Tantillo et al. 1998). This results in a large number of active site configurations, on the order of $10^{18}$ for the retro-aldolase and $10^{21}$ for the Kemp elimination catalyst. These active sites were then matched to a library of protein active sites selected from the PDB. Due to the large number of states to sample, highly efficient algorithms for matching the active site configurations to protein surfaces are needed. A technique developed in the field of computer vision known as geometric hashing was applied to this problem (Ladman et al. 1990; Zanghellini et al. 2006). By reducing active sites to a series of geometric descriptors, side-chain identities, and rotamers, it was possible to rapidly compare the transition state ensemble to the library of protein scaffolds. The hits from this search were energetically minimized (allowing for both backbone and side-chain flexibility) and subjected to local sequence redesign to allow for any second-sphere mutations to accommodate the new active site. For the retro-aldolase, nearly half of the 72 designs that were constructed showed catalytic activity. Rate enhancements of four to five orders of magnitude were observed. For the Kemp elimination catalyst, the successful designs were further optimized through directed evolution.

### FLUORESCENCE

Another example of computationally designed function is green fluorescent protein (GFP), a protein originally isolated from the *Aequorea victoria* jellyfish. The fluorophore in the protein is formed by an autocatalytic reaction in the core of the protein involving three amino acids. Different conjugation states of the rearranged amino acids result in different absorption/emission properties. The formation of the fluorophore is sensitive to the local structure and sequence, making it possible to modulate fluorescence by mutation (Jain and Ranganathan 2004). Computational design was used to bias the combinatorial libraries for amino-acid choices that preserve fluorescence while increasing the diversity of emission wavelengths over traditional random libraries generated by error prone PCR and related methods (Treynor et al. 2007). A novel blue fluorescent protein was found with this approach showing favorable photophysical properties (Mena et al. 2006).

### SOLUBILITY

Another property of practical importance is protein solubility. In some cases, it is desirable to increase protein solubility to achieve higher protein concentrations or improve bioavailability. Alternatively, one may wish to mutate protein surface amino acids to promote crystallization through specific interprotein contacts (Bauman et al. 2008). Tuning solubility is very challenging for computational protein design. Protein aggregation often involves association of misfolded protein intermediates that are challenging to characterize with sufficient atomic resolution for structure-based design. Modeling solubility also requires effective force fields and methods for computing protein-solvent interactions, which are still being developed (see Chapters 10 and 12).

One area where computational design has been successfully applied to solubility is in the conversion of transmembrane proteins to water-soluble ones. Based on the analysis of a small number of high-resolution transmembrane protein structures, Rees and Eisenberg hypothesized that membrane- and water-soluble proteins share chemically similar cores, and differ primarily in the hydrophobicity of the protein surface (Rees et al. 1989). As such, it should be possible to convert some proteins from one class to the other simply by altering the hydrophobicity of the protein surface (Choma et al. 2000; Mitra et al. 2002; Slovic et al. 2003). This approach was taken in the water solubilization of the bacterial potassium channel, KcsA (Slovic et al. 2004). Using SCADS (statistical computationally assisted design strategy), exposed positions on KcsA were mutated to optimize the solvent interaction energy equivalent to that of water-soluble proteins. After three generations of design and characterization, WKS-3 (water-soluble KcsA 3) formed a helical homotetramer and specifically bound agitoxin2, a channel blocking peptide, indicating that the tertiary structure of WSK-3 was retained. An automated approach to water-solubilizing transmembrane proteins will facilitate structure determination and allow facile screening compound libraries for binding without the logistical difficulties of doing biochemistry and biophysics in a membrane environment.

## CONCLUSION

In this chapter, we highlighted a number of successful applications of computational protein design. As computational methods continue to improve, it will be possible to design increasingly sophisticated targets. Another important future direction is the integration of molecular design with laboratory evolution, developing complementary approaches that take advantage of the strengths of each method.

## REFERENCES

Bauman, J.D., K. Das, W.C. Ho, M. Baweja, D.M. Himmel, A.D. Clark, Jr., D.A. Oren, P.L. Boyer, S.H. Hughes, A.J. Shatkin, and E. Arnold. 2008. Crystal engineering of HIV-1 reverse transcriptase for structure-based drug design. *Nucleic Acids Res* 36(15): 5083–92.

Benson, D.E., M.S. Wisz, and H.W. Hellinga. 2000. Rational design of nascent metalloenzymes. *Proc Natl Acad Sci U S A* 97(12): 6292–7.

Benson, D.E., M.S. Wisz, W. Liu, and H.W. Hellinga. 1998. Construction of a novel redox protein by rational design: Conversion of a disulfide bridge into a mononuclear iron-sulfur center. *Biochemistry* 37(20): 7070–76.

Bolon, D.N., and S.L. Mayo. 2001. Enzyme-like proteins by computational design. *Proc Natl Acad Sci U S A* 98(25): 14274–9.

Choi, E.J., and S.L. Mayo. 2006. Generation and analysis of proline mutants in protein G. *Protein Eng Des Sel* 19(6): 285–9.

Choma, C., H. Gratkowski, J.D. Lear, and W.F. DeGrado. 2000. Asparagine-mediated self-association of a model transmembrane helix. *Nat Struct Biol* 7(2): 161–6.

Clarke, N.D., and S.M. Yuan. 1995. Metal search: A computer program that helps design tetrahedral metal-binding sites. *Proteins* 23(2): 256–63.

Cristian, L., P. Piotrowiak, and R.S. Farid. 2003. Mimicking photosynthesis in a computationally designed synthetic metalloprotein. *J Am Chem Soc* 125(39): 11814–5.

de Lorimier, R.M., J.J. Smith, M.A. Dwyer, L.L. Looger, K.M. Sali, C.D. Paavola, S.S. Rizk, S. Sadigov, D.W. Conrad, L. Loew, and H.W. Hellinga. 2002. Construction of a fluorescent biosensor family. *Protein Sci* 11(11): 2655–75.

Desjarlais, J.R., and T.M. Handel. 1995. De novo design of the hydrophobic cores of proteins. *Protein Sci* 4(10): 2006–18.

Desmet, J., M. Demaeyer, B. Hazes, and I. Lasters. 1992. The dead-end elimination theorem and its use in protein side-chain positioning. *Nature* 356(6369): 539–42.

Fisinger, S., L. Serrano, and E. Lacroix. 2001. Computational estimation of specific side chain interaction energies in alpha helices. *Protein Sci* 10(4): 809–18.

Giver, L., A. Gershenson, P.O. Freskgard, and F.H. Arnold. 1998. Directed evolution of a thermostable esterase. *Proc Natl Acad Sci U S A* 95(22): 12809–13.

Gromiha, M.M., J. An, H. Kono, M. Oobatake, H. Uedaira, and A. Sarai. 1999. ProTherm: Thermodynamic database for proteins and mutants. *Nucleic Acids Res* 27(1): 286–8.

Guruprasad, K., and S. Rajkumar. 2000. Beta-and gamma-turns in proteins revisited: A new set of amino acid turn-type dependent positional preferences and potentials. *J Biosci* 25(2): 143–56.

Hellinga, H.W., J.P. Caradonna, and F.M. Richards. 1991. Construction of new ligand binding sites in proteins of known structure. II. Grafting of a buried transition metal binding site into *Escherichia coli* thioredoxin. *J Mol Biol* 222(3): 787–803.

Hellinga, H.W., and F.M. Richards. 1991. Construction of new ligand binding sites in proteins of known structure. I. Computer-aided modeling of sites with pre-defined geometry. *J Mol Biol* 222(3): 763–85.

Hellinga, H.W., and F.M. Richards. 1994. Optimal sequence selection in proteins of known structure by simulated evolution. *Proc Natl Acad Sci U S A* 91(13): 5803–07.

Hilvert, D. 2000. Critical analysis of antibody catalysis. *Annu Rev Biochem* 69: 751–92.

Jain, R.K., and R. Ranganathan. 2004. Local complexity of amino acid interactions in a protein core. *Proc Natl Acad Sci U S A* 101(1): 111–6.

Jiang, L., E.A. Althoff, F.R. Clemente, L. Doyle, D. Rothlisberger, A. Zanghellini, J.L. Gallaher, J.L. Betker, F. Tanaka, C.F. Barbas, 3rd, D. Hilvert, K.N. Houk, B.L. Stoddard, and D. Baker. 2008. De novo computational design of retro-aldol enzymes. *Science* 319(5868): 1387–91.

Jiang, L., B. Kuhlman, T. Kortemme, and D. Baker. 2005. A "solvated rotamer" approach to modeling water-mediated hydrogen bonds at protein–protein interfaces. *Proteins* 58(4): 893–904.

Jiang, X., H. Farid, E. Pistor, and R.S. Farid. 2000. A new approach to the design of uniquely folded thermally stable proteins. *Protein Sci* 9(2): 403–16.

Kaplan, J., and W.F. DeGrado. 2004. De novo design of catalytic proteins. *Proc Natl Acad Sci U S A* 101(32): 11566–70.

Klein-Seetharaman, J., M. Oikawa, S.B. Grimshaw, J. Wirmer, E. Duchardt, T. Ueda, T. Imoto, L.J. Smith, C.M. Dobson, and H. Schwalbe. 2002. Long-range interactions within a nonnative protein. *Science* 295(5560): 1719–22.

Klemba, M., K.H. Gardner, S. Marino, N.D. Clarke, and L. Regan. 1995. Novel metal-binding proteins by design. *Nat Struct Biol* 2(5): 368–73.

Korkegian, A., M.E. Black, D. Baker, and B.L. Stoddard. 2005. Computational thermostabilization of an enzyme. *Science* 308(5723): 857–60.

Kumar, S., C.J. Tsai, and R. Nussinov. 2000. Factors enhancing protein thermostability. *Protein Eng* 13(3): 179–91.

Ladman, Y., J.T. Schwartz, and H.J. Wolfson. 1990. Affine invariant model-based object recognition. *IEEE Transactions on Robotics and Automation* 6(5): 578–89.

Larson, S.M., J.L. England, J.R. Desjarlais, and V.S. Pande. 2002. Thoroughly sampling sequence space: Large-scale protein design of structural ensembles. *Protein Sci* 11(12): 2804–13.

Lazar, G.A., J.R. Desjarlais, and T.M. Handel. 1997. De novo design of the hydrophobic core of ubiquitin. *Protein Sci* 6(6): 1167–78.

Lazaridis, T., and M. Karplus. 1999. Effective energy function for proteins in solution. *Proteins* 35(2): 133–52.

Lee, C., and M. Levitt. 1991. Accurate prediction of the stability and activity effects of site-directed mutagenesis on a protein core. *Nature* 352(6334): 448–51.

Lippow, S.M., K.D. Wittrup, and B. Tidor. 2007. Computational design of antibody-affinity improvement beyond in vivo maturation. *Nat Biotechnol* 25(10): 1171–6.

Looger, L.L., M.A. Dwyer, J.J. Smith, and H.W. Hellinga. 2003. Computational design of receptor and sensor proteins with novel functions. *Nature* 423(6936): 185–90.

Looger, L.L., and H.W. Hellinga. 2001. Generalized dead-end elimination algorithms make large-scale protein side-chain structure prediction tractable: Implications for protein design and structural genomics. *J Mol Biol* 307(1): 429–45.

Lopez de la Paz, M., E. Lacroix, M. Ramirez-Alvarado, and L. Serrano. 2001. Computer-aided design of beta-sheet peptides. *J Mol Biol* 312(1): 229–46.

Marshall, S.A., G.A. Lazar, A.J. Chirino, and J.R. Desjarlais. 2003. Rational design and engineering of therapeutic proteins. *Drug Discov Today* 8(5): 212–21.

Matthews, B.W., H. Nicholson, and W.J. Becktel. 1987. Enhanced protein thermostability from site-directed mutations that decrease the entropy of unfolding. *Proc Natl Acad Sci U S A* 84(19): 6663–7.

Mayo, S.L., B.D. Olafson, and W.A. Goddard. 1990. Dreiding—a generic force-field for molecular simulations. *J Phys Chem* 94(26): 8897–909.

Mayor, U., N.R. Guydosh, C.M. Johnson, J.G. Grossmann, S. Sato, G.S. Jas, S.M. Freund, D.O. Alonso, V. Daggett, and A.R. Fersht. 2003. The complete folding pathway of a protein from nanoseconds to microseconds. *Nature* 421(6925): 863–7.

Meiler, J., and D. Baker. 2006. ROSETTALIGAND: Protein-small molecule docking with full side-chain flexibility. *Proteins* 65(3): 538–48.

Mena, M.A., T.P. Treynor, S.L. Mayo, and P.S. Daugherty. 2006. Blue fluorescent proteins with enhanced brightness and photostability from a structurally targeted library. *Nat Biotechnol* 24(12): 1569–71.

Mitra, K., T.A. Steitz, and D.M. Engelman. 2002. Rational design of "water-soluble" bacteriorhodopsin variants. *Protein Eng* 15(6): 485–92.

Mok, K.H., L.T. Kuhn, M. Goez, I.J. Day, J.C. Lin, N.H. Andersen, and P.J. Hore. 2007. A pre-existing hydrophobic collapse in the unfolded state of an ultrafast folding protein. *Nature* 447(7140): 106–09.

Munoz, V., and L. Serrano. 1995a. Elucidating the folding problem of helical peptides using empirical parameters. II. Helix macrodipole effects and rational modification of the helical content of natural peptides. *J Mol Biol* 245(3): 275–96.

Munoz, V., and L. Serrano. 1995b. Elucidating the folding problem of helical peptides using empirical parameters. III. Temperature and pH dependence. *J Mol Biol* 245(3): 297–308.

Nemethy, G., S.J. Leach, and H.A. Scheraga. 1966. The influence of amino acid side chains on the free energy of helix-coil transitions. *J Phys Chem* 70: 998–1004.

Pinto, A.L., H.W. Hellinga, and J.P. Caradonna. 1997. Construction of a catalytically active iron superoxide dismutase by rational protein design. *Proc Natl Acad Sci U S A* 94(11): 5562–7.

Prajapati, R.S., M. Das, S. Sreeramulu, M. Sirajuddin, S. Srinivasan, V. Krishnamurthy, R. Ranjani, C. Ramakrishnan, and R. Varadarajan. 2007. Thermodynamic effects of proline introduction on protein stability. *Proteins* 66(2): 480–91.

Rees, D.C., L. DeAntonio, and D. Eisenberg. 1989. Hydrophobic organization of membrane proteins. *Science* 245(4917): 510–13.

Regan, L., and N.D. Clarke. 1990. A tetrahedral zinc(II)-binding site introduced into a designed protein. *Biochemistry* 29(49): 10878–83.

Reichmann, D., Y. Phillip, A. Carmi, and G. Schreiber. 2008. On the contribution of water-mediated interactions to protein-complex stability. *Biochemistry* 47(3): 1051–60.

Reina, J., E. Lacroix, S.D. Hobson, G. Fernandez-Ballester, V. Rybin, M.S. Schwab, L. Serrano, and C. Gonzalez. 2002. Computer-aided design of a PDZ domain to recognize new target sequences. *Nat Struct Biol* 9(8): 621–7.

Rothlisberger, D., O. Khersonsky, A.M. Wollacott, L. Jiang, J. DeChancie, J. Betker, J.L. Gallaher, E.A. Althoff, A. Zanghellini, O. Dym, S. Albeck, K.N. Houk, D.S. Tawfik, and D. Baker. 2008. Kemp elimination catalysts by computational enzyme design. *Nature* 453(7192): 190–195.

Saiki, R.K., D.H. Gelfand, S. Stoffel, S.J. Scharf, R. Higuchi, G.T. Horn, K.B. Mullis, and H.A. Erlich. 1988. Primer-directed enzymatic amplification of DNA with a thermostable DNA polymerase. *Science* 239(4839): 487–91.

Schweiker, K.L., A. Zarrine-Afsar, A.R. Davidson, and G.I. Makhatadze. 2007. Computational design of the Fyn SH3 domain with increased stability through optimization of surface charge charge interactions. *Protein Sci* 16(12): 2694–702.

Scott, K.A., D.O. Alonso, S. Sato, A.R. Fersht, and V. Daggett. 2007. Conformational entropy of alanine versus glycine in protein denatured states. *Proc Natl Acad Sci U S A* 104(8): 2661–6.

Serov, A.E., E.R. Odintzeva, I.V. Uporov, and V.I. Tishkov. 2005. Use of Ramachandran plot for increasing thermal stability of bacterial formate dehydrogenase. *Biochemistry (Mosc)* 70(7): 804–08.

Shifman, J.M., and S.L. Mayo. 2002. Modulating calmodulin binding specificity through computational protein design. *J Mol Biol* 323(3): 417–23.

Shifman, J.M., and S.L. Mayo. 2003. Exploring the origins of binding specificity through the computational redesign of calmodulin. *Proc Natl Acad Sci U S A* 100(23): 13274–9.

Shimaoka, M., J.M. Shifman, H. Jing, J. Takagi, S.L. Mayo, and T.A. Springer. 2000. Computational design of an integrin I domain stabilized in the open high affinity conformation. *Nat Struct Biol* 7(8): 674–8.

Slovic, A.M., H. Kono, J.D. Lear, J.G. Saven, and W.F. DeGrado. 2004. Computational design of water-soluble analogues of the potassium channel KcsA. *Proc Natl Acad Sci U S A* 101(7): 1828–33.

Slovic, A.M., C.M. Summa, J.D. Lear, and W.F. DeGrado. 2003. Computational design of a water-soluble analog of phospholamban. *Protein Sci* 12(2): 337–48.

Stemmer, W.P. 1994. Rapid evolution of a protein in vitro by DNA shuffling. *Nature* 370(6488): 389–91.

Su, A., and S.L. Mayo. 1997. Coupling backbone flexibility and amino acid sequence selection in protein design. *Protein Sci* 6(8): 1701–07.

Szilagyi, A., and P. Zavodszky. 2000. Structural differences between mesophilic, moderately thermophilic and extremely thermophilic protein subunits: Results of a comprehensive survey. *Structure* 8(5): 493–504.

Tantillo, D.J., J. Chen, and K.N. Houk. 1998. Theozymes and compuzymes: Theoretical models for biological catalysis. *Curr Opin Chem Biol* 2: 743–50.

Trevino, S.R., S. Schaefer, J.M. Scholtz, and C.N. Pace. 2007. Increasing protein conformational stability by optimizing beta-turn sequence. *J Mol Biol* 373(1): 211–8.

Treynor, T.P., C.L. Vizcarra, D. Nedelcu, and S.L. Mayo. 2007. Computationally designed libraries of fluorescent proteins evaluated by preservation and diversity of function. *Proc Natl Acad Sci U S A* 104(1): 48–53.

van den Burg, B., and V.G. Eijsink. 2002. Selection of mutations for increased protein stability. *Curr Opin Biotechnol* 13(4): 333–7.

Ventura, S., M.C. Vega, E. Lacroix, I. Angrand, L. Spagnolo, and L. Serrano. 2002. Conformational strain in the hydrophobic core and its implications for protein folding and design. *Nat Struct Biol* 9(6): 485–93.

Villegas, V.V., A.R. Viguera, F.X. Aviles, and L. Serrano. 1995. Stabilization of proteins by rational design of alpha-helix stability using helix/coil transition theory. *Fold Des* 1(1): 29–34.

Vogt, G., and P. Argos. 1997. Protein thermal stability: Hydrogen bonds or internal packing? *Fold Des* 2(4): S40–46.

Walsh, C. 2001. Enabling the chemistry of life. *Nature* 409(6817): 226–31.

Walsh, S.T., V.I. Sukharev, S.F. Betz, N.L. Vekshin, and W.F. DeGrado. 2001. Hydrophobic core malleability of a de novo designed three-helix bundle protein. *J Mol Biol* 305(2): 361–73.

Wang, C., P. Bradley, and D. Baker. 2007. Protein–protein docking with backbone flexibility. *J Mol Biol* 373(2): 503–19.

White, G.W., S. Gianni, J.G. Grossmann, P. Jemth, A.R. Fersht, and V. Daggett. 2005. Simulation and experiment conspire to reveal cryptic intermediates and a slide from the nucleation-condensation to framework mechanism of folding. *J Mol Biol* 350(4): 757–75.

Wisz, M.S., C.Z. Garrett, and H.W. Hellinga. 1998. Construction of a family of $Cys_2His_2$ zinc binding sites in the hydrophobic core of thioredoxin by structure-based design. *Biochemistry* 37(23): 8269–77.

Zanghellini, A., L. Jiang, A.M. Wollacott, G. Cheng, J. Meiler, E.A. Althoff, D. Rothlisberger, and D. Baker. 2006. New algorithms and an *in silico* benchmark for computational enzyme design. *Protein Sci* 15(12): 2785–94.

# 17 Modulating Protein Interactions by Rational and Computational Design

*Jonathan S. Marvin and Loren L. Looger*

## CONTENTS

Exponential growth in the number of solved protein structures and the development of user-friendly structure analysis software has led to a greater understanding of the biophysical basis of protein structure and function, and the principles of molecular recognition. In the 1980s, researchers ambitiously postulated that with this information, the rational design of proteins would not be far off (Pabo 1983). Early efforts to alter the specificity of enzymes by inspection of the structure were rewarded with successful redesigns (Bone et al. 1989; Clarke et al. 1989a; Clarke et al. 1989b; Craik et al. 1985; Scrutton et al. 1990; Wilks et al. 1988), and it looked as if the ability to alter protein function at will was near.

Unfortunately, it was soon discovered that design efforts involving several simultaneous changes to a protein are often too complex to be solved by inspection of the structure alone. This results from two factors. First, subtle side-chain rearrangements in response to even a single mutation can have profound effects on the biophysical properties and structure of a protein. Qualitative or intuitive analysis is thus insufficient for predicting the effect of a mutation on the affinity of a protein–protein complex. For example, alanine-scanning mutagenesis is often used to identify critical "hot-spot" residues at protein–protein interfaces (Clackson et al. 1995). As expected, the most frequent result from truncation of a side chain in a naturally evolved interface is a decrease in affinity. However, mutations to alanine sometimes result in higher-affinity complexes (Jin et al. 1992; Schreiber and Fersht 1993; Shields et al. 2001). These unusual increases in affinity typically arise from interfaces with greater stereochemical complementarity being created by the rearrangement of neighboring side chains to occupy the void left by the truncated residue. Qualitative analysis cannot predict these rearrangements [but modern, advanced computational techniques can (Kortemme and Baker 2002)].

The second factor limiting design by inspection is the astronomical number of possible mutations to consider. For optimization of $n$ residues in a protein, there are $20^n$ possible naturally occurring amino-acid sequences. Attempting to optimize the multiple residues that compose a protein–protein interface quickly results in billions of possible sequences to be considered. Diversity-oriented and computational protein design techniques address the inadequacies of design by inspection by sampling large numbers of possible amino-acid sequences, and selecting the best variants through a physical or virtual fitness function.

Over the last few decades, as the field of computational design has matured, the complexity of the problems being addressed has grown significantly. This is mainly due to the advancement of software and knowledge so that more accurate descriptors of interatomic forces and more powerful algorithms for searching sequence space can be applied. As a result, the field of computational protein design has moved from the prediction of simple geometric arrangements of side chains to predicting intricate stereochemical complementarity and even catalytic functionality in proteins. Furthermore, designs are no longer being made simply because they represent readily testable hypotheses, but because they can have practical utility, and are even being included in therapeutic applications.

## GEOMETRY-BASED DESIGN

### METHODOLOGY

There are two significant limitations to all computational design efforts: (1) the extent to which the design criteria/theory match reality and (2) the speed with which one can make and test proteins. In the 1990s, when computational protein design was in its infancy, fitness functions to accurately describe molecular interactions and algorithms to search the astronomical number of potential designed proteins were too primitive for the design of complex stereochemical surfaces. Furthermore, the process of making genetic constructs and expressing and purifying proteins was laborious and required significant expertise. The combination of these limitations (physical models, search algorithms, computational power, and actual design and testing of proteins) necessitated that early design experiments be limited in scope to, for example, the design of metal-binding sites.

It is estimated that approximately half of all known proteins bind metal ions or metal-containing cofactors (Thomson and Gray 1998). With the wealth of structural information available as to how metals are chelated by proteins (Castagnetto et al. 2002), it is relatively straightforward to describe a metal-binding site in simple geometric terms (Figure 17.1). The free electron pairs from suitable side chains (His, Cys, Met, Asp, Glu) must be positioned in a specified geometry to form a primary coordination sphere (PCS) of overlapping orbitals with the metal (Chakrabarti 1989; Chakrabarti 1990a; Chakrabarti 1990b; Chakrabarti 1990c). Building a more elaborate secondary coordination sphere of hydrogen bonds to stabilize the geometry of the PCS can enhance the affinity and stability of the binding site (Christianson 1991).

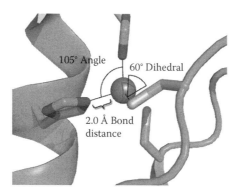

**FIGURE 17.1 (see color insert following page 178)**    Parameters for geometric definition of a metal-binding site. The primary coordination sphere is defined by the distances between the coordinating atoms and the metal, the angles around the metal (e.g., the angle formed by $N\varepsilon_{His}$-$Zn^{2+}$-$N\varepsilon_{His}$), and the dihedrals that define the relationship of the coordinating side chains to each other (in this example, the dihedral formed by the $C\beta_{Cys}$-$S\gamma_{Cys}$ bond relative to the $Zn^{2+}$-$N\varepsilon_{His}$ bond, about the $S\gamma_{Cys}$-$Zn^{2+}$ axis). Typically, the parameters are represented as a range of values. If an ideal $N$-$Zn^{2+}$-$N$ angle is 109.5°, then the designer could define an acceptable angle as being from 100° to 120°.

Besides the simplicity of their binding sites, transition metals possess unique spectroscopic signatures that are highly dependent on their coordination geometry, which makes the actual testing of the designs relatively easy.

Different specific computational methods for writing metal site searching algorithms have been described (Clarke and Yuan 1995; Hellinga et al. 1991; Hellinga and Richards 1991). They involve the same overall strategy, but with slight methodological differences. First, the protein backbone is scanned one amino-acid position at a time to determine where metal-chelating side chains can be placed without backbone clashes. Then combinations of those residues are analyzed to determine if they can create a primary coordination sphere with near-ideal geometry for the entire

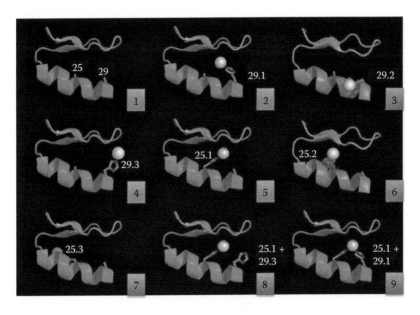

**FIGURE 17.2 (see color insert following page 178)** Snapshots of a $His_4$-$Zn^{2+}$ binding site search algorithm in progress. For simplicity, analysis of only two positions is illustrated. Frame 1: Individual amino-acid positions are evaluated for their ability to participate in a $His_4$-$Zn^{2+}$ binding site. At this point, residues 25 and 29 are being considered. Frames 2, 3, 4: Position 29 is being considered as a potential zinc-coordinating histidine residue. Frame 2: Rotamer #1 is accepted because it does not clash with the backbone. Frame 3: Rotamer #2 is rejected because it clashes with the backbone. Frame 4: Rotamer #3 of position 29 is, like rotamer #1, accepted. Frames 5, 6, 7: Position 25 is being considered as a potential zinc-coordinating histidine residue. Frame 5: Rotamer #1 of position 25 is accepted. Frames 6: Histidine rotamer #2 is rejected because the zinc clashes with the backbone. Frame 7: Rotamer #3 is accepted. Frames 8, 9: Combinations of sterically acceptable rotamers are queried for their potential to form the prescribed geometry (defined in Figure 17.1). Frame 8: The combination of 25.1 + 29.3 is rejected because it fails the geometric requirements. Frame 9: The combination of 25.1 + 29.1 meets the geometric criteria and is accepted. While only a few snapshots of the process are shown here, the search algorithm processes all positions to find candidate sites, and then checks all permutations of candidate sites for geometric compatibility. The output is a table of mutations predicted to form the desired geometry, which can be rank ordered by nearness to ideal geometry.

metal complex (Figure 17.2). Much of the effort in writing a geometry-based design program is focused on computational efficiency, because in an $n$-amino acid protein, there are approximately $n^4$ possible permutations of positions to be considered to design a tetracoordinating site (excluding rotameric possibilities).

### EXAMPLES

The use of computational design to create novel metal-binding sites has been thoroughly reviewed (Lu et al. 2001), but a few of the founding publications deserve to be mentioned here to illustrate the process. In one, a previously engineered four-helix bundle protein was used as a framework for identifying potential $His_2Cys_2$ metal-binding sites (Regan and Clarke 1990). To demonstrate that $Zn^{2+}$ binds the protein, simple gel filtration was used, and showed coelution of $Zn^{2+}$ and the protein. $Zn^{2+}$ was shown to stabilize the protein against chemical denaturation in circular dichroism. As mentioned previously, transition metals often have spectroscopic signatures that vary with the geometry of their coordination. $Zn^{2+}$ is spectroscopically silent, but binding of the similar cation $Co^{2+}$ to the designed protein yielded a spectroscopic signature consistent with metal coordination in a (thiolate)$_2$(imidazolate)$_2$ environment.

In another report, a distorted tetrahedron comprising one cysteine, one methionine, and two histidines was designed in thioredoxin to recreate a "blue-copper" site (Hellinga et al. 1991); the intended geometry was likewise confirmed by the distinct spectroscopic signature of the copper-protein complex. Unfortunately, neither design was confirmed by protein structure determination methods. Actual structural confirmation of an intended design was not achieved until 1995 when a $His_3Cys$ tetrahedral $Zn^{2+}$-binding site was engineered into the B1 domain of protein G (Klemba et al. 1995). The crystal structure of the designed site closely matched that of the model, and this structure was used to expand the design to include a secondary coordination sphere of Asp, Asn, Glu, or Gln to stabilize the positions of the metal-coordinating histidine residues while retaining the desired coordination geometry, resulting in a 10-fold increase in affinity (Marino and Regan 1999).

Geometry-based design can be expanded to include more than single atom/ion targets. An intriguing technique from the Lai group uses a combination of database mining and geometric search criteria to graft key residues from one protein onto unrelated proteins (Liang et al. 2000a; Liang et al. 2000b). The debut application of this technique used experimental mutagenesis data and the crystal structure of erythropoietin (EPO) in complex with its receptor (EPO-R) to identify three key residues in EPO assumed to be necessary and sufficient for binding. The Protein Data Bank (http://www.rcsb.org) was searched for proteins that could host these three residues, resulting in 15 candidate scaffolds. One of those proteins (rat pleckstrin homology domain of phospholipase C-∂1, PLC∂1-PH) was chosen for expression and purification in *Escherichia coli*. By grafting just these three residues from EPO onto PLC∂1-PH, the latter acquired the ability to bind EPO-R with 24 nM affinity (EPO binds EPO-R with 0.13 nM affinity) (Liu et al. 2007).

## STEREOCHEMISTRY-BASED DESIGN

### Methodology

Geometry-based approaches dominated the field of computational design until the mid-1990s, when a number of significant achievements made stereochemically complex problems addressable. More sophisticated semiempirical molecular mechanical potentials ("force fields") increased the accuracy of predictions, while new search algorithms and more powerful computers increased the combinatorial complexity that could be covered. Force fields typically comprise functions to quantify the steric, hydrogen bonding, and electrostatic interactions between pairs of atoms, a solvation term that quantifies the hydrophobic effect, and a variety of statistical or *ad hoc* terms.

While elementary descriptions of these interactions have been available since the mid-1980s (Brooks et al. 1983; Weiner et al. 1984), it has recently been discovered that seemingly minor, but critical adjustments are required to make accurate distinctions among highly similar structures. (For a comprehensive review of force field development, see Park et al. 2004.) One such adjustment is modification of the Lennard-Jones potential—a simple mathematical model of the London dispersion (van der Waals) force that represents the equilibrium balance of long-range attraction and short-range repulsion between atoms—to decrease the overemphasis of minor clashes. This can be done by decreasing the van der Waals radii (Dahiyat et al. 1997), stretching the van der Waals well (Looger et al. 2001), or using a linear (instead of $r^{12}$) repulsive term (Kortemme et al. 2002).

Substantial improvements to the quantitation of hydrogen bond energies have also been made. New angle dependence terms have been formulated by analyzing ideal hydrogen bond geometry (Dahiyat et al. 1997), and an empirically guided hydrogen bond energy function has been created by structural bioinformatics analysis of hydrogen bonds in high-resolution structures combined with quantum mechanical modeling (Kortemme et al. 2003). Furthermore, the contribution of water-mediated hydrogen bonds has been described quantitatively (Jiang et al. 2005). In some designs, a "hydrogen bond inventory" has been included to penalize nonsatisfaction of this important term (Looger et al. 2003). The electrostatic component of atomic interactions has often been modulated by varying the dielectric constant to account for observed differences between electrostatic interactions at the surface or in the core of proteins (for review, see Schutz and Warshel 2001). The distance-dependence term can be modified to more heavily weight shorter-range interactions (Selzer and Schreiber 1999), and recently, a pairwise-decomposable Poisson-Boltzmann function has been developed (Marshall et al. 2005). Another significant change to the potential function has been the addition of an empirical term that gives some amino acids preference over others, based on a statistical analysis of the backbone-dependent internal free energies of amino-acid rotamers (Kortemme and Baker 2002; Kuhlman and Baker 2000). Finally, quantum mechanics has been used to better estimate rotamer energies (Renfrew et al. 2007). Together, these modifications of the force field are leading to more complicated, but more accurate, quantitation of biophysical associations.

Stereochemistry-based design requires not only accurate force fields, but also methods for finding a combination of amino acids to produce a properly complementary surface from an astronomically large number of possible sequences. For the most part, three algorithms have been used in recent years: dead-end elimination (DEE), simulated annealing (SA, including Metropolis Monte Carlo), and fast and accurate side-chain topology and energy refinement (FASTER). DEE is an exact, deterministic algorithm that can quickly and drastically reduce the combinatorial complexity of the inverse folding problem. DEE in its original (Desmet et al. 1992) and generalized forms (Goldstein 1994; Gordon and Mayo 1998; Looger and Hellinga 2001) uses the pairwise-decomposability of the potential function to place upper and lower bounds on the energies of single side chains, removing those that provably cannot be members of the global minimum energy conformation (GMEC) (Figure 17.3). FASTER (Desmet et al. 2002) is a modified greedy algorithm that has been a major driving force in the radical redesign of some proteins. It functions by iteratively optimizing single protein positions or position clusters, in the context of a fixed combination of rotamers at the remaining positions. SA is an older technique, but many of the most impressive feats of computational protein design have resulted from its use to search sequence space (Baker 2006). SA searches a fitness landscape via a biased random walk: Energetically favorable moves are accepted, and unfavorable ones are accepted with a probability that decreases as the computation proceeds over time. The slower the rate of decrease (the "annealing"), the more likely the algorithm is to find an optimal or near-optimal state.

The advent of both accurate force fields *and* advanced methods for searching sequence space has made the design of complex stereochemical surfaces possible. Following, we highlight some of the more significant achievements in computational modulation of protein structure and interactions. This is not an exhaustive list, but rather an example of the diversity of problems that have been addressed.

## EXAMPLE: INTEGRAL MEMBRANE PROTEINS

One of the first principles discovered by analysis of protein structures is that proteins fold such that the interior is primarily hydrophobic, while the exterior is primarily hydrophilic. Transmembrane (TM) proteins are an exception to this rule, as the exteriors of the membrane-embedded portion match the hydrophobic environment of the membrane. Many TM proteins (especially G-protein coupled receptors) are key drug targets, playing essential roles in biological processes including signal transduction, ion conductance, and small molecule transport; thus they are of significant commercial interest. The design of new drugs would be greatly assisted by the ability to produce TM membrane proteins more reliably and with higher yield. In theory, one could make a water-soluble version of a transmembrane protein by simply redecorating the exterior facing hydrophobic side chains with hydrophilic ones. However, it is

| Position n | n Self energy | Position p | n–p Interaction energy | Position q | n–q Interaction energy | Position r | n–r Interaction energy |
|---|---|---|---|---|---|---|---|
| Rotamer $j$ | −2 | Rotamer $p_1$ | −2 | Rotamer $q_1$ | 5 | Rotamer $r_1$ | −2 |
| | | Rotamer $p_2$ | −1 | Rotamer $q_2$ | 1 | Rotamer $r_2$ | −2 |
| | | Rotamer $p_3$ | 4 | Rotamer $q_3$ | 2 | Rotamer $r_3$ | 0 |
| | | | | Rotamer $q_4$ | 3 | Rotamer $r_4$ | −3 |
| | | | | Rotamer $q_5$ | 2 | | |
| Rotamer $k$ | −1 | Rotamer $p_1$ | −3 | Rotamer $q_1$ | 0 | Rotamer $r_1$ | −6 |
| | | Rotamer $p_2$ | −1 | Rotamer $q_2$ | −5 | Rotamer $r_2$ | −5 |
| | | Rotamer $p_3$ | −4 | Rotamer $q_3$ | −3 | Rotamer $r_3$ | −7 |
| | | | | Rotamer $q_4$ | −2 | Rotamer $r_4$ | −6 |
| | | | | Rotamer $q_5$ | −2 | | |

$$\text{Sum(Best Energies for } n_j) = (-2 + -2 + 1 + -3) = -6$$
$$\text{Sum(Worst Energies for } n_k) = (-1 + -1 + 0 + -5) = -7$$

**FIGURE 17.3 (see color insert following page 178)** Example of dead-end elimination (DEE). The basic principle of the DEE algorithm is that some rotamers can be proven never to be in the GMEC, by identification of another rotamer to which it would always be favorable to switch. In the simplest form, rotamer $j$ at position $n$ ($n_j$) can be eliminated by rotamer $k$ at position $n$ ($n_k$) if the best energy conformation of $n_j$ is worse than the worst energy conformation of $n_k$. This calculation is easy to perform. First, a pair-wise matrix of energies is tabulated. In this table, the individual interaction energies of $n_j$ and $n_k$ with each possible rotamer at every other position is calculated, along with the self energy (the energy of $n_j$ or $n_k$ with the backbone template). In this example, there are three other positions: $p$, $q$, and $r$. Then the best possible total energy for $n_j$ is calculated, regardless of the interaction energies among $p$, $q$, and $r$. The worst possible total energy for $n_k$ is calculated, also regardless of the interactions among $p$, $q$, and $r$. In this example, $n_k$ eliminates $n_j$ from the GMEC, because the energetically worst combination of $p$, $q$, $r$ for $n_k$ ($p_2$, $q_1$, $r_2$) is still better than the best combination of $p$, $q$, $r$ for $n_j$ ($p_1$, $q_2$, $r_4$). The computational power of DEE can be seen by the fact that this table required only pairwise 24 energies to be calculated, whereas enumeration would require 120 (2*3*5*4) energies to be calculated.

more easily said than done, as mutation of the many residues to even the surface of a protein can lead to significant destabilization. DeGrado and colleagues have used computational design to resurface the exterior of two proteins with sufficient hydrophilic character to promote water solubility, while minimizing the disruptive actions of mutation to the global structure. Their redesigns of the potassium channel KcsA (Slovic et al. 2004) and phospholamban (Slovic et al. 2003) were based on either crystal structures or models assembled from individual subunits of the wild-type proteins. Of course, the "catch-22" is that the structure in the native environment is needed to apply computational design and produce a water-soluble version, but this requires a large quantity of the protein. Hopefully, with a combination of predictive techniques for protein folding and homology modeling, it may be possible to design solubilized versions of membrane proteins for which there are no available structures (Roosild and Choe 2005).

## Example: Nucleic Acids

Unlike nucleic acid hybridization, which can be predicted from the simple Watson-Crick base-pair recognition code, the sequence specificity of DNA-binding proteins cannot be easily predicted from the primary sequence of the protein. This is because recognition interfaces in proteins are constructed from complex stereochemical surfaces influenced by a combination of van der Waals, hydrogen bonding, solvation, and electrostatic forces. In theory, altering DNA-binding specificity in a protein is little different from altering its specificity for a small molecule, a peptide, or another protein. The limiting step is parameterization of the interactions between protein atoms and DNA atoms (Havranek et al. 2004). Baker and colleagues tested these parameters by altering the DNA recognition sequence specificity of the endonuclease I-*Mso*I (Ashworth et al. 2006). They redesigned the specificity of the enzyme by first screening *in silico* for base changes that would disrupt DNA binding by the wild-type enzyme. Two base substitutions (one each on the "left" and "right" sides of the symmetry dyad) were predicted to ablate binding to the wild-type protein. Computational redesign of the amino acids surrounding these bases suggested that just two mutations to the protein (K28L and T83R) could restore binding (Figure 17.4). Cleavage specificity, activity, and binding affinity of the redesigned protein are on par with the wild-type protein for its cognate sequence, and the crystal structure of the complex superimposes well with the predicted structure.

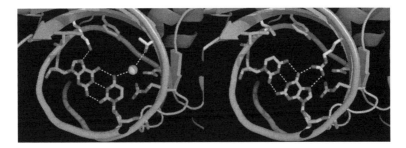

**FIGURE 17.4 (see color insert following page 178)** Designed DNA-binding protein (Ashworth et al. 2006). The left panel shows the structure of the wild-type (MsoI) enzyme in complex with wild-type DNA (from 2FLD.pdb). The G-C base pair is shown in the center, and important side chains from the protein are shown as sticks. The right panel shows the designed enzyme in complex with the mutant DNA (from 1M5X.pdb). The G-C base pair has been mutated to C-G. Notice that the Lys and Thr in the wild-type (yellow) that form hydrogen bonds with the DNA bases are mutated to Val and Arg in the designed enzyme to provide new van der Waals and hydrogen bonds.

## EXAMPLE: PEPTIDES

Computational design can also be used to alter the interaction between full-length proteins and small peptides. Calmodulin (CaM) is a calcium-modulated protein expressed in all eukaryotic cells that binds and regulates a large number of other proteins (many of which are enzymes) in response to fluctuating intracellular calcium levels. CaM binds up to four $Ca^{2+}$ ions and then undergoes a conformational change that greatly increases its affinity for various peptides (which are helical peptide domains of target proteins). This conformational change, when coupled with fluorescent proteins such as GFP, has been exploited to create genetically encoded calcium indicators (GECIs), which are very useful for monitoring calcium flux in neurons (Kotlikoff 2007). However, there are potential problems with CaM-based GECIs binding to endogenous cellular proteins, and it would thus be desirable to create a more specific CaM-peptide pairing. It has been hypothesized that the broad target specificity of CaM may be the result of its conformational flexibility, and that the abundance of methionines within the CaM-peptide interface makes for an accommodating binding site. Shifman and Mayo aimed to increase the specificity of CaM for one of its peptides, smooth muscle myosin light chain kinase (smMLCK), by redesigning CaM to minimize the free energy of the CaM-smMLCK complex (Shifman and Mayo 2002; Shifman and Mayo 2003). The affinity of the redesigned CaM (with eight mutations) for smMLCK is similar to that of the wild-type protein, and the affinity for other peptides is reduced by up to 86-fold.

Palmer and colleagues pursued a more complicated strategy for altering the specificity of the CaM-smMLCK interaction (Palmer et al. 2006). They first introduced "knobs" or "bumps" into the peptide and calculated free energies of interaction with wild-type CaM to determine whether binding of wild-type CaM to the mutant peptide would be destabilized compared to binding of the wild-type peptide. Then with the most destabilizing "knob" mutation in the peptide, four or five of the surrounding residues of CaM were redesigned (with "holes") to accommodate the protrusion, resulting in variants that form specific cognate pairs

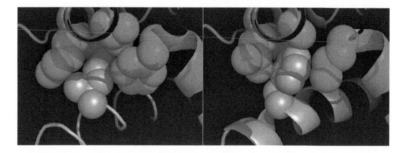

**FIGURE 17.5 (see color insert following page 178)**   Design of increased specificity for calmodulin (green) and a peptide (pink) (Palmer et al. 2006). The left panel shows the wild-type structure (from 1CDL.pdb). The right panel shows a model of the redesigned interface. On the peptide, Ile 14 has been mutated to Phe, which would form large steric clashes with the wild-type calmodulin. Three mutations were introduced in CaM (F19L, V35A, M36L) to create a compensatory "hole."

(Figure 17.5). This approach demonstrates the key principle of negative design: working not just to stabilize the target state (like the preceding example), but also to destabilize nontarget states.

## EXAMPLE: PROTEINS

The promiscuity of CaM is an exception among proteins. As a general rule, protein–protein interactions are remarkably specific and of high affinity (on the order of nM). This results from the evolution of stereochemically complementary surfaces interacting with each other. Typically, any mutation at the interface to one partner of the pair ablates or at least severely diminishes binding. Redesigning a protein–protein pair (A:B) by mutating both partners such that the mutants interact with each other (A*:B*) but not the parent proteins (A*:B or A:B*) requires not just stabilization of the target state (A*:B*), but also destabilization of alternative states (such as A*:B, A:B*, A*:A*, B*:B*, etc.). The redesign of the bacterial DNase colicin E7 and its tightly bound inhibitor Im7 (immunity protein 7) is another excellent example of negative design (Kortemme et al. 2004). Three residues on the E7 DNase and nine residues on the Im7 inhibitor, which are buried at the interface and form side chain–side chain contacts, were selected for mutagenesis. Computational screening identified mutations that maximized the computed free energy difference between the cognate (target state) and noncognate pairs (alternative states). The newly designed cognate pairs bound each other, and inhibition by the redesigned Im7 was observed, but there was still significant affinity between noncognate pairs (only 30-fold less than the designed cognate pairs). By extending the design process to include backbone flexibility and designing a novel hydrogen bond network, that difference in affinity was increased to 300-fold (Joachimiak et al. 2006). This is a great improvement, but still far from natural protein–protein pairs, for which noncognate affinity is negligible. From this design cycle, another principle of protein–protein interaction was elucidated. The structure of one of the designed pairs reveals that in some cases, even though direct side chain–side chain contacts were predicted to replace water-mediated hydrogen bonds, the waters were bound so tightly that the side chain was displaced instead. Since explicitly solvated side chains (Jiang et al. 2005) greatly expand the size of the rotamer library, this suggests that in future designs, retaining some tightly bound water molecules may be beneficial.

## EXAMPLE: SMALL MOLECULES

Small molecules possess broad functional group diversity, and often have greater degrees of internal flexibility than many other classes of molecules. As such, they can be difficult to parameterize for both drug and receptor design calculations (Gane and Dean 2000). They also have translational and rotational degrees of freedom, adding significant combinatorial complexity to the design problem. Substantial progress has been made in the computational design of small molecule binding sites, mainly by re-engineering existing binding pockets to accommodate different ligands. The family of bacterial periplasmic sugar- and amino acid–binding proteins makes an ideal scaffold for the redesign of small molecule binding sites (Dwyer and Hellinga 2004).

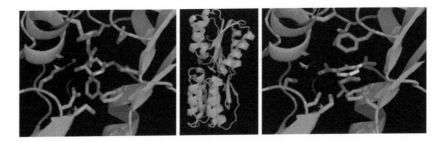

**FIGURE 17.6 (see color insert following page 178)**   Design of ribose-binding protein to TNT-binding protein (Looger et al. 2003). The center panel shows the overall structure of closed, ligand-bound ribose-BP (from 2DRI.pdb). Like all periplasmic binding proteins, the ligand-binding site is located in a cleft between two globular domains (top and bottom) separated by a hinge. In the unbound state, the two domains open up like a Venus flytrap. The left panel shows a close-up of the ribose (white) in the binding pocket with pertinent side chains shown in sticks. The right panel shows a model of TNT and the appropriate compensatory mutations required to turn ribose-BP into TNT-BP. Notice the π-sandwiching of the TNT ring by the two phenylalanine rings and the hydroxyl side chains properly positioned to form hydrogen bonds with the nitro-groups.

These binding proteins (BPs) consist of two domains connected by a hinge, with a ligand-binding site located at the interface between the two domains. They switch between a ligand-free open conformation and a ligand-bound closed conformation in a Venus flytrap–like way. This mechanism of binding results in bound ligands being in an environment that resembles a protein interior, with the residues that compose the binding site positioned at the cleft between the two domains. This characteristic has allowed the computational redesign of maltose-BP, glucose-BP, arabinose-BP, ribose-BP, glutamine-BP, and histidine-BP to bind $Zn^{2+}$ (using geometry-based design), trinitrotoluene (TNT, an explosive and pollutant, see Figure 17.6), L-lactate, serotonin, and pinacolyl methyl phosphonic acid (PMPA, a nerve agent surrogate) instead of their wild-type sugar or amino-acid ligands (Allert et al. 2004; Dwyer et al. 2003; Looger et al. 2003; Marvin and Hellinga 2001). However, the designed receptors have shown evidence of destabilization relative to the wild-type proteins. This suggests that the energetically favorable contacts made between the mutated side chains and the target ligand come at the expense of interactions with other residues in the protein. Quantifying these subtle yet important interactions will be a significant challenge for the design of small molecule binding proteins in the future.

## CATALYSIS

### EARLY EFFORTS

Perhaps the most rigorous test of computational protein engineering is the design of functional enzymes. Catalysis by enzymes is a complex process; it requires binding substrate(s), orienting the appropriate chemically reactive groups in the correct geometry, stabilizing the transition state to facilitate the chemical reaction, and releasing the product. As engineering a novel enzyme function is a very ambitious goal, it makes sense that initial attempts to design catalysts focused on the relatively simple target of oxygen-metal chemistry, and demonstration that the metal center can undergo cycles of reduction and oxidation when interacting with reactive oxygen species (Benson et al. 2002; Benson et al. 2000; Benson et al. 1998; Di Costanzo et al. 2001). Once it was demonstrated that the chemistry of oxygen could be controlled, designs progressed to include prototypes of organic substrate-binding sites, and used reactive oxygen locally within the active site, instead of releasing diffusible oxygen radicals. Using a previously designed four-chain heterotetrameric *di*-iron-binding helical bundle (Summa et al. 2002), DeGrado and colleagues modeled 4-aminophenol into the core of the protein, with its phenolic oxygen bridging the $Fe^{2+/3+}$ ions. With a few mutations to better accommodate this new substrate, they were able to achieve saturating kinetics with reaction rates 1000-fold above background (Kaplan and DeGrado 2004).

Moving beyond simple oxygen chemistry was made possible by the aforementioned advances made in the design of complex stereochemical surfaces. Since enzymes are believed to function by lowering the energy of the reaction transition state, the most tangible computational approach was to design proteins that bind transition state analogs, mimicking the approach taken in the development of catalytic antibodies (Hanson et al. 2005). In the first example of this strategy, Bolon and Mayo modeled a high-energy state intermediate of histidine-catalyzed *p*-nitrophenyl acetate (PNPA) hydrolysis as a series of transition-state/side-chain hybrid rotamers (Bolon and Mayo 2001). They searched *E. coli* thioredoxin for positions where the chimeric catalytic side-chain/substrate rotamer could be positioned without clashing with the backbone, and optimized the neighboring residues to accommodate the substrate. In theory, this should stabilize the transition state, reduce the energy of activation for the hydrolysis reaction, and enhance the catalytic rate. Substrate affinity is explicitly included in the design due to the stereochemical similarity between the transition state and the substrate. Their best "proto-enzyme" was constructed by mutation of a single residue to a "catalytic histidine" and mutation of three other residues to accommodate the substrate. It exhibited saturation kinetics with a $k_{cat}/k_{uncat}$ of 180, on the order of the first catalytic antibodies, but far short of the rate enhancements characteristic of natural enzymes.

## ADVANCED ENZYME DESIGN: COMBINING GEOMETRY-BASED AND STEREOCHEMISTRY-BASED DESIGN

The Baker group recently reported a substantial improvement to the process of designing enzymes (Rothlisberger et al. 2008). Unlike the method of Bolon and Mayo, in which the transition state side chain was forced into the chosen scaffold (thioredoxin), the Baker group used more advanced computational techniques to find an ideal scaffold. Making their experiments even more significant, they chose to catalyze a reaction for which there is no naturally occurring dedicated enzyme [although many proteins serendipitously catalyze this reaction (James and Tawfil 2001)]: the Kemp elimination—a model reaction for proton transfer from carbon. To find the ideal scaffold protein, they first modeled the transition state with geometric parameters describing the proper positioning of a general base and specific hydrogen bond donors. The next step was to find transition state placements in a large set of protein scaffolds that were consistent with the catalytic geometry using hashing algorithms (a computational tool to speed up data comparison) (Zanghellini et al. 2006). Once those sites (and scaffold proteins) were identified, residues surrounding the catalytic side chains and transition state were optimized with the simulated annealing-based Rosetta software (Meiler and Baker 2006). A total of 59 designs in 17 different scaffolds were selected for experimental characterization. Eight of those designs had measurable activity in a preliminary screen. For each of those, mutation of the catalytic base abolished activity, suggesting that the observed rate enhancement resulted from the designed active site. The best designs displayed saturation kinetics with a $k_{cat}/k_{uncat} \sim 10^5$, which is significantly greater than that previously achieved. Furthermore, the crystal structure of one of the designs and the predicted model are virtually superimposable (although no small molecule is present in the crystal structure), validating the design protocol (rmsd of 0.32 Å for the active site backbone and 0.95 Å for the active site side chains).

The most impressive feat in computational enzyme design to date (also from the Baker group) is the design of an enzyme that performs a multistep retro-aldol reaction (Jiang et al. 2008). The first step in their design process was defining the reaction pathway with a lysine as the active site nucleophile (Figure 17.7). They proposed four different catalytic motifs—three used combinations of different side chains to act as general acids/bases and one used an explicitly modeled bound water molecule to assist in the water elimination step. Then large ensembles of distinct 3D representations of these composite active sites (comprising the catalytic motif, key intermediates, and transition states) were created. The catalytic side chains in each scaffold protein (many scaffolds were tested) were then screened for the optimum rotamers and the ensuing position of the composite transition state was recorded in a hash table that was searched for the transition state (TS) positions compatible with

**FIGURE 17.7 (see color insert following page 178)** Definition of the Retro-Aldol reaction. First, nucleophilic attack of lysine on the ketone of the substrate forms an imine intermediate, and a water molecule is eliminated. Then carbon-carbon bond cleavage is triggered by deprotonation of the β-alcohol and release of the fluorescent ketone (blue). Finally, the enzyme is returned to its prereactive state by nucleophilic attack by water on the imine. Adapted from Jiang et al. 2008.

all catalytic constraints. When a match between the catalytic side chains and the TS was found, the remaining surrounding residues (not in the catalytic definition) were redesigned to optimize TS binding affinity. They ended up constructing and testing 70 designed proteins from 10 different scaffolds with between eight and 20 mutations each. The designs using the catalytic motif with a bound water molecule were the most successful, with rate enhancements up to four orders of magnitude over the uncatalyzed reaction. This truly remarkable achievement has significant implications for the development of enzymes *ab initio*. There is an infinitesimal probability that, as the result of random mutagenesis alone, three or four residues in a protein could simultaneously be appropriately positioned for catalysis. But once some detectable activity is reached by computational enzyme design, directed evolution could be used to optimize the surrounding residues by subtly changing the context of the active site.

## APPLICATIONS OF COMPUTATIONALLY DESIGNED PROTEINS

### BIOSENSORS

Computational design of ligand-binding sites is a particularly useful development for building new biosensors. Biosensors transduce microscopic binding events into macroscopically observable signals. Traditionally they have been developed by identifying a naturally occurring protein that binds the target analyte, discovering a suitable macroscopic signal, and building a detector that reports the concentration of the detected analyte to the user. While this strategy has produced many useful biosensors, each one is unique and requires substantial development time and optimization. Clearly, it would be useful to develop a modular system in which the binding site and the signaling site can be changed separately without destroying the communication between them, so that the development of new biosensors would require redesign of only one component—the analyte binding site. Many interesting platforms have been suggested (Hellinga and Marvin 1998), but periplasmic binding proteins, introduced previously in this chapter, have been the most thoroughly explored.

With the aim of creating a modular biosensor platform, *E. coli* maltose–binding protein was converted into a fluorescent maltose sensor by attaching a small molecule fluorophore to a region that is allosterically coupled to the ligand-binding site (Marvin et al. 1997). The maltose-binding site was redesigned to bind $Zn^{2+}$ while retaining the fluorescent response to ligand binding, indicating that, in theory, new biosensors could be created in a modular fashion by altering the binding specificity of the binding pocket (Marvin and Hellinga 2001). [Surprisingly, it was later determined by x-ray crystallography and small-angle x-ray scattering that the designed $Zn^{2+}$ binding site does not seem to form in its intended geometry (Telmer and Shilton 2005)]. Building on that work, ribose-binding protein was successfully converted into a zinc receptor with a secondary coordination sphere included in the initial designs (Dwyer et al. 2003). Once the method of stereochemistry-based design was available, ribose-binding protein (and other members of the periplasmic binding protein family) were designed into sensors for many other molecules (described previously in this chapter) (Allert et al. 2004; Looger et al. 2003). The redesigned ribose-binding proteins (zinc-BP and TNT-BP) were expressed in a bacterial strain containing a chimeric two-component signal transduction pathway connecting ligand binding in RBP to transcriptional activation of a reporter gene (β-galactosidase). This resulted in unique ligand-mediated synthetic signal transduction pathways. In theory, if the reporter gene were green fluorescent protein (GFP) or luciferase, then the engineered bacteria would glow in response to the analyte, making a truly biological biosensor.

## THERAPEUTIC PROTEINS AND ANTIBODIES

### Cytokine Design

Now that the field of computational design of protein–protein interactions is becoming more established, reports of its application to therapeutically relevant proteins are surfacing. One example of computational design in a therapeutic setting is the inactivation of TNF-α signaling by rationally designed dominant-negative TNF-α variants (Steed et al. 2003) using Xencor's Protein Design Automation (PDA) software (Filikov et al. 2002; Hayes et al. 2002), which is based on the work of Dahiyat and Mayo (Dahiyat and Mayo 1997). TNF-α is a homotrimeric cytokine that activates the inflammation pathway upon binding TNF receptor-1 (TNF-R1), and is implicated in a wide range of diseases including rheumatoid arthritis and inflammatory bowel syndrome (Chen and Goeddel 2002). Analysis of a homology model of the cytokine-receptor complex indicates that there is significant steric separation between residues of TNF-α that form the homotrimeric interface and those that form the binding interface with TNF-1R. The PDA software predicted one or two nonimmunogenic mutations that disrupt interactions with TNF-1R while preserving the structural integrity of the TNF variants and their ability to homotrimerize (and ideally heterotrimerize with endogenous TNF). In preclinical trials its ability to block soluble TNF activity in cell-based assays and two mouse arthritis models was established (Zalevsky et al. 2007). While this is a good example of computational design in a therapeutic setting, and is advantageous over other TNF-α inhibitors, destroying one interaction while preserving another distal one is not terribly challenging, and perhaps could have been achieved by inspection alone.

More impressive is the redesign of the TNF-related apoptosis inducing ligand (TRAIL). TRAIL is a potential anticancer drug that selectively induces apoptosis in cancer cells by interacting with the death receptors DR4 and DR5 (Jin et al. 2004; Kelley and Ashkenazi 2004) and is in Phase II trials for non-Hodgkin's lymphoma and non–small cell lung cancer (Genentech). It also binds decoy receptors that do not induce apoptosis (DcR1 and DcR2). Quax and colleagues computationally evaluated the effect of a single amino-acid mutation at 34 different positions of TRAIL on its interaction energy with each of the four receptors (Van Der Sloot et al. 2006). Seven mutations at five positions were predicted to confer selectivity for DR5. Of the various combinations of these mutations, one variant (with two mutations) showed no measurable affinity for DR4 or Dc1, and 20-fold reduced affinity for Dc2. In apoptosis assays, cells expressing DR5 were susceptible to killing by the mutant TRAIL. DR5-selective TRAIL variants have also been identified by phage display (Kelley et al. 2005); by using negative selection to remove variants that bind the other receptors

(DR4, Dc1, and Dc2), researchers at Genentech were able to isolate DR5-selective variants (which have more mutations than the computationally designed variants) that also induced apoptosis. Interestingly, the two solutions sets, computational and diversity-based, are nonoverlapping.

## Antibody Affinity Maturation

A common application of diversity-based engineering is affinity maturation of antibodies by phage display (Sidhu and Koide 2007), ribosome display (Lipovsek and Pluckthun 2004), and yeast display (Gai and Wittrup 2007) (reviewed in Marvin and Lowman 2005). Although these methods have proven robust, they are limited in the maximum genetic diversity they can search; a library of $6 \times 10^{23}$ Fab variants, even if present as single protein molecules, would weigh 50 kg! *In silico*, there are no such limitations. This has enabled computational design, with improved modeling of water and electrostatic interactions (Levy et al. 2003), to improve the affinity of antibodies beyond the achievements of diversity-oriented engineering (Lippow et al. 2007). The affinity of cetuximab (Erbitux®, ImClone Systems) for EGFR was improved 10-fold from 0.5 nM to 0.05 nM by mutations that modify the electrostatic interactions on the periphery of the binding interface; affinity of D44.1 (a common test antibody) for lysozyme was improved 100-fold from 4.4 nM to 0.04 nM. A similar approach was used to improve the affinity of an anti-VLA1 antibody from 10 nM to 1 nM (Clark et al. 2006). As computational design becomes more widely available, and its potential to create more efficacious antibodies is explored, it will eventually be adopted by companies that focus on therapeutic proteins.

## Surpassing the Technical Limitation of the Lab

The true value of computational design is its ability to surpass the technical limitations of the various diversity-oriented design approaches. Perhaps the best example is the design of antibody Fc variants with enhanced effector function (Lazar et al. 2006), for which no diversity-oriented approach is viable. Antibody-dependent cell-mediated cytotoxicity (ADCC), a key effector function that determines the clinical efficacy of some monoclonal antibodies, is mediated primarily through a set of closely related Fcγ receptors with both activating and inhibitory activities (excellently reviewed in Desjarlais et al. 2007). It has been shown by standard site-directed mutagenesis (Shields et al. 2001) and glycoform engineering (Shields et al. 2002) that the affinity of interaction between Fc and certain FcγRs correlates with cytotoxicity in cell-based assays. Of particular significance, higher affinity for FcγRIIIa (CD16) correlates with higher ADCC. Phage display, a powerful tool for antibody affinity maturation, could in principle be used to identify mutations that increase FcγRIIIa affinity, were it not for the critical role of the *N*-linked carbohydrate at N297. Deglycosylation of intact IgG decreases its affinity for FcγRIIIa by 20-fold, and deglycosylation of an Fc fragment decreases its affinity to below detectable levels (Radaev et al. 2001). Because *E. coli* bacteria lack the molecular machinery necessary for proper glycosylation of a displayed Fc fragment, and because full length IgGs are too large to be displayed, the only method for increasing the affinity of an IgG for FcγRIIIa used to date has been trial-and-error site-directed mutagenesis (Shields et al. 2001).

Lazar and coworkers at Xencor overcame this technical hurdle by using the crystal structure of the Fc/FcγRIIIa complex (Radaev et al. 2001) and their PDA software to predict mutations that would provide more favorable interactions at the Fc/FcγRIIIa interface (Lazar et al. 2006). Hedging their bets, they made and tested hundreds of combinations of those variants in high-throughput binding assays. They identified variants with over 100-fold enhancement of *in vitro* effector function, and this resulted in increased cytotoxicity in an *in vivo* preclinical model. This approach to optimization of FcR binding has enabled Xencor to form partnerships with MedImmune, Genentech, Boehringer Ingelheim, and Human Genome Sciences to incorporate (presumably) these mutations into their antibody-based products.

## CONCLUSIONS

### LOOKING BACK

The field of computational protein design has progressed rapidly in the last few decades. Like all new technologies, early efforts focused on achieving noticeable results with any system that could be used reliably. Designing metal-binding sites was an attractive target since such sites can be described in simple geometric terms, and near exhaustive searches could be performed within a reasonable period considering the processing power of computers at the time. Since many transition metals have spectroscopic signatures when chelated by particular ligands in specific geometries, testing the designs was straightforward. As the field progressed, geometry-based designs began to shift focus toward creating proteins with technological applications, most notably in the development of biosensors.

Advances in algorithms for searching sequence space and improved force fields to describe atomic interactions have made stereochemistry-based design of complex surfaces possible. Again, early efforts focused on tractable problems—having a strong assay for protein activity or binding often dictated which protein was used as a template. With new reports of progressively more intricate designs being published every year, computational design is starting to be seen as a realistic alternative or complement to diversity-oriented approaches. Furthermore, the software for designing proteins is becoming more accessible to the general scientific community. Many of the software packages used in the 1990s and early 2000s were (and still are) command line/script-based programs that require expert training and extensive programming knowledge to set the design parameters correctly.

### LOOKING FORWARD

The field of computational design is changing. Some of the programs are available for download or are accessible online. However, the design process is, for the most part, still not an interactive experience. This is likely to change in the near future. PyMol (http://www.pymol.org) is an open-source structure visualization program that runs on all main platforms and has a very intuitive interface. In addition to providing built-in tools for visualizing mutagenesis, plug-ins for various applications are available. As software becomes more user-friendly, computational protein design is likely to become more accessible to biochemists with less computational experience. With the RosettaDesign server (http://rosettadesign.med.unc.edu) (Liu and Kuhlman 2006) being accessible via a PyMol plug-in, the future could see protein engineers visualize a structure, pick the residues they want to mutate with a mouse, choose optimization criteria, and have a design returned to them shortly thereafter. Reliable "point-and-click" protein design will surely reduce the time it takes to design/improve/modify proteins, but the real value of computational design is in its ability to surpass the physical and technical limitations of diversity-oriented design. When those designs improve the quality of life of people, success will be real.

## REFERENCES

Allert, M., Rizk, S. S., Looger, L. L. and Hellinga, H. W. 2004. Computational design of receptors for an organophosphate surrogate of the nerve agent soman. *Proc Natl Acad Sci U S A* 101: 7907–12.

Ashworth, J., Havranek, J. J., Duarte, C. M., Sussman, D., Monnat, R. J., Jr., Stoddard, B. L. and Baker, D. 2006. Computational redesign of endonuclease DNA binding and cleavage specificity. *Nature* 441: 656–9.

Baker, D. 2006. Prediction and design of macromolecular structures and interactions. *Philos Trans R Soc Lond B Biol Sci* 361: 459–63.

Benson, D. E., Haddy, A. E. and Hellinga, H. W. 2002. Converting a maltose receptor into a nascent binuclear copper oxygenase by computational design. *Biochemistry* 41: 3262–9.

Benson, D. E., Wisz, M. S. and Hellinga, H. W. 2000. Rational design of nascent metalloenzymes. *Proc Natl Acad Sci U S A* 97: 6292–7.

Benson, D. E., Wisz, M. S., Liu, W. and Hellinga, H. W. 1998. Construction of a novel redox protein by rational design: Conversion of a disulfide bridge into a mononuclear iron-sulfur center. *Biochemistry* 37: 7070–76.

Bolon, D. N. and Mayo, S. L. 2001. Enzyme-like proteins by computational design. *Proc Natl Acad Sci U S A* 98: 14274–9.

Bone, R., Silen, J. L. and Agard, D. A. 1989. Structural plasticity broadens the specificity of an engineered protease. *Nature* 339: 191–5.

Brooks, B. R., Bruccoleri, R. E., Olafson, B. D., States, D. J., Swaminathan, S. and Karplus, M. 1983. CHARMM: A program for macromolecular energy, minimization, and dynamics calculations. *J Comput Chem* 4: 187–217.

Castagnetto, J. M., Hennessy, S. W., Roberts, V. A., Getzoff, E. D., Tainer, J. A. and Pique, M. E. 2002. MDB: The Metalloprotein Database and Browser at The Scripps Research Institute. *Nucleic Acids Res* 30: 379–82.

Chakrabarti, P. 1989. Geometry of interaction of metal ions with sulfur-containing ligands in protein structures. *Biochemistry* 28: 6081–5.

Chakrabarti, P. 1990a. Geometry of interaction of metal ions with histidine residues in protein structures. *Protein Eng* 4: 57–63.

Chakrabarti, P. 1990b. Interaction of metal ions with carboxylic and carboxamide groups in protein structures. *Protein Eng* 4: 49–56.

Chakrabarti, P. 1990c. Systematics in the interaction of metal ions with the main-chain carbonyl group in protein structures. *Biochemistry* 29: 651–8.

Chen, G. and Goeddel, D. V. 2002. TNF-R1 signaling: A beautiful pathway. *Science* 296: 1634–5.

Christianson, D. W. 1991. Structural biology of zinc. *Adv Protein Chem* 42: 281–355.

Clackson, T. and Wells, J. A. 1995. A hot spot of binding energy in a hormone-receptor inter-face. *Science* 267: 383–6.

Clark, L. A., Boriack-Sjodin, P. A., Eldredge, J., Fitch, C., Friedman, B., Hanf, K. J., Jarpe, M., Liparoto, S. F., Li, Y., Lugovskoy, A., Miller, S., Rushe, M., Sherman, W., Simon, K. and Van Vlijmen, H. 2006. Affinity enhancement of an in vivo matured therapeutic antibody using structure-based computational design. *Protein Sci* 15: 949–60.

Clarke, A. R., Atkinson, T. and Holbrook, J. J. 1989a. From analysis to synthesis: New ligand binding sites on the lactate dehydrogenase framework. Part I. *Trends Biochem Sci* 14: 101–05.

Clarke, A. R., Atkinson, T. and Holbrook, J. J. 1989b. From analysis to synthesis: New ligand binding sites on the lactate dehydrogenase framework. Part II. *Trends Biochem Sci* 14: 145–8.

Clarke, N. D. and Yuan, S. M. 1995. Metal search: A computer program that helps design tetrahedral metal-binding sites. *Proteins* 23: 256–63.

Craik, C. S., Largman, C., Fletcher, T., Roczniak, S., Barr, P. J., Fletterick, R. and Rutter, W. J. 1985. Redesigning trypsin: Alteration of substrate specificity. *Science* 228: 291–7.

Dahiyat, B. I., Gordon, D. B. and Mayo, S. L. 1997. Automated design of the surface positions of protein helices. *Protein Sci* 6: 1333–7.

Dahiyat, B. I. and Mayo, S. L. 1997a. De novo protein design: Fully automated sequence selection. *Science* 278: 82–7.

Dahiyat, B. I. and Mayo, S. L. 1997b. Probing the role of packing specificity in protein design. *Proc Natl Acad Sci U S A* 94: 10172–7.

Desjarlais, J. R., Lazar, G. A., Zhukovsky, E. A. and Chu, S. Y. 2007. Optimizing engagement of the immune system by anti-tumor antibodies: An engineer's perspective. *Drug Discov Today* 12: 898–910.

Desmet, J., De Maeyer, M., Hazes, B. and Lasters, I. 1992. The dead-end elimination theorem and its use in protein side-chain positioning. *Nature* 356: 539–42.

Desmet, J., Spriet, J. and Lasters, I. 2002. Fast and accurate side-chain topology and energy refinement (FASTER) as a new method for protein structure optimization. *Proteins* 48: 31–43.

Di Costanzo, L., Wade, H., Geremia, S., Randaccio, L., Pavone, V., DeGrado, W. F. and Lombardi, A. 2001. Toward the de novo design of a catalytically active helix bundle: A substrate-accessible carboxylate-bridged dinuclear metal center. *J Am Chem Soc* 123: 12749–57.

Dwyer, M. A. and Hellinga, H. W. 2004. Periplasmic binding proteins: A versatile superfamily for protein engineering. *Curr Opin Struct Biol* 14: 495–504.

Dwyer, M. A., Looger, L. L. and Hellinga, H. W. 2003. Computational design of a $Zn^{2+}$ recep-tor that controls bacterial gene expression. *Proc Natl Acad Sci U S A* 100: 11255–60.

Filikov, A. V., Hayes, R. J., Luo, P., Stark, D. M., Chan, C., Kundu, A. and Dahiyat, B. I. 2002. Computational stabilization of human growth hormone. *Protein Sci* 11: 1452–61.

Gai, S. A. and Wittrup, K. D. 2007. Yeast surface display for protein engineering and charac-terization. *Curr Opin Struct Biol* 17: 467–73.

Gane, P. J. and Dean, P. M. 2000. Recent advances in structure-based rational drug design. *Curr Opin Struct Biol* 10: 401–04.

Goldstein, R. F. 1994. Efficient rotamer elimination applied to protein side-chains and related spin glasses. *Biophys J* 66: 1335–40.

Gordon, D. B. and Mayo, S. L. 1998. Radical performance enhancements for combinatorial optimization algorithms based on the dead-end elimination theorem. *J Comput Chem* 19: 1505–14.

Hanson, C. V., Nishiyama, Y. and Paul, S. 2005. Catalytic antibodies and their applications. *Curr Opin Biotechnol* 16: 631–6.

Havranek, J. J., Duarte, C. M. and Baker, D. 2004. A simple physical model for the prediction and design of protein-DNA interactions. *J Mol Biol* 344: 59–70.

Hayes, R. J., Bentzien, J., Ary, M. L., Hwang, M. Y., Jacinto, J. M., Vielmetter, J., Kundu, A. and Dahiyat, B. I. 2002. Combining computational and experimental screening for rapid optimization of protein properties. *Proc Natl Acad Sci U S A* 99: 15926–31.

Hellinga, H. W., Caradonna, J. P. and Richards, F. M. 1991. Construction of new ligand binding sites in proteins of known structure. II. Grafting of a buried transition metal binding site into Escherichia coli thioredoxin. *J Mol Biol* 222: 787–803.

Hellinga, H. W. and Marvin, J. S. 1998. Protein engineering and the development of generic biosensors. *Trends Biotechnol* 16: 183–9.

Hellinga, H. W. and Richards, F. M. 1991. Construction of new ligand binding sites in proteins of known structure. I. Computer-aided modeling of sites with pre-defined geometry. *J Mol Biol* 222: 763–85.

James, L. C. and Tawfil, D. S. 2001. Catalytic and binding poly-reactivities shared by two unrelated proteins: The potential role of promiscuity in enzyme evolution. *Protein Sci* 10: 2600–07.

Jiang, L., Althoff, E. A., Clemente, F. R., Doyle, L., Rothlisberger, D., Zanghellini, A., Gallaher, J. L., Betker, J. L., Tanaka, F., Barbas, C. F., 3rd, Hilvert, D., Houk, K. N., Stoddard, B. L. and Baker, D. 2008. De novo computational design of retro-aldol enzymes. *Science* 319: 1387–91.

Jiang, L., Kuhlman, B., Kortemme, T. and Baker, D. 2005. A "solvated rotamer" approach to modeling water-mediated hydrogen bonds at protein–protein interfaces. *Proteins* 58: 893–904.

Jin, H., Yang, R., Fong, S., Totpal, K., Lawrence, D., Zheng, Z., Ross, J., Koeppen, H., Schwall, R. and Ashkenazi, A. 2004. Apo2 ligand/tumor necrosis factor-related apoptosis-inducing ligand cooperates with chemotherapy to inhibit orthotopic lung tumor growth and improve survival. *Cancer Res* 64: 4900–05.

Jin, L., Fendly, B. M. and Wells, J. A. 1992. High resolution functional analysis of antibody-antigen interactions. *J Mol Biol* 226: 851–65.

Joachimiak, L. A., Kortemme, T., Stoddard, B. L. and Baker, D. 2006. Computational design of a new hydrogen bond network and at least a 300-fold specificity switch at a protein–protein interface. *J Mol Biol* 361: 195–208.

Kaplan, J. and DeGrado, W. F. 2004. De novo design of catalytic proteins. *Proc Natl Acad Sci U S A* 101: 11566–70.

Kelley, R. F., Totpal, K., Lindstrom, S. H., Mathieu, M., Billeci, K., Deforge, L., Pai, R., Hymowitz, S. G. and Ashkenazi, A. 2005. Receptor-selective mutants of apoptosis-inducing ligand 2/tumor necrosis factor-related apoptosis-inducing ligand reveal a greater contribution of death receptor (DR) 5 than DR4 to apoptosis signaling. *J Biol Chem* 280: 2205–12.

Kelley, S. K. and Ashkenazi, A. 2004. Targeting death receptors in cancer with Apo2L/TRAIL. *Curr Opin Pharmacol* 4: 333–9.

Klemba, M., Gardner, K. H., Marino, S., Clarke, N. D. and Regan, L. 1995. Novel metal-binding proteins by design. *Nat Struct Biol* 2: 368–73.

Kortemme, T. and Baker, D. 2002. A simple physical model for binding energy hot spots in protein–protein complexes. *Proc Natl Acad Sci U S A* 99: 14116–21.

Kortemme, T., Joachimiak, L. A., Bullock, A. N., Schuler, A. D., Stoddard, B. L. and Baker, D. 2004. Computational redesign of protein–protein interaction specificity. *Nat Struct Mol Biol* 11: 371–9.

Kortemme, T., Morozov, A. V. and Baker, D. 2003. An orientation-dependent hydrogen bonding potential improves prediction of specificity and structure for proteins and protein–protein complexes. *J Mol Biol* 326: 1239–59.

Kotlikoff, M. I. 2007. Genetically encoded $Ca^{2+}$ indicators: Using genetics and molecular design to understand complex physiology. *J Physiol* 578: 55–67.

Kuhlman, B. and Baker, D. 2000. Native protein sequences are close to optimal for their structures. *Proc Natl Acad Sci U S A* 97: 10383–8.

Lazar, G. A., Dang, W., Karki, S., Vafa, O., Peng, J. S., Hyun, L., Chan, C., Chung, H. S., Eivazi, A., Yoder, S. C., Vielmetter, J., Carmichael, D. F., Hayes, R. J. and Dahiyat, B. I. 2006. Engineered antibody Fc variants with enhanced effector function. *Proc Natl Acad Sci U S A* 103: 4005–10.

Levy, R. M., Zhang, L. Y., Gallicchio, E. and Felts, A. K. 2003. On the nonpolar hydration free energy of proteins: Surface area and continuum solvent models for the solute-solvent interaction energy. *J Am Chem Soc* 125: 9523–30.

Liang, S., Li, W., Xiao, L., Wang, J. and Lai, L. 2000a. Grafting of protein–protein interaction epitope. *J Biomol Struct Dyn* 17: 821–8.

Liang, S., Liu, Z., Li, W., Ni, L. and Lai, L. 2000b. Construction of protein binding sites in scaffold structures. *Biopolymers* 54: 515–23.

Lipovsek, D. and Pluckthun, A. 2004. In-vitro protein evolution by ribosome display and mRNA display. *J Immunol Methods* 290: 51–67.

Lippow, S. M., Wittrup, K. D. and Tidor, B. 2007. Computational design of antibody-affinity improvement beyond in vivo maturation. *Nat Biotechnol* 25: 1171–6.

Liu, S., Liu, S., Zhu, X., Liang, H., Cao, A., Chang, Z. and Lai, L. 2007. Nonnatural protein–protein interaction-pair design by key residues grafting. *Proc Natl Acad Sci U S A* 104: 5330–5.

Liu, Y. and Kuhlman, B. 2006. RosettaDesign server for protein design. *Nucleic Acids Res* 34: W235–8.

Looger, L. L., Dwyer, M. A., Smith, J. J. and Hellinga, H. W. 2003. Computational design of receptor and sensor proteins with novel functions. *Nature* 423: 185–190.

Looger, L. L. and Hellinga, H. W. 2001. Generalized dead-end elimination algorithms make large-scale protein side-chain structure prediction tractable: Implications for protein design and structural genomics. *J Mol Biol* 307: 429–45.

Lu, Y., Berry, S. M. and Pfister, T. D. 2001. Engineering novel metalloproteins: Design of metal-binding sites into native protein scaffolds. *Chem Rev* 101: 3047–80.

Marino, S. F. and Regan, L. 1999. Secondary ligands enhance affinity at a designed metal-binding site. *Chem Biol* 6: 649–55.

Marshall, S. A., Vizcarra, C. L. and Mayo, S. L. 2005. One- and two-body decomposable Poisson-Boltzmann methods for protein design calculations. *Protein Sci* 14: 1293–1304.

Marvin, J. S., Corcoran, E. E., Hattangadi, N. A., Zhang, J. V., Gere, S. A. and Hellinga, H. W. 1997. The rational design of allosteric interactions in a monomeric protein and its applications to the construction of biosensors. *Proc Natl Acad Sci U S A* 94: 4366–71.

Marvin, J. S. and Hellinga, H. W. 2001. Conversion of a maltose receptor into a zinc biosensor by computational design. *Proc Natl Acad Sci U S A* 98: 4955–60.

Marvin, J. S. and Lowman, H. B. 2005. Antibody humanization and affinity maturation using phage display. In *Phage Display in Biotechnology and Drug Discovery*, ed. S. S. Sidhu, 493–528. New York: Marcel Dekker, Inc.

Meiler, J. and Baker, D. 2006. ROSETTALIGAND: Protein-small molecule docking with full side-chain flexibility. *Proteins* 65: 538–548.

Pabo, C. 1983. Molecular technology. Designing proteins and peptides. *Nature* 301: 200.

Palmer, A. E., Giacomello, M., Kortemme, T., Hires, S. A., Lev-Ram, V., Baker, D. and Tsien, R. Y. 2006. $Ca^{2+}$ indicators based on computationally redesigned calmodulin-peptide pairs. *Chem Biol* 13: 521–30.

Park, S., Yang, X. and Saven, J. G. 2004. Advances in computational protein design. *Curr Opin Struct Biol* 14: 487–94.

Radaev, S., Motyka, S., Fridman, W. H., Sautes-Fridman, C. and Sun, P. D. 2001. The structure of a human type III Fcgamma receptor in complex with Fc. *J Biol Chem* 276: 16469–77.

Radaev, S. and Sun, P. D. 2001. Recognition of IgG by Fcgamma receptor. The role of Fc glycosylation and the binding of peptide inhibitors. *J Biol Chem* 276: 16478–83.

Regan, L. and Clarke, N. D. 1990. A tetrahedral zinc(II)-binding site introduced into a designed protein. *Biochemistry* 29: 10878–83.

Renfrew, P. D., Butterfoss, G. L. and Kuhlman, B. 2008. Using quantum mechanics to improve estimates of amino acid side chain rotamer energies. *Proteins* 71(4), 1637≠46.

Roosild, T. P. and Choe, S. 2005. Redesigning an integral membrane K+ channel into a soluble protein. *Protein Eng Des Sel* 18: 79–84.

Röthlisberger, D., Khersonsky, O., Wollacott, A. M., Jiang, L., Dechancie, J., Betker, J., Gallaher, J. L., Althoff, E. A., Zanghellini, A., Dym, O., Albeck, S., Houk, K. N., Tawfik, D. S. and Baker, D. 2008. Kemp elimination catalysts by computational enzyme design. *Nature* May 8; 453(7192): 164–6.

Schreiber, G. and Fersht, A. R. 1993. Interaction of barnase with its polypeptide inhibitor barstar studied by protein engineering. *Biochemistry* 32: 5145–50.

Schutz, C. N. and Warshel, A. 2001. What are the dielectric "constants" of proteins and how to validate electrostatic models? *Proteins* 44: 400–17.

Scrutton, N. S., Berry, A. and Perham, R. N. 1990. Redesign of the coenzyme specificity of a dehydrogenase by protein engineering. *Nature* 343: 38–43.

Selzer, T. and Schreiber, G. 1999. Predicting the rate enhancement of protein complex formation from the electrostatic energy of interaction. *J Mol Biol* 287: 409–19.

Shields, R. L., Lai, J., Keck, R., O'Connell, L. Y., Hong, K., Meng, Y. G., Weikert, S. H. and Presta, L. G. 2002. Lack of fucose on human IgG1 N-linked oligosaccharide improves binding to human Fcgamma RIII and antibody-dependent cellular toxicity. *J Biol Chem* 277: 26733–40.

Shields, R. L., Namenuk, A. K., Hong, K., Meng, Y. G., Rae, J., Briggs, J., Xie, D., Lai, J., Stadlen, A., Li, B., Fox, J. A. and Presta, L. G. 2001. High resolution mapping of the binding site on human IgG1 for Fc gamma RI, Fc gamma RII, Fc gamma RIII, and FcRn and design of IgG1 variants with improved binding to the Fc gamma R. *J Biol Chem* 276: 6591–604.

Shifman, J. M. and Mayo, S. L. 2002. Modulating calmodulin binding specificity through computational protein design. *J Mol Biol* 323: 417–23.

Shifman, J. M. and Mayo, S. L. 2003. Exploring the origins of binding specificity through the computational redesign of calmodulin. *Proc Natl Acad Sci U S A* 100: 13274–9.

Sidhu, S. S. and Koide, S. 2007. Phage display for engineering and analyzing protein interaction interfaces. *Curr Opin Struct Biol* 17: 481–7.

Slovic, A. M., Kono, H., Lear, J. D., Saven, J. G. and DeGrado, W. F. 2004. Computational design of water-soluble analogues of the potassium channel KcsA. *Proc Natl Acad Sci U S A* 101: 1828–33.

Slovic, A. M., Summa, C. M., Lear, J. D. and DeGrado, W. F. 2003. Computational design of a water-soluble analog of phospholamban. *Protein Sci* 12: 337–48.

Steed, P. M., Tansey, M. G., Zalevsky, J., Zhukovsky, E. A., Desjarlais, J. R., Szymkowski, D. E., Abbott, C., Carmichael, D., Chan, C., Cherry, L., Cheung, P., Chirino, A. J., Chung, H. H., Doberstein, S. K., Eivazi, A., Filikov, A. V., Gao, S. X., Hubert, R. S., Hwang, M., Hyun, L., Kashi, S., Kim, A., Kim, E., Kung, J., Martinez, S. P., Muchhal, U. S., Nguyen, D. H., O'Brien, C., O'Keefe, D., Singer, K., Vafa, O., Vielmetter, J., Yoder, S. C. and Dahiyat, B. I. 2003. Inactivation of TNF signaling by rationally designed dominant-negative TNF variants. *Science* 301: 1895–8.

Summa, C. M., Rosenblatt, M. M., Hong, J. K., Lear, J. D. and DeGrado, W. F. 2002. Computational de novo design, and characterization of an A(2)B(2) diiron protein. *J Mol Biol* 321: 923–38.

Telmer, P. G. and Shilton, B. H. 2005. Structural studies of an engineered zinc biosensor reveal an unanticipated mode of zinc binding. *J Mol Biol* 354: 829–40.

Thomson, A. J. and Gray, H. B. 1998. Bio-inorganic chemistry. *Curr Opin Chem Biol* 2: 155–8.

van der Sloot, A. M., Tur, V., Szegezdi, E., Mullally, M. M., Cool, R. H., Samali, A., Serrano, L. and Quax, W. J. 2006. Designed tumor necrosis factor-related apoptosis-inducing ligand variants initiating apoptosis exclusively via the DR5 receptor. *Proc Natl Acad Sci U S A* 103: 8634–9.

Weiner, S. J., Kollman, P. A., Case, D. A., Singh, U. C., Ghio, C., Alagona, G., Profeta, S. and Weiner, P. 1984. A new force field for molecular mechanical simulation of nucleic acids and proteins. *J Am Chem Soc* 106: 765–84.

Wilks, H. M., Hart, K. W., Feeney, R., Dunn, C. R., Muirhead, H., Chia, W. N., Barstow, D. A., Atkinson, T., Clarke, A. R. and Holbrook, J. J. 1988. A specific, highly active malate dehydrogenase by redesign of a lactate dehydrogenase framework. *Science* 242: 1541–4.

Zalevsky, J., Secher, T., Ezhevsky, S. A., Janot, L., Steed, P. M., O'Brien, C., Eivazi, A., Kung, J., Nguyen, D. H., Doberstein, S. K., Erard, F., Ryffel, B. and Szymkowski, D. E. 2007. Dominant-negative inhibitors of soluble TNF attenuate experimental arthritis without suppressing innate immunity to infection. *J Immunol* 179: 1872–83.

Zanghellini, A., Jiang, L., Wollacott, A. M., Cheng, G., Meiler, J., Althoff, E. A., Rothlisberger, D. and Baker, D. 2006. New algorithms and an *in silico* benchmark for computational enzyme design. *Protein Sci* 15: 2785–94.

# 18 Future Challenges of Computational Protein Design

*Eun Jung Choi, Gurkan Guntas, and Brian Kuhlman*

## CONTENTS

With the recent advances in the field of computational protein design, we have seen many successful examples of *de novo* designs, including novel protein–protein and protein-DNA interactions, as well as enzymes capable of catalyzing novel reactions. However, the rate of success remains low, which suggests that our understanding of the sequence-structure-function relationship is far from complete. There are many formidable challenges that need to be overcome before we can reliably design new protein molecules at will. Here we review three topics that will be important in the future progress of computational protein design: backbone flexibility, negative design, and experimental approaches.

## BACKBONE FLEXIBILITY

The initial computational protein design methods were developed to accept a pre-determined backbone structure as an input and keep it fixed throughout the design process. This was necessary because it lowered the complexity of the search space by eliminating the need to consider the degrees of freedom for the backbone and as a result decreased the computation time. Also the energy function did not need to discriminate between energetically favorable backbones with unfavorable ones, so by utilizing it together with rotamer libraries one could disregard covalent interactions from the energy function altogether. Although fixed-backbone design methods have been very successful at various applications (Kuhlman and Baker 2004; Lippow and Tidor 2007), they have severe limitations (Figure 18.1). One is their tendency to predict false negatives. By keeping the backbone fixed, the energy landscape for potential sequences becomes more rugged and less physical. Certain residue combinations, which may be compatible with slight changes in the backbone, would be considered sterically incompatible (Desjarlais and Handel 1999). Another problem arises when applying the fixed-backbone protein design methods to a *de novo* backbone. Although a natural protein backbone is known to be the ground state or near ground state conformation for a natural sequence, the designability of a *de novo* backbone structure is not known. To select a designable backbone, you may have to sample many different local backbone conformations and find an energetically favorable structure for some sequence(s) to fold into, which necessitates the consideration of explicit backbone flexibility (Harbury et al. 1998; Kuhlman et al. 2003).

These limitations with fixed-backbone protein design have been pointed out since the beginning of the field (Vasquez 1996). Many attempts have been made to incorporate backbone flexibility into computational protein design methods by utilizing different approaches that either simplify the backbone structure, use an ensemble composed of different backbone structures and design each structure individually, sample multiple backbones by optimizing the backbone and the side chains simultaneously, or a combination of these methods. Different methods resulted in different levels of success. Descriptions of various flexible-backbone protein design methods with their pros and cons will be given in this section.

### PARAMETERIZATION OF STRUCTURES

Algebraic parameterization of regular secondary structures or folds can introduce backbone flexibility while minimizing the degree of freedom of the backbone. Parametric representation of supercoiled helices was used by Harbury and coworkers to design a *de novo* right-handed coiled coil trimer and a tetramer (Harbury et al. 1995; Harbury et al. 1998; Plecs et al. 2004). Although the crystal structures of the trimer and the tetramer showed high correlation to the design, the method was restrictive in that the parameterization was used to search for backbone coordinates that satisfied a predetermined amino-acid sequence. Su and Mayo also utilized secondary structure parameters for α/β proteins to redesign the core residues of protein G β1 domain (Su and Mayo 1997; Ross et al. 2001). They did not see any drastic differences between the design sequences obtained using backbones that

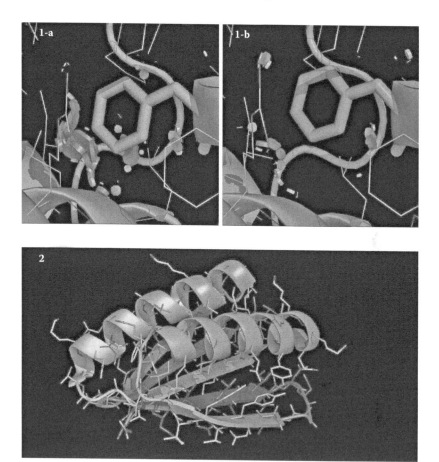

**FIGURE 18.1 (see color insert following page 178)**   The fallacy of fixed-backbone protein design. (1) Mutation of residue 59 of 434 cro from Leu to Phe is disruptive and is not allowed by fixed-backbone design (Desjarlais and Handel 1999) (a). When the backbone is allowed to move the clashes seen with fixed backbone are alleviated (b). (2) Crystal structure of Top7, the first *de novo* design of a novel backbone fold. The design of Top7 showed that to select a designable backbone, backbone flexibility is necessary. Clashes are shown as red disks. All figures were generated using PyMOL (Delano 2002).

deviated from the parent backbone and concluded that the protein design method is robust enough to tolerate significant amount of perturbation to the backbone. However, the nuclear magnetic resonance (NMR) structure of a design with a large translational perturbation of the helix along the sheet axis had backbone that was closer to the parent backbone than the designed. A recent study by Fu and coworkers used normal mode calculations for helices to parameterize the backbone of Bcl-$x_L$ (Fu et al. 2007). Based on a study that shows that backbone movement on a helix can be mostly captured by three low energy modes, they generated multiple backbones using normal mode analysis. Self-consistent mean field method was used to prune the rotamer library and then sequences were designed onto the backbone using a Monte Carlo procedure. By using this protocol they designed peptides that bind to Bcl-x$_L$ with nanomolar affinity. The limitation of this method is that the normal modes used have to capture most of the structural variation; thus guidance from pre-existing structures might be necessary to produce designable backbones. Parameterization of structures is a simple way to decrease the degree of freedom for backbone modeling but some of the general limitations are that it does not allow explicit backbone flexibility at every position of the protein and that it may not be generally applicable to complex motifs with nonsymmetrical folds or to irregular structural changes.

## ENSEMBLE APPROACH

Various methods have been attempted that incorporate backbone flexibility into protein design by using an ensemble of structures with slight differences in the backbone. In the ensemble approach, starting backbone structures are generated from multiple x-ray structures, from different models in a single NMR structure, from Monte Carlo perturbation runs, or from multiple snapshots along a molecular dynamics simulations trajectory. Then each individual structure in the ensemble is designed using the fixed backbone assumption. For the search algorithm, mean field methods have been applied to many designs that use the ensemble approach. The sequence information from the large numbers of designed structures allows you to calculate the specific probabilities and entropies of residues at each position in the context of all other residues and the backbones. Koehl and Delarue thus optimized the backbone and side chains simultaneously in loop designs of bovine pancreatic trypsin inhibitor by using the self-consistent mean field method with the ensemble approach (Koehl and Delarue 1995). Although in the loop designs a relatively small set of backbone conformations (10 or less) from short protein segments (five residues or less) were used, it was suggested that this method could be applicable to full sequence protein design problems. Kono and Saven utilized an entropy-based statistical method analogous to the self-consistent mean field method and 21 model backbones from an NMR structure to sample the sequence space of Protein L (Kono and Saven 2001). Using 30 backbone structures generated by a Monte Carlo simulation, Kraemer-Pecore and coworkers simulated the full sequence design of the WW domain (Kraemer-Pecore et al. 2003). In this study, each member of the ensemble was designed individually before an exhaustive search of all rotamers was carried out at each position with all other positions fixed. The energies from these runs were used to calculate the probability of all amino acids at each position. When the amino acids with the highest calculated probabilities were selected to produce a designed WW domain, the designed protein had the correct fold but was less stable than the native WW domain. Design of a large ensemble of backbone variants using a distributed computing network has also been explored (Larson et al. 2002; Larson et al. 2003). Generation of a large ensemble of 100 structural variants of the target structure was produced by a Monte Carlo procedure, and fixed backbone protein design was conducted on each of the individual structures by the Genome@home distributed computing system. The results of such calculations were used for fold recognition for structural and functional genomics. A general restriction with the ensemble approach is that a limited number of backbone conformation needs to be specified in advance.

## Simultaneous Optimization of Sequence and Structure

Methods incorporating backbone flexibility by optimizing the backbone and side chains simultaneously might represent a more accurate relaxation of local structures because of cross talk between the two components during optimization. The self-consistent mean field method used in the loop design mentioned previously, where the mean field of both the backbone and the side chains are simultaneously considered, is an example of this method. Another approach to this method is the use of molecular dynamics to allow backbone flexibility synchronously with side-chain adjustments. However, during side-chain repacking simulations the protein structure could get trapped in a local minimum. To prevent this from happening simulated annealing protocols are used but this also poses problems because explicit waters and overall protein folds can be distorted. Riemann and Zacharias suggested the use of a potential scaling molecular dynamics method to overcome these issues (Riemann and Zacharias 2005). The smooth rescaling of the potential during simulation allows the lowering of energy barriers, while minimizing the distortion of the protein fold and explicit waters. Using initial structures with arbitrary perturbed buried side chains and backbone, the potential scaling molecular dynamics method resulted in a better side-chain prediction compared to a fixed backbone side-chain packing algorithm (SCRWL3.0). A third approach of simultaneous backbone–side chain optimization is to iterate between backbone deformation and sequence optimization at all positions in the protein. The iterative strategy was adopted by Desjarlais and Handel, who used genetic algorithm for backbone perturbation and sequence selection on a pool of backbone structures obtained by altering random phi and psi angles of the parent template (Desjarlais and Handel 1999). Typically, a refinement step follows in the end where Monte Carlo is used for rotamer optimization and small backbone movements. Using this method, they designed a core variant of 434 cro protein with a mutation known to be incompatible with fixed-backbone design and a melting temperature slightly lower than the wild type. Unfortunately, the stability measurements of various 434 cro and T4 lysozyme protein variants showed that the predictive power of the flexible-backbone method is worse than those of the fixed-backbone method. Interestingly, in side-chain structure prediction comparison, the more the backbone deviates from that of wild type, the better the flexible-backbone prediction becomes compared to the fixed-backbone prediction. A similar result was obtained by Yin and coworkers, who compared the experimental $\Delta\Delta G$ with the computational $\Delta\Delta G$ for five proteins and their mutants using either a flexible- or a fixed-backbone prediction (Yin et al. 2007). They observed that in three out of five cases, the fixed-backbone prediction gave better correlation compared to flexible-backbone prediction method, although flexible-backbone method gave slightly better prediction if the mutations are classified into different types. In general, there is a concern with flexible-backbone design methods where an increased number of false positives are outputted by being too permissive in the prediction of mutations. As a result, mutations that are calculated to be energetically favorable in a flexible-backbone design often turn out to be destabilizing.

Another example of iterative backbone and side-chain design is the method used in the design of a novel protein fold (Kuhlman et al. 2003). Because the designed protein adopts a nonpreexisting fold, it is considered to be a more rigorous test of the flexible backbone design method. First, a pool of backbone models was created by using fragments from the PDB database that had the secondary structure of interest. Each of these backbones was used for design. Then based on the resulting sequence, the backbone was optimized using a Monte Carlo minimization procedure. Iterations of the sequence and backbone optimization were carried out to produce a novel $\alpha/\beta$ fold protein that folded into the topology of interest with atomic accuracy and was stable. An interesting observation from this study is that the energies of the designs from the initial starting backbone structures had worse energies compared to natural proteins, corroborating the fact that not all backbones are designable and that optimizing the backbone for the sequence is a critical step in the flexible-backbone design procedure. The ensemble approach, however, does not optimize the backbone and sequence simultaneously, but rather the backbone of the design is determined prior to sequence design. The same iterative flexible-backbone protocol was used to design a novel 10-residue loop on the tenascin protein (Hu et al. 2007). A detailed study of this iterative method, subsequently carried out using native protein structures and sequences (Saunders and Baker 2005), showed that in order to construct a designable backbone, it is important to sample a larger structural space around the starting structure. This was achieved by utilizing a high temperature Monte Carlo melting procedure, torsional minimization, and systematic substitution of the fragments that cause the greatest disruption in the global structure. However, when this modified iterative flexible backbone design method was used, there was no energy discrimination between the correct backbone configuration and the incorrect one. A similar difficulty was reported by Desjarlais and Handel, who observed no significant correlation when predicting the stability of proteins using the Amber/OPLS potential to represent the backbone (Desjarlais and Handel 1999). Although the current energy function may be sufficient to design a backbone, there is a clear need for improvement in the current energy function for a more reliable backbone selection (Bradley et al. 2005). Recent fold prediction studies also show that selection of good models cannot be achieved by considering energy alone (Bradley and Baker 2006). In this study, clustering based on structural similarities was more robust than using energies. In the loop design of tenascin, designs were filtered based not only on energy but also on solvent accessible surface area pack score and the number of unsatisfied hydrogen bonds (Hu et al. 2007). In the study by Fung and coworkers, root mean squared deviation (RMSD) from the template structure was used to rank structures (Fung et al. 2008). These results show that the development of energy functions, which can accurately predict a favorable backbone from an unfavorable one, is still needed as we migrate from fixed- toward flexible-backbone design. Furthermore, other methods to rank good structures may be helpful at times, since the energy function is currently not accurate enough for some applications. An interesting conclusion from the Saunders and Baker study (Saunders and Baker 2005) is that the occupancy of the sequence space of iterative flexible-backbone design overlaps more with those of the natural homologs compared to the ensemble approach. This shows that the iterative approach does not result in increased false positives, which are often seen with flexible-backbone design.

## THE FUTURE OF FLEXIBLE-BACKBONE PROTEIN DESIGN

Although the successful flexible-backbone design results described in the previous section reflect the extent to which the field has progressed, there is still room for improvement, as can be seen from the inconsistent results of sequence recovery and the modest correlation between computational and experimental stability of known proteins. The complexity of the flexible backbone design is still considered to be too large to obtain complete coverage, and extensive conformational sampling of structurally diverse populations is critical for its success. Although conformational sampling has led to many successful results, proper sampling of the conformational space still remains the primary bottleneck for accurate structure prediction (Bradley et al. 2005). This is probably the case for flexible-protein design as well, since conformational sampling does not guarantee identification of an optimal solution over the backbone/sequence search space. The line that separates inverse folding and structure prediction has gotten very vague in the recent years. Utilization of structure prediction protocols in protein design and vice versa are common, and there are also examples where structure prediction is used to evaluate the design result (Bradley et al. 2004; Hu et al. 2007).

Despite many significant achievements in the past incorporating backbone flexibility into protein design, myriad challenges lie ahead. The infinite complexity of the sequence and structural space and the limited accuracy of the current energy functions are some of these grand challenges, and resolving these issues would require novel approaches. One approach to reduce complexity might be allowing not random but "realistic" movement in the backbone, such as the application of "backrub" motions of natural proteins to simulate backbone flexibility (Davis et al. 2006). A precise energy representation of the backbone is not necessary to describe the "backrub" motion, as it is characterized by low energy and results in only favorable backbone. It is also local with no motion beyond two residues, which is appropriate for protein design, since the goal is not to sample a completely different fold but to allow small deviations from the parent backbone to accommodate for sequences and conformations that would be incompatible with fixed backbone (Smith and Kortemme 2008). Novel algorithms for flexible-backbone design that restrict the complexity are surfacing, such as the development of dead-end elimination (DEE) for flexible backbone (Georgiev and Donald 2007). Georgiev and Donald suggest reducing the complexity of the problem by allowing flexibility in both backbone and side chains but utilizing DEE algorithms to prune the rotamers not in the GMEC in the context of flexible backbone and minimized rotamers. Fung and coworkers have developed a new algorithm that identifies an optimal backbone via a continuum template and NMR structure refinement (Fung et al. 2007; Fung et al. 2008). Continuous values of $C_\alpha - C_\alpha$ distances and dihedral angles within a preset boundary were considered and by using this method they showed that the sequences of the designed β-defensins recapitulate the sequences found in nature to a large degree. However, the energy function used for side-chain selection is based on the $C_\alpha - C_\alpha$ distance and not on explicit consideration of side-chain rotamers. The accuracy in the prediction of structure, energetics, and design is important, especially with flexible backbone incorporation. In-depth study of the successful results found in the literature as well as of the various failures not reported is therefore a necessary step toward developing a reliable and accurate flexible-backbone protein design method.

## NEGATIVE DESIGN

Positive design aims to design a molecule to perform a desired function. Negative design seeks to ensure the specificity of function by minimizing unintended side effects. Biological molecules and systems exhibit great specificity precisely because nature is proficient in both positive and negative design. Computational protein modeling today is still in the process of developing and optimizing positive design, with most efforts focused on engineering macromolecules to perform specific target functions. To that end, recent years have seen significant progress in macromolecule structural modeling and recognition. In particular, protein sequences that fold into novel structures, bind novel ligands (e.g., small molecules, nucleic acids), and catalyze nonbiological reactions have been designed (Dahiyat and Mayo 1997; Kuhlman et al. 2003; Looger et al. 2003; Ashworth et al. 2006; Jiang et al. 2008; Rothlisberger et al. 2008). The core of positive design used in these problems involves setting a target structure and optimizing a sequence to stabilize it. This single-state approach, however, does not explicitly consider the possibility of the designed sequence to adopt other structures in the conformational space, including the unfolded state. This oversight is addressed by negative design, which interrogates each designed sequence against multiple target structures to ensure that it does not favor alternate conformations (Figure 18.2). In so doing, negative design allows the selection of sequences with structural uniqueness as well as favorable free energy of folding.

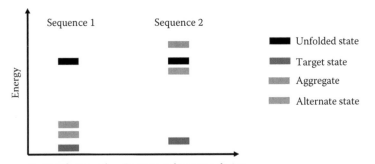

FIGURE 18.2 (see color insert following page 178)  The energy diagram of different states for two hypothetical designed sequences. Sequence 1 is designed using positive design only, whereas Sequence 2 utilizes both positive and negative design. Although the energy of the target structure is lower for Sequence 1, Sequence 2 is preferable as the energy difference between the target and alternative states is larger.

The set of alternate structures vary depending on the situation. For example, for the inverse-folding problem, that is, protein design, one might consider the unfolded state, oligomers, aggregates, and other stable folds as potential competing states. When designing protein–protein interactions, structures that represent competing protein–protein complexes would constitute potential alternate structures. In all cases, the goal is to search the sequence space for a unique and specific structure-sequence pair. In the following, we will examine examples in which negative design has been successfully applied together with positive design.

## PROTEIN–PROTEIN INTERACTIONS

Protein–protein interactions within the cell govern the signaling networks that are necessary to sustain life. Understanding and designing key protein interactions will thus help research in protein therapeutics and cell biology. Most studies that involve negative design have aimed at designing specific protein–protein interactions. These studies can be broadly classified into three categories based on differences in their strategy: computational algorithms that perform explicit multistate design, prescreening for sequences that destabilize alternate states, and experimental negative selections.

### Explicit Multistate Design

Although specificity in protein–protein interactions has been designed in many labs, only a few studies involved computational algorithms that explicitly penalize sequences that form undesired complexes. In order for such a multistate strategy to be successful, the structures of the undesired complexes as well as the target structure must be known or set beforehand. In the most recent example, Bolon and coworkers compared the success of conventional positive design algorithm versus explicit negative design in achieving specificity (Figure 18.3) (Bolon et al. 2005). They selected the same four residues in each chain of the SspB homodimer and first redesigned them for improved stability. This positive design strategy optimizes the sequence for each chain so that the mutants formed a stable heterodimer. The sequences of the designed amino acids were FAFI and LALI, compared to the wild-type sequence LAYV, and the FAFI/LALI complex was predicted to be more stable than LAYV/LAYV, FAFI/FAFI, and LALI/LALI homodimers. Hence, the calculation seemed to suggest that positive design alone is sufficient to design a desired state while avoiding alternate states. However, experimental studies showed that the homodimers were almost as stable as the heterodimer and all oligomers were more stable than wild-type homodimer. Overall, computational design was good at stabilizing a target state, but did not perform well in predicting how destabilizing these mutations were for the alternate homodimer states. In this study, the main chains of each polypeptide were kept fixed, and mutations that caused steric overlap in the models were penalized. In reality proteins adapt to destabilizing mutations through relaxation. This result, therefore, also underscores the importance of incorporating backbone flexibility into the protein design algorithms. In a second round of calculation, the authors used explicit negative design to select for sequences that optimized the energy difference between the preferred heterodimer and the sum of two alternate homodimers. *In silico*, the designed sequences for both chains, LSLA

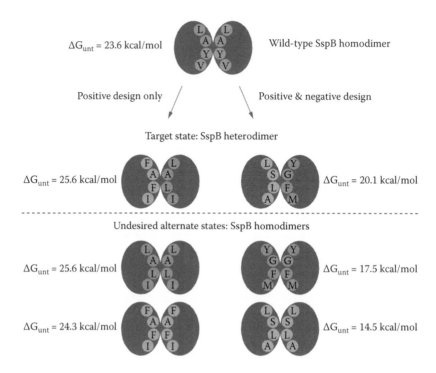

**FIGURE 18.3 (see color insert following page 178)** Effect of explicit multistate design on specificity and stability (Bolon et al. 2005). Positive design produces stable but not specific heterodimers, whereas inclusion of negative design results in less stable but specific heterodimers.

and YGFM, formed a heterodimer that was significantly more stable than the alternate homodimers (LSLA/LSLA and YGFM/YGFM), but was less stable than the wild-type homodimer. Experimental results confirmed that indeed they exclusively formed a heterodimer, LSLA/YGFM, albeit with lower stability than the wild-type homodimer, suggesting that there is a tradeoff between specificity and stability. This study shows that explicit consideration of alternate states can be critical for negative design, and the sole consideration of stability as in the case of positive design might not result in improved specificity.

Summa and coworkers designed a heterotetrameric ($A_2B_2$) diiron-binding helical bundle (Summa et al. 2002). For their design, two alternate topologies (heterotetramer configurations with different monomer arrangements) were explicitly disfavored. Although many other undesired arrangements might have been considered, authors concluded that other topologies would not be populated due to the energetic cost of exposing nonpolar residues and destabilization of metal-binding sites. Once the target and competing structures were determined, the surface residues were mutated to charged residues that preferentially stabilized the target structure. The energy difference between target and competing states was maximized for each considered competing state separately. Circular dichroism and analytical ultracentrifugation analysis showed that at neutral pH the designed peptides are by themselves unable to form homotetramers, whereas they exhibited strong helical content and formed a heterotetramer when mixed in equimolar proportion. Cobalt binding and ferrooxidase activity tests suggested that the designed topology agreed well with the target structure.

Havranek and Harbury implemented an algorithm to optimize coiled-coil sequences that preferentially form homodimers or heterodimers (Havranek and Harbury 2003). In their work, in addition to the target state, they explicitly disfavored three competing states: alternate oligomer (heterodimer disfavored if the target is homodimer, or vice versa), the unfolded state, and the aggregates. Rather than stabilizing the target state alone, sequence selection was performed to maximize transfer free energy from the target state to the three competing states. Eight designed sequences (four homodimer and four heterodimer targeting sequences) were experimentally characterized and in each case the sequence was found to prefer the target oligomeric state. In another study, Keating and coworkers redesigned BBAT2, a previously designed 21-residue α/β mini protein that forms a homotetramer, to form a heterotetramer by maximizing the energy gap between the homotetramer and heterotetramer states (Ali et al. 2005). They also modeled the unfolded state explicitly and maximized the energy difference between the target and unfolded states. This negative design strategy yielded two sequences that were less stable than the wild-type homotetramer but had a strong preference for heterotetramer formation.

Given the target and alternate structures, multistate design does quite well in selecting sequences optimized for specificity. Computational algorithms that apply multistate design to specificity in protein–protein interactions are expected to be particularly useful if the target and competing structures share the same interface. The model systems studied in the aforementioned examples are small oligomeric complexes with small binding interfaces. Even with the relatively smaller size of sequence space searched, adequate rotamer sampling and computational workload coming from flexible backbone are often bottlenecks in the selection of designs with high specificity. Finally, it will be interesting to see whether the tradeoffs between stability and specificity, as in Bolon and coworkers (Bolon et al. 2005), are a result of the fact that naturally existing interfaces are near optimal or stem from the shortcomings of our computational modeling tools.

## Prescreening for Destabilizing Mutations

An alternative approach to multistate design is first to prescreen for sequence changes that destabilize an existing interaction. For example, if a promiscuous "Protein A" binds both "Protein B" and "Protein C" using the common surface, and one wants to design a cognate A-B interaction while keeping the sequence "C" constant, the explicit multistate design might require searching for A-B mutants to create a large "free energy of binding" difference between the A-B and A-C interfaces. However, unlike explicit multistate design that would require a simultaneous analysis of the complex structures, predicting the mutations in "A" to increase the free energy of A-C binding may only take a simple calculation, or even a visual inspection, of the A-C natural interface. In other words, the "negative design" part of the problem might easily be solved by placing amino acids at the A-C interface that are highly likely to be destabilizing. As the interfaces for both complexes are similar, it is likely that "Protein A" mutants will also have low affinity against "Protein B." In that case, "Protein B" can be redesigned to compensate for the loss of binding due to the mutations in "Protein A." Therefore, a multistate negative design project has been converted to a much simpler positive design task.

This computational "second-site suppressor" strategy has been applied to design cognate "Colicin E7–Immunity protein Im7" binding pairs (Figure 18.4) (Kortemme et al. 2004) . The authors first analyzed the effect of E7 mutations on the wild-type interface between E7 and Im7 and introduced two mutants that were predicted to most destabilize the bound complex. Next, they applied positive design to search for complementary Im7 mutants that were predicted to have high affinity for the mutant E7s. This procedure resulted in the design of an Im7 mutant that had 40-fold higher affinity for its cognate E7 mutant than the wild-type E7 protein, suggesting that prescreening for destabilizing mutants is a quick way to solve the negative design problem. However, the specificity came with a loss in stability, as the affinity between cognate mutant pair was several orders of magnitude lower than wild-type pair. The residual affinity between the noncognate wild-type E7 and mutant Im7 pair and the trade-off between specificity and stability have evolutionary implications. For example, the proteins that promiscuously bind multiple proteins might have used moderately destabilizing mutations as a way to acquire additional partners. Similarly, evolution of promiscuity (Humphris and Kortemme 2007) in multispecific proteins may also be a consequence of soft negative design where nature evolved a

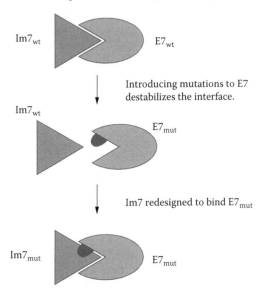

**FIGURE 18.4 (see color insert following page 178)**   Computational second-site suppressor strategy (Kortemme et al. 2004). A specific Im7-E7 interaction was designed by introducing interface-destabilizing mutations to E7 followed by the redesign of Im7 interface residues to compensate for the loss of binding. The interaction between $Im7_{mut}$ and $E7_{mut}$ was 40-fold stronger than $Im7_{mut}$-$E7_{wt}$ interaction, but several orders of magnitude weaker than $Im7_{wt}$-$E7_{wt}$ pair.

subset of a particular protein's interface with its partner in order to accommodate a second partner while ensuring that the sequence changes preserve the interactions with the original partner. It should also be noted that most of the destabilizing sequence subspace was not accessed in the study, which together with the conformational constraints due to fixed-backbone assumption may have prevented the full optimization of specificity.

## Experimental Approaches

Combinatorial protein engineering and directed evolution are useful tools to engineer specificity. As a positive design strategy, a library of protein mutants can be screened against an immobilized target for enhanced affinity using *in vitro* display (i.e., phage, bacterial, yeast, ribosome, mRNA) technologies or *in vivo* selections. If the design requires that the selected mutant should not bind to another competitor protein, then the competitor protein may be added to the medium during the selection to introduce negative design.

In a recent example, Mason and coworkers applied such a specificity screen to the design of specific coiled coils (Mason et al. 2007). Protein-fragment complementation assay (PCA) utilized in the study is an *in vivo* genetic selection for protein–protein interactions. In this system, the genes for two proteins that are interrogated for binding are separately fused to the genes encoding the two halves of the murine-DHFR (dihydrofolate reductase) enzyme and coexpressed in the same cell. If binding occurs between the proteins of interest, DHFR fragments are brought together and their association restores DHFR enzymatic activity, whereas if the two proteins do not bind, these DHFR fragments are unable to complement each other. Since DHFR activity is required for cell growth, only the cells expressing interacting proteins will be able to survive in selective media. Using this conventional PCA strategy, Mason and coworkers selected cFos and cJun mutants that form heterodimers with wild-type cJun and cFos, respectively (Mason et al. 2007). However, the selected peptides also formed homodimers and nonspecific heterodimers with their parent peptides. Based on thermal denaturation experiments, the melting temperatures of selected cJun mutant (cJun\*)-cFos, cJun\*-cJun, cJun\*-cJun\* complexes were 44°C, 57°C, and 66°C, respectively. In order to enable negative design, they coexpressed a competitor peptide to select for more specific heterodimers. A second cJun mutant (cJun\*\*) thus selected formed significantly less homodimers and nonspecific heterodimers, while the stability of its complex with cFos peptide remained the same. The measured melting temperatures for cJun\*\*-cFos, cJun\*\*-cJun, cJun\*\*-cJun\*\* complexes were 44°C, 23°C, and 29°C, respectively.

Experimental selection schemes that enable the engineering of specific protein–protein interactions are in place. In the future, it may be possible to combine the power of high-throughput screen with computational tools containing negative design to optimize specificity from computer-designed protein libraries.

## PROTEIN DESIGN

A longstanding goal in protein science is the ability to design sequences that adopt a desired structure. This inverse folding problem has been studied by several labs and resulted in the computational design of sequences that fold into naturally existing (Dahiyat and Mayo 1997) and nonnatural (Kuhlman et al. 2003) folds. To model the protein folding process, Yue and Dill used 2D lattice simulations where each residue in a 14-mer peptide was assigned a hydrophobic or polar residue (Yue and Dill 1992). The simulation suggested that hydrophobic amino-acid propensity should be enough to drive the collapse of the polypeptide, but should not be so excessive to allow alternate conformations. Yue and coworkers subsequently used lattice algorithms to fold rationally designed 48-mer peptides (Yue et al. 1995). That most sequences had more than one predicted structure pointed to the possibility that a designed sequence may adopt several different conformations.

Although its importance is generally recognized, it is not clear how one should apply negative design to inverse-folding problem. Negative design requires destabilization of alternate structures using a multistate approach, but the misfolded structures are typically not well characterized. The challenges of dealing with misfolded proteins are further exemplified by the fact that several natural proteins are able to adopt different folds by only a few mutations. For example, introducing two mutations in P22 Arc repressor has been shown to cause its structure to switch from an α-β into all helical fold (Cordes et al. 2000). Similarly, Alexander and coworkers mutated seven residues of β1 domain of Protein G to amino acids that are at the same positions of another similar size domain of Protein G (Alexander et al. 2007). These mutations resulted in β1 domain changing its fold from mostly β-sheet into all α-helical. These studies suggest that highly similar sequences may adopt significantly different folds.

The ultimate goal of negative design is to create an energy gap between the desired native state and other misfolded conformations. One way to achieve this would be to improve the positive design algorithms such that the probability of an alternate conformation having a lower energy than the design is extremely low. However, in studies where single- and multistate simulations were compared (Havranek and Harbury 2003; Bolon et al. 2005), designing for stabilization of a target state did not yield specific sequence-structure pairs. An alternative is to optimize the Z-score that is defined as the difference of energy between the target structure and an ensemble of randomly picked structures. Using Z-score optimization, Jin and coworkers redesigned a 47-residue α-helical bundle (Jin et al. 2003). They performed folding and unfolding simulations on random sequences to create snapshots of the denatured states. During the sequence design, the set of the denatured models served as the reference to optimize the Z-score. The new sequence had no similarity to the wild-type sequence, and based on CD and NMR studies, the designed protein folded cooperatively.

An alternative solution that might potentially be useful is to utilize protein structure prediction algorithms to verify that the designed sequences do not fold into structures other than the target. Recently, significant progress has been made in the development of structure-prediction algorithms (Bradley et al. 2005). In structure prediction, the metric of success is usually the RMSD of the $C_\alpha$ coordinates between the predicted model and the final structure. Although accuracy is inconsistent, for proteins with fewer than 85 residues, predictions have approached the true structures with RMSDs as low as 0.8 Å. Structure prediction would not only serve as a filter for foldability, but also might give insight as to the designability of a given backbone.

Similarly, docking simulations to predict protein–protein binding modes have the potential to evaluate whether a sequence finds exactly the intended structure or competing bound-complex structures.

Incorporating negative design during inverse protein folding is a challenging problem. To our knowledge, there is no experimental screen that can resolve various potential folds unless the structural differences are coupled to differences in function. From a computational standpoint, multistate design is promising and has been applied to design a sequence that is able to adopt two different folds (Ambroggio and Kuhlman 2006). However, the lack of knowledge about competing folds precludes multistate design to be useful for the protein design problem. It appears that protein design will benefit most from postdesign filters that use structure prediction and possibly molecular dynamics simulations for validation.

## EXPERIMENTAL APPROACHES FOR HIGH-THROUGHPUT DESIGN SAMPLING

Despite recent advances, the success rate for computer-based protein designs is well below 100%. Therefore, it is important to be able to experimentally characterize as many designs as possible. Typically, this requires cloning the gene for the designed polypeptide in an expression vector to allow subsequent microbial expression and purification for *in vitro* characterization. This one-protein-at-a-time approach is expensive, tedious, and time consuming. There are two approaches that in the near future might potentially alleviate these experimental bottlenecks: (1) massively parallel protein fabrication platforms and (2) construction and high-throughput screening of computationally designed protein libraries

### HIGH-THROUGHPUT PROTEIN FABRICATION

The first approach is to develop a high-throughput system to automate the traditional process of sample preparation, including oligonucleotide synthesis, gene assembly, sequencing, expression, purification, and bioassay.

The success of high-throughput gene synthesis hinges on the cost, efficiency, and extent of parallelization in oligonucleotide synthesis. In practice, the majority of custom oligonucleotides are kept shorter than 100 bases due to the imperfect chemical coupling between bases. Therefore, assembling genes coding for more than 35 amino acids requires multiple overlapping oligonucleotides. For long oligonucleotides, additional purification steps may also be required to recover the correct sequence, which increases the overhead in constructing the full-length gene product.

Alternatively, long DNA sequences can be assembled from shorter oligonucleotides by PCR alone (Stemmer et al. 1995) or PCR in conjunction with hybridization driven ligation (Smith et al. 2003). Since high-throughput synthesis of shorter oligonucleotides with high purity and low cost is essential to the success of these approaches, developing strategies to synthesize and purify oligonucleotides in parallel has attracted much attention. Recently, Zhou and coworkers reported the miniaturization of oligonucleotide synthesis using microarray chips (Zhou et al. 2004).

Despite the small production scale, the sequences can be easily amplified by PCR to improve the yield. Indeed, Tian and coworkers (Tian et al. 2004) designed programmable DNA microchips where they first used PCR to amplify the oligonucleotides and then used hybridization to select for the oligonucleotide with the right sequence. Furthermore, the designed microchips were able to perform gene assembly using a single-step PCR reaction.

Although microchip-based DNA synthesis has a potential for parallel processing, laboratories are also developing automated, 96-well platforms to use oligonucleotides from conventional sources. Cox and coworkers reported a procedure for using PCR to assemble genes in the 96-well format with automated liquid handling (Cox et al. 2007). This approach is particularly appropriate for probing protein function by combinatorial alanine scanning and for testing different computer-based designs with closely related sequences. The method, however, requires molecular cloning of the assembled genes followed by transformation into an expression host in order to eliminate errant mutations by a genetic selection, which highlights the need for cheap and highly pure oligonucleotides to simplify downstream processing.

The accuracy of the assembled genes can be verified by high-throughput sequencing (Margulies et al. 2005), which has currently reached a capacity of about $10^8$ nucleotides in a single run. The last step in the procedure requires high-throughput *in vitro* transcription/translation to obtain microgram scale quantities of designed protein for testing stability, catalytic activity, or binding affinity, provided that an appropriate bioassay is available.

## DESIGN AND HIGH-THROUGHPUT SELECTION OF COMPUTATIONALLY DESIGNED PROTEIN LIBRARIES

The second strategy for optimizing the screening of designed proteins involves creating a biased library of genes that best represents an ensemble of rationally designed proteins. The library of proteins is then expressed and screened for members with the desired property. In order to create a biased library one must decide what sequence positions will be varied and which amino acids will be allowed at each position. A variety of approaches have been used to choose the optimal set of amino acids at each position.

Rationally designed libraries have been used to probe the determinants of protein folding and stability. Realizing that naturally existing beta-sheet and α-helices often exhibit repeating patterns of hydrophobicity, Kamtekar and coworkers designed libraries of sequences with the hydrophobic pattern that would favor the formation of helical bundles (Kamtekar et al. 1993). Surprisingly, these libraries were rich in the number of molecules that formed compact, α-helical structures. In another study, proteins that folded cooperatively were isolated from a library of 80–100 residue proteins that were encoded by only three amino acids: leucine, glutamine, and arginine (Davidson and Sauer 1994; Davidson et al. 1995). Therefore, it appears that screening libraries of proteins with appropriate amino-acid type propensity and pattern might be an efficient way of selecting well-folded molecules.

Multiple sequence alignments (MSA) have been used to create profiles of amino acids at each position within a family of closely related proteins, which are then used to improve protein stability. Schellenberger and coworkers, for example, developed "combinatorial consensus mutagenesis" to improve the stabilities of β-lactamase and variable domain of an antibody (Amin et al. 2004; Roberge et al. 2006).

In another demonstration of the utility of bioinformatics analysis coupled with high-throughput screening, the structural analysis of antibody-antigen complexes shows a preference for the tyrosine and serine residues at the interface (Amin et al. 2004; Roberge et al. 2006). Based on this observation, a combinatorial library of FN3 mutants containing tyrosine and serine mutations in the complementary determining region was constructed and screened by phage display to optimize binding against arbitrary targets (Koide et al. 2007). The screen successfully identified the mutants that bound their targets with a dissociation-constant in the low nanomolar range.

Apart from the strategies that use clever sequence selection methods, libraries that are created based on explicit computational simulations also hold promise as a discovery tool. Kono and coworkers used mean field simulations to develop a probabilistic sequence theory that optimizes the site-specific amino-acid distributions of a protein with a known structure (Kono and Saven 2001). Rather than testing individual sequences *in silico* for folding, the theory directly calculates the probability of a site being occupied with a particular amino acid. Based on this theory, Park and coworkers designed a small library in order to improve the stability of a three-helix-bundle protein (Park et al. 2006). Concurrently, there have been other studies that used structure-based scoring functions in order to create more efficient libraries for the sole purpose of directed evolution. Using a site entropy function, Voigt and coworkers correlated the mutations that catalytically improved subtilisin and T4 lysozyme to their tolerance to mutations at those positions (Voigt et al. 2001). The authors suggest that their computational method to predict mutation-tolerant sites will allow significant reductions in the searched sequence space. Further, Voigt and coworkers developed an algorithm, SCHEMA, that computationally evaluates recombinations of homologous proteins in order to identify which crossovers between two proteins would lead to stable chimeric proteins (Voigt et al. 2002). In the last few years there have been a few studies that took advantage of full-atom computational design in order to decrease the experimental search space in search of novel function, for example, fluorescence and catalysis. Although these algorithms do not model function directly, they are useful when designing libraries as they can prescreen the vast, experimentally inaccessible sequence space for mutants that are least likely to disrupt the structure of the target protein. For example, Hayes and coworkers used "protein design automation" and Monte Carlo sampling to create an amino-acid profile at each of 19 residue positions within 5 Å of the five active site residues of β-lactamase (BLA) (Hayes et al. 2002). Following elimination of certain amino acids due to their low frequency of occurrence, they expressed a library of 200,000 mutants in cells. When the mutant proteins were screened for cefotaxime resistance, the best clones conferred 1280-fold greater cefotaxime resistance to the cells than wild-type BLA. Mena and coworkers constructed libraries of blue fluorescent mutants following full-atom simulations that discarded destabilizing mutations (Mena et al. 2006). Although mutants selected from the library were not significantly improved in terms of emission intensity, one mutant had 40-fold higher photobleaching half-life and enhanced quantum yield compared to wild type. In support of computationally designed libraries, Treynor and coworkers compared the frequency of active GFP mutants in a designed library versus a randomly mutated library (Treynor et al. 2007). They observed that the former was significantly richer in the frequency of active members as the mutants are prescreened for stability.

## GENE ASSEMBLY

When computation is used to design libraries, an amino-acid profile is first created for each position of a scaffold and degenerate oligonucleotides encoding all or most of the amino acids specified for each mutated position are synthesized and combined to assemble a library of genes. As gene assembly reactions generate a large library of mutants, the cost and time of synthesis per clone is low. However, the use of degenerate oligonucleotides also has several technical challenges. First, the designed sequences must share a minimum level of sequence homology for successful PCR extension. Second, the use of degenerate codons does not allow the exact representation of the desired amino acids with desired frequencies. As a result, either desired residues might be left out or undesired residues might have to be included at any given position (Mena and Daugherty 2005). One may avoid this misrepresentation at a mutated position by using multiple primers during gene assembly, each encoding a single designed amino acid. Alternatively, one may achieve more accurate representation by using the mix-and-split strategy during oligonucleotide synthesis (Glaser et al. 1992). However, given the current cost of oligonucleotides these solutions will be expensive if there are multiple positions that are close to each other and have the same misrepresentation issue. Third, although theoretically it is straightforward to find an optimal nucleotide composition for any desired amino-acid distribution (Wang and Saven 2002), synthesizing oligonucleotides with wobble codons to match the exact amino-acid distributions is expensive as the nucleotide composition at each chemical conjugation step will be arbitrary and hence will result in increased labor cost. One possible solution was proposed by Park and coworkers (Park et al. 2005), in which the DNA synthesizer was programmed to dispense an arbitrary mix of the four standard bases A, C, G, and T to construct a collection of degenerate oligonucleotides that best reproduce the desired amino-acid probabilities.

So far, rationally designed libraries were mostly constructed using low-resolution sequence selection methods that do not interrogate sequences in atomistic detail using computer simulations. The next step is to investigate whether using full-atom simulations will lead to more true positives in the library. However, the few studies that addressed this question assumed a rigid protein backbone, raising a possibility of false negatives (Hayes et al. 2002; Mena et al. 2006; Treynor et al. 2007). Including flexible-backbone simulations may potentially offer more accurate sequence diversity in the future. Screening computationally designed libraries might be a faster way to uncover the flaws in the computational algorithms. For example, the libraries designed using different scoring-function parameters and weights or search methods (e.g., dead-end elimination, Monte Carlo, genetic algorithm) might help to optimize the procedure used in evaluating structures. Another possibility is to use computation together with directed evolution to design or improve function. Computational modeling may be accurate enough to generate designs that are structurally sound but a few mutations away from the optimum functional fitness. In that case, error-prone PCR and DNA shuffling may help explore additional mutations that computational design has missed.

## CONCLUDING REMARKS

In this chapter, we summarized three issues that need to be addressed in order to increase the scope and accuracy of computational protein modeling and design. First, in using flexible-backbone methods, searching the conformational space more efficiently will potentially result in more accurate representation of protein biophysics. Second, negative design will have to be included into the modeling process to be able to generate sequences that are specific to the intended function. Finally, high-throughput experimental sampling of computational predictions will constitute a useful dataset and speed up the engineering objectives. Although many different ideas, including the ones mentioned previously, have been considered to incorporate or improve these methods, there is still much room for improvement. As mentioned previously, not only are the successes found in the literature important for continuous development but studying and learning from the many failures not reported is necessary toward progress in the field of computational protein design.

## REFERENCES

Alexander, P. A., Y. He, Y. Chen, J. Orban and P. N. Bryan. 2007. "The design and characterization of two proteins with 88% sequence identity but different structure and function." *Proc Natl Acad Sci U S A* 104: 11963–8.

Ali, M. H., C. M. Taylor, G. Grigoryan, K. N. Allen, B. Imperiali, et al. 2005. "Design of a heterospecific, tetrameric, 21-residue miniprotein with mixed alpha/beta structure." *Structure* 13: 225–34.

Ambroggio, X. I. and B. Kuhlman. 2006. "Computational design of a single amino acid sequence that can switch between two distinct protein folds." *J Am Chem Soc* 128: 1154–61.

Amin, N., A. D. Liu, S. Ramer, W. Aehle, D. Meijer, et al. 2004. "Construction of stabilized proteins by combinatorial consensus mutagenesis." *Protein Eng Des Sel* 17: 787–93.

Ashworth, J., J. J. Havranek, C. M. Duarte, D. Sussman, R. J. Monnat, Jr., et al. 2006. "Computational redesign of endonuclease DNA binding and cleavage specificity." *Nature* 441: 656–9.

Bolon, D. N., R. A. Grant, T. A. Baker and R. T. Sauer. 2005. "Specificity versus stability in computational protein design." *Proc Natl Acad Sci U S A* 102: 12724–9.

Bradley, P. and D. Baker. 2006. "Improved beta-protein structure prediction by multilevel optimization of nonlocal strand pairings and local backbone conformation." *Proteins* 65: 922–9.

Bradley, P., B. Kuhlman, G. Dantas and D. Baker. 2004. "Predicting protein structures accurately—response." *Science* 304: 1596–7.

Bradley, P., K. M. Misura and D. Baker. 2005. "Toward high-resolution de novo structure prediction for small proteins." *Science* 309: 1868–71.

Bradley, P., K. M. S. Misura and D. Baker. 2005. "Toward high-resolution de novo structure prediction for small proteins." *Science* 309: 1868–71.

Cordes, M. H., R. E. Burton, N. P. Walsh, C. J. McKnight and R. T. Sauer. 2000. "An evolutionary bridge to a new protein fold." *Nat Struct Biol* 7: 1129–32.

Cox, J. C., J. Lape, M. A. Sayed and H. W. Hellinga. 2007. "Protein fabrication automation." *Protein Sci* 16: 379–90.

Dahiyat, B. I. and S. L. Mayo. 1997. "De novo protein design: Fully automated sequence selection." *Science* 278: 82–7.

Davidson, A. R., K. J. Lumb and R. T. Sauer. 1995. "Cooperatively folded proteins in random sequence libraries." *Nat Struct Biol* 2: 856–64.

Davidson, A. R. and R. T. Sauer. 1994. "Folded proteins occur frequently in libraries of random amino acid sequences." *Proc Natl Acad Sci U S A* 91: 2146–50.

Davis, I. W., W. B. Arendall, D. C. Richardson and J. S. Richardson. 2006. "The backrub motion: How protein backbone shrugs when a sidechain dances." *Structure* 14: 265–74.

Desjarlais, J. R. and T. M. Handel. 1999. "Side-chain and backbone flexibility in protein core design." *J Mol Biol* 290: 305–18.

DeLano, W. L. 2002. *The PyMol Molecular Graphics System.* DeLano Scientific, Palo Alto, CA.

Fu, X., J. R. Apgar and A. E. Keating. 2007. "Modeling backbone flexibility to achieve sequence diversity: The design of novel alpha-helical ligands for Bcl-xL." *J Mol Biol* 371: 1099–117.

Fung, H. K., C. A. Floudas, M. S. Taylor, L. Zhang and D. Morikis. 2008. "Toward full-sequence de novo protein design with flexible templates for human beta-defensin-2." *Biophys J* 94: 584–99.

Fung, H. K., M. S. Taylor and C. A. Floudas. 2007. "Novel formulations for the sequence selection problem in de novo protein design with flexible templates." *Optimization Methods & Software* 22: 51–71.

Georgiev, I. and B. R. Donald. 2007. "Dead-end elimination with backbone flexibility." *Bioinformatics* 23: I185–94.

Glaser, S. M., D. E. Yelton and W. D. Huse. 1992. "Antibody engineering by codon-based mutagenesis in a filamentous phage vector system." *J Immunol* 149: 3903–13.

Harbury, P. B., J. J. Plecs, B. Tidor, T. Alber and P. S. Kim. 1998. "High-resolution protein design with backbone freedom." *Science* 282: 1462–7.

Harbury, P. B., B. Tidor and P. S. Kim. 1995. "Repacking protein cores with backbone freedom: Structure prediction for coiled coils." *Proc Natl Acad Sci U S A* 92: 8408–12.

Havranek, J. J. and P. B. Harbury. 2003. "Automated design of specificity in molecular recognition." *Nat Struct Biol* 10: 45–52.

Hayes, R. J., J. Bentzien, M. L. Ary, M. Y. Hwang, J. M. Jacinto, et al. 2002. "Combining computational and experimental screening for rapid optimization of protein properties." *Proc Natl Acad Sci U S A* 99: 15926–31.

Hu, X., H. Wang, H. Ke and B. Kuhlman. 2007. "High-resolution design of a protein loop." *Proc Natl Acad Sci U S A* 104: 17668–73.

Humphris, E. L. and T. Kortemme. 2007. "Design of multi-specificity in protein interfaces." *PLoS Comput Biol* 3: e164.

Jiang, L., E. A. Althoff, F. R. Clemente, L. Doyle, D. Rothlisberger, et al. 2008. "De novo computational design of retro-aldol enzymes." *Science* 319: 1387–91.

Jin, W., O. Kambara, H. Sasakawa, A. Tamura and S. Takada. 2003. "De novo design of foldable proteins with smooth folding funnel: Automated negative design and experimental verification." *Structure* 11: 581–90.

Kamtekar, S., J. M. Schiffer, H. Xiong, J. M. Babik and M. H. Hecht. 1993. "Protein design by binary patterning of polar and nonpolar amino acids." *Science* 262: 1680–5.

Koehl, P. and M. Delarue. 1995. "A self consistent mean field approach to simultaneous gap closure and side-chain positioning in homology modelling." *Nat Struct Biol* 2: 163–70.

Koide, A., R. N. Gilbreth, K. Esaki, V. Tereshko and S. Koide. 2007. "High-affinity single-domain binding proteins with a binary-code interface." *Proc Natl Acad Sci U S A* 104: 6632–7.

Kono, H. and J. G. Saven. 2001. "Statistical theory for protein combinatorial libraries. Packing interactions, backbone flexibility, and the sequence variability of a main-chain structure." *J Mol Biol* 306: 607–28.

Kortemme, T., L. A. Joachimiak, A. N. Bullock, A. D. Schuler, B. L. Stoddard, et al. 2004. "Computational redesign of protein–protein interaction specificity." *Nat Struct Mol Biol* 11: 371–9.

Kraemer-Pecore, C. M., J. T. Lecomte and J. R. Desjarlais. 2003. "A de novo redesign of the WW domain." *Protein Sci* 12: 2194–205.

Kuhlman, B. and D. Baker. 2004. "Exploring folding free energy landscapes using computational protein design." *Curr Opin Struct Biol* 14: 89–95.

Kuhlman, B., G. Dantas, G. C. Ireton, G. Varani, B. L. Stoddard, et al. 2003. "Design of a novel globular protein fold with atomic-level accuracy." *Science* 302: 1364–8.

Larson, S. M., J. L. England, J. R. Desjarlais and V. S. Pande. 2002. "Thoroughly sampling sequence space: Large-scale protein design of structural ensembles." *Protein Sci* 11: 2804–13.

Larson, S. M., A. Garg, J. R. Desjarlais and V. S. Pande. 2003. "Increased detection of structural templates using alignments of designed sequences." *Proteins* 51: 390–6.

Lippow, S. M. and B. Tidor. 2007. "Progress in computational protein design." *Curr Opin Biotechnol* 18: 305–11.

Looger, L. L., M. A. Dwyer, J. J. Smith and H. W. Hellinga. 2003. "Computational design of receptor and sensor proteins with novel functions." *Nature* 423: 185–90.

Margulies, M., M. Egholm, W. E. Altman, S. Attiya, J. S. Bader, et al. 2005. "Genome sequencing in microfabricated high-density picolitre reactors." *Nature* 437: 376–80.

Mason, J. M., K. M. Muller and K. M. Arndt. 2007. "Positive aspects of negative design: Simultaneous selection of specificity and interaction stability." *Biochemistry* 46: 4804–14.

Mena, M. A. and P. S. Daugherty. 2005. "Automated design of degenerate codon libraries." *Protein Eng Des Sel* 18: 559–61.

Mena, M. A., T. P. Treynor, S. L. Mayo and P. S. Daugherty. 2006. "Blue fluorescent proteins with enhanced brightness and photostability from a structurally targeted library." *Nat Biotechnol* 24: 1569–71.

Park, S., H. Kono, W. Wang, E. T. Boder and J. G. Saven. 2005. "Progress in the development and application of computational methods for probabilistic protein design." *Computers & Chemical Engineering* 29: 407–21.

Park, S., Y. Xu, X. F. Stowell, F. Gai, J. G. Saven, et al. 2006. "Limitations of yeast surface display in engineering proteins of high thermostability." *Protein Eng Des Sel* 19: 211–7.

Plecs, J. J., P. B. Harbury, P. S. Kim and T. Alber. 2004. "Structural test of the parameterized-backbone method for protein design." *J Mol Biol* 342: 289–97.

Riemann, R. N. and M. Zacharias. 2005. "Refinement of protein cores and protein-peptide interfaces using a potential scaling approach." *Protein Eng Des Sel* 18: 465–76.

Roberge, M., M. Estabrook, J. Basler, R. Chin, P. Gualfetti, et al. 2006. "Construction and optimization of a CC49-based scFv-beta-lactamase fusion protein for ADEPT." *Protein Eng Des Sel* 19: 141–5.

Ross, S. A., C. A. Sarisky, A. Su and S. L. Mayo. 2001. "Designed protein G core variants fold to native-like structures: Sequence selection by ORBIT tolerates variation in backbone specification." *Protein Sci* 10: 450–4.

Rothlisberger, D., O. Khersonsky, A. M. Wollacott, L. Jiang, J. DeChancie, et al. 2008. "Kemp elimination catalysts by computational enzyme design." *Nature* 453: 190–5.

Saunders, C. T. and D. Baker. 2005. "Recapitulation of protein family divergence using flexible backbone protein design." *J Mol Biol* 346: 631–44.

Smith, C. A. and T. Kortemme. 2008. "Backrub-like backbone simulation recapitulates natural protein conformational variability and improves mutant side-chain prediction." *J Mol Biol* 380: 742–56.

Smith, H. O., C. A. Hutchison, 3rd, C. Pfannkoch and J. C. Venter. 2003. "Generating a synthetic genome by whole genome assembly: phiX174 bacteriophage from synthetic oligonucleotides." *Proc Natl Acad Sci U S A* 100: 15440–5.

Stemmer, W. P., A. Crameri, K. D. Ha, T. M. Brennan and H. L. Heyneker. 1995. "Single-step assembly of a gene and entire plasmid from large numbers of oligodeoxyribonucleotides." *Gene* 164: 49–53.

Su, A. and S. L. Mayo. 1997. "Coupling backbone flexibility and amino acid sequence selection in protein design." *Protein Sci* 6: 1701–7.

Summa, C. M., M. M. Rosenblatt, J. K. Hong, J. D. Lear and W. F. DeGrado. 2002. "Computational de novo design, and characterization of an A(2)B(2) diiron protein." *J Mol Biol* 321: 923–38.

Tian, J., H. Gong, N. Sheng, X. Zhou, E. Gulari, et al. 2004. "Accurate multiplex gene synthesis from programmable DNA microchips." *Nature* 432: 1050–4.

Treynor, T. P., C. L. Vizcarra, D. Nedelcu and S. L. Mayo. 2007. "Computationally designed libraries of fluorescent proteins evaluated by preservation and diversity of function." *Proc Natl Acad Sci U S A* 104: 48–53.

Vasquez, M. 1996. "Modeling side-chain conformation." *Curr Opin Struct Biol* 6: 217–21.

Voigt, C. A., C. Martinez, Z. G. Wang, S. L. Mayo and F. H. Arnold. 2002. "Protein building blocks preserved by recombination." *Nat Struct Biol* 9: 553–8.

Voigt, C. A., S. L. Mayo, F. H. Arnold and Z. G. Wang. 2001. "Computational method to reduce the search space for directed protein evolution." *Proc Natl Acad Sci U S A* 98: 3778–83.

Wang, W. and J. G. Saven. 2002. "Designing gene libraries from protein profiles for combinatorial protein experiments." *Nucleic Acids Res* 30: e120.

Yin, S., F. Ding and N. V. Dokholyan. 2007. "Modeling backbone flexibility improves protein stability estimation." *Structure* 15: 1567–76.

Yue, K. and K. A. Dill. 1992. "Inverse protein folding problem: Designing polymer sequences." *Proc Natl Acad Sci U S A* 89: 4163–7.

Yue, K., K. M. Fiebig, P. D. Thomas, H. S. Chan, E. I. Shakhnovich, et al. 1995. "A test of lattice protein folding algorithms." *Proc Natl Acad Sci U S A* 92: 325–9.

Zhou, X., S. Cai, A. Hong, Q. You, P. Yu, et al. 2004. "Microfluidic PicoArray synthesis of oligodeoxynucleotides and simultaneous assembling of multiple DNA sequences." *Nucleic Acids Res* 32: 5409–17.

# Index

Printed and bound by CPI Group (UK) Ltd, Croydon, CR0 4YY

24/10/2024

01779064-0007